SERIES ON SEMICONDUCTOR
SCIENCE AND TECHNOLOGY

Series Editors

H. Kamimura   R. J. Nicholas   R. H. Williams

SERIES ON SEMICONDUCTOR
SCIENCE AND TECHNOLOGY

Low-Dimensional Semiconductors

Materials, Physics, Technology, Devices

M. J. Kelly

University of Surrey

CLARENDON PRESS · OXFORD
1995

Oxford University Press, Walton Street, Oxford OX2 6DP

Oxford New York
Athens Auckland Bangkok Bombay
Calcutta Cape Town Dar es Salaam Delhi
Florence Hong Kong Istanbul Karachi
Kuala Lumpur Madras Madrid Melbourne
Mexico City Nairobi Paris Singapore
Taipei Tokyo Toronto
and associated companies in
Berlin Ibadan

Oxford is a trade mark of Oxford University Press

Published in the United States
by Oxford University Press Inc., New York

A catalogue record for this book is available from the British Library

Library of Congress Cataloging in Publication Data
Kelly, M. J. (Michael J.), 1949–
Low-dimensional semiconductors : materials, physics, technology,
devices / M. J. Kelly.
(Series on semiconductor science and technology ; 3)
Includes bibliographical references and index.
1. Semiconductors—Surfaces. 2. One-dimensional conductors.
3. Crystal growth. 4. Electron gas. I. Title. II. Series.
QC611.6.S9K45   1995   537.6'22—dc20   94-47918

ISBN 0 19 851781 5 (Hbk)
ISBN 0 19 851780 7 (Pbk)

Typeset by Pure Tech India Ltd.,
Printed in Great Britain on acid-free paper by
Bookcraft (Bath) Ltd, Midsomer Norton

ANN and CONSTANCE

Preface

This book was prepared to form the basis of a set of lectures given to first-year graduate students at the University of Surrey at Guildford in 1993, and builds upon a much shorter and optional course of lectures given previously to final year undergraduates and MSc students at the University of Cambridge. In both cases, it was my aim to present the results of the exciting semiconductor science and technology research that has been in progress since about 1970. In that year, Esaki and Tsu published their paper introducing the concept of a superlattice, and semiconductor science has not been the same since.

In preparing for the course of lectures, I was aware of a large number of books containing collected research review papers written by practising specialists (see Appendix 1 for a partial list). Every year the proceedings of several summer schools are published, with updates on the various subjects. The most impressive and orderly collection of the key research material in one volume was published in 1991 by Claude Weisbuch and Borge Vinter, and contains an extensive list of references. A number of books are already available at the introductory level, but with a heavy bias towards some particular aspect of the subject. There is a lack of a suitable introductory book that would allow students in physics, electronic engineering, and materials science to progress from elementary courses in quantum mechanics and solid state physics to the position of being able to digest the material and appreciate the nuances in research texts.

The subject matter of this book—heterojunctions, multilayers, and microstructures in semiconductors—takes up the energies of approximately half of contemporary semiconductor physicists, as judged by the papers presented at the principal international conference on the physics of semiconductors. Although we shall be concerned in the main with III–V semiconductors, the principles are now increasingly being applied to mainstream silicon, amorphous semiconductors, and exotic semiconductor materials.

In 1984 I was given the opportunity by Cyril Hilsum, Derek Roberts, and Dennis Scotter to establish a Superlattice Research Group at the Hirst Research Centre of the General Electric Company. I have had the great satisfaction of seeing several items of new physics in multilayer semiconductor structures find their way into commercially successful microwave components. In addition, the Group sponsored some research at the Cavendish Laboratory in Cambridge, the outcome of which helped to focus our in-house activities. Our close interactions with the Product Division at Lincoln (formerly

Marconi Electronic Devices Limited, now GEC Plessey Semiconductors) reinforced the fact that not every new physics idea leads to a new and world-beating device. I learnt much of what follows here from my colleagues at these three centres, including Richard Davies, Adrian Long, Peter Beton, Nigel Couch, Mike Kearney, Steve Andrews, Richard Syme, and Gareth Ingram at GEC, Mike Pepper, Geb Jones, Haroon Ahmed, Mike Stobbs, and many research students at Cambridge, and Sean Neylon and Ian Dale at Lincoln. Claude Weisbuch and Borge Vinter at Thomson-CSF were stimulating partners in an Esprit project. My colleagues at the University of Surrey, from the Vice-Chancellor Tony Kelly on, have been generous hosts while I have been writing this book. The text has been read by a number of colleagues who made valuable suggestions, including Peter Beton, John Shannon, David Woolf, Andrew Norman, and others. I thank Sheila Rudman for the graphics. The final version is mine.

Surrey   M. J. K.
May 1995

Contents

Introduction

There is a contrast between semiconductor science and technology as practised before and after about 1970. During the 1930s silicon was established as a modern semiconductor when A. H. Wilson* applied the Bloch energy band theory (see Chapter 1) to establish the distinctive role of an energy gap. The modern theory of the electronic structure of dopants was completed in the 1950s by J. M. Luttinger and W. Kohn. During the late 1940s and early 1950s, the silicon bipolar transistor acting as a solid state amplifier was invented and refined at the Bell Laboratories. The realization of the field-effect transistor in the 1960s, together with the appreciation by R. Noyce and J. Kilby that resistors and capacitors could be fabricated using silicon, led to the integrated circuit. Arguably, this last has had as great an impact on contemporary life as any other single piece of science and technology. Since 1970, the progress of silicon technology has been relentless, requiring a logarithmic scale on which to display progress whether as increased computational power per unit chip area or reduced cost per transistor. The change has been one of continued evolution and refinement, with hundredfold reductions in the key feature sizes of individual transistors (from small fractions of a millimetre in the 1960s to fractions of a micrometre today). The increasing precision with which one defines volumes of n- and p-doping in bipolar transistors, the thickness of the gate oxide in field-effect transistors, and the delineation of features on the surface of the chip all represent recognizable advances on the state of the art 20 years ago. The semiconductor at the heart of these chips remains single-crystal silicon.

In 1970, Leo Esaki and Raymond Tsu published a paper entitled 'Superlattice and negative differential conductivity in semiconductors' (*IBM Journal of Research and Development*, **14**, 61–5). They described a new physical effect—negative differential conductivity (i.e. a regime of decreasing current under increasing bias voltage)—that should be present if only one could produce a new type of semiconductor crystal, a superlattice consisting of precisely periodic repetitions of a sequence of layers of at least two different semiconductors. The energy of electrons (the so-called band structure introduced in Chapter 1) in silicon and other materials is obtained by solving an appropriate Schrödinger equation with a potential that shares the underlying periodicity of the semiconductor lattice; with the artificial

---

*At a meeting in Oxford in 1986 to celebrate his 80th birthday, I heard Sir Alan Wilson recall the initial scepticism that met his ideas. Professor J. D. Bernal reported the increasing conductivity of a piece of silicon that he was purifying, convinced that he had a metal! (A beryllium crucible perhaps!)

periodicity of a superlattice, a new energy band structure is generated. Esaki and Tsu predicted novel electrical properties which set off a world-wide research effort, devoted at first to the preparation of the superlattices, using crystal growth techniques described in Chapter 2. We have now reached the stage where certain semiconductor multilayers can be prepared atomic layer by atomic layer with independent control over both the composition of the semiconductor and the level of doping in each layer. The many ramifications of this research on multilayer semiconductor structures are described in this book. Ironically, the negative differential conductivity in the superlattice remained illusive until very recently (and the effect was surprisingly complex to justify beyond their initial simple analysis). Now it has been observed, but it is a delicate phenomenon and is likely to find little practical application (Chapter 9). However, the spin-offs from the advances in crystal growth, and in the techniques for preparing small semiconductor structures (described in Chapter 3), have had an enormous impact both on semiconductor physics (Chapters 4–15 inclusive) and on virtually every non-silicon semiconductor device (Chapters 16–19 inclusive). Indeed, contemporary research suggests a major impact on future silicon technology (Chapter 14).

The new physics described in this book comes from one important consideration—length scale. If one examines solid state physics textbooks for an understanding of, say, conductivity, one encounters the theory established by Ludwig Boltzmann 120 years ago. There is an underlying assumption of three length scales, each of which are widely separated in magnitude: the diameter of the atom or molecule a, the mean free path λ of an electron between collisions or scattering events, and the size (L) of the sample. We normally assume $a \ll \lambda \ll L$. The structure of the Boltzmann theory for conductivity theory is similar to the kinetic theory of gases, where the size of the atom/molecule is much smaller than the mean free path between collisions, which in turn is much smaller than the gas container. If one can assume the clear separation of length scales, all kinds of averages and associated approximations can be invoked; for example, the density of a gas has a meaning in a large container but not in a very fine capillary. Further, density gradients are well defined and can be removed by net diffusive motion of the atoms/molecules in the gas. In the semiconductor microstructures we shall encounter, this clear separation of length scales can no longer be assumed. Under the condition that the size of the sample in one spatial direction is only a few atomic layers, quantum mechanics plays an important role in describing the energy levels for electrons. The motion of the electrons may be ballistic over the sample in this spatial direction, rather than diffusive as in macroscopic solids. The concepts that appear in Chapters 4–15 relate to electrons that may be confined in one or more spatial directions to within a few atomic layers; this is called a low-dimensional system, and its general properties are outlined in Chapter 4, with some more specific examples being described in Chapters 5 and 6. There are strong modifications to both the electrical and optical properties of these semiconductor microstructures. It is quite easy to place extremely large electric fields which excite the electrons into the so-called hot electron regime (Chapter 7). Specifically quantum-mechanical phenomena, such as tunnelling and resonant tunnelling through thin barriers, are observed and are described in Chapters 8 to 11. In statistical mechanics, one is usually allowed to introduce averaged properties of N particles (atoms, electrons, etc.) whenever $1 << \sqrt{N} << N$. In some of the semiconductor microstructures we shall encounter, the results of our electrical or optical measurements will be determined by a few tens or hundreds of electrons, and we shall be in a regime intermediate between macroscopic solids and microscopic (i.e. atomic) systems. The name mesoscopic (meso denoting middle or midway) is used to describe this regime where conventional averaged quantities (density, conductivity, etc.) cannot be defined (see Chapter 12). We eventually reach the microscopic limit of one or a few electrons entering the key regions of our microstructures, and their electronic charge dominates the electrical and optical properties in a way that is not possible in larger systems. It will be apparent from Chapter 2 why the III–V compound semiconductors feature strongly in our discussion. The preparation of very high quality multilayer single crystals is relatively easy using gallium arsenide and the ternary alloy aluminium gallium arsenide with any gallium-to-aluminium ratio; the materials have the same lattice constant to within a small fraction of a per cent. With increasing boldness on the part of the crystal growers, an ever wider combination of materials is now being prepared and used in both physics studies and device development. Chapters 13 and 14 are devoted to these. The properties, other than electronic and optical, of semiconductor microstructures can be interesting, but have not been widely investigated, so that the excitement is still to come (see Chapter 15).

In Chapters 16–19, I describe the impact of heterojunctions and semiconductor multilayers on a wide range of electronic and optical devices. This in part reflects my satisfaction at seeing the introduction of new generations of devices whose superior performance (efficiency, low noise, insensitivity to ambient temperature, etc.) can be traced back to a subset of the physics and technology of the previous chapters. Indeed, it allow us to claim not only that we have already been at the heart of the continued advances in other than silicon technology, but also that we may be about to have a major impact on the future of silicon technology itself. The final two chapters are devoted respectively to the future of mainstream semiconductor science and to radical alternatives. In addition, the lessons from semiconductors are being applied more widely, and in Chapter 21 some reference is made to recent progress with the use of metallic multilayers to enhance various magnetic properties for applications in memories, motors, etc. The other reason for the discussion of devices is that the expense of the infrastructure for a well-founded laboratory for the fabrication of advanced semiconductor microstructures is of order £5 million or $10 million, when one includes crystal growth and lithography. This expense is met only as long as it is justified by the returns from sales of advanced devices and systems.

The themes in this book have an important lesson for materials science more generally. One name coined for much of the research in semiconductor multilayers is 'band-structure engineering'. With control over the composition and level of doping in each atomic layer during the growth of multilayer semiconductor structures, we are able to produce crystals with the desired electronic and optical properties, i.e. we are engineering materials. As the techniques are applied to an ever wider range of materials, metals, dielectrics, superconductors, and combinations of these with each other and with semiconductors, we are starting to appreciate the ultimate limits to the properties of crystalline materials. This will become a reality once methods for preparing multilayers are fully integrated with methods for patterning the layers during preparation, as hinted in Chapter 3.

Modern semiconductor science and technology is fertile ground for teaching, as it brings together the quantum theory of materials, novel physics in the electronic and optical properties of solids, the engineering of small structures, and the design of high performance electronic, photonic, and optoelectronic systems. This book is aimed at students of physics, electronic engineering, and materials science. The only prerequisite here is familiarity with the elementary quantum mechanics and solid state physics of metals and semiconductors that is encountered in a first course in these subjects. The treatment attempts to be introductory, comprehensive, and phenomenological. The structure of individual chapters, highlighting materials, technology, theory, physics, or device issues, necessarily means that some material (e.g. the quantum well) is reintroduced in the different contexts. This duplication could have been avoided if a narrower focus of readership was intended, but it is hoped that the breadth is a virtue. Many interesting details within the overall picture have been omitted, together with most of the more advanced theoretical formulations and details of finer experiments. It should lead naturally into the texts listed in Appendix 1, and from there to the papers that appear in the leading European, Japanese, and American research journals and conference proceedings.

1 Resumé of bulk semiconductor physics

1.1 Introduction

In this chapter, we collect some of the key basic facts about the bulk properties of the semiconductors silicon, gallium arsenide, and aluminium arsenide that form the background material for the rest of this book. The material is selected solely on the basis of its use in subsequent chapters. If this were a book on solid state physics, such as those written by Kittel (1986), Ashcroft and Mermin (1976), or Madelung (1978), or on bulk semiconductors, such as those by Smith (1978), Seeger (1991), or Harrison (1980), much greater detail would be appropriate. We refer readers to these texts in the search of that greater detail. In particular, the terms in italics in this chapter have precise definitions which can be obtained from these texts.

At the outset, we refer readers to Appendix 2, where we have collected details of the relevant energy, length, and time scales for modern semiconductor physics.

1.2 Crystal structure

At room temperature and pressure, the element silicon (Si), from group IV in the periodic table, takes up the same structure as carbon in its diamond form. Gallium arsenide (GaAs) and aluminium arsenide (AlAs) both take up the zinc blende structure; these are both III–V semiconductors, given the group III origin of Ga and Al and the group V origin of As. The two crystal structures are shown in Fig. 1.1. The cubic lattice constants are 0.543095 nm for Si, 0.56533 nm for GaAs and 0.56611 nm for AlAs. In each case the atoms are tetrahedrally coordinated. Both crystal structures can be considered as *face-centred cubic lattices* with two atoms per unit cell (identical atoms in the case of Si). The *reciprocal lattice* is *body-centred cubic*, and in Fig. 1.2 the first *Brillouin zone* is shown, together with the nomenclature of the *principal symmetry points and lines*.

1.3 Energy band structure

The Schrödinger equation for the many-ion many-electron problem of a solid is exceedingly complex, but there is a well-established lore for simplifying such an equation (Ashcroft and Mermin 1976; Kittel 1986; Madelung 1978; Harrison 1980). The ions are regarded as rigidly fixed at their lattice sites, and any one of the *valence electrons* is considered to move in a potential formed by the ions and all the other electrons; this is the one-electron approximation.

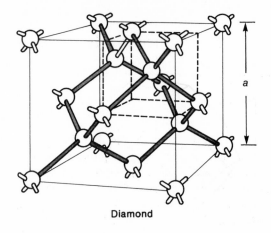

Diamond

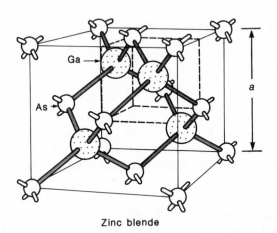

Zinc blende

Fig. 1.1 The diamond and zinc blende crystal structures for Si and GaAs.

The effective one-electron potential $V(\mathbf{r})$ is periodic, sharing the periodicity of the underlying lattice. If the potential has value zero everywhere in a large volume V (and infinity outside), the Schrödinger equation is

$$(-\hbar^2/2m)\nabla^2\psi = E\,\psi(\mathbf{r}),$$

where $\hbar = h/2\pi$ is the reduced Planck constant and m is the electron mass, with plane wave solutions $\psi(\mathbf{r}) = \exp(i\mathbf{k}\cdot\mathbf{r})/\sqrt{V}$ and energy

$$E = E(\mathbf{k}) = \hbar^2 k^2/2m.$$

It follows that $\hbar\mathbf{k}$ has the dimensions of momentum. *Bloch's theorem* for a periodic one-electron potential $V(\mathbf{r})$, states that the eigenfunctions (Bloch functions) $\psi_n(\mathbf{r})$ of the one-electron Schrödinger equation

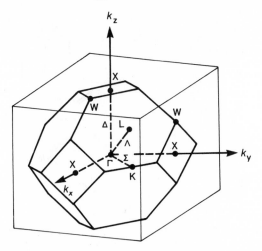

Fig. 1.2 The first Brillouin zone of the diamond and zinc blende structures.

$$[-(\hbar^2/2m^*)\nabla^2 + V(r)]\psi_n(r) = E_n(k)\,\psi_n(r)$$

are of the form of the same plane waves $\exp(i k \cdot r)$ multiplied by another function $u_n(k,r)$ sharing the same underlying periodicity as the potential:

$$\psi_n(r) = \exp(i k \cdot r)u_n(k,r),$$

where it is assumed that the function $u_n(k,r)$ is normalized over the volume V. The relation between k, a crystal momentum represented by a point in the first Brillouin zone, and the associated energy $E_n(k)$ is known as the *energy band structure*, or just the band structure, and is the starting point for describing most of the electrical and optical properties of semiconductors. (The integer n refers to a countable set of solutions, just like the n in the $\sin(n\pi x/L)$ solutions for electrons in a one-dimensional potential well of length L.) The detailed construction of the $V(r)$ for Si, GaAs, and AlAs is a matter of continuing research and refinement, but the resulting band structures are reasonably well known from calculations and have been corroborated by relevant experiments (Cohen and Chelikowsky 1988). These band structures are shown in Fig. 1.3. The multiple bands emerge from the different values of n. With eight valence electrons per unit cell in each of the three semi-conductors, precisely four bands are occupied, allowing for a spin degeneracy of two for each electron state. In the language of chemical bonds (Harrison 1980), the orbitals between the atoms form four bonds shown schematically as the rods between adjacent atoms in Fig. 1.1. There are precisely four outer electrons per Si atom and eight electrons per formula unit for GaAs and AlAs so that the *bonding* (symmetric) combinations of orbitals are fully occupied and the *antibonding* (antisymmetric) combinations are empty.

The following features should be noted; their fuller significance will become apparent through this and subsequent chapters.

Resumé of bulk semiconductor physics

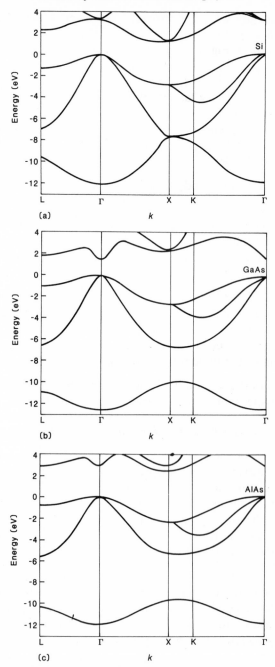

Fig. 1.3 Simple energy band structures calculated for for Si, GaAs, and AlAs using the same approximations (e.g. excluding the spin–orbit interaction) showing the principal features on the same energy scale. (G. P. Srivastava, University of Exeter, private communication.)

1. There is a clear *gap* in energy between filled valence band states (at negative energies) and empty conduction bands; note that in the high-symmetry directions shown in Fig. 1.3, *the upper valence bands are doubly degenerate*. The energy gap is the signature of a semiconductor, and takes values of 1.12 eV for Si, 1.42 eV for GaAs, and 2.23 eV for AlAs, at room temperature, and slightly larger values at low temperatures. The data on the crystal structures, electronic energy bands, electronic transport, and optical phenomena, and some information on the properties of defects, have been collected for group IV and III–V semiconductors and systematically tabulated (Madelung 1991), often as a function of temperature.

2. The lowest conduction band is *singly degenerate*, and at the Γ point has the symmetry of s-type orbitals at each site,

3. The lowest energy in the conduction band is at Γ for GaAs but away from Γ in the case of both Si (where it is 85 per cent of the way to the X point in the ⟨100⟩ direction, of which there are six different points in the Brillouin zone) and AlAs (at the X point of which there are three different points, with the pairs on opposite sides being equivalent),

4. The highest energy states for the valence bands are at Γ in all three cases, and in the absence of higher-order effects (*spin–orbit interactions* and strain that we discuss further in Chapter 13) are threefold degenerate having a representation of the three p-type orbitals,

5. An expansion of the conduction band energies about the minimum energies (at $k = k_0$) is approximately parabolic, and is written as

$$E(k) = E(k_0) + \hbar^2(k - k_0) \overleftrightarrow{(1/m^*)}(k - k_0)/2.$$

This expression defines the *inverse effective mass tensor*. In the case of GaAs, the effective mass is $m_* = 0.067m_e$, where m_e is the free-electron mass, and the constant energy surfaces are spherical in k-space about Γ. For silicon, the constant energy surfaces are ellipsoidal, characterized by a longitudinal effective mass $m_l = 0.916m_e$ and two transverse masses $m_t = 0.19m_e$. AlAs is not as well characterized as Si or GaAs, but it is thought that $m_l = 1.1m_e$ and $m_t = 0.19m_e$, giving a *density-of-states mass* $3\sqrt{m_l m_t^2}$ of $0.26m_e$.

6. The expansion of the valence-band states about Γ is more complicated (see Appendix 3 for a detailed treatment of valence band structure of relevance to bulk materials and the thin layers that we encounter subsequently) as the symmetry dictates that the functional form is not spherical but rather of the form

$$E(k) = E(k_0) + \hbar^2\sqrt{[B(k_x^4 + k_y^4 + k_z^4) + C(k_x^2 k_y^2 + k_y^2 k_z^2 + k_z^2 k_x^2)]},$$

if we write $k - k_0 = (k_x, k_y, k_z)$, where B and C are constants which have been tablulated by Madelung (1991). Nevertheless, there are spherical

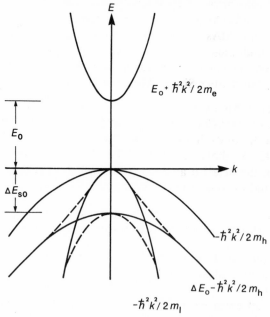

Fig. 1.4 The schematic band structure near the energy gap in greater detail, showing increasingly sophisticated approximations to the valence bands. (After Harrison 1980.)

approximations with effective masses. Indeed, the detailed forms of the valence bands for all three semiconductors are shown schematically in Fig. 1.4, and the effective masses of the light, heavy, and split-off hole bands for the three materials at low temperature are given in Table 1.1.

7. GaAs is a *direct-gap* material, as the highest valence band states and lowest conduction band states are at the same point Γ in the Brillouin zone. In contrast, both Si and AlAs are *indirect-gap* semiconductors. This innocent difference is an important reason why GaAs has been researched. Excita-

Table 1.1

Effective masses for Si, GaAs, and AlAs and spin split-off hole band separation Δ,

	Si	GaAs	AlAs
Electrons	$m_l = 0.916m_e$ $m_t = 0.19m_e$	$0.067m_e$	$m_l = 1.1m_e$ $m_t = 0.19m_e$
Holes	$m_l = 0.153m_e$ $m_h = 0.537m_e$, $m_{so} = 0.234m_e$	$m_l = 0.082m_e$ $m_h = 0.51m_e$, $m_{so} = 0.15m_e$	$m_l = 0.15m_e$, $m_h = 0.76m_e$ $m_{so} = 0.24m_e$
Δ (eV)	0.04	0.26	0.30

Electrons: longitudinal m_l; transverse m_t
Holes: light m_l; heavy m_h; split-off m_{so}.

From Madelung 1991.

tions of electrons between the valence band and conduction band extrema can take place in GaAs with automatic conservation of momentum *k*, while in both Si and AlAs some other momentum-containing process (e.g. *phonons*) or multiple-electron effects are required. For this reason, as we shall see later, GaAs can act as an optical device (light-emitting diode or laser diode), but neither Si nor AlAs can be used.

8. At an energy of about 0.3 eV above the Γ conduction band minimum in GaAs, there are '*satellite*' *minima* at the L points, and minima at a similar energy at the X points. Similarly, the Γ conduction band minimum in AlAs is about 0.42 eV above the X minima. These higher-energy minima are important in the detailed analysis of tunnelling structures as described in Chapter 8, and in some microwave devices described in Chapter 17.

1.4 Theory of donor (acceptor) levels and equilibrium carrier statistics

In a perfect (*intrinsic*) semiconductor at zero temperature, there is no conductivity as the valence bands are full and the conduction bands are empty. We now describe the process of *doping* which is used to achieve conduction. If an extra electron is placed in GaAs, it occupies the lowest conduction band state at Γ. If one of the Ga atoms in GaAs is changed into Si, creating what is called a *donor*), there is an extra positive charge on the ion but an extra electron as well. This process results in a material that is described as n-doped (where n stands for the negatively charged mobile carriers that result). Luttinger and Kohn (1955) have shown that there is a particularly simple description for the energetics and electronic structure of that extra electron, namely that of a modified hydrogen atom. Although the detailed calculations are intricate but straightforward, one can account for all the complications of the periodic potential and the Bloch functions if one replaces the bare electron mass in the Schrödinger equation by the effective mass (or the inverse effective mass tensor for non-spherical energy band minima as in Si and AlAs), and one screens the bare Coulomb potential of a proton (hydrogen ion) in free space with the static dielectric constant. The relative permittivities of Si, GaAs, and AlAs are 11.9, 13.1, and 10.1 respectively. The situation for GaAs is shown schematically in Fig. 1.5(a). The *binding or ionization energy* (a half Rydberg) is modified from the hydrogenic ionization energy (13.6 eV) by a factor $m^*/(\varepsilon_s)^2$, which is about 4×10^{-4} for GaAs, and the effective Bohr radius a^* is increased from the hydrogenic value of $a_0 = 0.052$ nm by a factor of $\varepsilon_s/m^* \approx 200$. Furthermore, the wavefunction for this extra donor electron, instead of being the Bloch function (i.e. a plane wave multiplying a periodic function), becomes an envelope function. $F(r)$ multiplying the same periodic function. $F(r)$ comes from solving the modified hydrogenic Schrödinger equation with a Coulomb potential for a single positive charge, but screened by the semiconductor dielectric constant:

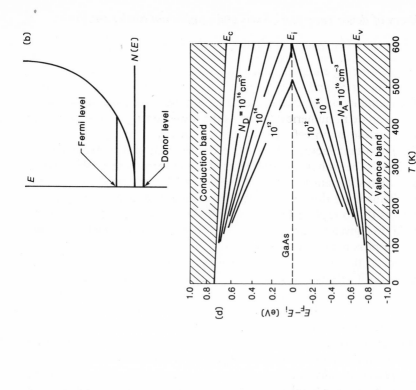

(a) Hydrogenic envelope wavefunction

Conduction band edge

Extra donor potential

Atomic potentials

Hydrogenic level

(b)

E

Fermi level

Donor level

$N(E)$

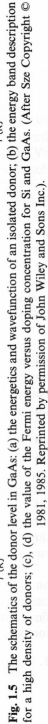

(c)

$N_D = 10^{18} \text{ cm}^{-3}$

10^{16}

10^{14}

10^{12}

E_c

E_i

Conduction band

Si

10^{12}

10^{14}

10^{16}

$N_A = 10^{18}$

E_v

Valence band

$E_F - E_i$ (eV)

T (K)

(d)

$N_D = 10^{16} \text{ cm}^{-3}$

10^{14}

10^{12}

E_c

E_i

Conduction band

GaAs

10^{12}

10^{14}

$N_A = 10^{16} \text{ cm}^{-3}$

E_v

Valence band

$E_F - E_i$ (eV)

T (K)

Fig. 1.5 The schematics of the donor level in GaAs: (a) the energetics and wavefunction of an isolated donor; (b) the energy band description for a high density of donors; (c), (d) the value of the Fermi energy versus doping concentration for Si and GaAs. (After Sze Copyright © 1981, 1985. Reprinted by permission of John Wiley and Sons Inc.).

$$- [(\hbar^2/2m^*)\nabla^2 + e^2/(4\pi\varepsilon_s\varepsilon_0|\,r\,|)]F(r) = EF(r),$$

where E is the binding energy referred to above.

The situation for Si and AlAs is more complicated because of the multiplicity of the conduction band minima. Although the underlying physics is the same, there is an extra numerical factor as the lowest energy states are similar to that of a sixfold or threefold hydrogenic molecule (Luttinger and Kohn 1955). The important factor is that twice the thermal energy per degree of freedom kT (k is the Boltzmann constant) at room temperature (about 0.025 eV) is more than sufficient for thermal ionization of the extra electron on the donor site. Thus if we dope any of our three semiconductors with electron-rich impurities (Si to replace Ga in GaAs or Al in AlAs, or As to replace a Si atom in Si) to a given concentration, we can assume that at room temperature we have an approximately equal concentration of electrons occupying states in the bottom of the conduction band, which are free to move (Fig. 1.5(b)), leaving fixed positively charged donor sites behind. At high concentrations ($\geqslant 3 \times 10^{24}$ m^{-3}), not all donors are charged, and one must resort to careful measurements of the carrier concentrations (as described in Chapter 2).

A similar theory is available for the *acceptor* levels that arise if electron-deficient atoms are used to replace atoms of the host lattice (i.e. B in the case of Si, or Si on an As site in GaAs), but again there is the added complication of the *multiplicity of valence bands*. The excitations here are *holes* (electrons missing from the valence bands), and the acceptor sites are charged negatively by the electrons trapped at the acceptor sites after excitation from the valence bands. In this case we refer to p-type doping, because of the effectively positive mobile carriers (the holes).

Note that if we were subsequently to deal with bulk semiconductors in this text, we would consider in much greater detail than we do below the effect of finite temperature and the exponentially small probability of the thermal excitation of an electron from the valence band into a conduction band state (see treatments by Kittel (1986), Ashcroft and Mermin (1976), Madelung (1978), and Sze (1981)). This consideration proves less significant in the modern theory of semiconductors where, in most cases, we dope the semiconductor sufficiently that it usually shows metallic conduction behaviour at room and at cryogenic temperatures.

The following brief description of the calculation of the *Fermi level* E_F in an n-doped semiconductor, the energy which separates the occupied from the empty states at low temperatures, or the states of 50 per cent occupation at finite temperatures, in both cases with the assumption that all donors are ionized, is sufficient for our purposes. There are three ingredients:

1. the probability of occupation of electron states of energy E which is determined by the Fermi–Dirac distribution

$$f_{\mathrm{FD}}(E) = \frac{1}{\exp[(E - E_F)/k_B T] + 1};$$

2. the Fermi energy E_F determined by the solution of the implicit equation

$$N = \int_{E = E_c}^{\infty} f_{FD}(E) n_e(E) \, dE$$

which says that the number of electrons per unit volume N is obtained by integrating from the edge of the conduction band E_c to infinity the product of the probability $f_{FD}(E)$ that the states at energy E are occupied and the density of those states $n_e(E)$ in energy per unit volume;

3. the density of states $n_e(E)$ given by (see Chapter 4 for further details)

$$n_e(E) = 1/(2\pi^2)(2m*/\hbar)^{3/2}\sqrt{(E - E_c)}.$$

We repeat here that N is usually given by the doping level. The results for E_F versus N are obtained numerically and are shown above in Figs 1.5(c) and 1.5(d). If we have heavily doped material, as indeed is the usual case in this book, we have to retain the $+1$ term in the denominator of the Fermi–Dirac distribution. In this regime, a semiconductor is described as *degenerate*.

Semiconductors are referred to as *non-degenerate* if the Fermi energy, as we proceed to calculate now, is in the gap and more than about $5kT$ away from either band edge. In the case of an intrinsic semiconductor, where all the carriers come from excitation across the bandgap (of energy E_g), the $+1$ term in the denominator of the expression for $f_{FD}(E)$ above can be ignored and the integration can be performed analytically, giving an intrinsic concentration of electrons of

$$n = 2(m_e^* k_B T / 2\pi\hbar^2)^{3/2} \exp(-E_C/kT) \exp(E_F/kT)$$

(defn)
$$\equiv N_C \exp[-(E_C - E_F)/kT]$$

We can repeat this exercise for acceptor levels and the density of mobile holes in the valence bands (from the band edge at E_V), as well as for obtaining the intrinsic concentration for holes as

$$p = 2(m_h^* k_B T / 2\pi\hbar^2)^{3/2} \exp(-E_F/kT) \exp(E_V/kT)$$

(defn)
$$\equiv N_V \exp[-(E_F - E_V)/kT]$$

In this special case, we have two relations for the product of the carrier concentrations and the position of the Fermi energy which depend on temperature only:

(defn)
$$n_i^2 = 4(k_B T / 2\pi\hbar^2)^3 (m_e^* m_h^*)^{3/2} \exp(-E_g/kT) \equiv N_c N_V \exp(-E_g/kT).$$

$$E_F = E_g/2 + 0.75 kT \ln(m_h*/m_e)$$

These statistics apply to undoped material, as often used in optical experiments. In the modern multilayer structures described below, some layers are heavily doped and others are not doped at all. The general results in Figs 1.5(c) and 1.5(d) apply to the former case, but also to undoped layers if, as often occurs, the *carriers* (the mobile electrons or holes from donors or acceptors) cross from one layer to another.

1.5 The electrical, optical, thermal, and mechanical properties

In this section we consider a homogeneous piece of semiconductor which may be doped.

1.5.1 Electrical properties

If a low electric field is applied to a piece of semiconductor, the carriers (electrons or holes) drift in the field with a velocity v_d, (the drift velocity) which is proportional to the electric field E, where the constant of proportionality is defined as the *drift mobility μ*:

$$v_d = \mu E$$

The units for μ are $m^2 V^{-1} s^{-1}$. The *Drude theory* of conductivity (Ashcroft and Mermin 1976; Kittel 1986; Madelung 1978) gives $\sigma = ne^2 \tau / m$, and so $\mu = e\tau / m^*$, reinforcing the notion that long times τ between scattering events of light mass particles correspond to high conductivity. There are texts (Seeger 1991; Ridley 1993) with extensive sections devoted to the microscopic theories of μ, and we shall encounter some aspects of these theories in latter chapters, as low dimensionality (in Chapter 4), for example, can modify the value of μ. Crudely, one can see the mobility μ as a measure of scattering processes that stop the carriers from accelerating indefinitely in response to an electric field. Indeed, the conductivity of a semiconductor which is heavily n- or p-doped is given by

$$\sigma = en\mu_n \qquad \text{or} \qquad \sigma = ep\mu_p$$

respectively. Mobilities at room temperature and for modest levels of doping 10^{20}–10^{24} m^{-3} are given in Table 1.2.

The various scattering processes that contribute to the mobility enter (via *Mattheissen's rule*) as additive contributions to $1/\mu$, and include the effects of

Table 1.2 Electron and hole mobilities

	Si	GaAs	AlAs
Electron mobility $(m^2 V^{-1} s^{-1})$	0.15	0.9	0.03
Hole mobility $(m^2 V^{-1} s^{-1})$	0.035	0.04	0.02

From Madelung 1991.

scattering of impurities, including the charged donor or acceptor sites, (and naturally their concentration), the various types of lattices, vibrations, etc. Furthermore, these scattering processes are usually dependent on temperature, and indeed a field dependence of the mobility sets in as the electric fields are raised (see Chapter 7). Again, if a mixture of mobile electrons and holes occupy the same region of space, as in p–n junctions (Chapter 2) or bipolar transistors (Chapter 16), it is possible for an electron and hole to *recombine* as the electron in the conduction band reoccupies the hole state in the valence band, with the excess energy being emitted as light or via some other process that ends up heating the lattice. This process alters both the n and the τ in the expressions for conductivity and mobility.

Finally, we note that there are important semiclassical equations of motion, as used to discuss transport in solids. This topic will be introduced in Chapter 4.3.1.

1.5.2 Optical properties

The momentum of light of energies typically encountered in semiconductors is very small compared with that of electrons or holes. The process of optical absorption can be viewed as the vertical excitation of an electron in the band structure from an occupied initial state to an unoccupied final state. The optical power P absorbed at a given energy $\hbar\omega$ per unit volume Ω defines the *frequency-dependent conductivity* as

$$P/\Omega = \langle E{\cdot}\sigma(\omega){\cdot}E \rangle$$

where E is the strength of the electric field from the light waves. The rate of absorption is calculated by the *Fermi golden rule*, and in turn the probability of excitation includes a *matrix element* that is a measure of the 'allowedness' of the excitation. *Selection rules* based on the symmetry of the initial and final wavefunctions with respect to each other, and the polarization ε of the electric field vector of the light, are involved, just as in atoms. We also need the Fermi occupation functions to ensure filled initial states (from band n) and empty final states (from band n'). We obtain

$$\sigma(\omega) \sim \sum_{n,\,n',\,k} f(E_n\mathbf{k})[1 - f(E_{n'k})]\delta(E_{n'k} - E_{nk} - \hbar\omega)\langle \psi_{n'k}|\,\varepsilon{\cdot}\nabla\,|\psi_{nk}\rangle\langle \psi_{nk}|\,\varepsilon{\cdot}\nabla\,|\psi_{n'k}\rangle.$$

The complication of considering the initial and final states leads to a definition of a *joint density of states* that examines the whole Brillouin zone for pairs of initial and final states separated by the correct energy. The optical absorption of Si and GaAs is shown as a function of optical energy in Fig. 1.6. (There are no reliable data for AlAs.)

The features in the optical absorption contain much information that can be related back to the band structure.

1. Because of the direct-gap nature of GaAs, optical absorption across the gap at Γ is allowed, thus providing an initial confirmation of the estimate

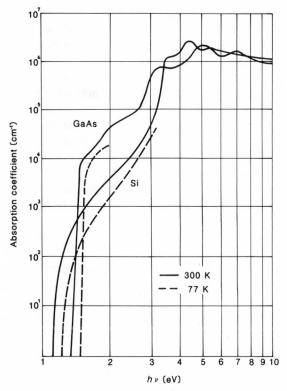

Fig. 1.6 The optical absorption of pure Si and GaAs at 77 K(– – –) and 300 K (—). The vertical axis is a measure of the inverse distance over which light is attenuated. (After Sze Copyright © 1981. Reprinted by permission of John Wiley and Sons Inc.).

of the direct gap. Furthermore, because of the $\sqrt{E}$ form of the density of states of both the conduction and valence bands away from the bandgap, the optical absorption has a similar $\sqrt{E}$ form.

2. In contrast, the lowest energy 'vertical' transition in both Si and AlAs is away from Γ. However, if we invoke *quanta of lattice vibrations* ('phonons', introduced in Section 1.5.3 and in subsequent chapters), we can allow a process by which an electron is excited optically at the same time as a phonon is generated or absorbed. The process satisfies conservation of energy and momentum, but it is weak because of the need to invoke the secondary momentum-conserving process. Thus it is possible to obtain optical absorption across an indirect gap in a semiconductor, although it has a different energy dependence.

3. A careful examination of the band structure allows identification of regions where conduction and valence bands are roughly parallel over significant volumes of the Brillouin zone. This should lead to peaks in the optical absorption. Furthermore, there are analytic singularities (*van Hove singu-*

larities), in the density of states arising from the form of energy bands near the faces, edges, and corners of the Brillouin zone, and the effects of these can in turn appear as features in the optical absorption.

4. The onset of optical absorption in GaAs at low temperature has an extra peak superimposed on the $\sqrt{E}$ absorption edge predicted above. This comes from an important correction to the 'one-electron' approximation used to establish the band structure in the first place. If an electron in the valence band is given an energy just slightly less than that of the energy gap, it can be excited in such a way that a *bound electron–hole pair* (an *exciton*) is formed. This is precisely the modified hydrogen atom problem referred to above in the case of donors, except that the positive hole in the valence band is now thought of as a positive ion. The *cross-section* for this particular optical absorption process is strong at low temperatures. Because the mass of the hole is much lighter than that of any ion, the reduced mass that appears in the ionization energy (or Rydberg constant) is much smaller. Indeed, the binding energy of an exciton in GaAs is approximately 5 meV (order of kT at 50 K), and so a feature that dominates the absorption near the bandgap at low temperatures is reduced to a modest feature at 300 K. There is much interest in excitons in multilayer semiconductor systems (see Chapters 4, 10, and 18).

1.5.3 Thermal properties

The introductory texts on solid state physics (Ashcroft and Mermin 1976; Kittel 1986; Madelung 1978) include as a worked example the wave-like modes of vibration of a line of atoms (each of mass M and a distance a apart) with adjacent atoms connected by springs (of stiffness K), and show that modes of wavevector q, with an amplitude that varies as $\exp(iqx)$, have an energy $\hbar\omega = 2\hbar\sqrt{(K/M)} |\sin(qa/2)|$. Hence the so-called *phonon spectrum* (the $\omega - q$ relationship) is plotted.

An extension of this simple model is possible when the atoms alternate in mass; in this case there are two types of modes, one similar to that encountered above (the so-called *acoustic mode*) where adjacent atoms tend to be moving in the same direction and a second type (the so-called *optic mode*) where adjacent atoms tend to be moving against each other.

The whole problem can be extended further to cover Si, GaAs, and AlAs in three dimensions. In addition to the acoustic and optic modes, we now have to consider whether the amplitude of the motion of the atoms is in the same direction as the wavevector q of the wave propagation (as in longitudinal modes) or at right angles (transverse modes), or whether some more complex relationship exists as is the case in directions other than those of high lattice symmetry. The phonon spectra for our three key materials are shown in Fig. 1.7.

Whereas there are well-developed semiclassical theories of electronic transport (see Chapter 4, Section 4.3) that relate back to band structure, the theory of thermal transport bears a much more complex relationship to the phonon

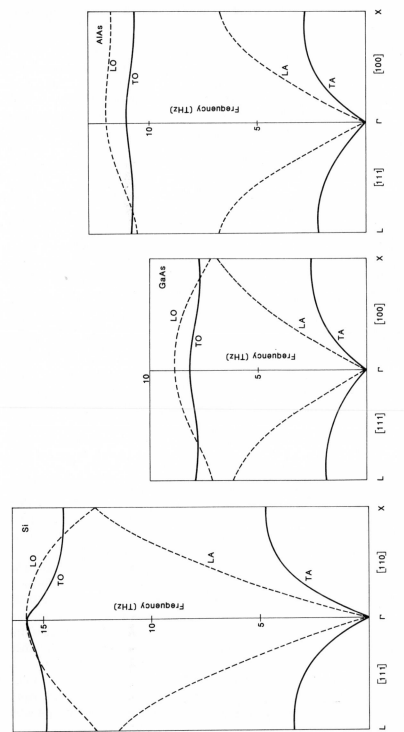

Fig. 1.7 The phonon spectra of Si, GaAs, and AlAs. Note that the similarity of the latter follows from scaling the effects of the different masses of Ga and Al with respect to As. (After Molinas-Mata *et al.* 1993, and M. Cardona, Max Planck Institute, Stuttgart, private communication.)

spectra, particular at other than low temperatures. The *thermal conductivity* at low temperatures can be written as

$$K = C_v v \lambda / 3$$

where C_v is the specific heat at constant volume and is given by

$$C_v = (2\pi^2/5)k_B^4 T^3/v^3,$$

where v is an appropriately averaged speed of sound and λ is the phonon mean free path. The averaged sound velocity has contributions from both transverse and longitudinal phonon modes:

$$3/v^3 = 2/v_t^3 + 1/v_l^3$$

where v_t and v_l are the transverse and longitudinal sound velocities in the semiconductor (given by the limiting slopes of the phonon spectrum in the regime where $\omega = vq$). There are interesting cases of pure materials where λ is determined by the size of the sample, but for doped materials much of the heat is carried by electrons (see Chapter 15). The stiffness of a solid (which gives rise to steep phonon spectra and high speeds of sound) translates to a high thermal conductivity at low temperatures, where the T^3 dependence is widely observed. Phonons obey *Bose–Einstein* statistics, and their number is not conserved. Indeed, phonon–phonon scattering is an important practical ingredient in thermal conductivity at elevated temperatures (Ziman 1960). Above about 77–100 K, the generation and absorption of optical phonons is an added complication; from their flat dispersion curves, they are slow moving and they have a short lifetime, decaying into multiple acoustic phonons. In terms of the thermal conductivity, optic phonons are not as effective as acoustic phonons in transporting heat (Fig. 1.8). However, optical phonons are an important mechanism by which energy from electrons is initially given to the lattice.

1.5.4 Mechanical and miscellaneous properties

A section on the mechanical properties of Si and GaAs is included here for completeness. In practice, GaAs is more brittle than Si, and this implies that extra care is needed in handling the latter during the fabrication and use of III–V semiconductor microstructures. The technology for the fabrication of Si wafers of diameter up to 8 inch is more advanced (in large part because over 90 per cent of the volume of semiconductor materials used is Si) than that of GaAs, where 3–4 inches is the maximum. AlAs is not prepared as wafer material, but is grown in thin layers on GaAs.

Si is a stable material, with a thin (about 2 nm) native oxide film acting as a protection against water and some chemicals. A thin oxide provides some protection for GaAs, although this is more susceptible to chemical attack than is the silicon oxide. AlAs is hygroscopic, and layers of this material must be capped with thin layers of GaAs for protection.

One property of GaAs that is important for its applications, in both very-high-frequency microwave devices and high-speed circuits, is its availa-

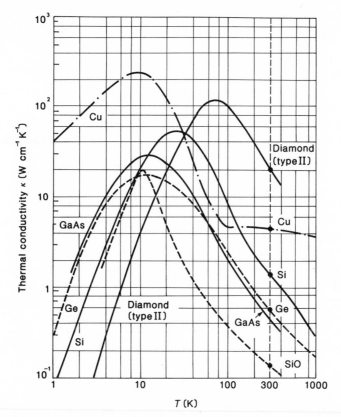

Fig. 1.8 The thermal conductivities of Si and GaAs compared with other device and circuit materials. (After Sze © 1981 Reprinted with permission of John Wiley and Sons Inc.).

bility as a semi-insulating substrate which maintains this property even after the various processing steps, including severe heat treatment, required to fabricate devices. This means that materials with a very high resistivity are available as substrates. In Si, there are a number of processing steps that induce low levels of free carriers, and these respond to high-frequency electric fields as unwanted *parasitic impedances*. Electrically active substrates interfere with the operation of devices that are prepared on one surface. The higher mobility of electrons in GaAs compared with that in Si means that GaAs finds applications in very-high-speed circuits, further assisted by the absence of parasitic impedances.

1.6 Other bulk semiconductors

1.6.1 *The III–V and II–VI semiconductors*

The discussion so far has concentrated on Si, GaAs, and AlAs. The first is the dominant semiconductor material. The second has a key role in the operation of optical devices (including lasers) and also in some microwave devices; in both instances aspects of the GaAs band structure (direct-gap and satellite valleys), not available in Si, are exploited. AlAs is included in the description above, as it plays an important role in the multilayer structures and devices described throughout this text.

There are many other III–V and II–VI semiconductors, and Chapter 14 is devoted to some of the novel physics that can be achieved with these materials, although of lesser overall device relevance at the present time. Figure 1.9 shows Si, Ge, and the binary members of the various compound semiconductors plotted on axes of lattice constant and energy gap. This diagram is useful in subsequent chapters, as lattice matching of pairs of materials is a great help in the preparation of high-quality single crystals containing single or repeated interfaces between different materials. The almost vertical line between GaAs and AlAs shows that it should be relatively easy to prepare crystals containing multilayers of GaAs and any of the $Al_xGa_{1-x}As$ alloys (see Section 1.6.2). Indium phosphide is a convenient substrate (although it is even more brittle than GaAs), and it is possible to prepare $In_{0.53}Ga_{0.47}As$ a lattice-matched overlayer. In Chapter 14 we shall refer to the CdTe–HgTe materials system, which has been widely investigated

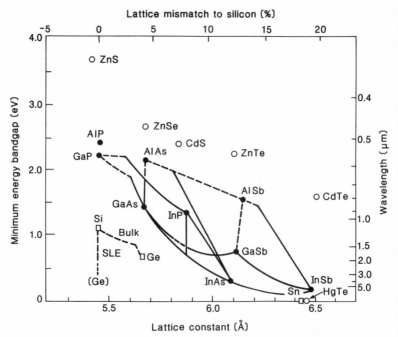

Fig. 1.9 The various semiconductors plotted as a function of lattice constant and energy gap. The full lines indicate a direct-gap semiconductor and the broken lines indicate an indirect-gap semiconductor, while the two curves related to Si refer to bulk unstrained material and the strained layer epitaxial (SLE) material. (After Bean 1990 from S. M Sze (ed) © 1990. Reprinted with permission of John Wiley and Sons Inc.)

for its optical properties in the infrared. The vertical axis on Fig. 1.9 shows the energy gap, and much of this book is devoted to exploiting the different energy gaps in adjacent semiconductor layers that are separated by atomically abrupt interfaces. The lines show the lattice constant–energy gap relationship for the ternary alloys, so that the line between GaAs and AlAs encompasses all the $Al_xGa_{1-x}As$ alloys. What is also shown is the distinction between direct-gap (full lines) and indirect-gap (broken lines) semiconductors, with the kinks in the curves joining AlAs and InAs, and AlSb and InSb, for example, indicating where the indirect-gap alloy starting from AlAs changes to a direct-gap alloy approaching InAs and InSb. The details associated with Si–Ge alloys are discussed further in Chapter 14.

1.6.2 Bulk ternary $Al_xGa_{1-x}As$ alloys

The ternary alloy system $Al_xGa_{1-x}As$ plays an important role throughout this book, because GaAs and AlAs form a solid solution over the entire composition range ($0 \leqslant x \leqslant 1$) with very little variation (<0.15 per cent) of the lattice constant. These facts, plus others to be described in the next chapter, mean that semiconductor multilayers of very high quality with different x values can be prepared with relative ease. The most exploited aspect of the $Al_xGa_{1-x}As$ alloys is their band structure, and the most important features are summarized in Fig. 1.10(a). At $x = 0$ (i.e. for GaAs), we have a direct-bandgap semiconductor with a room temperature bandgap of 1.42 eV.

With reference to Fig. 1.3(b), the bandgaps at the high symmetry points on the Brillouin zone faces (X in the ⟨100⟩ direction and L in the ⟨111⟩ direction) are 1.86 eV and 1.72 eV respectively. From optical experiments on alloys with increasing substitution of Ga with Al, the bandgap of the alloy is seen to rise, as indeed do the bandgaps at the X and L points but with a smaller slope. At $x = 1$, AlAs is an indirect-gap semiconductor with the minimum energy separation at the X point. At somewhere about $x = 0.45$ there is a crossover from direct to indirect-gap structure as shown. We anticipate one of the key questions in Chapter 2, namely how the bandstructures of GaAs and one of the $Al_xGa_{1-x}As$ alloys line up when perfect single crystals of the two are brought together. Electrical transport measurements have been performed on these interfaces and the results are summarized in Fig. 1.10(b). We defer to the next chapter a discussion of the accuracy and interpretation of transport and optical measurements, but note the differences between the two which typify this problem.

Other properties of the $Al_xGa_{1-x}As$ bulk alloys that are of interest and have been measured include the phonon spectra, which we describe in Chapter 15; there is evidence for vibrational modes associated with GaAs and AlAs bonds in the alloys. For thermal, elastic, and other properties, interpolations are often made between the properties of GaAs and AlAs (where these latter are known). The transport properties of $Al_xGa_{1-x}As$ near the indirect–direct crossover are not well known.

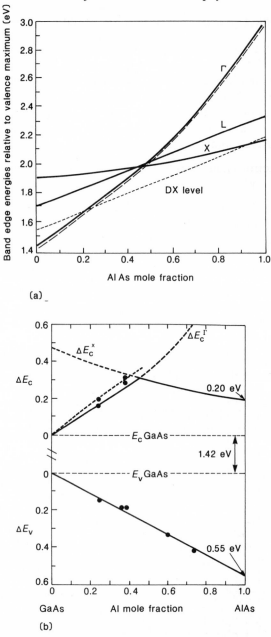

Fig. 1.10 (a) Optical data on the bandgaps at Γ, X, and L in bulk GaAs and the bulk $Al_xGa_{1-x}As$ alloys (after Pearton and Shah 1990): the broken line following the Γ curve gives the energy of shallow (hydrogen-like) donor levels, while that marked DX refers to the energy level of deep defects in $Al_xGa_{1-x}As$. (b) Electrical data on multilayers as interpreted in terms of the relative alignment of the energy bands in GaAs and the $Al_xGa_{1-x}As$ alloys (after Solomon 1986).

1.7 Velocity–field characteristics and high-field phenomena

The linear Ohm's law regime of the electrical properties of doped semiconductors was outlined in Section 1.5.1. In practice, nearly all useful devices operate under conditions of high electric fields in some key part of the device. For this reason, the relationship between the *drift velocity* and the applied electric field is an important feature in the design of devices. The velocity–field characteristics for electrons and holes in Si and electrons in GaAs are shown in Fig. 1.11. Several points should be noted.

1. For both electrons and holes in Si, ohmic behaviour persists up to fields of 100 kV m^{-1}, with the mobility of electrons being greater than that of holes.

2. In the case of Si, the electron and hole mobilities saturate at fields of the order of 1000 kV m^{-1}. A mechanism is operating by which the energy input from the electric field into the carriers is matched by energy loss (in practice to the lattice). The carriers are accelerating along the band structure, and once their energy with respect to the Fermi level exceeds that of an optic phonon (about 60 meV in Si and about 36 meV in GaAs) the emission of optic phonons is an efficient process for taking energy from the carriers. The saturated drift velocity is an important quantity in the design and simulation of devices.

3. The behaviour of electrons in GaAs is unusual, as the velocity–field curve reaches a peak before saturating at a lower value of velocity. There is a

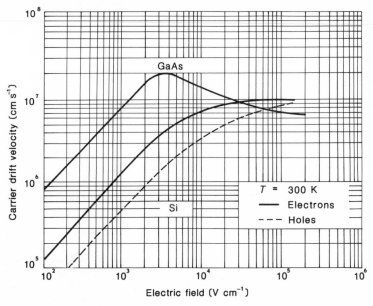

Fig. 1.11 The velocity–field characteristics in Si and GaAs at 300 K: —electrons; – – –holes. (After Sze © 1981. Reprinted with permission of John Wiley and Sons Inc.).

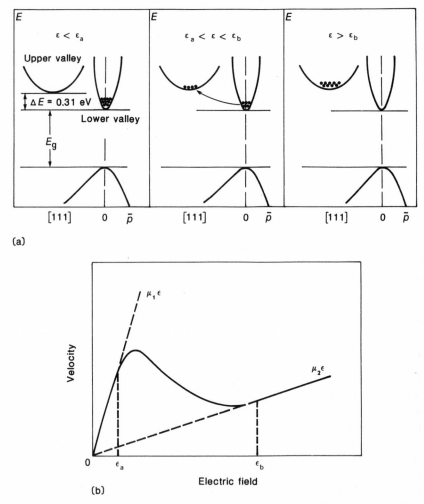

Fig. 1.12 (a) The two-valley origin of negative differential mobility (velocity) in GaAs, showing the increasing transfer of electrons to high-mass low-mobility satellite valleys with increasing field; (b) the resulting velocity–field characteristic. (After Sze © 1981 Reprinted with permission of John Wiley and Sons Inc.).

regime of negative differential velocity, which translates to a negative differential conductivity which is the basis of microwave oscillator devices (see Chapter 17). The explanation of this behaviour is shown schematically in Fig. 1.12. In low fields, the electrons are all in the Γ valley, characterized by a light effective mass m^*, and a low density of states, so that the low-field electron mobility $\mu = e\tau/m^*$ is high. The low density of (final) states implies a relatively long time between scattering events. At modest fields, some of the carriers acquire sufficient energy to *transfer* to the satellite (L, X) valleys, which have a heavy effective mass and a high

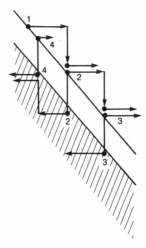

Fig. 1.13 Avalanching in semiconductors under very high fields as very highly excited electrons use their excess energy to create extra electron–hole pairs. (After Sze © 1981 Reprinted with permission of John Wiley and Sons Inc.).

density of states for scattering the electrons, i.e. a lower mobility. At higher fields most of the carriers are in the satellite valleys, resulting in the reduced velocity.

4. Figure 1.11 does not indicate what occurs at the highest fields, i.e. above about 30 000 kV m^{-1} or 30 V μm^{-1}. Under these conditions a number of electrons are accelerated high into the conduction bands, to the extent that it is energetically feasible for them to excite another electron from the valence band across the gap into the conduction band. This is a process that multiplies the number of carriers, as shown schematically in Fig. 1.13, and results in avalanching of current. If the avalanche is controlled (for example by having very high fields over relatively short distances, or by operating only for short time intervals), it can be exploited; otherwise, it is merely the same thing as the onset of dielectric breakdown. A number of electronic devices operate relatively close to, or under, the conditions of avalanche breakdown, and we shall examine them in Chapters 15–19.

1.8 The drift–diffusion model of electrical transport

Although we shall encounter a frequent need to go beyond it in this book, the *drift–diffusion model* of electrical transport has served semiconductor physicists and device designers well. We assume that the scale of a semiconductor sample is sufficiently large that densities of electrons n and holes p and their spatial and temporal gradients are well defined. Under these conditions, we define contributions to current densities J from carriers (electrons and holes) drifting in applied electric fields and from net diffusion against concentration gradients:

$$J_{\text{tot}} = J_{\text{n}} + J_{\text{p}},$$

where

$$J_{\text{n}} = e\mu_{\text{n}}nE + eD_{\text{n}}\nabla n,$$
$$J_{\text{p}} = e\mu_{\text{p}}pE - eD_{\text{p}}\nabla p$$

and e is the magnitude of the electronic charge. The local values of the electron and hole densities are determined by continuity equations that take into account the rates G and U of generation and recombination of carriers and their outdiffusion from a given position:

$$\partial n/\partial t = G_{\text{n}} - U_{\text{n}} + \nabla \cdot J_{\text{n}}/e$$

$$\partial n/\partial t = G_{\text{p}} - U_{\text{p}} + \nabla \cdot J_{\text{p}}/e.$$

The generation processes may include the effects of optical excitation or ionization of a donor or acceptor. The *recombination* is usually described by a relaxation time model of the form

$$U_{\text{n}} \sim (n_{\text{p}} - n_{\text{po}})/\tau_{\text{n}}$$

where n_{p} is the actual minority carrier density (i.e. electrons in p-type material), which differs from the equilibrium value n_{p0}. The time scale for the relaxation is given by τ_{n}, which is in the nanosecond to picosecond range.

In some of our low-dimensional structures, the length scales of interest are too small for this formalism to be valid in key parts of the structure, even though it may suffice to describe currents in the regions that lead into and out of those key parts.

1.9 Summary

We have summarized those elements of the physics of bulk semiconductors, concentrating on Si, GaAs, and AlAs, that provide necessary background material for addressing the more recent research results in semiconductor science and technology. References are given to texts that contain further details about solid state and semiconductor physics. In particular, these texts should be used to clarify the technical terms given in italics in this chapter.

References

Ashcroft, N. W. and Mermin, N. D. (1976). *Solid state physics*. Holt, Rinehart, Winstone, New York.

Bean, J. C. (1990). Materials and technology. In *High-speed semiconductor devices* (ed. S. M. Sze), pp. 13–55. Wiley, New York.

Cohen, M. L. and Chelikowsky, J. R. (1988). *Electronic structure and optical properties of semiconductors*. Springer-Verlag, Berlin.

Harrison, W. A. (1980). *Electronic structure and the properties of solids*. Freeman, San Francisco.

Kittel, C. (1986). *Introduction to solid state physics*. (6th edn). Wiley, New York.

Luttinger, J. M. and Kohn, W. (1955). *Physical Review*, **97**, 869–83.

Madelung, O. (1978). *Introduction to solid state theory*. Springer-Verlag, Berlin.

Madelung, O. (1991). *Semiconductors: group IV elements and III–V compounds*. Springer-Verlag, Berlin.

Molinas-Mata, P., Shields, A. J. and Cardona, M. (1993). Phonons and internal stresses in IV–IV and III–V semiconductors: the planar bond–charge model. *Physical Review B*, **47**, 1866–75.

Pearton, S. J. and Shah, N. J. (1990). Heterostructure field-effect transistors. In *High-speed semiconductor devices* (ed. S. M. Sze), pp. 283–334. Wiley, New York.

Ridley, B. K. (1993). *Quantum processes in semiconductors* (3rd edn). Oxford University Press.

Smith, R. A. (1978) *Semiconductors* (2nd edn). Cambridge University Press.

Seeger, K. (1991). *Semiconductor physics: an introduction* (5th edn). Springer-Verlag, Berlin.

Solomon, P. M. (1986). Three-part series on heterojunction transistors. In *The physics and fabrication of microstructures and microdevices* (ed. M. J. Kelly and C. Weisbuch). Springer-Verlag, Berlin.

Sze, S. M. (1981). *Physics of semiconductor devices* (2nd edn). Wiley, New York.

Sze, S. M. (1985). *Semiconductor devices: physics and technology*. Wiley, New York.

Ziman, J. M. (1960) *Electrons and phonons*. Oxford University Press.

2 III–V Semiconductor homojunctions and heterojunctions

2.1 Introduction: ever-sharper interfaces in semiconductor devices

In the Introduction, we referred to the hundredfold reduction in the lateral feature size of transistors in integrated circuits over the period from the late 1960s to the present day. This reduction has been accompanied by a similar scaling down of vertical dimensions of the devices, i.e. in the thicknesses of the individual layers (see Dennard *et al.* (1974) and Chapter 21 below). Indeed, if one takes the p–n–p bipolar transistor from its earliest fabrication to the present day, new technologies that have allowed the ever sharper internal interfaces of devices to be achieved have appeared at intervals of approximately 10 years.

The original transistors were made in the 1950s by alloying in the species to make a p-type layer in a host n-type material, and then overalloying to produce a new n-type layer near the surface. This technique is exceptionally crude by today's standards. The interface between the n- and p-type material is uncontrolled over thicknesses of micrometres. Furthermore, the interface follows the melt front of the alloying species.

The technique of ion implantation was introduced in the 1960s (see Chapter 3 for more details). The ions of the dopant material are fired at the host wafer with doses and energies that allow layers of the required doping to be formed at given depths beneath the surface. This technique has the advantage of producing interfaces that are parallel to the surface of the wafer, and the shape of the implanted layer can be controlled to some extent by using a range of implant ions and a range of energies. Lower-energy implants of heavier ions produce layers nearer the surface. The ions are generally deposited in other than host lattice sites (often in interstitial sites) and the implantation process damages the integrity of the lattice. An annealing schedule (subjecting the sample to a short duration but high-temperature excursion) allows the crystalline order to be restored, with the implanted ions taking up lattice sites to behave as donors or acceptors (the so-called 'activation' of the implants). It is possible to control the interfaces between different levels of implant to within about 0.1 µm, and the total depth to which implants can conveniently be driven is of order of 2–5 µm. Ion implantation remains a technology in widespread use in the production of semiconductor devices, both Si and GaAs.

During the 1960s and 1970s, an initial series of epitaxial (epi meaning upon, taxis meaning arrangement) growth techniques were introduced, in particular liquid phase and vapour phase epitaxy. The emphasis here is on the over-growth of crystals on a suitable substrate, continuing with the same crystal structure while the composition of the liquid or of the vapour that supplies the material to the growing crystal is changed to reflect the desired doping of the layers being grown. Liquid phase epitaxy is similar to conventional crystal pulling, except that the melt is being changed during the growth of the crystal. Depending on the rate of crystal growth and the speed with which the liquid or vapour can be changed (without setting up turbulent patterns that might be replicated in the growing crystal), it is possible to prepare interfaces that are abrupt on the scale of about $0.01\,\mu m$, with a pair of interfaces having a separation as small as 0.1–0.2 μm. Note that changes to the liquid or vapour may not now be confined to doping, but can also include changes in chemical composition. For the first time, one can contemplate the design of structures with interfaces between (say) GaAs and AlGaAs. These techniques are still used in volume production of devices (e.g. double heterojunction lasers). Note also that layers of great total thickness (tens of micrometres) are possible. For further information see Nakajima (1985) on liquid phase epitaxy and Beuchet (1985) on vapour phase epitaxy.

During the 1980s, two newer epitaxial techniques (invented and refined in the 1970s) were introduced which are in widespread use today for the preparation of III–V semiconductor multilayers for both physics studies and device fabrication. These techniques are molecular beam epitaxy (MBE) and metal–organic chemical vapour deposition (MOCVD). With these it is possible to control both the chemical composition and the level of doping down to thicknesses that approach an atomic monolayer, and certainly to within less than one nanometre ($0.001\,\mu m$). MBE and MOCVD thus approach the absolute limits of control for preparing layered semiconductor structures. We discuss these two crystal growth techniques in the next section, and in the subsequent section we outline some of the further refinements to them that are at the level of initial investigation.

2.2 Molecular beam epitaxy and metal organic chemical vapour deposition

MBE and MOCVD are highly controlled forms of evaporation and of cracking respectively as techniques for presenting material to the surface of a semiconductor substrate that continues the growth of a semiconductor crystal. The schematic heart of the two processes is shown in Fig. 2.1, while the photographs in Fig. 2.2 give an indication of the complexity of the overall systems that result once all the relevant techniques for controlling the environment of the crystal growth are integrated.

2.2.1 Molecular beam epitaxy

An MBE machine consists of a stainless steel vessel of diameter approximate-ly 1 m which is kept under ultrahigh vacuum (10^{-11} torr) by a series of pumps.

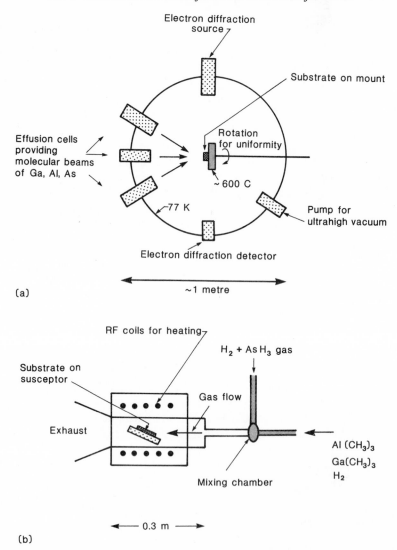

Fig. 2.1 A schematic diagram of (a) the growth chamber of an MBE machine and (b) the growth cell of an MOCVD reactor.

On one side, a number of cells (typically eight) are bolted onto the chamber. These (Knudsen) cells are of some complexity, again to control the processes for which they are responsible. Inside each a refractory material boat contains a charge of one of the elemental species (Ga, Al, or As) for growth of the semiconductor, Si (for n-type doping), and Be or B (for p-type doping). Each boat is heated so that a vapour is obtained which leaves the cell for the growth chamber through a small orifice. The vapour is accelerated by the pressure differential at the orifice and forms a beam that crosses the vacuum chamber

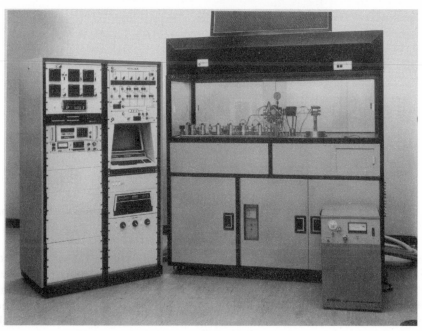

Fig. 2.2   (a) An MBE machine and (b) an MOCVD reactor. (After GEC-Marconi Ltd, Hirst Research Centre, Borehamwood, UK.)

to impinge on a (GaAs) substrate which is mounted on a holder controlled from the opposite side of the chamber. The flux rate is controlled by the temperature within the Knudsen cell. The control and monitoring of the fluxes from the different cells ensure that approximately a monolayer's worth of molecular beam species impinge on the substrate in 1 s. Thus the growth rates are typically 1 $\mu m\,h^{-1}$. Shutters in front of the orifices can be opened and closed in less than 0.1 s, and so combinations of Ga and As beams are used for GaAs growth, while the ratio of Ga to Al flux can be varied to produce the species for growing $Al_xGa_{1-x}As$ alloys. It is quite common to have two silicon cells, each set to deliver either high or low levels of doping—there is an unacceptably long thermal time constant associated with changing the Knudsen cell temperature quickly enough to increase the Si flux by (say) a factor of 100. The opening and closing of different shutters determines the multilayer structure that is grown in terms of both semiconductor composition and doping profile.

The substrate holder contains a heater, as the quality of the grown crystal is a sensitive function of the substrate temperature (about 580 °C and 630 °C is best for GaAs and AlAs). Models of the growth process suggest that atoms (Ga or Al) or molecules (As_2 or As_4) stick to the surface, but that it is hot enough for the species to migrate small distances to find correct lattice sites. The sticking coefficient of the group V elements is often lower than of the group III, so that the partial pressures of the various species are altered to compensate for this. Because of the size of the cells and the growth chamber, the substrate holder rotates about its central axis to average out any geometric variations in the incident flux density coming from the cells at the different angles. In contrast with the high temperatures of the Knudsen cells and substrate, the surrounding vessel is cooled to 77 K by liquid nitrogen. This ensures that any material in the atomic or molecular beams that misses the target on the first pass sticks to the surrounding vessel and is not able to reflect and make a second pass, so upsetting the growth pattern.

One advantage of MBE is the possibility of adding on a range of *in situ* techniques for monitoring the growth. The techniques include glancing-angle high-energy electron reflections to monitor the crystalline integrity of the growing surface (diffraction streaks of oscillating sharpness and brightness appear and fade as successive layers are grown). There are mass spectrometers to monitor the fluxes of the atomic and molecular species. Such monitoring is in addition to the routine growth of thick multilayers (i.e. effectively bulk samples) for the purpose of calibrating the growth rates, the alloy composition of multilayers, and the doping levels.

What the schematic diagram in Fig. 2.1(a) does not show, but Fig. 2.2(a) does, is the preparation chamber (the horizontal cylinder that adjoins the growth chamber from behind the sample holder). In this chamber, wafers can be introduced in batches of 10 or more and be subjected to heating, electron beam irradiation, and other forms of cleaning prior to introduction into the ultrahigh vacuum system of the growth chamber. In addition, the preparation chamber may have extra diagnostic equipment (X-ray crystallography, secondary ion mass spectroscopy, etc.) which is not needed during growth, but invaluable for a relatively rapid, but sophisticated, analysis of grown multilayers. These techniques are described further in Section 2.4.

The thermodynamics and kinetics of MBE crystal growth have been the subject of widespread investigation. Some of the findings are listed below.

1. In practice an overpressure of the group V material is required for stoichiometric growth of GaAs and the prescribed $Al_xGa_{1-x}As$ alloys.

2. The morphology of an interface between, say, GaAs and an $Al_xGa_{1-x}As$ alloy is strongly dependent on the growth conditions; it has become popular to interrupt the growth for a few seconds to allow the surface species to form as large area layers without steps.

3. The incorporation of dopant species is often not straightforward. They tend to ride on the growing surface, so that a nominally 1 nm layer of Si-doped GaAs might in fact be several nanometres thick; growth interruption is often of some help in keeping doping layers very thin.

4. Crystal growers have become increasing adventurous. Within limits set by the thermal time-constants of the Knudsen cells and the crystal growth rate, it is possible to grade the composition of $Al_xGa_{1-x}As$ layers with, say, x increasing linearly between two different values (0 and 0.35 over 50 nm).

5. It is possible to grow thin layers of materials that are not well lattice matched. The overgrowing layers tend to distort in a tetragonal fashion, retaining the substrate surface crystallography as a template but expanding or contracting the interlayer spacing; beyond a certain 'critical' layer thickness (see Chapters 13 and 14) the stored strain energy is sufficient to force the overlayer to break up into a heavily dislocated layer.

6. The range of binary, ternary, and even quaternary semiconductor materials and epitaxial multilayers that are available is increasing all the time, with metals, dielectrics, and other materials being prepared (see Chapter 21).

7. Typical layers take of the order of an hour to grow, although some thicker layers (about 10 μm) may take longer. Production-scale MBE machines are now available that grow on about 10 wafers at a time, and are used for making high-electron-mobility transistors (see Chapter 13).

MBE has taken the lead in research laboratories. Its disadvantages include its reliance on complex ultrahigh-vacuum techniques. Detailed procedures are required to re-establish the ultrahigh vacuum each time modifications must be made to the MBE chamber, including the replenishing of the material in the Knudsen cells. The advantages include the availability of elaborate *in situ* diagnostics. Further details of MBE growth techniques can be found in books of research reviews edited by Parker (1985) and by Chang and Ploog (1985).

2.2.2 Metal–Organic Chemical Vapour Deposition

MOCVD, which is also known as organometallic chemical vapour deposition (OMCVD), takes place in a glass reactor, typically about 0.3 m long and about 0.1 m in diameter. In that reactor, a heated substrate sits at an angle to a laminar flow of gas. Radiofrequency inductive heating is used to achieve substrate temperatures comparable to those for MBE growth, although research is aimed at being able to use ever lower substrate temperatures while maintaining high-quality growth. The gases that pass over the GaAs substrate are a mixture of hydrogen as a carrier gas, organometallic precursors for the group III elements (usually trimethyl gallium and/or trimethyl aluminium) and arsine as the source of As. The basic chemical reaction is

$$Ga(CH_3)_3 + AsH_3 \rightarrow 3CH_4 + GaAs,$$

with an analogous reaction for producing AlAs. The relative ratio of the organometallic precursors in the incoming gas mixture determines the chemical composition of the grown layers. Dimethyl zinc ($Zn(CH_3)_2$) or silane (SiH_4) are used to provide the dopant species.

The growth of multilayers by MOCVD involves an elaborate premixing of the required gases (some of which have to be preheated to exert a sufficient vapour pressure) in a premixing chamber, with care being taken to achieve uniform mixing without disturbing a laminar flow. The pressure inside the reactor may be near or somewhat below atmospheric pressure, but there is no question of high vacuum conditions.

Although the growth chamber of an MOCVD reactor is much smaller and less expensive than the MBE equivalent, the total costs of installing and running MBE and MOCVD growth systems are comparable. What is gained in simplicity in MOCVD is offset by the intricacy of the gas handling, and in particular the rigorous safety standards that are demanded with the use of the particularly toxic gases involved. Negative-pressure rooms, alternative venting systems in the case of malfunction or of explosive (catastrophic) failure, and the routine disposal of the excess gases are all expensive.

The level of *in situ* diagnostics is quite primitive in the case of MOCVD, and much greater reliance is placed on post-growth characterization than is the case with MBE.

The growth rate attained with MOCVD is typically about 10 times that of MBE. MOCVD is more widely used and favoured as the growth technique in the production of commercial devices. The interfaces between adjacent layers are not proven to be as abrupt as with MBE. The technology is a recognizable improvement over vapour phase epitaxy. The system is easily scaled up to grow tens of wafers at a time. MOCVD-grown material, and AlAs in particular, suffers from the incorporation of carbon as a donor level (from any incompleteness of the cracking process leaving carbon-based radicals) and of oxygen (as a contaminant of trimethyl aluminium). These impurities cause some difficulties in the growth of the very-high-performance microwave devices described in Chapter 17, where detailed device models may not incorporate these materials, or even if they do, only in ill-defined concentrations. However, the best semiconductor laser devices have been made routinely with MOCVD rather than MBE.

Some of the most remarkable crystals have been grown with MOCVD. If one is prepared to go to great lengths with the gas-flow controllers and package up small bursts of material for growth of single layers or fractions of a layer, a high degree of control can be achieved. In particular the growth of alternative half-layers of GaAs and AlAs to form a vertically propagating superlattice has been achieved by MOCVD, where as the MBE equivalent is known as a serpentine lattice because it wavers owing to lack of adequate control over the growth (see Fig. 2.3 for a comparison).

For more details of the science and technology of MOCVD growth see Razeghi (1989).

2.3 Recent developments in crystal growth

The principles behind both MBE and MOCVD growth were established in the 1970s and refined in the 1980s, since which time further developments have taken place. Once control over growth in the depth direction has been achieved that approaches atomic layer precision, there is no further room for improvement. Nevertheless there are several areas of growth research whose benefits will be realized during the 1990s.

Chemical beam epitaxy (CBE) or gas source MBE (GSMBE) is the name given to a method of crystal growth (Tsang 1989) that combines the ultrahigh-vacuum advantages of MBE with the gas-source advantages of MOCVD. There is no need to bring a vacuum up to atmosphere to replenish the stocks of material for growth, and there is a greater control over the gases to the extent that a wider range of graded composition layers are feasible.

The main advances have concerned the patterning of layers during their growth, and several techniques have been investigated. The most primitive is growth through masks that allow selected area deposition. This works on relatively large-scale structures, but shadowing effects limit the resolution of features of the smallest areas that can be grown. Submicrometre lateral features are ruled out by fringing edge effects. If there is to be a future for hybrid GaAs-on-Si circuits, then large areas of GaAs epitaxial layers suitable for whole circuits must be prepared on a Si substrate. Substrates are being patterned before growth. V-grooves have been used, and their infilling with overgrown multilayers produces structures with narrow (less than $0.1\,\mu m$) linear features in the bottom of the grooves (see Fig. 2.4). Attempts to modify the local surface temperature, and hence the local growth rate, have been performed using focused laser and electron beams. Some success has been achieved, but the degree of control seems insufficient to date for widespread applications.

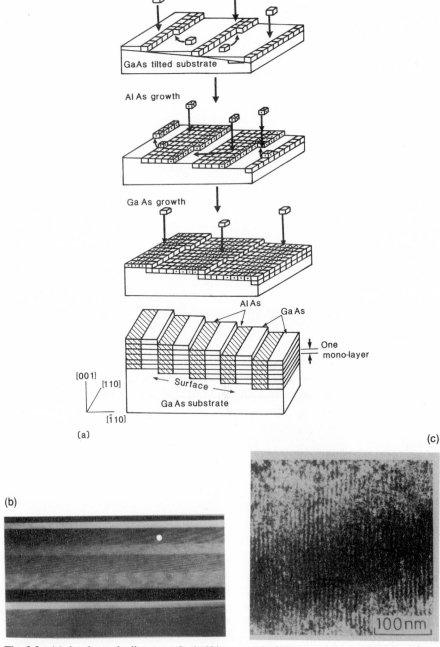

Fig. 2.3 (a) A schematic diagram of a half-layer vertical propagating superlattice, and the superlattice as realized by (b) MBE growth (after Petroff *et al.* 1992) and (c) MOCVD growth (after Fukui 1992).

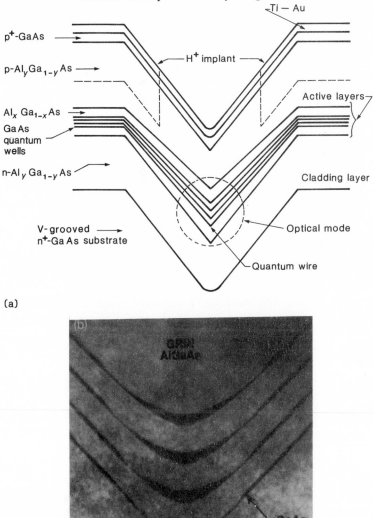

Fig. 2.4 The growth of multilayers for a quantum wire laser on V-grooved substrates: (a) the overall multilayer structure; (b) a transmission electron microscope cross-section of the quantum wells and quantum wires. (After Kapon *et al.* 1992.)

One promising technique is the incorporation of focused ion beams (FIBS) within an MBE machine. Several groups have reported success. An ion beam column works on principles similar to that of an electron beam column (as used in electron microscopes or lithography machines (see Chapter 3)). A liquid metal ion source produces a tip less than 0.01 μm in diameter from which ionized metal atoms emerge. These can be focused to *c*.50 nm spots and steered over the surface of a wafer during growth, or during an interruption in growth, implanting the ions into the surface with either high or low net energy. To date, Ga ions have been used to induce damage, and so turn conducting volumes into insulating volumes, while Sn ions (and soon Si ions) have been used to produce local volumes of n-doped material. To date, fairly coarse structures have been proven, but this FIB–MBE system seems the first likely to produce genuinely three-dimensional engineered crystals in useful volumes (Jones *et al.* 1992).

2.4 Characterization of semiconductor multilayers

Having specified to a crystal grower a sequence of layers by their thickness, chemical composition, and doping levels, it is most important at a research stage and later at a production stage to be able to verify that the layers as grown are as specified. In this section we describe briefly some of the routine screening techniques, as well as some of the more elaborate techniques for qualifying layers (i.e. checking that they are as specified). In practical terms, there is often a trade-off between a rapid turnround of results using a relatively crude technique, but one which can feed back information in time to correct the following wafer that is grown, and obtaining results in great detail suitable for accurate modelling of the physics and device results.

2.4.1 Electrical characterization

The simplest method for checking the doping levels in bulk semiconductors is the Hall measurement. In the experiment shown in Fig. 2.5, an electric field is induced along the *y*-axis by the Lorentz force (ev_xB_z) exerted by a magnetic

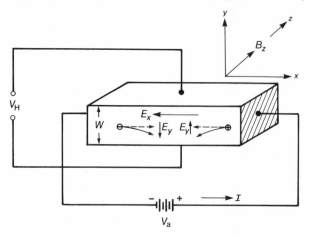

Fig. 2.5 Hall measurement of the transverse field established by a magnetic field applied at right angles to a current.

field in the z direction when electrons are moving with a drift velocity in the x direction (Sze 1981; Seeger 1991):

$$E_y = R_H J_x B_z,$$

where $R_H = (r/e)(p - b^2 n)/(p + bn)^2$, with $b = \mu_n/\mu_p$ and $r = \langle \tau^2 \rangle/\langle \tau \rangle^2$ is the Hall coefficient in the case of bipolar conduction. The theory for r has been worked out for various types of scattering process (τ is the mean free time between carrier collisions), with values of 1.18 for lattice scattering and 1.93 for impurity scattering. In practical cases, where one is dealing with dominant hole or electron conduction, one obtains R_H from the measured E_y and obtains the carrier concentration directly as $R_H = r(-1/en)$ if $n \gg p$ and $R_H = r(1/ep)$ if $p \gg n$. The size and sign of the Hall coefficient gives the concentration and type of the dominant carrier.

In the case of low-dimensional electron systems, particularly the single two-dimensional electron gas of Chapter 5, the above results remain true if the magnetic field is placed in a direction perpendicular to the plane of the electron gas.

The situation is less simple if there are spatial variations in the doping profiles, and two techniques are used to extract the carrier concentration as a function of depth, namely capacitance–voltage profiling and secondary ion mass spectrometry. In the former, use is made of the *depletion approximation* (see Section 2.5.1 for further details) to show that the concentration $N^*(x)$ of majority carriers at the edge of a depletion region can be found from a differential capacitance–voltage measurement as

$$N^*(x) = \frac{-C^3}{(\varepsilon_0 \varepsilon_s A^2 e)} \frac{1}{dC/dV}$$

This technique has been used to obtain good information on $N^*(x)$ through a multilayer (Fig. 2.6). See Thomas (1991) for more details.

Secondary ion mass spectrometry (SIMS) involves removing material from a multilayer structure using a beam of high-energy ions (i.e. sputtering) and a mass analysis of the species that come from the flat centre of the crater. This technique is particularly useful for multilayers and has the advantage of a large dynamic range. The alloy composition and the doping profile can be measured simultaneously (see Fig. 2.7 for an example). The caveats are as follows.

1. SIMS measures all the dopant ions, whether activated or not, so that calibration runs are required, particularly for highly doped material (i.e. more than $3 \times 10^{24} \, m^{-1}$ where only partial activation of donors occur).

2. The depth resolution can be good if care is taken to perform the experiments very slowly, and is capable of revealing repeated 1.5 nm-thick layers of AlAs only 4 nm apart and buried 0.5 μm below the surface of an otherwise GaAs structure.

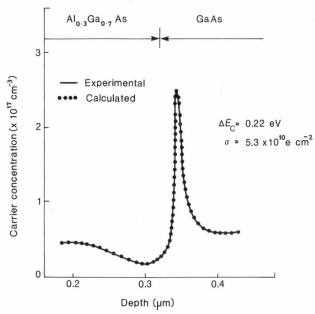

Fig. 2.6 A *C–V* plot used to extract doping details of a heterojunction using techniques described in the text. (After Thomas 1991.)

3. SIMS can measure the doping profile, but unfortunately has some difficulty with the detection of Si in AlGaAs materials as AlH has a mass that cannot be differentiated from Si, and there is always some residual hydrogen, even from pump oil in the evacuated chamber in which the experiments are performed.

  SIMS data have always been of great use in the verification of the design of practical device multilayers. For more details on the SIMS technique, see Sykes (1989).

2.4.2 *The optical characterization*

Photoluminescence, electroluminescence, and photoreflectance spectroscopy have all proved useful in the qualification of certain multilayer structures. The first two techniques involve the examination of the wavelength- or energy-selected emission of light from a structure when the structure has first been excited by light or an electron beam respectively. The last technique involves changes of reflectivity at different energies as small electric fields are applied to the solid. All three provide information about optical transitions, and we shall see in subsequent chapters (and in particular Chapter 10) the way in which thin layers have optical transitions modified by quantum-size effects including the shape of heterojunction interfaces (atomically abrupt changes of composition, or changes over two or more atomic layers, with or without lateral steps in the interface plane).

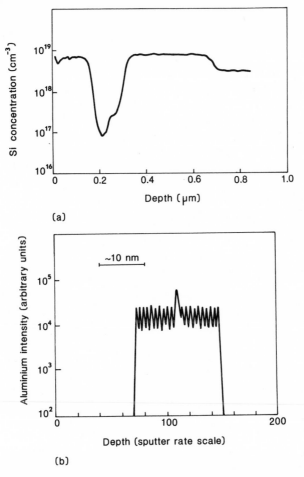

Fig. 2.7 A SIMS profile of doping and composition of a multilayer structure. The midpoint of the Al profile is coincident with the step in doping density in the low-doped region. (After Davies *et al.* 1989.)

Photoluminescence (PL) spectra are obtained by flooding a solid with high-energy light which is absorbed, exciting some electrons which eventually relax back to their original energies. These excited electrons may emit light from the fully excited or any partially relaxed states. The structure in the PL spectra as a function of energy can be used to infer the position of various energy levels—in particular the bandgaps of the different layers—while the linewidths of the PL features can be interpreted in terms of uniformity and absence of fluctuations. PL as a diagnostic technique has a number of advantages: it is non-destructive, fairly simple to implement, able to give a rapid turnround of information to crystal growers, and capable of use in a wafer-mapping mode (i.e. checking out the uniformity of layers in a manufacturing environment). The disadvantages include complications that arise if one has (as is often the case) an active device structure with heavily doped contact layers on either side; such doped layers tend to reabsorb and redistribute the luminescent energy. The qualitative results produced are particularly useful for many structures of optical devices (lasers etc.). In some cases luminescence from the substrate can mask effects from the overgrown multilayers. Changes to the PL spectra under the influence of temperature changes, applied stress, magnetic fields, etc. all provide further insights into the thicknesses, compositions, and uniformity of multilayers. Photolumines-cence excitation (PLE) spectroscopy involves the detection of light at a fixed energy (generally near a prominent band-edge feature, e.g. the lowest bandgap), while the excitation energy is monochromatic and is swept in energy. The amplitudes of features in the PLE spectra provide further information about the cross-section of the optical absorption processes involved in the relaxation of the excited electrons. Typical PL and PLE data for a range of thin layers of GaAs between thicker AlGaAs layers is shown in Fig. 2.8.

Electroluminescence is a similar optical emission process, but one where the initial excitation of electrons in the multilayer is brought about by high-energy injected electrons (as in an electron microscope) that are localized in space.

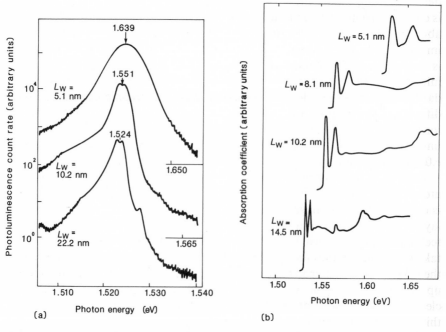

Fig. 2.8   (a) PL spectra for wells of thickness (A) 5.1 nm, (B) 10.2 nm, and (C) 22.2 nm thickness with the curves offset for clarity, and (b) the PLE spectra from GaAs quantum wells. (After Weisbuch and Vinter 1991.) The relationship of peak energy to well width is treated in more detail in Chapter 10.

This technique is less widely used as a routine characterization, but is a useful adjunct to the electron microscopy described in the next section.

Photoreflection spectroscopy is another specialized technique where the optical reflection is monitored as a function of light energy while the sample is subject to a time-varying electric field. The in-phase signal is a measure of the way that optical transitions vary under electric fields (as we shall see in further detail in Chapters 4 and 10). The changes are most prominent when associated with energies at bandgaps, zone edges, etc., and their energy can be used to infer structure of layers within a multilayer stack. Further descriptions of these optical methods as applied to characterizing semiconductor multilayers are given by Delalande and Voos (1986) and Clausen *et al.* (1990).

2.4.3 Transmission electron microscopy and structural characterization

In its various forms, transmission electron microscopy (TEM) has proved the most useful analytical tool during the development of new physics and new devices in III–V semiconductor multilayers. High-energy (*c.* 50 keV) electrons diffract and scatter off the ion core potentials of solids as they pass through approximately 0.1 μm thick layers. The electrons can be collected in a number of ways (in diffracted beams, with fixed amounts of energy loss, etc.), and each way provides complementary information about samples. TEM is used to determine both the structure and chemistry on an atomic scale. The technique is currently being refined in the hope of obtaining doping profiles and of being able to image individual atomic steps at interfaces. In the so-called lattice-imaging mode, the Fourier transform of the Fourier transform of the sample is displayed, and columns of atoms are 'imaged', from which structural information can be read off. In a lower-magnification mode (the so-called dark-field technique) selected diffraction spots are recombined and imaged; differences show up in, for example, the relative scattering from Al and Ga ions. This technique provides an image that gives the thickness, planarity, uniformity, and overall integrity of layers over linear distances of order *c.*0.1 μm (Fig. 2.9).

TEM gives the most vivid picture of semiconductor multilayers down to the atomic scale. However, it has a number of disadvantages. First, the technique is destructive. Clear images come from transmission of electrons through thin layers (*c.*0.1 μm maximum), and elaborate and time-consuming thinning techniques are required for the preparation of suitable samples. One must take care that the thinning process does not introduce its own structure into the sample—this is a problem with strained layers, and with AlAs which takes up water and oxygen very effectively. A recent rapid technique involves cleaving a wafer twice at right angles and sending the electrons through the thin section at the very corner; however, this gives information at only one location in a wafer, with no possibility of knowing whether it is typical. Properly thinned samples allow information to be obtained along a line across the surface of a wafer that is *c.*0.1 μm long. This precludes the use of TEM in any form of wafer-mapping mode.

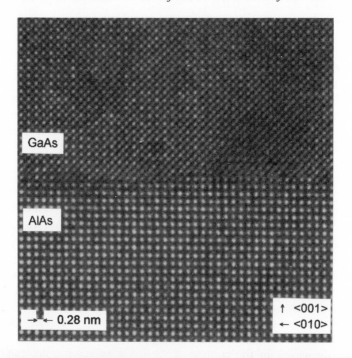

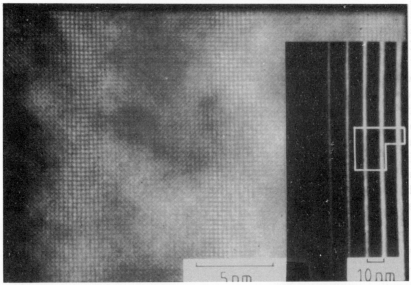

Fig. 2.9 Different TEM images from semiconductor multilayers. Different diffraction conditions can be used to accentuate the crystalline perfection at near-atom-scale resolution, or the well thicknesses and uniformity based on chemical composition contrast. (T. Walther and W. M. Stobbs respectively, University of Cambridge, private communication.)

The really detailed information from TEM that is useful for physics and device studies comes from the comparison of TEM images (often using defocusing of the image as an extra variable) with the predictions from increasingly elaborate simulations of the behaviour of electron beams in the putative multilayer samples; the thicknesses and compositions of each layer are used as fitting variables. Information from thick, well-characterized, and uniform calibration layers is built into the simulation packages. At present, layer thicknesses can be determined down to less than ±0.5 nm for layers anywhere from 1 nm thickness upwards, and composition (as in the fraction of Al in AlGaAs layers) can be determined to within ±2 per cent. TEM has monitored the improvement in quality of MBE and MOCVD layers from the earliest days of these growth techniques, and has been used throughout in the analysis of defects whether contained in the starting wafer surface (including deliberate V-grooves as shown in Fig. 2.4) or arising during the growth. More recently, TEM has provided useful information about the preservation of the quality of semiconductor multilayers as they are processed into devices (being subjected to thermal, chemical, and other treatments (see Chapter 3)). TEM is also used to monitor changes in the multilayers that reflect the ageing of devices, and indeed in the analysis of device failure itself. The structural and chemical information provided by TEM has made it the single most important analytical tool available. Its inability to handle dopant levels remains a significant drawback in the context of device development projects. Futher information on this technique can be found in Stobbs (1986), Britton *et al.* (1987), and Ross (1988).

X-ray diffraction was used earlier in this century to verify the crystal structure of common solids, and is used today in determining the structure of exceptionally complex biological molecules. The types of multilayers that we encounter are also suited to X-ray analysis, and in some laboratories the technique is used as a complement or an alternative to TEM. The angular position and the strength of many diffracted beams are used to reconstruct the multilayer system that gave rise to them. The accuracy approaches that of TEM in the direction perpendicular to grown layers, but X-rays do not have the same local sensitivity to variations in the plane of the layers. X-ray diffraction has the advantage of being non-destructive and can be used in a wafer-mapping mode. Fewster (1993) gives more details of X-ray diffraction in the present context.

2.4.4 The limitations on characterization

The development of multilayer physics and devices over the last 20 years has prompted the development of many of the analytic tools themselves. There remains a close symbiosis between the semiconductor science and materials analysis communities. The situation is ever-evolving. The questions that physicists ask about the semiconductor structure often cannot be answered, and the implications of some of the data from analytical studies are not always incorporated into the physical models. For those developing practical devices, the performance of some of which depend very sensitively on the thickness, uniformity, and position of critical layers, no single analytical technique suffices on its own and the information from many techniques must be integrated. Several projects with precisely this aim are in progress (see Chapter 8, Section 8.5, for an example).

We encounter the following problem with some of the structures described in Chapter 11. Electrons are tunnelling down a conducting column of GaAs only about 0.1 μm in diameter and through thin $Al_xGa_{1-x}As$ barrier layers. If the electron has a wavelength of *c*.20 nm (see Appendix 2 and Chapter 3), how precisely does the electron wavefunction sample the barriers and any lateral steps across the interfaces? Further refinements to all the electrical, optical, and structural analytical techniques described above are necessary to provide inputs to an adequate answer to this question. The science–materials analysis symbiosis will continue.

2.5 Basic electronic structure of homojunctions and heterojunctions

In this section we consider the electrostatics associated with the principal types of atomically abrupt interfaces that we shall encounter. We consider homojunction interfaces between n^+- and n^--GaAs, and between n- and p-GaAs. Here n^+, n, and n^- refer to heavy, normal, and light doping (typically $c.10^{24} m^{-3}$ $10^{23} m^{-3}$ and less than $10^{22} m^{-3}$. We shall also consider a heterojunction between GaAs and AlGaAs alloy, both in the general case of uniform doping and the special and important case where the doping is confined to the alloy (known as modulation doping).

2.5.1 The p–n homojunction in GaAs

The p–n junction is treated in standard textbooks (Ashcroft and Mermin 1976; Sze 1981; Kittel 1986) and we summarize the results here, referring the reader to these texts for the details. The basic physics is contained in Fig. 2.10: some of the electrons from the n region fill hole states in the p region, and we are left with fixed donor and acceptor changes that generate an electric field and build up a potential. With no applied bias, there is no current flowing through the sample and the Fermi energy is constant throughout the sample. The built-in potential eV_{bi} ensures that there is no net diffusion current. This allows us to write (with reference to Fig. 2.10)

$$eV_{bi} = E_g - (eV_n + eV_p)$$

where V_n and V_p are the Fermi energies relative to the band edges deep inside the respective materials. If the equilibrium carrier densities there are n_{n0} and p_{p0}, we can use the results in the previous chapter to write

$$E_g = kT \ln(N_C N_V/n_i^2) \qquad eV_n = kT \ln(N_C/n_{n0}) \qquad eV_p = kT \ln(N_V/p_{p0}),$$

so that

$$eV_{bi} = kT\ln(n_{n0}p_{p0}/n_i^2) + kT\ln(N_A N_D/n_i^2)$$

where N_A and N_D are the acceptor and donor densities on either side and n_i is the intrinsic carrier concentration.

In thermal equilibrium the electric field far from the junction must vanish, so that we have a charge neutrality condition (in fact we have built up a dipole layer) in the vicinity of the p–n junction which can be expressed in terms of the thicknesses of the exposed layers of donor and acceptor ions:

$$N_A x_p = N_D x_n.$$

To obtain the overall thickness $W = (x_n + x_p)$ of the p–n junction, we set up and solve the Poisson equation, first for the electric fields

$$\mathcal{E}(x) = -eN_A(x + x_p)/\varepsilon_s \text{ for } x_p \leqslant x \leqslant 0$$

and

$$\mathcal{E}(x) = -\mathcal{E}_m + eN_D x/\varepsilon_s = (eN_D/\varepsilon_s)(x - x_n) \text{ for } 0 \leqslant x \leqslant x_n$$

where $\mathcal{E}_m = eN_D x_n/\varepsilon_s = eN_A x_p/\varepsilon_s$, and then for the potential $V(x)$ and the built-in potential V_{bi}, bearing in mind the limiting values and boundary conditions implicit in Fig. 2.10,

$$V(x) = \mathcal{E}_m(x - x^2/2W) \qquad V_{bi} = \mathcal{E}_m W/2 = \mathcal{E}_m(x_n + x_p)/2$$

We can eliminate $\mathcal{E}_m$ and arrive at a value of W in terms of the built-in potential:

$$W = \sqrt{[(2\varepsilon_s/e)(1/N_A + 1/N_D)V_{bi}]}.$$

In the case of a one-sided junction, where the density of either the donors or acceptors is much higher, the thickness reduces to

$$W = \sqrt{[(2\varepsilon_s/eN_B)V_{bi}]}$$

where N_B is the smaller of N_A and N_D. There is a further refinement: if one corrects the charge density on either side from its assumed value by virtue of dopant ions to include the majority carriers left in the junction region, a minor correction applies as the built-in voltage is reduced by an amount $2kT/e$, and one can introduce a Debye length L_D characterized by the majority dopant and temperature, from which the more general depletion distance is given by

$$W = L_D\sqrt{[2(V_{bi}/kT - 2)]}, \quad L_D = \sqrt{[\varepsilon_s kT/e^2 N_B]}$$

As an aside here we derive the equation used earlier in the capacitance–voltage method of determining the doping density at an interface. The differential capacitance per unit area for an abrupt one-sided junction can be defined in terms of changes in the charge Q_c in the depletion layers by incremental changes of $\pm V$ to the built-in potential as

Space charge

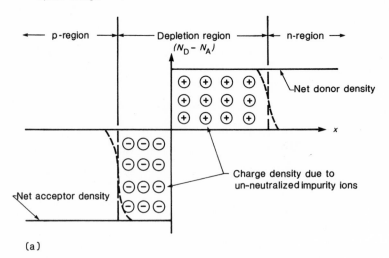

(a)

Built-in electric field

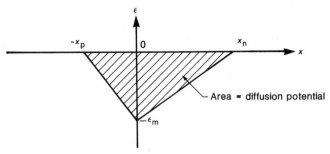

(b)

Potential variation

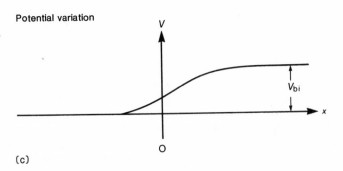

(c)

Energy band diagram

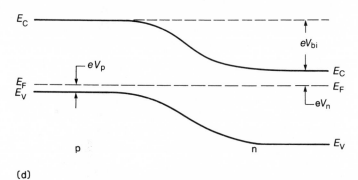

(d)

Fig. 2.10 The space charge, built-in electric field, potential variation, and energy band diagram at a p–n junction. (After Sze © 1981 Reprinted with permission of John Wiley and Sons Inc.).

$$C = \frac{dQ_c}{dV} = \frac{d(eN_B W)}{d(eN_B W^2/2\varepsilon_s)} = \frac{\varepsilon_s}{W} = \frac{\sqrt{[\varepsilon_s N_B/2]}}{\sqrt{[V_{bi} \pm V - 2kT/e]}}$$

which can be rewritten as

$$1/C^2 = 2(L_D^2/\varepsilon_s^2)(\beta V_{bi} \pm \beta V - 2),$$

where $\beta = 1/kT$, from which

$$d(1/C^2)/dV = (2/(e\varepsilon_s N_B)) \quad (= L_D^2 \beta/\varepsilon_s^2).$$

Although the result derived here is for a special case of uniform $N(x) = N_B$, it does apply more generally, with $d(1/C^2)/dV = 2/(e\varepsilon_s N(W))$.

In order to derive the current–voltage characteristics of a p–n junction, the abrupt depletion model is used, the Boltzmann approximation for the carrier densities is assumed valid everywhere, the minority injected carrier densities remain small, and no generation occurs in the depletion layer so that hole and electron currents are equal. From these assumptions, the drift–diffusion model can be used to derive the distribution of holes in the n region in an electric field E and vice versa:

$$-(p - p_{n0})/\tau_p - \mu_p E \partial p_n/\partial x + D_p \partial^2 p_n/\partial x^2 = 0.$$

Here p_{n0} is the hole density in the n region ($p_{n0} = p_{p0} \exp(-eV_{bi}/kT)$), μ_p is the hole mobility, D_p is the diffusion constant of holes, and τ_p is the ambipolar lifetime (of holes in n-type material). We concentrate on the steady state solutions. Outside the depletion regions, where there is no field, the solutions are simple:

$$p_n - p_{n0} = p_{n0}[\exp(eV/kT) - 1]\exp[-(x - x_n)/L_p]$$

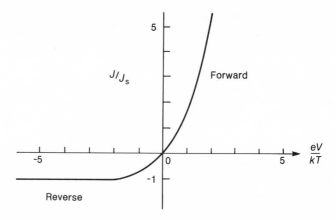

Fig. 2.11 The current–voltage characteristics of a p–n junction diode, showing the low limiting forward resistance and high limiting reverse bias resistance.

where $L_p = \surd(D_p \tau_p)$. Thus at the boundaries of the depletion region (where the field has gone to zero) we have

$$J_p = -eD_p \partial p_n / \partial x|_{x_n} = (eD_p p_{n0} / L_p)\,[\exp(eV/kT) - 1].$$

The sum of the electron and hole currents gives the total current:

$$J = J_n + J_p = J_s [\exp(eV/kT) - 1]$$

where

$$J_s = eD_p p_{n0} / L_p + eD_n n_{p0} / L_n$$

L_n and L_p are diffusion lengths for electrons and holes in gradients of carrier distributions. The form of the current–voltage characteristic is shown in Fig. 2.11. There is a low resistance in the forward bias direction and a high resistance in the reverse bias direction. Note also that there is an exponentially strong temperature dependence, which can present problems in some practical applications (see the discussion in Chapter 17).

2.5.2 The n^+–n^- homojunction in GaAs

In a number of more complex structures (e.g. the planar-doped barrier in the next section, and the tunnelling structures in Chapter 8), there are interfaces between GaAs layers that are heavily and lightly doped. We can infer from the limits of heavily and lightly doped materials the form of the conduction band edge since, in the absence of bias, there is a well-defined Fermi energy (strictly chemical potential, but this terminology is not widely used in the semiconductor literature) separating filled and empty states throughout the structure. The situation is shown schematically in Fig. 2.12 for a symmetric n–i–n structure. Asymmetry complicates the analysis with a built-in potential without adding new physics. To a first approximation, a dipole layer is

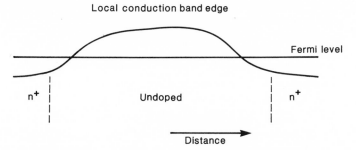

Fig. 2.12 Homojunctions between heavily and lightly doped GaAs in a n–i–n structure, showing the spatial variation of the bottom of the conduction band for a symmetric structure under no bias with a constant Fermi level.

formed as some electrons 'spill over' from the heavily doped to the lightly doped layer. We are interested in the total charge and the length scale associated with this spillover, and subsequently in the changes in these when a small bias is applied. We cannot use the simplest depletion approximation as we did for the p–n junction, as there is no neutralizing fixed charge to form a localized dipole layer. However, the Debye length scale L_D of the previous section can be used. All the electrostatics is based on mobile charge diffusing into the low-doped region from the high-doped layer. First there is the Poisson equation for the potential $\phi(x)$ from which we also regard $e\phi(x)$ as the local bottom of the conduction band from which the kinetic energy of electrons are measured:

$$\partial^2 \phi(x)/\partial x^2 = (e/\varepsilon_s)[N(x) - \rho(x)].$$

Here $N(x)$ is the distribution of positive ions dropping from a high value to a low value at $x = 0$, and $\rho(x)$ is the mobile charge in the conduction band. The second equation is a summation over the occupied states in the conduction band ($g(\varepsilon)$ is the density of states (see Chapter 4)) to obtain the charge density, i.e.

$$\rho(x) = \int \frac{g(\varepsilon)\,d\varepsilon}{1 + \exp[\varepsilon - \mu(x) + e\phi(x)/kT]}$$

Under zero bias $\mu(x) = E_F$ throughout the structure. These equations must be solved numerically and self-consistently in the practical cases of interest to us. Well into the lightly doped region, where the term +1 in the denominator can be neglected (the so-called Boltzmann approximation) and integrals can be performed analytically (see Section 2.6, Appendix 4, and Luryi (1990)), leading to a current linear in bias for small biases, crossing over to a space-charge-limited V^2 dependence once the bias exceeds about $10\ kT/e$. However, it has been found that in the detailed simulation of the current–voltage characteristics in device-type structures, there is no alternative to a full numerical solution. The n^+–n^- interface is not encountered in isolation, but rather a pair are used back to back. In later sections we encounter tunnel barriers etc. within the i layer, but the physics of an n^+–i–n^+ structure (i = intrinsic or very lightly doped) structure has been worked out, and we shall encounter this in Chapter 8.

2.5.3 The ideal semiconductor heterojunction

Without enquiring at this stage into the detailed physics of electrons in various bonds or bands, let us consider the gross properties of a heterojunction between two semiconductors, both of which, for convenience, we assume to be direct-gap materials (e.g. GaAs and $Al_xGa_{1-x}As$, with $x < 0.4$). If we consider separate pieces of semiconductor, i.e. not in intimate contact, and we set aside some details associated with electronic structure specifically associated with free surfaces, we shall have to consider bandgaps, electron affinities (the energy of an extra electron with respect to the vacuum level, i.e. the bottom of the conduction band), and work functions (i.e. the Fermi level, which is the energy from which one electron could be taken). The situation might be as in Fig. 2.13.

The principles on which the band alignments are made are as follows: (i) there is no distortion of the energy bands of the constituent semiconductors right up to the interface (this cannot be correct, but will be discussed subsequently); (ii) in the absence of an external bias, there is a Fermi energy common to the entire structure; (iii) there may be some displacement of mobile charge in order to achieve the alignments. Even these assumptions do not account for the relative discontinuity of the conduction and valence band edges at the interfaces: how much of the discontinuity appears in the conduction band edges and how much in the valence band edges? This question is of considerable theoretical and experimental subtlety, and we return to it in Section 2.8 and in Chapter 13, Section 13.2.

These heterojunctions are pervasive throughout this book in the context of both physics investigations and device applications. Various aspects of their electronic, optical, elastic, and other properties are discussed in subsequent chapters.

2.5.4 The metal–semiconductor interface

The metal–semiconductor interface and the types of electrical contacts that result have been widely investigated (Rhoderick and Williams 1988). If a metal is deposited on a clean semiconductor under ultrahigh-vacuum conditions, electrons from the metal fill surface electron states in the semiconductor and form an electrostatic barrier to further motion of electrons between the metal and semiconductor; this is known as the Schottky barrier, after the physicist who first identified it. The height of the barrier is related to the amount of charge transfer, and the thickness is defined by the Debye length or depletion width in the semiconductor. The detailed physics and technology for producing reliable and reproducible Schottky barriers has been a continuing problem for 40 years. The nature of the electrical contact is intimately

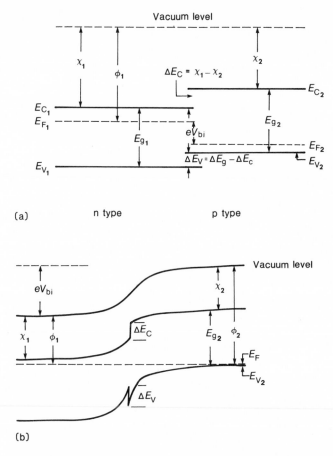

Fig. 2.13 The band diagrams of (a) an n-type and a p-type semiconductor in different materials, and (b) the abrupt heterojunction formed from them. In (a) the energy gaps E_{g}, conduction and valence band edges E_{v} and E_{c}, electron affinities χ, and work functions ϕ correspond to bulk values of the separate materials. In the heterojunction charge is transferred (as in a p–n junction) with the resulting built-in potential eV_{bi}. The Fermi level is a constant for the structure when not under applied bias, and the other quantities vary as shown. (After Pearton and Shah 1990 from S. M. Sze © 1990. Reprinted with permission of John Wiley and Sons Inc.).

associated with the atomic-scale morphology of the metal–semiconductor interface, including the effects of local oxide layers, metal grain boundaries, and defects in the semiconductor. Where there is strong alloying of the metal and semiconductor, and the metal (is an alloy which) contains dopant species for the semiconductor, the effective Schottky barrier can be small and very thin and the I–V characteristics at room temperature are linear; this is the ohmic contact. The efforts of scientists and technologists are devoted to reducing the resistance of this contact.

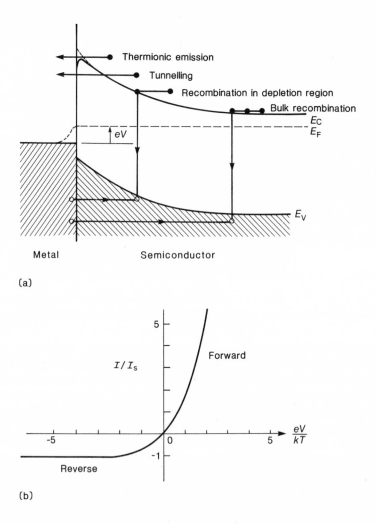

Fig. 2.14 (a) The Schottky barrier and (b) its current–voltage characteristics (note the similarity with the p–n junction in Fig. 2.10). The two principal mechanisms by which electrons enter the metal under forward bias are (1) thermionic emission from the barrier (the dominant process) and (2) tunnelling through the top of the barrier. (After Rhoderick and Williams 1988.)

Figure 2.14 shows a schematic diagram of the energy band profile for a metal–semiconductor interface, showing the two most prominent current mechanisms, thermionic emission over the top of the barrier, and tunnelling through it. The probability of finding electrons at an energy E above the Fermi energy of the semiconductor varies as $\exp(-E/kT)$, and thermionic emission results from collecting those electrons travelling towards the metal and having sufficient energy to exceed the barrier. The net current for bias V is of the form

$$I_{s \to m} = \text{Area} \times A^{**} T^2 \exp(-e\phi_B/kT)[\exp(+eV/kT) - 1] = I_s \exp(+eV/kT) - 1]$$

where ϕ_B is the Schottky barrier height, is shown in Fig. 2.14, along with the *I–V* characteristics. The Richardson constant is given by

$$A^{**} = 4\pi e m^* k^2/h^3$$

where all symbols have their usual meanings. Note the strong temperature dependence of the current. The form is similar to the p–n junction, except that the physics for the saturation current I_s is quite different. Under bias the value of ϕ_B hardly changes; this is not the case with the new planar doped barrier diode described in the next section as a replacement for the Schottky diode. Practical Schottky diodes, where other current mechanisms (tunnelling or recombination) apply, are fitted to a form with a factor $\exp(-eV/nkT)$, where the factor n is called an ideality factor. For diodes whose properties are dominated by thermionic emission, $n < 1.1$.

2.6 The planar doped barrier

We are now in a position to describe one multilayer structure in GaAs consisting of an n–i–p–i–n doping profile, as shown in Fig. 2.15(a) along with the energy band profile under a small, but finite, bias in Fig. 2.15(b). In Chapter 17 we describe the success of this structure as a replacement for a metal–semiconductor (Schottky) diode in applications as a microwave mixer or detector. The principle of this structure is simple. A thin layer of heavily p-doped material (say about 4 nm of about $10^{24} \, m^{-3}$) is placed within a layer of intrinsic (undoped) GaAs, which in turn is clad with heavily n-doped GaAs as contact layers. The acceptors in the thin layer each capture an electron, creating a potential barrier for the flow of electrons. In the absence of any bias, we can define the Fermi energy in the n+-GaAs layers and recognize that there is no electric field in these layers. We can use Gauss's theorem to give the change in electric field on passing through the negatively charged p layer of doping–thickness product $N_A t$ per unit area:

$$E_R - E_L = eN_A t/\varepsilon_s.$$

If for now we ignore the subtleties of the spillover charge and the depletion charges at the two n–i interfaces (Fig. 2.15(a)), the doping–thickness product $N_A t$ and the lengths of the two intrinsic layers (L_1 and L_2 in Fig. 2.15) are sufficient to determine the entire shape of the barrier. The asymmetry is given by the ratio L_1/L_2, and the barrier height ϕ is obtained by solving Gauss's theorem:

$$\phi = eN_A t L_1 L_2/[\varepsilon_s(L_1 + L_2)].$$

In practice the Poisson equation has been solved with the correct carrier statistics equations throughout the structure to allow for spillover and depletion, but the simple geometrical argument above accounts for over 90

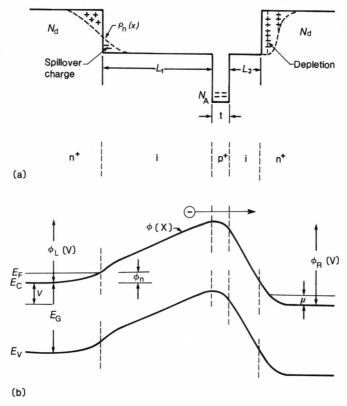

Fig. 2.15   (a) The doping profile and (b) the electrostatic potential profile of the planar doped barrier. (After Kearney and Dale 1990.)

per cent of the barrier height. While this agreement might seem satisfactory, the principal current mechanism for this structure is thermionic emission over the electrostatic barrier which has an exponential dependence on barrier height. Quantitative agreement can be achieved between theoretical simulations and the experimental data if one accounts for the detailed shape of the potential at the two n–i interfaces and throughout the finite thickness of the p-doped layer.

The current–voltage characteristics can be obtained in a manner similar to that for the p–n junction, or indeed for Schottky diodes. The potential energies $\phi_L(V)$ and $\phi_R(V)$ are obtained by solving the Poisson equation for the potential profile. With the Boltzmann approximation, we write the electron charge densities on the left- and right-hand sides of the barrier as

$$\rho_L(x) = N_D\exp[-e\phi(x)/kT], \qquad \rho_R(x) = N_D\exp\{[eV - e\phi(x)]/kT\}$$

where we have assumed that the charge density in the contact regions is N_D and $\phi(x = -\infty) = 0$. If the potential maximum lies a distance z into the p^+-doping layer, the electric field within the left-hand intrinsic layer is

$$E_L = eN_Az/\varepsilon_s$$

The Poisson equation and its first integral within the left-hand n region are

$$\partial^2\phi/\partial x^2 = (e/\varepsilon_s)[N(x) - \rho(x)]$$

and

$$\partial\phi/\partial x = \sqrt{[2N_D/\varepsilon_s]}\sqrt{[e\phi + kT\exp(- e\phi/kT) - kT]}$$

respectively.

The electric field at the n–i interface must be continuous (i.e. equal to E_L), and if the spillover charge is small we neglect the exponential term. Together these determine the value of the potential step at the interface:

$$e\phi_n = (e^2N_A^2z^2/2\varepsilon_sN_D) + kT.$$

The potential drop over the intrinsic arm is eE_LL_1, and the potential from the start of the p^+ layer a further distance z into the peak is

$$\delta\phi = eN_Az^2/2\varepsilon_s.$$

Summing all these terms gives the left-hand barrier:

$$\phi_L = (eN_Az/\varepsilon_s)[L_1 + (N_Az/2N_D) + kT\varepsilon_s/(e^2NAz) + z/2].$$

Note how ϕ_L is determined by the distance z of the peak potential from the left-hand edge of the p^+ layer. We can obtain ϕ_R if we replace z by $t-z$ and L_1 by L_2 in the expression for ϕ_L, and with the zero of ϕ_R offset by $-eV$. Since we can write

$$\phi_R(z) - \phi_L(z) = eV(z),$$

the variable z forms a useful parameterization of the whole barrier profile as a function of bias. In the particular case of $L_1 = L_2 = 0.2\ \mu m$, and $N_D = N_A = 10^{18}\text{cm}^{-3}$ the zero-bias barrier height is 530 meV, of which 500 meV is contributed by the first term (the L_1 factor) in the above expression for ϕ_L. The extra 30 meV is important in achieving a quantitative understanding of the I–V characteristics, with an exponential sensitivity to barrier height.

The net thermionic current is given by

$$I = \text{area} \times A^{**}T^2\exp(e\mu/kT)\{[\exp- \phi_L(V)/kT] - \exp[-\phi_R(V)/kT]\}$$

where A^{**} is the same effective Richardson constant for thermionic emission in GaAs as encountered in the Schottky diode and μ is the chemical potential (Fermi energy) due to the doping N_D in the contacts. This can be rewritten as

$$I = I_s[\exp(eV/kT) - 1]$$

where

$$I_s = \text{area} \times A^{**}T^2\exp\{[e\mu - \phi_R(V)]/kT\}.$$

In a Schottky diode, the barrier height is determined by the metal–semiconductor interface and is independent of bias, but this does not hold for planar doped barriers. The difficulties in fabricating practical Schottky diodes have been referred to: the device designer has only the metal, the semiconductor, and its doping at his disposal. In contrast, for the planar doped barrier in Fig. 2.15, the design variables include L_1, L_2, the doping–thickness product N_At in the thin p-layer, and the doping N_D in the n layers. This gives much greater flexibility in that one can design a multilayer structure to have specific current–voltage characteristic and allow much wider range of such characteristics to be specified. An example of this comes from the growth of test layers where only the doping–thickness product N_At was varied, keeping $L_1 = L_2$ and a fixed doping. Figure 2.16 shows the experimental results as the various lines, together with the results of simulations based on the solution of the Poisson equation and the above formulae for the I–V characteristics. There are no free parameters in the calculation! The only systematic deviation is at higher current levels, and we can attribute this in part to our neglect of the series resistance associated with the contact layers and in part to some

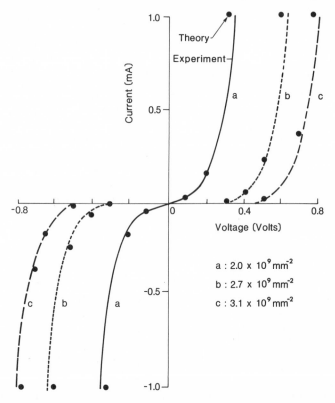

Fig. 2.16 *I–V* characteristics (● theory; — experiment) of a symmetric planar doped barrier diode with varying doping in the p-layer: curve A, $2.0 \times 10^9 \, \text{mm}^{-2}$; curve B, $2.7 \times 10^9 \, \text{mm}^{-2}$; curve C, $3.1 \times 10^9 \, \text{mm}^{-2}$. (After Kearney and Dale 1990.)

intervalley scattering (see Chapter 1) in the GaAs. We return to the planar doped barrier as a microwave device in Chapter 17.

The planar doped barrier is a first and simple example of the ability to tailor multilayer structures for a given end purpose in physics and/or device technology. This theme of tailoring will recur throughout this book.

2.7 The quantum well

If we sandwich a thin layer of GaAs (say, less than 10 nm thick) between two layers of $Al_xGa_{1-x}As$, (with, say, $x \approx 0.3$), we expect an energy band profile such as that in Fig. 2.17. Again deferring any detailed questions about the behaviour of electrons in the GaAs valence bands near the heterojunctions until later, we see that it is not possible to place an extra electron at the conduction band edge in the thin GaAs layer. If we try to confine a carrier in one dimension, there is a quantum kinetic energy price to pay, of magnitude $h^2/8m^*l^2$, and for $l = 10$ nm this value is about 50 meV in GaAs (see Chapter 4). Similarly, there is an equivalent energy for confining holes (but now involving the effective mass of the holes). Optical absorption is possible from the top of the valence band in GaAs to the bottom of the conduction band, as the underlying form of the periodic part of the Bloch function makes this transition allowable (from p-like to s-like states). If we adopt the shape of the electrons-in-a-box wavefunction as shown in Fig. 2.17, and remember the presence of the underlying Bloch functions (further details are given in Chapters 4 and 10), then the optical absorption of the electrons in a box from valence to conduction band states will have to include these extra quantum energies of confinement. We have already seen this in Fig. 2.8(b). The electrons and holes are still free to propagate within the GaAs plane; the quantum-size effects only apply in the direction perpendicular to the layer.

The quantum-well theme also recurs throughout this book as a method of tailoring the optical properties of semiconductors through the thickness of layers and the height of potential barriers.

2.8 Determining the bandgap offset

In all branches of solid state physics, key principles are often easily established and trends in properties are predictable with some precision. There comes a point where detailed *ab initio* theories are sought that give quantitative values of key properties. In bulk semiconductors, the value of the bandgaps, effective masses, and the energy separation of satellite valleys have preoccupied many theorists since the 1960s. The early successes were with semi-empirical theories, i.e. part calculation and part experimental input. Over the last decade, progress has been made in first-principles calculations for the bulk semiconductors (Cohen and Chelikowsky 1988). The equivalent problem in heterojunctions is the precise estimate of the ratio by which offsets (ΔE_v and ΔE_c) of the valence and conduction band edges take up the

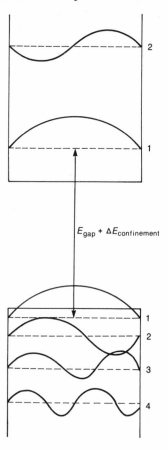

$E_{gap} + \Delta E_{confinement}$

Fig. 2.17 Schematic diagram of a quantum well showing bound states for electrons in the conduction band and holes in the valence band.

discontinuity ΔE_g in the bulk bandgap (see Fig. 2.13). The history of the simplest GaAs–AlGaAs system is revealing. The value of the bandgap for $Al_xGa_{1-x}As$ (for $x < 0.45$) at low temperatures is

$$E_g = 1.42 + 1.26x \text{ eV.}$$

Early optical experiments were used to obtain a best fit to the free parameter $\Delta E_c/\Delta E_v$ from a range of semiconductor quantum wells with different widths and alloy compositions in the cladding layers. The net result, which was used for a decade, was the 85 per cent–15 per cent rule (Dingle 1975). This value made a number of device ideas plausible, which were pursued but with disappointing results (see Chapter 19). Continuing examination of this problem has resulted in a consensus somewhere in the range between 60 per cent–40 per cent and 65 per cent–35 per cent; this may vary systematically with x, but this is not usually considered, and can be tailored by the doping profile at an interface (Martin 1989). The theorists freely admit that their calculations and models are insufficiently accurate, or the computation resources required are too great, to perform adequate first-principles calculations. To date, a series of experiments are used to try to infer the value of the bandgap offset by comparing optical and transport measurements with simulations of the experiments; examples of this will be seen in several chapters below. These comparisons are often quite difficult because of the number of extraneous factors that must be taken into account—contact resistances, corrections because of high doping densities, etc. The satisfactory resolution of this problem is still awaited. In the meantime, practical device designers continue as they always have: they use whatever is available and accepted as reliable, and supplement that with empirical data (Kroemer 1983).

References

Ashcroft, N. W., and Mermin, N. D. (1976). *Solid state physics.* Holt, Rinehart, Winston, New York.

Beuchet, G. (1985). Halide and chloride transport vapor-phase deposition of InGaAsP and GaAs. In *Semiconductors and semimetals*, Vol. 22A (eds R. K. Willardson and A. C. Beer), pp. 261–98. Academic Press, New York.

Britton, E. G., Alexander, K. B., Stobbs, W. M., Kelly, M. J., and Kerr, T. M. (1987). The atomic scale characterisation of multilayer semiconductor structures using TEM, *GEC Journal of Research*, **5**, 31–9.

Chang, L. L. and Ploog, K. (ed). (1985). *Molecular beam epitaxy and heterostructures.* Martinus Nijhoff, Dordrecht.

Clausen, E. M., Kapon, E., Tamargo, M. C. and Huang D. M. (1990). Cathodoluminescence imaging of patterned quantum well heterostructures grown on non-planar substrates by molecular beam epitaxy. *Applied Physics Letters*, **56**, 776–8.

Cohen, M. L. and Chelikowsky, J. R. (1988). *Electronic structure and optical properties of semiconductors.* Springer-Verlag, Berlin.

Davies, R. A., Bithell, E. G., Chew, A., Harris, P. G., Dineen, C., Kelly, M. J., *et al.* (1989). Correlation of electronic and analytical data from a superlattice tunnel diode. *Semiconductor Science and Technology*, **4**, 35–40.

Delalande, C. and Voos, M. (1986). Optical processes in semiconductor quantum wells. *Surface Science*, **174**, 111–19.

Dennard, R. H., Gaensslen, F. H., Yu, H. N., Rideout, V. L., Bassous, E. and LeBlanc, A. R. (1974). Design of ion-implanted MOSFET's with very small physical dimensions. *IEEE Journal of Solid-State Circuits*, **SC-9**, 256–68.

Dingle, R. (1975). Confined carrier quantum states in ultrathin semiconductor heterostructures. In *Festkorperprobleme (Advances in Solid State Physics)* Vol. 15, pp. 21–48. (ed H. J. Queisser). Pergamon-Vieweg, Braunschweig.

Fewster, P. (1993). X-ray diffraction from low dimensional structures. *Semiconductor Science and Technology*, **8**, 1915–34.

Fukui, T., Tsubaki, K., Saito, H., Kasu, M. and Honda, S. (1992). Fractional superlattices grown by MOCVD and their device applications. *Surface Science*, **267**, 588–92.

Jones, G. A. C., Ritchie, D. A., Linfield, E. H., Thompson, J. H., Hamilton, A. R. and Brown, K. (1992). UHV *in-situ* fabrication of three-dimensional semiconductor structures using a combination of particle beams. *Journal of Vacuum Science and Technology*, **B10**, 2834–7.

Kapon, E., Hwang, D. M., Walther, M., Bhat, R. and Stoffel, N. G. (1992). Two-dimensional quantum confinement in multiple wire lasers grown by OMCVD on V-grooved substrates. *Surface Science*, **267**, 593–600.

Kearney, M. J. and Dale, I. (1990). GaAs planar doped barrier diodes for mixer and detector applications, *GEC Journal of Research*, **8**, 1–12.

Kittel, C. (1986). *Introduction to Solid State Physics* (6th edn). Wiley, New York.

Kroemer, H. (1983). Heterostructure devices: a device physicist looks at interfaces. *Surface Science*, **132**, 543–76.

Luryi, S. (1990). Device building blocks. In *High-speed semiconductor devices.* (ed. S. M. Sze), pp. 57–136. Wiley, New York.

Martin, R. M. (1989). Comments on 'Can band offsets be changed controllably?' In *Band structure engineering in semiconductor microstructures* (ed. R. A. Abram and M. Jaros). NATO ASI Series B, Physics, Vol. 189, pp. 1–6. Plenum, New York.

Nakajima, K. (1985). The liquid-phase epitaxial growth of InGaAsP. In *Semiconductors and semimetals*, Vol. 22A (eds R. K. Willardson and A. C. Beer), pp. 2–94. Academic Press, New York.

Parker, E. H. C. (ed) (1985). *The technology and physics of molecular beam epitaxy.* Plenum, New York.

Pearton, S. J. and Shah, N. J. (1990). Heterostructure field-effect transistors. In *High-speed semiconductor devices* (ed. S. M. Sze.), pp. 283–334. Wiley, New York.

Petroff, P. M., Gaines, J., Tsuchiya, M., Simes, P., Coldren, L., Kroemer, H., *et al.* (1992). Band gap modulation in two dimensions by MBE growth of tilted superlattices and applications to quantum confinement structures. *Journal of Crystal Growth*, **95**, 260–5.

Razeghi, M. (1989). *The MOCVD challenge*. Adam Hilger, Bristol.

Rhoderick, E. H. and Williams, R. H. (1988). *Metal–semiconductor contacts.* Clarendon Press, Oxford.

Ross, F. M., Bithell, E. G. and Stobbs, W. M. (1988). The application of Fresnel fringe contrast analysis to measurements of composition profiles in GaAs/AlGaAs heterostructures. *Proceedings of AEM* (ed G. Lorimer), pp. 205–9. Institute of Metals, London.

Seeger, K. (1991). *Semiconductor physics: an introduction* (5th edn). Springer-Verlag, Berlin.

Stobbs, W. M. (1986). Recently developed TEM approaches for the characterisation of semiconductor heterostructures and interfaces. In *The physics and fabrication of microstructures and microdevices* (eds. M. J. Kelly and C. Weisbuch), pp. 136–49. Springer-Verlag, Berlin.

Sykes, D. E. (1989). Dynamical secondary ion mass spectroscopy. In *Methods of surface analysis* (ed. T. M. Walls), pp. 216–62. Cambridge University Press.

Sze, S. M. (1981). *Physics of semiconductor devices* (2nd edn). Wiley, New York.

Thomas, H. (1991). Characterisation of heterojunctions: electrical methods. In *Physics and technology of heterojunctions devices* (eds D. V. Morgan and R. H. Williams). Peter Peregrinus, London.

Tsang, W. T. (1989). From chemical vapour epitaxy to chemical beam epitaxy. *Journal of Crystal Growth*, **95**, 121–31.

Weisbuch, C. and Vinter, B. (1991). *Quantum semiconductor structures.* Academic Press, New York.

3 Fabrication technologies for semiconductor microstructures

3.1 Introduction

The last chapter began with a reference to the hundredfold decrease in the key feature sizes of the transistors that appear in memory chips from small fractions of a millimetre in the 1960s to fractions of a micrometre today. In Fig. 3.1, we show a section of part of a Si circuit and a high magnification photograph of the surface. In general, such diagrams and photographs do not do justice to the very tight tolerances that must be achieved before the devices work to full specification, but we should bear them in mind as we proceed. It is not sufficient to produce a small structure as the key part of a device, unless that structure can be repeated millions of times with a spread in properties that is a smaller fraction again of the structure of interest, for example 0.1 μm length gates with rather less than 10 per cent variation in that length. The purpose of this chapter is to describe the technologies used in the fabrication of semiconductor microstructures for physics and device studies. The techniques described here as used for research purposes are recognizable extrapolations of the technologies as used in the manufacture of today's integrated circuits (whether in Si or GaAs). These state-of-the-art fabrication techniques are available precisely because they are foreseen as essential in one or other strategy for the fabrication of future devices. The physics investigation of these microstructures, quite apart from its intrinsic interest, is meant to shed light on the various choices of technology that might be applied in the year 2000 and beyond.

Given a semiconductor substrate, usually containing an epitaxially grown multilayer, the processes required *en route* to the fabrication of a microstructure for optical or electronic investigation will include methods for (a) patterning the various layers, (b) altering the electronic or optical properties of selected volumes of the material, (c) making conducting or blocking electrical contacts, (d) adding or removing material from selected parts of the surface, and (e) protecting the free surfaces of the finished structures against unwanted chemical, electrical, or other form of attack. These technologies will be described in outline, with the emphasis on the final structures rather than the technology *per se*, or the finer points of the arguments between competing technologies.

3.2 Lithography and pattern transfer

The stages in the transfer of a pattern onto the surface are shown schematically in Fig. 3.2.

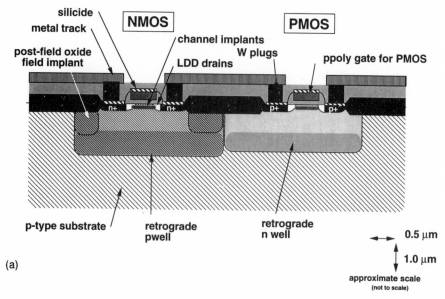

(a)

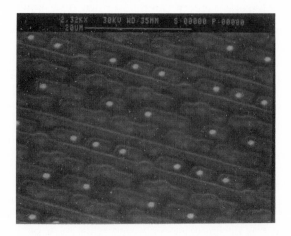

(b)

Fig. 3.1 (a) A cross-section of a modern transistor and (b) a surface view of part of a circuit. (Courtesy of GEC Plessey Semiconductors.)

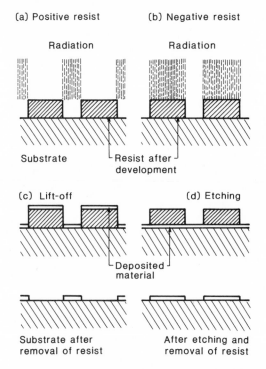

Fig. 3.2 The stages in pattern transfer as described in the text.

A number of organic materials, polymethylmethracrylate (PMMA) being one, are available in solution in the monomer form, and can be used to form a hard thin polymeric coating on the surface of a semiconductor. These materials play a key role in the fabrication of both small structures and integrated circuits. A liquid film can be formed on a planar surface (or near-planar as implied by some of the features in Fig. 3.1) of thickness in the 0.5–2 μm range. The film thickness can be controlled precisely by the concentration of the monomer in its solvent and the rate at which we spin the wafer in its own plane to remove excess solvent. After a short baking (say 10 min at 160 °C) the film is hard and polymerized to some extent. The key property of these so-called resist materials now comes into play. When exposed to light, electron beams, or some other form of attack, some (the positive resists) are broken into smaller organic units that are much more easily dissolved by solvents that do not attack the unexposed material. Others (negative resists) polymerize further and harden under the same conditions of exposure so that solvents subsequently remove those parts that have not been exposed. PMMA is in the former category, while most of the latter go under proprietary names. In addition, there are inorganic resists. Common salt is a positive resist while chalcogenide glasses (As–S, Ge–Se) are negative; the latter are used in research but not in device production. A typical developer of resist patterns is a 1:3 methyl isobutone ketone (solvent)–isopropal alcohol (non-solvent host) solution.

If the resist material is the template used to transfer a pattern onto a substrate, there are several (lithographic) techniques for defining accurate patterns in the resist. Optical lithography proceeds in one of two forms. An opaque mask has been prepared (typically a metal film on glass) with openings where the resist is to be exposed. Such a mask can be prepared by the electron beam lithography methods described below. The mask may be adjacent to the resist layer, so that one refers to contact masks, or it might form part of an optical projection system, so that the final pattern on the surface might be (say five times) smaller. Masks can be used many times. The light may come from powerful lamps, lasers, or even soft X-ray sources. It is clear that one practical limit to the smallest feature sizes that can faithfully be reproduced in resist is the wavelength of that light, i.e. the diffraction limit, with visible light, in the range 0.4–0.8 μm. Those familiar with diffraction patterns appreciate the side-lobes that appear in the Fresnel limit (the appropriate optics for describing contact printing). In fact, these can be used to good effect in that the apertures in the mask may contain not only the pattern to be exposed, but also a number of 'ghost' structures adjacent to the edges whose function is to interfere with the diffracted light from the main aperture in such a way as to sharpen up lines of intensity contrast in the resist. It is now thought that optical lithography using visible light is available down to about 0.15 μm feature sizes in lithography for integrated circuits (Grenville *et al.* 1991). If one is prepared to go to shorter wavelengths (using soft X-rays), one can reach *c.*0.01 μm features. Apart from these diffraction limits, there are also depth-of-field limits so that unevenness on the large scale across a wafer cannot be tolerated, nor can buckling or distortions to the mask, nor even the structure in the surface as a result of previous processing steps for making devices or structures.

Columns for controlling high-energy electron beams have been developed for high-resolution microscopes. The system can be turned around (Fig. 3.3(a)) and the computer-controlled lens optics can be used to steer an intense beam of electrons around the surface of a resist-covered wafer.

The higher the energy of the electron beam, the finer is the focusing and the smaller is the spot on the surface. The limitation to this process is reached at 0.02 μm when the primary electron beam interacts once with the resist and forms a cascade of secondary electrons with sufficient energy to break other bonds in the vicinity. The secondary electron process is also a limitation on the proximity with which two separate features can be written without the intervening resist being weakened as well (Fig. 3.3(b)). If one turns to inorganic resists, which rely on direct sublimation of the resist in an intense electron beam, the smallest features are 20 times smaller at *c.*1 nm.

In recent years focused ion beam columns have been investigated, so that one uses ions rather than electrons. Here the depth range of the species in the resist is less, but the lateral resolution is of order 0.01 μm.

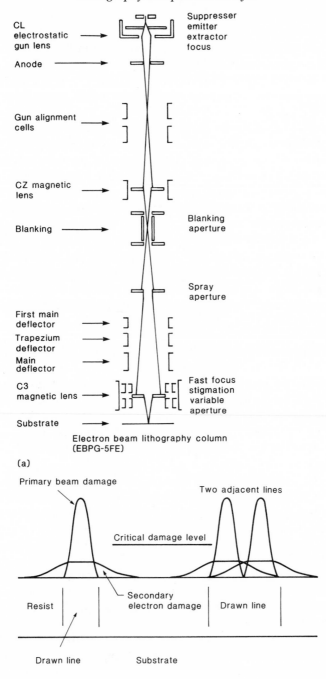

Electron beam lithography column
(EBPG-5FE)

(a)

Fig. 3.3 (a) An electron beam lithography column (courtesy of Leica Cambridge Ltd) and (b) factors relevant to drawing very fine but closely spaced structures.

In comparing optical (and X-ray) lithographies with electron (and ion) beam lithographies, the most obvious difference is in speed and throughput. A mask permits a single flash exposure, whereas the operation of the beam lithographies is tantamount to redefining the whole mask each time. Optical masks can be made (and altered and repaired) using electron beam techniques. (See Einspruch and Watts (1988) and Watts (1988) for more details on lithography.)

In many of the semiconductor microstructures of interest for physics investigations, coarse features (such as electrical contacts) might be made using optical lithography, keeping the electron beam techniques for defining the critically small features.

After developing (i.e. removing the weaker part of the resist material), we have reached the stage where a pattern in resist is left on the surface. That pattern is approximately as deep as the resist (allowing for slight chemical attack from solvents), and it is hoped that the solvents have been able to remove all the weakened resist material. (One cause of defects in devices and structures is the incomplete removal of resist.) It now remains to exploit this pattern in one of several ways. The exposed underlying material can be removed by various forms of etching. More material (metals, dielectric, or semiconductor) can be deposited by selected-area overgrowth. The electrical properties can be altered by ion implantation (including proton isolation). It is to these various further options that we turn. One challenge to keep in mind is how can (as many as 30) different processing steps be undertaken in perfect registry in order to make the transistors shown in Fig. 3.1?

3.3 Etching

There are two principal forms of etching—using wet chemicals or using (dry) plasmas. Each has advantages and disadvantages.

Wet chemistry has been used since the earliest days of integrated circuit manufacture. The range of dilute acid and alkaline materials used is quite wide. Their etching properties can also differ. Some attack the material in an almost isotropic manner. Others are very sensitive to crystal plane. For example NaOH attacks the {100} but not the {111} planes of silicon. A 1:10 citric acid–hydrogen peroxide mix reveals {221} and {111} planes in GaAs (see discussion of the fabrication of fine wires in Chapter 15). Wet chemical methods are simple, and at their most reliable when dealing with relative coarse and well-separated features as they rely on the etchant solution reaching and attacking all relevant areas uniformly. On the micrometre scale, one cannot be sure of the fluid flows with the possibility of vortices of high or low exposure of the etchant within confined regions. In some cases, wet chemistry provides beautiful coloration from interference fringes generated by the thin layers on the surface which can serve as a guide to the progress of the etching and as a suitable method for end-point detection (i.e. that the required etch depth has been achieved).

As the demand has continued for ever tighter control of features of ever increasing aspect (i.e. height-to-diameter) ratio, the move has been towards plasma etching. The structure to be etched is introduced to a chamber containing gases (chlorine, methane, etc.) in which chemically active radicals are generated by radiofrequency radiation. The radicals attack the exposed parts of the semiconductor and the resist differentially. Again, some etches are isotropic, while others are more crystal-plane specific. A body of lore is steadily being built up of the various conditions of temperature, radio-frequency power density, d.c. voltages applied to radicals, gas composition, and flow rates needed to achieve specific etch rates.

With multilayer structures, one has the extra freedom of incorporating an extra very thin layer into the design which has little effect on the subsequent device performance, but which has a very different etch rate from the adjacent layers so that etching can be stopped more easily. After all, one normally wants to remove just a certain prescribed amount of material; neither more nor less, and end-point detection is an important issue. Again, with the use of anisotropic etches in conjunction with resist materials, desirable undercut-ting profiles can be achieved. These can simplify subsequent processes. The condition of the etched surface and subsurface is a further important consideration. Wet chemistry leaves various residues, while any hydrogen atoms in any plasma process can enter the material and alter the doping profile to depths of 50 nm or more by trapping and neutralizing donor or acceptor sites. Again, these become serious problems when dealing with nanometre-scale structures.

The structures achievable with etching techniques are becoming more elaborate all the time. Very high aspect ratios, very steep sidewalls, and closely adjacent microstructures are all possible. In Fig. 3.4 we show examples of

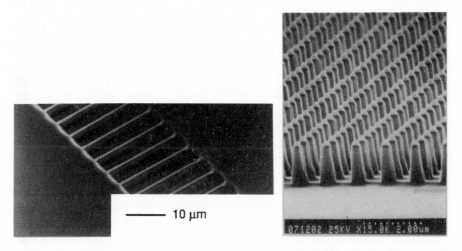

Fig. 3.4 Examples of structures produced by wet and dry etching. (After Hasko *et al.* 1993, and V. Law, University of Cambridge, respectively.)

structures etched in GaAs. For further details on etching techniques see Mukherjee and Woodward (1985), Schutz (1988), Ashby (1990), and Law *et al.* (1991).

3.4 Ion implantation

In the introduction to Chapter 2, it was pointed out that layered doping profiles could be placed with 0.1 μm accuracy in depth to depths of up to 2 μm by implanting the semiconductor with high-energy ions. The purpose of placing these doping profiles is either to create conducting layers of desired polarity (electron or hole transport) or to render some layers non-conducting (as with hydrogen implantation). Clearly with a *c.*2 μm thick patterned layer of resist, it is possible to obtain highly selective ion implantation, with sharp boundaries between implanted and unimplanted regions. The process of implantation, particularly with heavy ions, disrupts the lattice, and it becomes imperative to anneal the sample afterwards to give the ions the opportunity to take up lattice sites as the lattice heals itself. Some residual damage still remains. Furthermore, the annealing schedule itself may be sufficiently severe that a free surface of, say, GaAs might lose Ga in unacceptable quantities, so that protecting layers (such as of Si_3N_4) must be deposited to to protect the surface during the anneal, only to be removed afterwards before the next processing step.

The advent of multilayers adds one further range of possible structures. Some heavy ions, e.g. Se^+, have been shown to disrupt to GaAs lattice with lattice constant *c.*0.5 nm, but not a GaAs–AlGaAs superlattice with periodicity *c.*5 nm. Upon annealing, the superlattice can be restored along with the underlying lattice. TEM studies (Fig. 3.5) reveal residual damage after the anneal, and it may be that further annealing is required for optimum results (Bithell *et al.* 1990). One cannot anneal for too long at too high a temperature, as a uniform alloy rather than a composition-modulated structure will result. Another material, zinc, which is alloyed to p-GaAs when making ohmic contacts, has been shown to completely dissolve a GaAs–AlGaAs superlattice and hasten the formation of a uniform alloy (Laidig *et al.* 1981). It is yet to be seen, but would be valuable, with ion-implanted Zn.

One problem associated with ion implantation is the existence of channelling, whereby ions travel very great distances in certain tightly defined crystallographic directions relative to the lattice (one thinks of relatively open channels in the crystal structure that would assist the passage of high energy ions). In the case of GaAs–AlGaAs superlattices the modulated composition of Al and Ga does not alter the channelling directions, while in strained-layer and other heteroepitaxial systems there may be severe reduction in channelling. Another problem is the existence of a tail to the statistical spread of depths to which the implanted ions penetrate. There will always be some ions at greater depths; this has been shown to be a problem in some transistor structures where the implant damage is sufficient to destroy sensitive buried layers of material. (For more details on ion implantation, see Morgan and Eisen (1985) and Sealy (1990).)

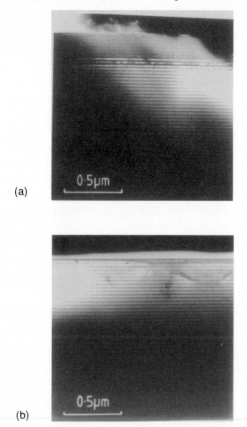

(a)

(b)

Fig. 3.5 Examples of Se⁺-ion-implanted multilayers showing (a) the disruption to the lattice and superlattice and (b) their recovery during annealing. (After Bithell *et al.* 1990.)

A combination of lithography and implantation allows selected volumes of a superlattice to be modified. Studies are in progress to see to what lateral extent the modification to a given layer spreads under the resist. Zn spreading seems very efficient while with other materials (such as Ga⁺ ions) a more faithful image of the resist pattern is preserved (Cibert *et al.* 1986). Figure 3.5 shows some of the structures made using ion implantation. One area that remains for investigation is the analogy of the process of very heavy implantation of oxygen into Si to form buried silicon oxides. Can one produce novel semiconductor microstructures by this technique (Oostra *et al.* 1993, Hunt *et al.* 1994)?

3.5 Metallization, dielectric deposition, etc.

The principal form of selected-area metal deposition is known as 'lift-off'. A uniform film of metal is deposited by evaporating the metal (or alloy) over the entire patterned wafer surface. In areas where the resist has been removed, the metal makes contact with the underlying surface. Otherwise, it coats the resist. A subsequent powerful solvent is used to remove the remaining resist and 'lift off' that metal which is covering it. An effective metallization relies on good adhesion of the metal overlayer to the semiconductor, and for this some form of annealing may be required before lift-off. The formation of good fine-geometry metal gates for transistors (see Chapters 5, 6, and 14) is helped by any undercutting caused by the etching process, as there is no join between the metal of the gate and the metal over the resist (Fig. 3.6(a)). In those cases where undercutting is absent, one must rely on a relatively thin layer of metal on side-walls to break during the lift-off procedure.

A further refinement is often required in the case of very narrow (short) gates. In very-high-speed devices, it may be necessary for the gate to have a short (*c*.0.1 µm) length of contact with the surface, and even if the metal were *c*.1 µm high, the resistance of the gate would be too high (in terms of

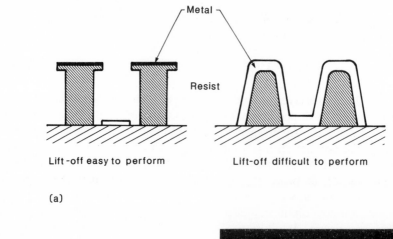

(a)

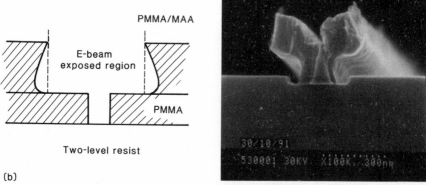

(b)

Fig. 3.6 Resist schemes for metallization: (a) the use of undercutting of resist to ease the formation of narrow gates; (b) the use of double-layer resists to form mushroom gates. (After Cameron *et al.* 1993, as part of the ESPRIT 'MONOFAST' project.)

any *RC* time constants associated with high-speed device operation (see Chapter 16). This problem is overcome by forming a metal line of a mushroom shape (Fig. 3.6(b)), which is obtained if a two-level resist, each level with different electron beam sensitivity, is used and two sets of exposures are applied to provide a mushroom cross-section of removed resist before the metal is deposited.

The metal–semiconductor contact may need to be either ohmic, as is the case if current is to be conducted into or out of a particular device or structure, or blocking or Schottky, if a bias applied to the metal is to be used subsequently to vary the potential profile within a semiconductor (as in transistors in Chapters 5, 6, and 16). In the former case, some form of heat treatment to alloy the metal overlayer into the surface of the semiconductor will normally be required to ensure a good low-resistance contact. This situation has been helped in some cases by the detailed choice of a semiconductor multilayer structure; it is possible to form non-alloyed ohmic contacts by using In-containing GaAs layers, with which In forms low resistance contacts. Selected-area low-energy ion implants may also be used to lessen the severity of the alloying schedule. Schottky contacts are more determined by the choice of metallization.

Traditional materials used for ohmic contacts to n-type GaAs include Au–Ge–Ni alloys (usually deposited in layers of 5–10 nm thickness, with perhaps a 100 nm overlayer of gold) which have a low melting point. The Ge helps to form heavily doped layers adjacent to the contact, the Ni alloys easily with the GaAs, and the Au gives overall adhesion of the contact to the surface and renders the contact strong for bonding with contact wires. For special cases one might use Ag or Pd instead of Au. The contacts to p-type material usually include Zn in place of Ni and Ge. One of the real challenges of contemporary research, which we shall encounter subsequently, is the ability to make very shallow ohmic contacts with high selectivity to one thin layer of a multilayer structure. A failure in this respect is holding back the development of the hot electron transistor (see Chapter 7). Where Schottky contacts are required, there is a recent tendency to use refractory metals including W and W–Si alloys. These materials resist chemical reactions, and only minor changes to their morphology and the electrical properties of the metal–semiconductor contact occur during any heat treatments required to activate implanted species or to form good ohmic contacts in other parts of the overall structure. The detailed metallurgy of ohmic and Schottky contacts has always been a fraught science; one is often dealing with five- or six-component alloy systems, and the formation of some intermetallic compounds is possible because of the highly non-equilibrium conditions present (gradients of the materials, local stress gradients, reduced time scales that prevent equilibration, a range of polycrystalline grain sizes, etc.) during the formation of metal–semiconductor contacts. The technology of the practical metal–semiconductor contact has been based on pragmatic development, with a few empirical rules governing contemporary practice. Further details on these subjects are given by Henisch (1984) and Rhoderick and Williams (1988).

The comments that apply to metal deposition and subsequent treatment also apply to the deposition of other materials, including dielectrics or other semiconductors. Dielectrics are used in optical devices and structures such as dielectric mirrors to reflect light or to separate metallic or other layers where large differences in voltages might occur. In the latter case one is utilizing the large dielectric breakdown strengths, which rely on pure crystalline materials without defect-containing grain boundaries. The art of depositing such layers has not yet been perfected, although mixed $(Ca, Ba)F_2$ layers are promising, as they can be lattice-matched to GaAs (Tu *et al.* 1986).

3.6 Selected-area growth and overgrowth

The possibility has recently been raised of linking two disparate semiconductor technologies on the same substrate. For example could one deposit GaAs of device-grade quality on top of parts of a silicon wafer, so that one could have Si transistor circuits performing high-speed signal-processing functions, the results of which are conveyed to optical devices (lasers etc.) on the GaAs-covered parts of the wafer? Apart from the extra concerns regarding the overall process route (see Section 3.8 below), the deposition of high-quality material is a first precondition. We return to strain-containing heteroepitaxy in Chapter 13, but limited successes can be reported. Whereas field-effect transistors are unipolar devices, the light-emitting devices are inherently bipolar, and it is the lifetime of the minority carriers in the GaAs that sets limits on this strategy. For further details see Chand (1990).

The other new form of research for physics and device structures is to prepare patterned substrates suitable for overgrowth. We saw an example in Fig. 2.4, where V-grooves were formed in a substrate surface. In this case lithography was used to form the overall surface size of the grooves, and wet chemical etching was used to form the particular shape of the grooves. However, all the resist has to be stripped off the wafer surface with a strong chemical solvent before the overgrowth is commenced. Furthermore, the patterned structure requires further cleaning (by electron beams or chemicals in the gas form, or by suitable plasmas) so that a clean surface is presented for the start of the growth, whether by MBE or MOCVD. There is only a limited ability to maintain precise control over the groove geometry during these processes, and the overgrowth process itself tends to smooth the structure (see Fig. 2.4 and Chapter 9, Section 9.6).

3.7 Passivation

Once a semiconductor structure is made, with electrical contacts connected and optical windows prepared as appropriate, it may still be necessary to complete one further stage of sample preparation, namely surface passivation, particularly of the semiconductor. This is a problem in the case of both high-frequency GaAs devices and II–VI IR sensing devices. In both cases, thin layers of native oxide or other materials that build up on the surface may be electrical active (e.g. slightly doped) or chemically active (so that the device is slowly eroded). In the interests of stable structures and devices, some overlayer of an inert material may be deposited to protect the surface by first saturating any spare chemical bonds and then forming a highly resistive overlayer. Suitable passivating materials include silicon nitride or silicon dioxide. Some epoxy resins may be used, although this is often to give a device a suitably robust environment. Subsequently the passivated device must perform in hostile environments, such as repeated cycling to low or high temperatures, being driven hard electrically, being subject to shock in highly accelerating systems (e.g. rockets), or operating in chemically active environments. Passivation is often an essential consideration in the development of stable and reliable devices (e.g. Botez 1990).

3.8 Process schedules

Having described above the range of operations applied to semiconductor multilayers *en route* to producing structures for physics experiments or practical devices, we can see that a range of aggressive treatments might be required. It is important that subsequent treatments do not alter the final effects of previous treatments. In this context, the design of process schedules is all important. It might prove necessary to perform particularly delicate operations at the very end of a process schedule, precisely to avoid difficulties. Alloying in shallow ohmic contacts is one example. The device might have to have the formation of more robust contacts, or the activation of ion-implanted species carried out earlier in the schedule. Sometimes the requirements are conflicting, so that extra steps have to be introduced into the schedule. This can occur when a surface has to be passivated with, say, a silicon nitride capping layer during the anneal schedule after ion implantation, so that As is not driven off an unprotected GaAs surface layer. The nitride might then need to be removed in part or completely before subsequent process steps are carried out. Each lithographic step must contain 'alignment marks' at the edge of the wafer, or whatever area is being patterned. The features associated with these marks must be graduated in size, so that tighter alignments can be made and continued with successive layers. One needs to align features to less than a tenth of the smallest critical feature size at any given level of processing. It is hoped that the most critical step (as in the gates of field-effect transistors in Chapters 5, 6, and 16) can be performed during the final steps and have a relatively less stringent alignment. This subject is not widely written about, but the principles established for Si VLSI (Hillenius 1988) apply more generally.

In this chapter, we have described some of the key steps required for making devices. Those who pursue particular device fabrications have to digest an enormous body of lore, as much of semiconductor device and structure fabrication remains a craft rather than a science. Those working on the design of experiments need to appreciate the complexities of structure and device fabrication, and the limitations imposed by it. One general feature of the research described elsewhere in this book is precisely the establishment of limitations on the future device performances because of the unattainable demands placed on the fabrication processes, for example unacceptable fluctuations occur in the size and electrical or optical properties of nominally identical microstructures.

References

Ashby, C. I. H. (1990). Etching of GaAs: overview. In *Properties of gallium arsenide* (2nd edn), pp. 653–4, and references cited therein. EMIS Datareviews Series 2. INSPEC, Institution of Electrical Engineers, London.

Bithell, E. G., Stobbs, W. M., Phillips, C. C., Ecclestone, R., and Gwilliam, R. (1990). Correlated transmission electron microscopy and photoluminescence studies of the Se^+–ion implantation of a GaAs/(Al, Ga)As multiple quantum well. *Journal of Applied Physics*, **67**, 1279–87.

Botez, D. (1990). GaAs as a light source. In *Properties of gallium arsenide* (2nd edn), pp. 703–12. EMIS Datareviews Series 2. INSPEC, Institution of Electrical Engineers, London.

Cameron, N. I., Ferguson, S., Taylor, M. R. S., Beaumont, S. P., Holland, M., Tronche, C., Soulard, M. and Ladbrooke, P. (1993). Selectively dry gate recessed GaAs MESFETs, HEMTs and MMICs. *Journal of Vacuum Science and Technology B*, **11**, 2244–8.

Chand, N. (1990). GaAs-on-Si: an overview. In *Properties of gallium arsenide* (2nd edn), pp. 459–79. EMIS Datareviews Series 2. INSPEC, Institution of Electrical Engineers, London.

Cibert, J., Petroff, P. M., Dolan, G. J., Pearton, S. J., Gossard, A. C., and English, J. H. (1986). Optically detected carrier confinement to one and zero dimensions in GaAs quantum well wires and boxes. *Applied Physics Letters*, **49**, 1275–7.

Einspruch, N. G. and Watts, R. K. (ed.). (1988). *Lithography for VLSI*. VLSI Electronics Physical Science Series, Vol. 16. Academic Press, New York.

Grenville, A., Hseih, R. L., von Bunau, R., Lee, Y.-H., Markle, D. A., Owen, G., and Pease, R. F. W. (1991). Markle–Dyson optics for 0.25 μm lithography and beyond. *Journal of Vacuum Science and Technology*, **B9**, 3108–12.

Hasko, D. G., Cleaver, J. R. A., Ahmed, H., Smith, C. G., and Dixon, J. E. (1993). Hopping conduction in a free-standing GaAs/AlGaAs heterostructure wire. *Applied Physics Letters*, **62**, 2533–6.

Henisch, H. K. (1984). *Semiconductor contacts*. Clarendon Press, Oxford.

Hillenius, S. J. (1988). VLSI process integration. In *VLSI Technology* (2nd edn) (ed. S. M. Sze) pp. 466–515. Wiley, New York.

Hunt, T. D., Reeson, K. J., Homewood, K. P., Teon, S. W., Gwilliam, R. M., and Sealy, B. J. (1994). Optical properties and phase transformations in α and β iron disilicide layers. *Nuclear Instruments and Methods in Physics Research B*, **84**, 168–71.

Laidig, W. D., Holonyak, N., Camras, M. D., Hess, K., Coleman, J. J., Dapkus, P. D. and Bardeen, J. (1981). Disorder of an AlAs/GaAs superlattice by impurity diffusion. *Applied Physics Letters*, **38**, 776–8.

Law, V. J., Ingram, S. G., Tewordt, M., and Jones, G. A. C. (1991). Reactive ion etching of GaAs using CH_4. *Semiconductor Science and Technology*, **6**, 411–3.

Morgan D. V. and Eisen F. H. (1985) Ion-implantation and damage in GaAs. In: *Gallium arsenide: Materials, devices and circuits* (ed. M. J. Howes and D. V. Morgan), pp. 161–94.

Mukherjee, S. D. and Woodward, D. W. (1985). Etching and surface preparation of GaAs for device fabrication. In *Gallium arsenide: Materials, devices and circuits* (ed. M. J. Howes and D. V. Morgan), pp. 119–60. Wiley, New York.

Oostra, D. J., Bulle-Lieuwma, C. W. T., Vandenhoudt, D. E. W., Felton, F. and Jans, J. C. (1993) $\beta FeSi_2$ in (111)Si and (001)Si formed by ion beam synthesis. *Journal of Applied Physics*, **74**, 4347–53.

Rhoderick, E. H. and Williams, R. H. (1988). *Metal–semiconductor contacts* (2nd edn). Clarendon Press, Oxford.

Schutz, R. J. (1988). Reactive plasma etching. In *VLSI technology* (2nd edn) (ed. S. M. Sze), pp. 185–232. Wiley, New York.

Sealy, B. J. (1990). Ion implantation of GaAs: overview. In *Properties of gallium arsenide* (2nd edn), pp. 685–8 and references cited therein. EMIS Datareviews Series 2. INSPEC, Institution of Electrical Engineers, London.

Tu, C. W., Ajuria, S. A., and Temkin, H. (1986). Broadband and high-reflectivity mirror using (Al,Ga) As/(Ca,Sr) F_2 multilayer structures grown using molecular beam epitaxy. *Applied Physics Letters*, **49**, 920–2.

Watts, R. K. (1988). Lithography. In *VLSI Technology* (2nd edn) (ed. S. M. Sze), pp. 141–83. Wiley, New York.

4 Low-dimensional physics

4.1 Length scales in modern solid state physics

In this chapter we introduce some of the key new physics principles that have dominated semiconductor physics research in the last two decades, namely the concepts associated with low-dimensional systems, i.e. systems that exhibit properties associated with less than three spatial dimensions. However, we start with an introduction to some of the length scales that are relevant to modern solid state physics. An appreciation of these length scales is an important prerequisite to a more detailed understanding of the new physics and when it is likely to apply in practice. Figure 4.1 shows, on a logarithmic scale extending over six orders of magnitude between 0.1 nm and 0.1 mm, an indication of lengths associated with a number of phenomena that occur in physics, technology, or everyday life

The lengths fall into four categories. First there are those measured on an atomic scale, associated with the deposition of atomic layers. The interlayer spacing in GaAs is now only approximately a quarter of the period of a regular semiconductor superlattice, which consists of double atomic layers of GaAs and AlAs. The thickest superlattices or other epitaxial structures grown (apart from exercises in extremes) are typically 10 μm thick. A second set of scales are set by achievements of lithography and technology at a given point in time. While the ultimate limits to certain lithographic processes are known (see Chapter 3), the practical implementation of lithography may not have yet reached, or even approached, these limits. A third set of lengths are set by physics principles (diameter of excitons, radius of cyclotron orbits, etc., as discussed in this and other chapters). A fourth and important set of length scales are not readily entered on Fig. 4.1; these relate to various lengths associated with scattering, coherence, etc. that are derived from the optical or electronic properties. In some ways they may be related to the quality of a particular sample, the material it is made of, and the precision of the fabrication route. Thus the low-field mobility of certain electron systems translates to macroscopic mean free paths of millimetre proportions over which electrons can remember their quantum phase, but this length may also depend on temperature, magnetic field, and other conditions. In this and subsequent chapters, we shall put more of these length scales together.

The main point to take from Fig. 4.1 is that we shall encounter an enormous richness of physical phenomena as we arrange for various length scales to have different relationships with each other, with one being much greater than, comparable with, or much less than another, or to make them vary systematically from one regime to another.

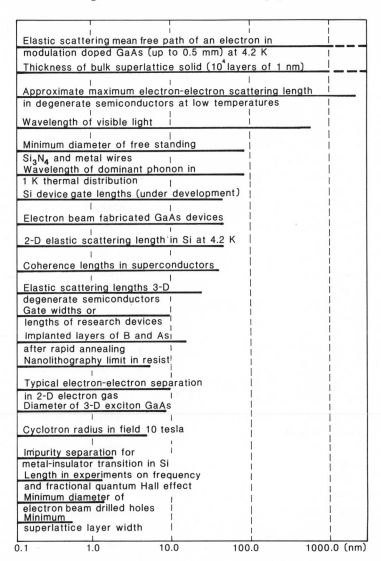

Fig. 4.1 Length scales in solid state physics. (After H. Ahmed 1986, updated in 1994.)

Despite the complexity of Fig. 4.1, there are some natural length, energy, and time scales on which many phenomena occur, and there are a number of convenient units in which the results of theory and experiment might be expressed. The nanometre, millielectronvolt, and the picosecond are the most important scales, and many of the important quantities appear as tens or hundreds of these units (see Appendix 2). We anticipate the results of the next section to note that the one-dimensional kinetic energy of confinement of carriers in a distance l expressed in nanometres in GaAs is

$E_0 = h^2/8ml^2 = 5.608/l^2$ eV (i.e. 56.08 meV for $l = 10$ nm as typical).

Again, the donor binding energy in bulk GaAs can be written as $E_D = \hbar/(2m^*a^{*2})$, and with $a^* = 10.34$nm, then $E_D = 5.316$ meV.

4.2 Dimensionality

4.2.1 Electron states in zero, one, two, and three dimensions

Consider a box of sides L_1, L_2 and L_3, and assume that there is zero potential energy for electrons inside and an infinite potential outside. The one-electron Schrödinger equation

$$[-\hbar^2/2m)\nabla^2 + V(r)]\psi_{lmn}(r) = E_{lmn}\psi_{lmn}(r)$$

has analytic solutions satisfying the condition $\psi_{lmn}(r) = 0$ on the boundaries:

$$\psi_{lmn}(r) = \sqrt{(8/L_1L_2L_3)} \sin(l\pi x/L_1) \sin(m\pi y/L_2) \sin(n\pi z/L_3)$$

with an energy (entirely kinetic since $V = 0$ within the box)

$$E_{lmn} = (\hbar^2\pi^2/2m)[(l/L_1)^2 + (m/L_2)^2 + (n/L_3)^2].$$

In practical solids, where the L_i are on the scale of millimetres, the quantum numbers (l, m, n) are very large, and the kinetic energy terms are very small in units of meV. Note that if $L_1 = 1$ mm, then $(\hbar^2\pi^2/2m)(1/L_1)^2$ takes the value 3.8×10^{-10} meV. In such a case, we consider the k vectors and eigenenergies

$$k = (l\pi/L_1, m\pi/L_2, n\pi/L_3)$$

and

$$E_{lmn} = E(k) = \hbar^2 k^2/2m$$

as being quasi-continuous, as the values of the integers (l, m, n) are all very large. In the special cases that are important here, we consider that some of the L_i may be only a few nanometres, and the l, m, n small integers.

First we consider the distinction between quasi-one-dimensional (Q1D) and one-dimensional (1D) systems. If there is strictly one dimension in the problem, we consider the limits where L_2 and L_3 tend to zero, and the 'zero' of energy is set by $(\hbar^2\pi^2/2m)[(m/L_2)^2 + (n/L_3)^2]$. In practice, we regard the solution to the one-electron Schrödinger equation as

$$\psi_l(x) = \sqrt{(2/L_1)}\sin(l\pi x/L_1)$$

(where we assume that L_1 is a large distance) with an energy

$$E_l = (\hbar^2\pi^2/2m)(l/L_1)^2 = E_l(k_x) = \hbar^2 k_x^2/2m$$

In a Q1D system, we consider finite, but small, values for L_2 and L_3. This means that we have a quasi-continuum of states in the one dimension, but separated and discrete sets of energy levels in the other two directions. In this case we write the energy as

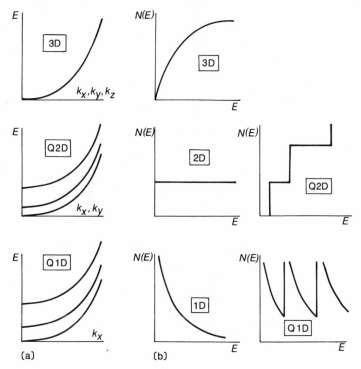

Fig. 4.2 (a) The energy bands and (b) the densities of states in energy in various dimensions and quasi-dimensions (see text).

$$E_{lmn} = (\hbar^2\pi^2/2m)\,[(l/L_1)^2 + (m/L_2)^2 + (n/L_3)^2]$$

$$= E_{lmn} = \hbar^2 k_x^2/2m + (\hbar^2\pi^2/2m)\,[(m/L_2)^2 + (n/L_3)^2]$$

where $L_1 > L_2, L_3$. There are similar sets of solutions in precisely two-dimensional (2D) systems where L_3 tends to zero, and quasi-two-dimensional (Q2D) systems where $L_1, L_2 \gg L_3$. In the quasi-dimensional systems, we use the term sub-band to apply to the band of continuous states having energies that vary as $\hbar^2 k^2/2m$ but starting off from some discrete energy E_{lm} (or E_n etc.) at $k = 0$ (Fig. 4.2(a)).

Instead of the boundary condition $\psi_{lmn}(r) = 0$, there is an alternative form widely used in solid state physics, namely the periodic boundary condition $\psi_{lmn}(r + (\theta_1 L_1, \theta_2 L_2, \theta_3 L_3)) = \psi_{lmn}(r)$ where $\theta_i = 0, \pm 1$. This allows for solutions of the form

$$\psi_{lmn}(r) = \sqrt{(1/L_1 L_2 L_3)}\,\exp(i k \cdot r),$$

where $k = (2l\pi/L_1, 2m\pi/L_2, 2n\pi/L_3)$, i.e. twice the spacing in k-space between adjacent solutions compared with the box solutions above, but now the integers can be positive or negative ($l, m, n, = 0, \pm 1, \pm 2, \ldots$). These solutions only apply in the direction where the L_i are large, and the box solution must be used to treat structures in the spatial directions/dimensions in which they are only a few nanometres thick. Note that the separation in k-value is twice that obtained with box boundary conditions, but the space of solutions is now twice as long in each dimension (from $-N$ to $+N$ instead of 1 to N if there are N unit cells in that direction). They are an equivalent method for counting the same number of eigenstates (Mandl 1988).

A most important derived quantity is the density of states (DOS) in energy, $n(E)$, defined such that $n(E)\delta E$ is the number of solutions to Schrödinger's equation in the energy interval between E and $E + \delta E$ (Ashcroft and Mermin 1976). In the cases where the L_i tend towards zero, the discreteness associated with the quantum numbers appears as discontinuities in the DOS. Totally discrete energy solutions give sharp energy levels, i.e. δ-functions of the form $\delta(E - E_{lmn})$ in what we can call a quasi-zero-dimensional (Q0D) system when all the L_i above are small (i.e. of the order of a nanometre). In order to take into account the quasi-continuous energy levels that follow when the L_i are macroscopic, we consider that the density of states in k-space, i.e. $dN(k)/dk$, is uniform, i.e. there is an interval of $2\pi/L_i$ between adjacent solutions, so that their density in k-space is $L_i/2\pi$ per spatial dimension. (Note here that we choose to use the periodic boundary conditions, but the results are the same if we follow the box boundary conditions.) From this we can write an implicit expression in the one-dimensional case of parabolic energy bands, namely that

$$
\begin{aligned}
n_1(E) = d(n)/dE &= [dN(k)/dk](dk/dE) \\
&= (L_1/2\pi)\,[1/\,(\hbar^2 k/m)] \\
&= (L_1/2\pi)\,(m/\hbar^2)\surd(\hbar/2mE) \\
&= (L_1/2\pi)\,\surd(m/2\hbar^2)\,(1/\surd E)
\end{aligned}
$$

In a Q1D system, we obtain a separate $1/\surd E$ branch in the DOS for each lateral discrete solution of Schrödinger's equation. We end up with a picture of the DOS such as in Fig. 4.2(b) of a single $1/\surd E$ branch in the strictly 1D case, and the multiple branch version in the Q1D case.

In the case of a strictly 2D system, the above arguments are repeated taking the two dimensions into account. Again, we write that

$$
d[n_2(E)]/dE = [dN(\mathbf{k})/d\mathbf{k}](d\mathbf{k}/dE).
$$

The DOS in $\mathbf{k}$-space is now $(L_x/2\pi)(L_y/2\pi)$, and the annular area in $\mathbf{k}$-space between solutions of energy E and $E + \delta E$ is now given for parabolic bands by $2\pi k\delta k = 2\pi(m/\hbar^2)\,\delta E$, leading to the expression

$$
n_2(E) = (A/4\pi^2)\,2\pi(m/\hbar^2) = Am/(2\pi\hbar^2),
$$

where A is the sample area, which is a constant function (in energy) starting from the bottom of the band—the $\mathbf{k} = \mathbf{0}$ value of energy. Again, in a Q2D system, there will be a separate branch for each of the discrete solutions in the third dimension where the L_3 is thought of as being on the nanometre scale. These forms of $n_2(E)$ are also shown in Fig. 4.2(b).

Finally, the case of three dimensions (usually treated in solid state physics textbooks) incorporates a density of solutions in 3D k-space of $V/8\pi^3$ and uses the fact that there is a shell of volume of $4\pi k^2 dk$ between spheres of radius k and $k + dk$, which in turn convert into appropriate energies of E and $E + dE$, leading to

$$n_3(E) = (V/8\pi^3)\,[2\pi(m/\hbar^2)]^{3/2}\,\sqrt{E}.$$

The $n_i(E)$ given above refer to the density in energy of solutions to the Schrödinger equation. In practice, each solution can accommodate two electrons (one of each spin), and it is this doubled quantity that appears in the various equations where the electrons are being counted in the analysis of optical and electrical data.

We note that in the presence of a magnetic field at right angles to a macroscopic two-dimensional system (say B_z with respect to the x–y plane) the eigensolutions to the Schrödinger equation change from $\psi \sim \exp[(ik_x x + k_y y)]$, with solutions separated by $2\pi/L_1$ in dk_x or by $2\pi/L_2$ in dk_y, to Gaussian-type solutions (of the form $\psi \sim \exp(-|r - r_0|^2/a_0^2)$ where $a_0^2 = \hbar/eB$ and r_0 is the centre of the circular orbit that the electron executes) (Landau and Lifshitz 1977). These solutions (known as Landau levels) have discrete energies $\hbar\omega_c(n + 0.5)$ where $\omega_c = eB/m$ with degeneracy eB/h per magnetic level per unit area and per spin. Note that this discreteness of solutions in the k_x–k_y plane means that a 3D system in a strong magnetic field has a spectrum similar to that of a Q1D system (Madelung 1978), while a 2D system in the same magnetic field has a spectrum of discrete levels as in a Q0D system.

In later chapters and in the research literature, we find the terms quantum dot, quantum wire, and quantum well used to describe systems which in fact are Q0D, Q1D, and Q2D respectively.

Why is the DOS such an important concept, particularly in low dimensional systems?

1. The functional form of the DOS, particularly near a band edge (e.g. the $\sqrt{E}$, constant, or $1/\sqrt{E}$ variation) has a profound influence on the transport and optical properties as we shall see below, and the raw data can often show immediately the dimensionality of the system under investigation.

2. The form of multiple sub-bands (from the different discrete solutions in the short transverse directions) can show up in transport experiments (as a change in conductivity or mobility) and in optical data (with extra features in the absorption) and gives an immediate indication of the size of the sample in that lateral dimension.

3. The above remarks apply to all materials, but another reason for the importance of the density of states comes from the particular structure of semiconductor valence bands in small structures: extra selection rules on

the polarization of light allow further investigation of small structures (see below and Chapter 10).

4. Many thermodynamic quantities are related closely to the density of states and the density of excited states. This is clear for both the Fermi energy and the total one-electron energy, and their dependence on temperature via the Fermi–Dirac distribution. It also applies (see Chapter 14) to the density of vibrational modes and the lattice specific heat.

5. Magnetotransport and magneto-optics in small structures also reflect the dimensionality.

6. The influence of fluctuation phenomena and transport and optical properties is a strong function of dimensionality.

7. The response to applied electric fields is a strong function of dimensionality.

The deviations from parabolic energy bands can easily be accounted for in practice in the DOS via a numerical evaluation of $1/|\nabla_k E(k)|$, instead of using dk/dE above, and the appropriate integral over k-space just as above (Ashcroft and Mermin 1976).

4.2.2 Practical definitions of dimensionality

If an electron is in a particular quantum state at an ambient temperature T, it may pick up an energy ΔE of order kT from thermal fluctuations. In an applied field F, it may acquire an energy $\Delta E = eF\lambda$, where λ is the mean free path for either elastic or inelastic scattering (i.e. scattering which does not or does involve a change in the electron energy). A practical definition of dimensionality follows if the only states into which an electron can scatter are determined by no change in the quantum numbers in one or more dimensions, or at most changes of one or more. Thus if ΔE is less than the energy between discrete energy differences in all three (two, one, or no) spatial dimensions, we are in a Q0D (Q1D, Q2D, or Q3D) situation. Note that this practical description also applies in the case of magnetic fields: a Q2D system can become a Q0D system in strong enough magnetic fields when $\hbar\omega_c > \Delta E$.

An equally relevant definition of dimensionality comes from considering lengths rather than energies. If the mean free path for some electron transport process is large compared with the lateral extent of a structure in some spatial dimension, that dimension is not available for transport.

4.3 Transport in various dimensions

In this section, we introduce much of the new physics that has been discovered and investigated in electronic transport experiments performed on low-dimensional structures. We start with a brief introduction to Boltzmann transport that has been used to explain conductivity in bulk solids, and proceed to discuss the various new effects encountered in systems that are small in one or more spatial dimensions and the corrections to the Boltzmann theory that they imply.

4.3.1 Semiclassical and Boltzmann transport

In conventional Boltzmann transport theory as applied to bulk solids, one starts with a distribution function $f(k,r,t)$ which describes the number of carriers in phase space of volume dr at r, and dk at k at time t. It must be assumed that the solid is infinite and homogeneous for r, k, and f to be well defined, and any electric magnetic or thermal fields applied to the solid induce only small changes from the value of f in equilibrium. By analogy with the continuity equations elsewhere (in gas dynamics, electrostatics, etc.), the Boltzmann equation for carriers in a semiconductor can be written as

$$df/dt = \partial f/\partial t + k \cdot \nabla_k f + r \cdot \nabla_r f = \partial f/\partial t |_{\text{coll}},$$

where the right-hand side refers to collisions suffered by the carriers with each other, with impurities, or with phonons etc., which have the effect of trying to return the carriers to equilibrium. In the semiclassical theory of transport in solids (Ashcroft and Mermin 1976) we assume that we can make up wavepackets of electrons with well-defined k, r, and we write

$$v = \dot{r} = (1/\hbar)\nabla_k E(k) \; (= \hbar k/m^*)$$

in which we relate the carrier motion to the band structure through the effective mass m^*. The collision term is exceptionally complex to derive with any real generality (Madelung 1978), and for our purposes a simple relaxation time approximation is used. We introduce a phenomenological time which we may allow to depend explicitly on energy, i.e. $\tau[E(k)]$, and write

$$\partial f/\partial t |_{\text{coll}} = -(f - f_0)/\tau[E(k)]$$

where f_0 is the local equilibrium value of the distribution function $f(k,r,t)$. We assume that collisions destroy all phase memory of the carriers. In the presence of electric fields E, thermal gradients $\nabla_r T$, and changes in chemical composition through varying chemical potential μ, we can write a spatially homogenous solution (Ashcroft and Mermin 1976):

$$f(k,t) = f_0(k) + \int_0^\infty dt' \exp\{-(t - t')/\tau[E(k)]\}(-\partial f/\partial E)v[k(t')]$$

$$\times \{-eE(t') - \nabla_r\mu - eE(t') - [E(k) - \mu]\nabla_r T(t')/T\}$$

showing how the distribution function of the system is now determined by its states at previous times, but with a lessening effect for time intervals that are long on the scale of τ. From this we obtain the steady state current density in materials with a uniform temperature and composition as

$$j = (-e/4\pi^3)\int f(k)v(k)\,\mathrm{d}k = \sigma E,$$

where

$$\sigma = (-e/4\pi^3)\int (-\partial f(k)/\partial E)v(k)v(k)\,\mathrm{d}k.$$

For free electrons and a single relaxation time τ, this equation gives $\sigma = ne^2\tau/m$. Note that this derivation applies in 1D, 2D, and 3D, and, via a summation over separate bands, to quasi-dimensional systems; only the dimension of the carrier density n varies. One obtains conductivities with units of $(\Omega\,\mathrm{m})^{-1}$ in 3D, Ω^{-1} per square in 2D, and $(\Omega\,\mathrm{m}^{-1})^{-1}$ in 1D, as the carrier density is expressed in m^{-3}, m^{-2}, and m^{-1} respectively. Note that the $-\partial f(k)/\partial E$ factor above implies that we must have the equivalent of a Fermi energy within the band; a completely filled or completely empty band or sub-band does not contribute to the conductivity. This solution does not admit any forms of inter-sub-band scattering. Appropriate variations of the above formulation can be used to obtain the Onsager relations between transport coefficients in magnetic fields, the various thermoelectric coefficients, and the electronic contribution to the thermal conductivity (Ashcroft and Mermin 1976), again in 3D, 2D, and 1D.

The underlying validity of the Boltzmann treatment relies on the sample size L being greater than the mean free path λ of the carriers, and in turn λ being rather larger than the interatomic spacing a, i.e. $L \gg \lambda \gg a$. The interest in low-dimensional structures is that one can examine the corrections to Boltzmann transport theory, as these conditions on length scales are systematically violated. When $L \sim a$, we obtain quantum corrections to energy levels as discussed above, while for $L \sim \lambda$, the carriers are ballistic and the randomizing caused by scattering processes is absent. These phenomena appear in several of the following chapters.

4.3.2 Disorder and small-scale effects

In Chapter 1 we noted the coexistence in space of electron states that extended over the whole of a macroscopic crystal (the Bloch functions), and those that were concentrated in one location (cf. the envelope function and the exponential fall-off in the probability of finding the extra electron near its parent donor ion). The subject of extended and localized states has been debated vigorously over the last three decades (Anderson 1958; Kramer *et al.* 1985; Lee and Ramakrishnan 1985). Semiconductor structures have been particularly useful, as dimensionality figures prominently in the formation of localized states, and lower-dimensional systems have been used to probe such states (see Chapter 5). In general one needs some form of disorder (like the donor ion disturbing the perfect lattice periodicity) to induce the localization. While there is coexistence in space of localized and extended states, there is no coexistence (at least at zero temperature) in energy, and states at a given energy are all either extended or localized. The immediate implication is that

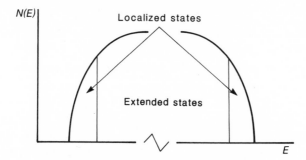

Fig. 4.3 The band edges of a disordered semiconductor system.

as the temperature is lowered, the conductivity tends to a constant, as in a metal, if the Fermi energy lies within a band of extended states, while it tends to zero if the Fermi energy appears in a regime of localized electron states. As with the isolated donor or acceptor, the disorder tends to affect the density of states and the nature of those states at the edges of bands. The situation is shown schematically in Fig. 4.3. Some of these issues are encountered later in this chapter and in Chapters 5 and 6.

The concept of disorder and its influence on conductivity is pervasive in the study of low-dimensional systems. The scattering time τ in the Boltzmann conductivity ($\sigma = ne^2\tau/m$) can be rewritten $\tau = l/v_f$ where $v_f = \hbar k_f/m$ is the Fermi velocity and l is the mean free path. The Boltzmann condition ($L \gg l$) is well satisfied when $k_f l \gg 1$. When $k_f l \sim 1$ (and $k_f \sim 2\pi/a$ where a is the interatomic spacing) the condition for the well-defined extended wave-functions for carriers is no longer met, so that we obtain a minimum metallic conductivity of order $1/(3\pi^2 a)(e^2/h)$ where a is the lattice spacing. The disorder that reduces l to atomic proportions may come from impurities or other crystal imperfections (which include alloying and interface roughness effects in heterojunctions). Note that the condition $k_f l \sim 1$ is more easily satisfied in semiconductors than in metals, given the lower carrier densities generally encountered.

We introduce a system that does not satisfy the conditions for the Boltzmann treatment above, and an argument originally due to Thouless (1977) which leads to the conclusion that there is a minimum metallic conductivity beyond which long enough and thin enough samples of *all* materials are insulating at low enough temperatures. This argument applies to, and has been tested in, metallic microstructures (Santhanam *et al.* 1984). We consider a metallic sample of length L and area A, in which the density of states per unit volume is dn/dE, i.e. a box like that introduced in Section 4.2. We estimate that individual energy states are sensitive to changes in their boundary conditions on a scale set by $\Delta E \sim h/\tau$, where τ is the time taken for an electron to diffuse to the end of the wire and sample the boundary, i.e. a time L^2/D, where D is the diffusion constant which is obtained from the conductivity and the density of states through the Einstein relation

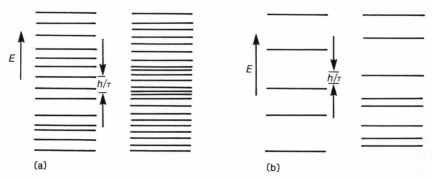

Fig. 4.4 The energy levels in two small pieces of material (see text): (a) $h/\tau >> dE/dN$; (b) $h/\tau << dE/dN$.

$(\mu = eD/kT)$. Since the number of transport electrons, i.e. those with energy within kT of E_F and moving in one direction, is $n = N(E_f)kT/2$, $\sigma = ne\mu$ implies $\rho = (0.5e^2)D(dn/dE)$. We can write

$$h/\tau = (h/e^2)(2/r)(dE/dN),$$

where r is the conventional resistance of the wire and dE/dN is the average separation between adjacent energy levels in the sample. If we now join together two similar but not identical lengths of wire, we can ask about the resistance of the combined system. If $h/\tau >> dE/dN$ (Fig. 4.4), all the energy levels of one sample interact with all the energy levels of the other, and the resulting calculation scales up to a sample of length $2L$. The process of adding further lengths of wire can be continued, eventually reproducing the conditions to which Ohm's law applies. However, if $h/\tau << dE/dN$, the electron energy levels only weakly interact and motion will involve electrons hopping from one piece of material to the next, a condition that is repeated each time another piece of wire is added, which leads to an exponential rather than linear increase in resistance with length. The condition $r > 2h/e^2 \sim 50\,k\Omega$ is met in very fine wires $\sqrt{A} \sim 30$ nm and $L \sim 100$ μm.

The Thouless type of argument has be taken rather further by a form of scaling argument (Lee and Ramakrishnan 1985). We define a dimensionless conductance $g = G/[e^2/h]$, which must scale with sample size; in Ohm's law regime we have $g(L) = \sigma L^{(d-2)}$. We define a further variable $\beta = d\ln(g)/d\ln(L)$, and note that $\beta(g) = d-2$ in Ohm's law regime, while it has the form $\beta(g) = \ln(g/g_c)$ in the Thouless example above. Note that if $\beta < 0$ the conductivity vanishes in the limit of large structures, and that this applies in all cases in 1D. In $2 + \varepsilon$ dimensions, we obtain the same type of results as for any analysis of the scaling of the 3D conductivity. The situation in 2D is marginal, and in fact all electron states are localized in two dimensions (Abrahams *et al.* 1979); indeed it was detailed physics of the type described in the next chapter that clarified this. The localization in 2D is actually weak, and often indistinguishable from non-localization in practical experiments.

The analysis of disorder just given is at zero temperature, and is within the one-electron picture. Many-electron effects, via electron–electron scattering, provide further (quantum-interference) corrections to the Boltzman conductivity. The details require rather more sophisticated many-body physics than is appropriate here, but rely on the DOS as a function of dimensionality, and the results are quoted without proof (Lee and Ramakrishnan 1985). The general result is that there is a correction to the conductivity of the form $-(2e^2/h\pi)(1/L^d)\,\Sigma_Q(1/Q^2)$, where the cut-off at low Q of an otherwise divergent summation is set by the sample size L and the cut-off at high Q is set by l. The corrections to the Boltzmann conductivity in the various dimensions from electron–electron scattering are given by

$$\Delta\sigma_{3D} = -(2e^2/3h\pi^3)(1/l - 1/L)$$

$$\Delta\sigma_{2D} = -(e^2/h\pi^2)\ln(L/l)$$

$$\Delta\sigma_{2D} = -e^2h\pi(L - l)$$

At finite temperatures one has to consider the inelastic scattering processes in which the electrons lose their memory of quantum phase, thus reducing the effects of quantum interference. In general, if the inelastic scattering process has a temperature dependence $\tau \sim T^{-p}$, then the relevant upper length of the sample is of the form $L_{th} = aT^{-p/2}$, since in diffusive processes $L_{th} = \sqrt{(D\tau_i)}$, and the resulting further conductivity corrections are

$$\Delta\sigma_{3D} = + (e^2/h\pi^3)(1/a)T^{p/2}$$

$$\Delta\sigma_{2D} = [p/2(e^2/h\pi^2)]\ln(T/T_0)$$

$$\Delta\sigma_{1D} = -(e^2/h\pi T^{-p/2})$$

The conductivity decreases with decreasing temperature, a sign that electrons are localized. Further treatments have also been used to cover the effects of magnetic field which alter the phase of the electron wavefunctions, and hence their ability to interfere. We return to these when discussing the data in subsequent chapters.

4.3.3 Transport in magnetic fields

An electron in perpendicular electric and magnetic fields (say B_z and E_x) moves under the Lorentz force

$$m\ddot{r} = \hbar\dot{k} = F = -e(E + v \times B)$$

for which the solution for the trajectory is a circular orbit (with angular frequency eB/m) superimposed on a uniform velocity of magnitude E/B in the direction perpendicular to both E and B. Thus electrons moving in a current in the x direction with the electric field develop a Hall field at right angles, and the Hall coefficient is just the ratio $E_y/J_xB_z = -1/ne$ for electrons or $1/ne$ for holes, as described in Chapter 1. This is the so-called cyclotron motion. If we introduce a phenomenological relaxation time τ, the equation above can be rewritten as

$$\hbar(d/dt + 1/\tau)\mathbf{k} = \mathbf{F}.$$

This equation can be solved in the form of Ohm's law to give a steady state solution of the form ($\sigma_0 = ne^2\tau/m$)

$$\mathbf{j} = \sigma \cdot \mathbf{E}.$$

where

$$\sigma = \frac{\sigma_0}{1 + (\omega\tau)^2}\begin{pmatrix} 1 & -\omega_c\tau & 0 \\ \omega_c\tau & 1 & 0 \\ 0 & 0 & 1 + (\omega_c\tau)^2 \end{pmatrix}$$

Note that the magnetic fields plays no role in the motion in the z direction. This form of solution is of interest in discussing the conductivity of electrons in various low-dimensional systems, where the strength and direction of the magnetic field can be used to probe the electron states.

We have already seen that the electron states are strongly modified in the presence of a magnetic field. In the case of a two-dimensional electron gas, the density of states is a set of sharp peaks broadened by the effects of disorder as discussed above, which is shown schematically in Fig. 4.5. If we can arrange for the density of carriers to vary, the conductivity (proportional to the density of states at the Fermi energy, as seen above) should oscillate as the Fermi level passes through the different Landau levels. The conditions for these oscillations to be observed are two fold: (a) $kT << \hbar\omega_c$, which guarantees that thermal effects are too small to destroy the effects of the magnetic field, and (b) $\omega_c\tau >> 1$, which allows many cyclotron orbits to be completed between scattering events. Clear oscillatory magnetotransport phenomena are only seen at low temperatures and in pure materials. We discuss the longitudinal transport and the so-called Shubnikov–de Haas oscillations in the next chapter.

By examining the form of the conductivity above, we see that we can write

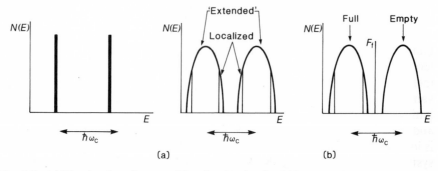

Fig. 4.5 (a) The density of states of Landau levels of a Q2D system in a magnetic field without and with some disorder; (b) the quantum Hall condition.

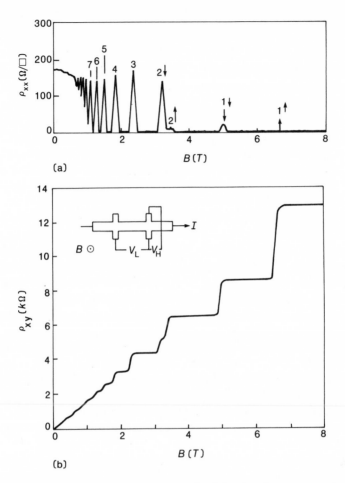

Fig. 4.6 The longitudinal and transverse conductivity for electrons constrained into two-dimensional motion in a strong magnetic field. (After Paalanen *et al.* 1982.)

$$\sigma_{xy} = -ne/B + \sigma_{xx}/\omega_c \tau.$$

Note that if we have a carrier density $n_s = ieB/h$ that allows precisely i Landau levels to be filled and we go to very low temperatures, the probability of an inelastic scattering process exciting an electron into the next Landau level vanishes as $\exp(-\hbar\omega_c/kT)$ and the value of τ increases exponentially. In the limit we are left with $\sigma_{xx} = 0$, as an integral number of Landau levels are filled, and σ_{xy} takes the value $-e^2 i/h$, a combination of fundamental constants which is independent of the material or device in which the two-dimensional electron system is measured. This is the quantum Hall effect (von Klitzing *et al.* 1980) (Fig. 4.6), which will be discussed in more detail in subsequent chapters, when the experimental conditions for varying the carrier density at fixed magnetic field (or varying the magnetic field at fixed carrier density) are explained.

4.3.4 Ballistic motion

In Boltzmann transport theory, we assume that electrons move in simple trajectories between scattering events. In recent years the conditions have been reached in which the ballistic motion of electrons between scattering events can be accessed experimentally (see Chapters 6 and 7 for further details). We consider here the consequences of ballistic motion in 1D, where we know from earlier results that, in 1D, the density of states varies as $1/\sqrt{E}$ from the bottom of the appropriate lateral sub-band. Consider now the situation shown in Fig. 4.7. We assume that an infinitesmal bias is applied along the channel, and that the channel is short enough that there is no scattering of the electrons during their motion, which we also assume occurs at zero temperature. The density of states of those electrons with positive-going velocity is just half the total, and electrons with energy E with respect to the bottom of the sub-band have a velocity given by $v = \sqrt{(2E/m^*)}$. Thus the forward-going differential current for electrons with energy between E and $E + dE$ is given by

$$dJ_+ = eN(E)v(E)dE/2.$$

Note that in 1D, there is no explicit energy dependence in the product $N(E)v(E)$, which is important below. The reverse-going current is the same in magnitude. The net total current is obtained by integrating over the energy states from zero to E_f in the forward-going direction and from eV to E_f in the reverse-going direction. Thus the net current is

$$J = J_+ - J_- = 2e^2 V/h \qquad \text{i.e. } \sigma = 2e^2/h$$

We arrive at a quantized conductance (not conductivity) again, this time in 1D and in the absence of a magnetic field. The derivation has been for one

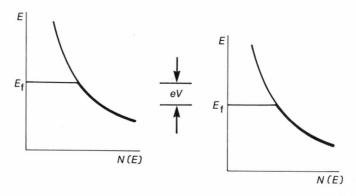

Fig. 4.7 The 1D density of states in a narrow channel offset by an applied bias, leading to the quantization of the 1D ballistic resistance as described in the text.

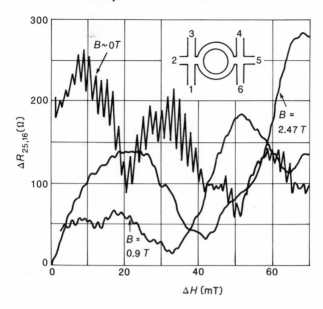

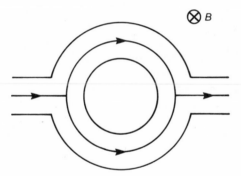

Fig. 4.8 The Aharonov–Bohm structure and magnetoresistance interference effects as electrons combine coherently after taking either arm and being subject to a different phase difference. (From Timp *et al.* 1989.)

sub-band, but the result is additive; as further sub-bands are introduced the conductance increases in steps of $2e^2/h$. We discuss refinements to this formula in the presence of high fields and artificial potential barriers in Chapter 6.

4.3.5 Aharonov–Bohm effect and quantum interference

Consider the special configuration in Fig. 4.8 that has proved of great experimental and theoretical interest: a Q1D channel splits into two and rejoins further down in such a way that an area A is enclosed by the loop. Provided that the structure has linear features which are small compared with any scattering processes that destroy the coherence of the phase, an electron wavefunction that passes one way round the loop can interfere with that part of its wavefunction that takes the other route. In particular, if a magnetic field threads the loop, an extra phase difference is picked up or lost depending on which sense the electron takes. The form of this phase difference is $\exp(\pm iHA/\phi_0)$, where H is the magnetic field intensity, and $\phi_0 = h/2e$ is the quantum of magnetic flux. The transmission of electrons from the left to the right of the structure in Fig. 4.8 depends on any interference of the electrons recombining after taking opposite arms of the loop, and this is seen experimentally as oscillations in the resistence. Again, we discuss the details in Chapter 6.

4.3.6 The Landauer–Buttiker formalism for describing resistance

The novel physics ideas introduced in the recent sections are far from those of the Boltzmann regime of $a_0 \ll \lambda \ll L$ (atomic scale $\ll$ mean free path $\ll$ sample size). One can ask for a more general definition of resistance appropriate to microscopic structures. The pervasive appearance of the conductance in units of e^2/h suggests that this is an appropriate starting point. Different resistances will emerge as different numbers of channels (equivalent to electron sub-bands) become available, and as the experimental conditions allow perfect, partial, or no transmission of electrons in each channel (Buttiker 1992).

The general derivation of resistance in a quantum sense comes from considering a multiterminal conductor as shown in Fig. 4.9. For convenience, we assume that it is a small structure and that each terminal j is characterized by a finite number M_j of one-dimensional sub-bands. The current entering from probe j and leaving via probe i can be analysed in terms of a transmission probability between each channel in each lead, written as

$$T_{ij} = \sum_{m=1}^{M_i} \sum_{n=1}^{M_j} T_{ij,\,mn} \quad (m, n \text{ the channel indices})$$

while that reflected from lead j due to carriers entering lead j is given by

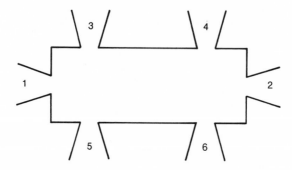

Fig. 4.9 The multiterminal conductor for the Landauer–Buttiker formalism.

$$R_{jj} = \sum_{m=1}^{M_j} \sum_{n=1}^{M_j} R_{jj,\,mn}$$

The specific R and T coefficients for each channel can be calculated as in quantum-mechanical calculations of transmission and reflection coefficients from potential barriers (as described in Chapters 6, 8, and 11). The reservoirs beyond the leads are assumed to be able to feed each of the channels equally and with a unit amplitude of current. If there is an imbalance in chemical potentials μ_i (or Fermi energy at low temperature, as induced by a small bias), there is a lowest μ_0 below which all electron states are occupied everywhere and cannot contribute to the current (there are no empty final states). Reservoir i can feed all the channels in the energy range $\mu_i - \mu_0$. Since the density of channels in energy multiplied by the velocity of carriers in the channel is a product independent of energy (cf. the $1/\sqrt{E}$ and $\sqrt{E}$ cancellation), the current fed by reservoir i is (cf. section 4.3.4)

$$I = (e/h)(\mu_i - \mu_0) M_i.$$

A quantum form of Kirchhoff's current laws gives the desired result. Of the current incident into probe i, a fraction R_{ii} is reflected and the net inward current is further reduced by those currents incident from other probes that transmit out of probe i. Thus

$$I_i = (e/h)\left[(M_i - R_{ii})\mu_i - \sum_{j \neq i} T_{ij}\mu_j \right].$$

Current conservation for probe i, taken by considering individual channels, can be written as

$$M_i = R_{ii} + \sum_{(j \neq i)} T_{ij}.$$

This allows us to obtain the following key result:

$$I_i = (e/h) \sum_{j \neq i} T_{ij}(\mu_i - \mu_j).$$

Then the two-terminal resistance, where both voltage $(\mu_1 - \mu_2)/e$ and current I are measured between terminals 1 and 2, is

$$R_{12,12} = (\mu_1 - \mu_2)/eI = (h/e^2)(1/T_{12}).$$

We can take this analysis further to obtain the four-terminal resistance, where the voltage $V_{mm} = (\mu_m - \mu_n)/e$ is measured between terminals m and n while the current goes from i to j, as

$$R_{kl,\,mn} = (h/e^2)[(T_{mk}T_{nl} - T_{ml}T_{nk})/D_{kn}].$$

Here D_{kn} is the subdeterminant of the matrix of all the transport coefficients derived from the equation

$$M_i = R_{ii} + \sum_{j \neq i} T_{ij}$$

but with row k and column n deleted. If we denote the determinant of the full matrix by Det, current conservation implies that

$$0 = \text{Det} = (M_i - R_{ii}) D_{ii} - \sum_{j \neq i} T_{ij} D_{ij}.$$

Note that this is just the equation for probe i with no current ($I_i = 0$), and so only the equilibrium solution is possible with all chemical potentials the same. This implies that all subdeterminants are equal $D_{ij} \equiv D$).

This formulation (named after R. Landauer and M. Buttiker who developed it) can also account for the various magnetotransport phenomena in small structures. The simple formulation does not include the effects of phase-breaking inelastic scattering processes, although appropriate generalizations are being sought. It is far removed from the Boltzmann theory of transport.

4.4 Optical properties in low-dimensional systems

4.4.1 One-electron effects

The basic form of optical absorption was introduced in Chapter 1, with a frequency-dependent conductivity giving the relationship between the power absorbed at that frequency and the intensity of the incident light. The conductivity has built into it a sum over all possible initial and final electron states separated by the energy $\hbar\omega$ (the so-called joint density of states), such that the initial state is filled and the final state is empty, and each contributing absorption process is multiplied by the appropriate optical matrix element. This is an application of the Fermi golden rule transition probability for the optical absorption:

$$W = (2\pi/h) \sum_{f,i} |\langle \psi_f | H_{\text{int}} | \psi_i \rangle|^2 \, \delta(E_f - E_i - \hbar\omega).$$

The same basic formulation applies to the optical properties of low-dimensional systems, but both the matrix elements (and any selection rules) and the joint density of states reflect the lower dimensionality. In bulk semiconductors, we distinguish between interband absorption (involving excitation of an electron from the valence band to the conduction band) and the intra-impurity-level excitations of donor levels which occur in the far infrared. There is a similar distinction to be drawn between interband and inter-subband excitations in low-dimensional systems. The key point about low-dimensional systems is that the feature size L_i in one or more dimensions can be controlled during the preparation of a sample, and in turn this can alter the optical properties since the energy level separations include contributions from the quantum energies of confinement (cf. $(\hbar^2\pi^2/2m)(m/L_2)^2$ in Section 4.2)

The optical matrix element can be obtained relatively simply using the nature of the wavefunctions involved (i.e. a product of the envelope functions discussed in Section 4.2, but multiplied by the full Bloch function at the energy band extremum). The interaction Hamiltonian is taken as $-e\boldsymbol{r} \cdot \boldsymbol{E}$ in the usual dipole approximation (Ashcroft and Mermin 1976), and the optical matrix element becomes

$$M \sim |\langle \psi_f | H_{int} | \psi_i \rangle| \sim | \int F_f(\boldsymbol{r}) \exp(-i\boldsymbol{k}\cdot\boldsymbol{r}) u_{ck0} \varepsilon \cdot \boldsymbol{r} F_i(\boldsymbol{r}) \exp(i\boldsymbol{k} \cdot \boldsymbol{r}) u_{vk0} d\boldsymbol{r}|.$$

We assume that the envelope functions are real, that subscripts c and v refer to Bloch functions from conduction and valence band extrema (at $\boldsymbol{k} = \boldsymbol{0}$), and that ε is the polarization vector of the light. As in the theory of 3D impurities (Luttinger and Kohn 1955), when the integral is taken first over unit cells of the semiconductor to handle the rapidly varying functions (the u_c, u_v terms), the result contains selection rules coming from the symmetries of the conduction and valence bands and the light polarization. The $\boldsymbol{k}$ terms are equal in the two exponential terms because of the vertical transition involving the negligible momentum of light. In general, if the selection rules permit a non-zero value, it is because of the atomic p-like and s-like symmetry of the u_c and u_v terms respectively. There is no further role to be played by the optical matrix element *per se* in determining interband optical absorption selection rules in spatially localized structures. The final integral of the form

$$M \sim \int F_f(\boldsymbol{r}) F_i(\boldsymbol{r}) d\boldsymbol{r}$$

implies that optical absorption can occur as long as there is overlap between the envelope functions themselves (cf. Fig. 4.10)

The result is different for inter-sub-band excitation in the infrared (i.e. where the excitation is from one confined state to an excited confined state in a quantum well). In this case the u_c term appears twice in the integral over a unit cell, and ε appears in the slower integral over the envelope functions

$$M \sim \int F_f(\boldsymbol{r}) \boldsymbol{\varepsilon} \cdot \boldsymbol{r} F_i(\boldsymbol{r}) d\boldsymbol{r}.$$

Because of the normalization of the envelope functions, the optical matrix element is the same expression in any dimensions (as long as we are considering only one-electron effects). Furthermore, the oscillator strength, defined as

$$f = (2m\omega/h) |\langle \psi_f | H_{int} | \psi_i \rangle|^2,$$

is comparable in 3D and 2D. As Weisbuch and Vinter (1991) make clear, the main effect of the reduction in dimensionality is to concentrate the oscillator strength from wide bands of energy into ever narrower bands.

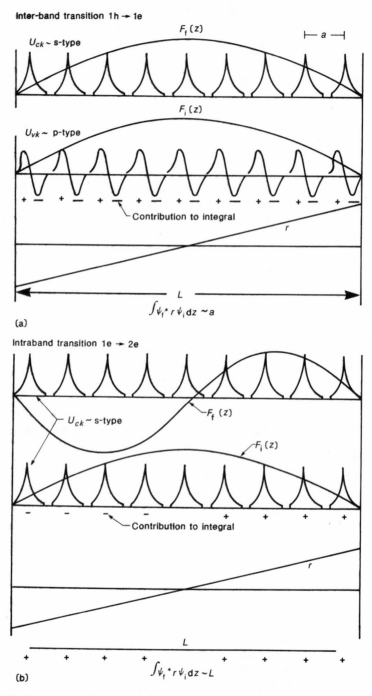

Fig. 4.10 Structure of wavefunctions for (a) interband and (b) inter-sub-band absorption of light. (After Weisbuch and Vinter 1991).

4.4.2 Excitonic effects

In practice, the optical absorption of GaAs/AlGaAs low-dimensional systems and microstructures at low temperatures is strongly modified by excitonic effects. Consider first the problem of a hydrogen atom in strictly two dimensions. This 2D problem is exactly soluble, although a little more difficult technically than in 3D, and the net result for the Rydberg constant is

$$R_{n,2D} = R_{3D}[1/(n - 0.5)^2],$$

giving a fourfold increase in 2D compared with 3D for $n = 1$. The effective Bohr radius in 2D is a factor $\sqrt{(3/8)}$ smaller than the 3D Bohr radius (Weisbuch and Vinter 1991).

The interest in the excitonic properties of low-dimensional systems is because the oscillator strength appears in narrow energy ranges (just below the threshold for interband absorption) and the binding energies are increased, thus reducing the effects of thermal ionization (even up to room temperature). The range of practical samples includes layers that are rather thinner than the 3D exciton diameter, and the scaling of the exciton behaviour towards 2D behaviour is easy to verify. Detailed calculations of the energy levels of excitons in quantum wells of finite thickness and finite potential-well depth are possible. Typical results are shown in Fig. 4.11. As wells of finite potential depth become thinner, an increasing fraction of the electron and hole wavefunctions appear in the barrier layer, and so one loses some of the extra degree of binding by virtue of confinement in Q2D.

4.4.3 Linear and non-linear optical effects

The optical properties in the presence of a strong electric field are also of great interest, and again the reason is simple and is shown schematically in Fig. 4.12. An electric field applied to a bulk semiconductor generates Airy-function-type solutions which contain an exponentially damped solution in the energy gap region. This in turn implies that the effective bandgap is reduced by an electric field; light could be absorbed from carriers originating from the region of damped wavefunctions that extend into the bandgap—the Franz–Keldysh effect (Seeger 1991). If an electric field is sufficient to allow a significant potential energy difference across an exciton diameter, the shape of the applied potential profile is as shown in Fig. 4.12, and there is a finite probability that the electron will tunnel out from the confining Coulomb potential. Thus excitons do not survive other than modest electric fields in bulk semiconductors.

The situation in a quantum well is quite different. The confining potential provided by the alloy barriers is on the scale of 300 meV rather than $c.5$ meV. Very large electric fields can be applied to a quantum well (as much as 40 times that in bulk material) and yet retain a sufficient overlap of electron and hole wavefunction that an exciton can be formed and survive. There is a modest reduction in the binding energy as the electron and hole move so as to polarize the exciton, but the quantum-well barrier is much larger. Detailed

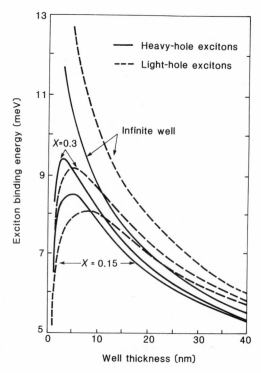

Fig. 4.11 Calculations of the electron–hole pair (exciton) binding energy in finite and infinite quantum wells. As the well width decreases, the exciton becomes more 2D-like with increased binding energy, until for very thin finite quantum wells the binding drops as the electron and hole confinement reduces (Greene *et al.* 1984)

results (see Chapter 10) show a quadratic dependence on applied electric field of the reduction in the excitonic absorption energy.

A general feature of small structures is that the regime of excitation intensity that induces a linear response is usually very small. In optical absorption in a quantum well, and more so in a quantum wire or quantum dot, there is a limited number of carriers to be excited, and the modification to the electrostatic potential caused by electrons residing in excited states can be appreciable. This means that non-linear optical response is much easier to obtain in microstructures. This non-linearity might come in the saturation of the absorption (i.e. no further absorption even if the incident power is increased) or in shifts in the energies between initial and final states that depend on the intensity of the light or combinations of the two.

4.4.4 Donor levels in reduced dimensions

The close similarity between the calculations for an exciton and a donor ion with its donor electron has been noted in Chapter 1. The binding energies of donors and acceptors are modified in microstructures because the spatial

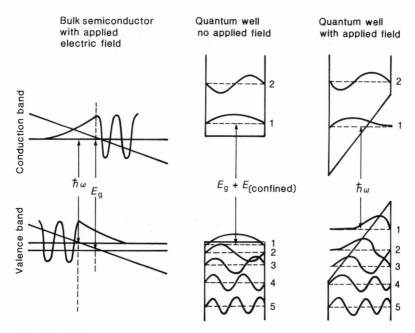

Fig. 4.12 The effect of an electric field on an exciton in a bulk semiconductor and in a quantum well. The blue shift in the optical absorption of a bulk semiconductor under an electric field is known as the Franz–Keldysh effect, while the corresponding phenomenon (on a larger scale) in quantum wells is known as the quantum-confined Stark effect.

uniformity of the energy band extrema can no longer be exploited; indeed, detailed calculations have to be performed, as the problem of a donor level with a Coulombic potential in the presence of a potential step does not lead to a separable and analytically soluble form for the Schrödinger equation. The case is comparable to that of the exciton in a quantum well, and needs a full numerical solution. The qualitative behaviour is quite predictable—as the electron wavefunction has an increasing probability of being in a barrier layer, so the energy of the bound states rises. One or two special cases can be obtained: a donor level at a heterojunction interface with an infinite step can have as valid solutions the $l = 1$ solutions of the problem of a bulk Coulomb potential as these solutions share the condition of having a zero value everywhere at the $z = 0$ plane that can be used as the interface.

4.5 Dimensional crossover

In this chapter we have dealt with structures that are 3D, 2D, 1D, or 0D, and the quasi-dimensionality variations. In subsequent chapters we shall see that experimental studies of the transition between 3D and 2D and between 2D and 1D have been popular. Structures where applied biases to Schottky gates have been used to modify the thickness of an electron system in one spatial direction from many times the Fermi wavelength (or scattering length, or one of the practical methods by which dimensionality is observed) to a small fraction of that value. The experiments seek out changes to the temperature dependence of the conductivity, the energy dependence (i.e. shape) of an optical absorption edge, etc. as the effective lateral dimension is varied.

4.6 Summary

The main theoretical aspects of the electronic structure, and the transport and optical properties of structures with a varying spatial dimensionality have been introduced. In the following two chapters we focus on two important heterojunction structures that have permitted exhaustive studies to be made of electronic structure, electronic and optical properties of electrons that are constrained into two and one spatial dimensions, and the transition between them, as well as the transition from three to two dimensions. The applications of some of the key topics in this chapter have been of commercial significance as well as of intrinsic scientific interest.

References

Abrahams, E., Anderson, P. W., Lee, P. A. and Ramakrishnan, T. V. (1979). Scaling theory of localisations: absence of quantum diffusion in two dimensions. *Physical Review Letters*, **42**, 673–6.
Ahmed, H. (1986). An integrated microfabrication system for low-dimensionality structures and devices. In *The physics and fabrication of microstructures and microdevices* (eds. M. J. Kelly and C. Weisbuch), pp. 435–42. Springer-Verlag, Berlin.
Anderson, P. W. (1958). Absence of diffusion in certain random lattices. *Physical Review*, **109**, 1492–1505.
Ashcroft, N. W. and Mermin, N. D. (1976). *Solid state physics*. Holt, Rinehart, Winston, New York.
Buttiker, M. (1992). The quantum Hall effect in open conductors. In *Semiconductors and Semimetals*, Vol. 35 (ed. M. Reed), pp. 191–277.
Greene, R. L., Bajaj, K. K. and Phelps, D. E. (1984). Energy levels of Wannier excitons in GaAs–Ga$_{1-x}$Al$_x$As quantum well structures. *Physical Review B*, **29**, 1807–12.
Kramer, B., Bergmann, G. and Bruynseraede, Y. (1985). *Localization, interaction and transport phenomena*. Springer-Verlag, Berlin.
Landau, L. D. and Lifshitz, E. M. (1977). *Quantum mechanics* (3rd edn). Pergamon, Oxford.
Lee, P. and Ramakrishnan, T. (1985). Disordered electronic systems *Reviews of Modern Physics*, **57**, 287–337.
Luttinger, J. M. and Kohn, W. (1955). *Physical Review*, **97**, 869–83.
Madelung, O. (1978). *Introduction to solid state theory*. Springer-Verlag, Berlin.
Mandl, F. (1988). *Statistical physics* (2nd edn). Wiley, New York.
Paalanen, M. A., Tsui, D. C. and Gossard, A. C. (1982). Quantised Hall effect at low temperatures. *Physical Review B*, **25**, 5566–9.

Santhanam, P., Wind, S. and Prober, D. E. (1984). One-dimensional localisation and superconducting fluctuations in narrow aluminium wires. *Physical Review Letters*, **53**, 1179–82.

Seeger, K., (1991). *Semiconductor physics: an introduction* (5th edn). Springer-Verlag, Berlin.

Thouless, D. (1977). Maximum metallic resistance in thin wires. *Physical Review Letters*, **39**, 1167–9.

Timp, G. Mankiewich, P. M., de Vegvar, P., Behringer, R., Cunningham, J. E., Howard, R. E., *et al.*, (1989). Suppression of the Aharonov–Bohm effect in the quantised Hall regime. *Physical Review B*, **39**, 6227–30.

Von Klitzing, K., Dorda, G., and Pepper, M. (1980), Realisation of a resistance standard based on fundamental constants. *Physical Review Letters*, **45**, 494–7.

Weisbuch, C. and Vinter, B. (1991). *Quantum semiconductor structures*. Academic Press, New York.

5 The two-dimensional electron gas

5.1 Introduction

In this chapter, we describe the two main systems that have driven research into low-dimensional physics over the last two decades. The metal–oxide–semiconductor field-effect transistor (MOSFET) was patented over 50 years ago, but the technology of oxidizing a Si surface in a way that left few remaining unsatisfied bonds (which have undesirable electrical activity) was perfected only in the 1960s, just in time for its exploitation in modern integrated circuitry where it has been the key active device (Nicollian and Brews 1982). High electron mobilities in GaAs–AlGaAs multilayers with selective doped layers were discovered in 1978 (Dingle *et al.* 1978), and this selective doping principle has become incorporated into the high-electron-mobility transistor (HEMT), at present the lowest noise transistor available, and used in amplifying weak signals, with the particular domestic application of receiving television signals from space. The commercial success of these two structures has increased the levels of research investment in multilayer structures more generally. In both structures, we have electrons that are 'trapped' at an interface, but are free to move (with metallic-like conduction properties) along the interface. The physics of these electron systems has been of wide interest, while attempts to improve the performance of these structures as devices have led to many of the new materials and device technologies that we now take for granted. Historically, the Si–SiO$_2$ system dominated research in the 1970s (Ando *et al.* 1982), while the GaAs–AlGaAs heterojunction dominated research in the 1980s. Although most of this book is about III–V epitaxial multilayers, experience with the Si MOSFET set the scene for much of the research that followed.

5.2 Practical realizations

5.2.1 The silicon MOSFET

Consider the MOS structure shown in Fig. 5.1(a). In effect we have a capacitor, with SiO$_2$ as the dielectric and p-type Si as one of the plates. In practice, the oxide is formed by heating the Si substrate in a moist oxygen atmosphere at temperatures and times (typically *c.*1000 °C for an hour) that allow the formation of a thin uniform SiO$_2$ layer which has few unsatisfied chemical bonds at the interface. The details of these growth schemes is a key and usually proprietorial aspect in fabricating high-performance transistors. In scaling down to small transistors (see Chapters 14 and 21) the precision in

Metal gate

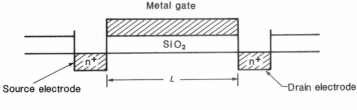

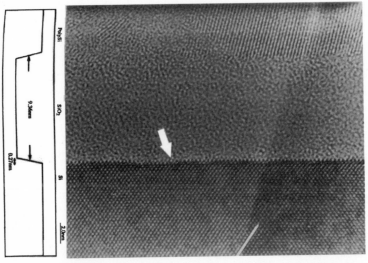

Fig. 5.1 The Si MOSFET: (a) multilayer capacitor structure including contacts to form the transistor; (b) TEM image of the Si–SiO₂ interface. (After W. M. Stobbs, University of Cambridge, private communication.)

the depth direction is all important. For our physics investigations the oxide thickness is of order $c.0.1$ μm, but state-of-the-art transistors require oxide thicknesses 10 times thinner. The application of a positive bias to the metal (the so-called gate in the transistor structure) draws electrons from the semiconductor into the vicinity of the Si–SiO₂ interface, forming a thin layer of n-type material, typically a few nanometres thick. In the bulk the Fermi level is just above the bottom of the conduction band, but a sufficient gate bias allows the energy bands near the surface to be bent downwards, first reducing the number of p-type carriers but eventually bringing the conduction band down below the Fermi level, allowing electron states to become populated. With a p-type substrate, the thin n-type layer means that the carrier density is locally inverted from the bulk value, and this thin electron layer is referred to as an inversion layer. Just beyond the region of the inversion layer, where the material is non-conducting, there is a depletion region of thickness approximately $\sqrt{(2\varepsilon_s \phi_d)/[e(N_A - N_D)]}$, where ϕ_d is the extent of the band-bending (equivalent to the built-in potential V_{bi} derived in

Chapter 1), and N_A and N_D are the number of acceptor and donors respectively, in which the number of mobile carriers has been reduced effectively to zero. The band-bending in the depletion layer comes from the residual fixed charges from ionized acceptors. Thus we have a thin sheet of electrons, free to move along the interface but electrically isolated from the p-type substrate. It is now possible to remove the oxide in various areas and to form n-type contacts to the semiconductor surface region. Any attempt to force electrons in through a (source) contact and out through a (drain) contact will be unsuccessful without the gate bias, as one or other of the n–p junctions will be reverse biased. A positive gate bias allows the inversion layer to form and easy conduction from source to drain to occur via entirely n-type regions. This influence of the gate bias on the source–drain conduction is at the heart of the action of the field-effect transistor: we refer to this technology as NMOS, since n-type conduction is involved. Of course it is possible to use a negative gate bias on a MOS capacitor with n-type Si and form a p-type conducting channel between p-type contacts, and hence there is a PMOS technology. Indeed, for reasons of low-power consumption, the two can be fabricated side by side on a chip (using ion implantation to control both the substrate and the local contact dopings), giving rise to CMOS (C = complementary) technology (Nicollian and Brews 1982).

The suggestion by Schrieffer (1957) that motion of inversion layer electrons might be quantized led to much of the subsequent research interest. The thickness of the conducting region is comparable to the Fermi wavelength for electrons or holes. Furthermore, the potential well formed by the bending conduction band has a number of bound states, and excitations in and between them are possible. We return to these topics in Section 5.3.

5.2.2 The modulation-doped (or selectively doped) heterojunction

In Chapter 2, we introduced the GaAs–AlGaAs heterojunction. The mobility of bulk GaAs is very high at 77 K, but is reduced by the onset of optic phonon scattering at higher temperatures, and by carrier freeze-out at lower temperatures. The details will be discussed below. At any temperature, the scattering of mobile carriers off the charged impurity states, from which they originated, is a significant contribution to the resistance of a piece of semiconductor. It is the main achievement of modulation-doped heterojunctions to reduce this Coulomb charge scattering to very small values.

The principle is simple. A crystal of ultrapure undoped GaAs is grown, followed by a thin layer (c.10–40 nm) of undoped $Al_xGa_{1-x}As$ alloy (with typically 20–30 per cent Al), and then a layer of moderately heavily n-doped AlGaAs (Fig. 5.2). In practice, a thin capping layer of GaAs completes the structure to prevent oxidation or other degradation of the AlGaAs layer. The electrons liberated from the donors by temperature, or by light, are free to move and may cross the heterojunction into the GaAs layer. From there they cannot easily return, as the potential barrier is typically about 0.2 eV high, i.e. typical of thermal energies at over 2000 K. The electrons are trapped in the GaAs by this barrier, but the residual Coulomb attraction keeps them at the interface, with the net formation of a dipole layer. We discuss the electrostatics in more detail in a later section, but the layer is typically about 10–20 nm thick. The electrons are free to move along the interface, but the Coulomb scattering is small because the electrons only see the overlapping tails of many distant Coulomb charge centres (Fig. 5.2). At very low temperatures and in very pure materials, where the Coulomb scattering would otherwise be large, it is found that the mobility of electrons at heterojunctions can exceed the values in low-doped bulk material by factors in excess of 1000. Indeed, the mean free path can approach fractions of a millimetre; over this distance, electrons retain their quantum phase memory. In the application of modulation-doped (selectively doped) heterojunctions in transistors operating at room temperature, the mobility of the electrons in the pure GaAs is typically double that of the doped GaAs used in metal-gate field-effect transistors. As we shall see in Chapter 16, this has two important implications for the performance of high-speed transistors: (i) the resistances are reduced, and with them the RC time constants, so that devices of a given size are faster, and (ii) the levels of noise generated by the device (usually from scattering processes) are much reduced. A large heterojunction potential step is clearly desirable, as one would like to drive the electrons parallel to the interface without them gaining sufficient kinetic energy to scatter back into the AlGaAs. This so-called real-space transfer (see Chapter 7) is to be contrasted with the k-space transfer to the satellite X and L valleys under high fields in GaAs, as described in Chapter 1. There are two problems (both related) that suggest that an Al composition of about 30 per cent is optimum in the GaAs–AlGaAs materials combination. Above 45 per cent Al, the alloy becomes an indirect gap semiconductor, with the X valley falling in energy as the Al fraction is increased, so that the maximum gap is at about 40 per cent. At lower concentrations, the theory of shallow donors, as applied to the alloy, breaks down. Above about 25 per cent Al an increasing fraction of the Si atoms form 'deep' levels, with higher activation energies and with a high sensitivity to light. Both these lead to unwanted side-effects in the performance of practical transistors, such as light or field dependences that reflect the previous history of the structure.

In both the Si MOSFET and the modulation-doped GaAs–AlGaAs heterojunction, we generate a two-dimensional electron gas (2DEG). The term gas is used as the effective density of carriers is low, typically 10^{15}–10^{16} m^{-2}, and so the typical interelectron spacing is large compared with the relevant effective Bohr radius. In the regimes of practical interest, the compressibility of the electron gas remains high, namely an increase in gate voltage and/or doping level in the AlGaAs results in a commensurate increase in the 2D carrier density. Later we shall encounter an unusual regime (in the presence of intense magnetic fields) where an incompressible electron liquid is formed, and a search for a Wigner crystal (an electron crystal) is underway (Stormer 1992).

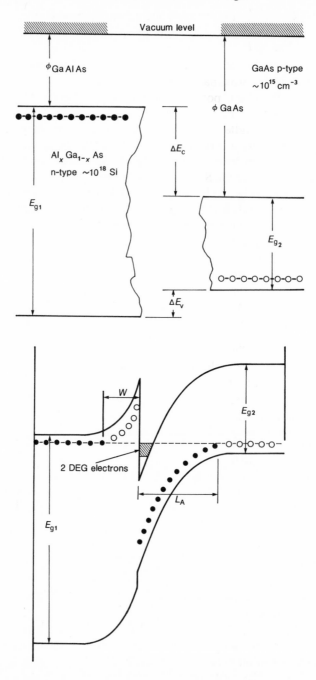

Fig. 5.2 The modulation-doped heterojunction showing the different electron affinities and the band-bending cause by charge transfer from the AlGaAs to GaAs when the materials are joined to produce a uniform Fermi level. (After Weisbuch and Vinter 1991.)

5.3 The theoretical description of the electron states

5.3.1 The Si MOSFET

There are many treatments of electron states in an n-channel MOSFET, ranging from crude approximations working on the assumption that electrons are trapped in a triangular potential well, through to complex numerical and self-consistent solutions of the Poisson and Schrödinger equations. In this section we adopt a middle course in treating many of the key physics issues in detail while retaining a formalism capable of analytic solutions.

We consider the schematic structure in Fig. 5.3. Under the inversion conditions, electrons are trapped near the Si–SiO$_2$ interface, and the thickness of the electron gas layer is constrained by the limited extent to which electrons can penetrate the oxide in one direction and into the bandgap of the semiconductor on the other. We adapt the envelope wavefunction theory that proved successful in treating impurity levels in Chapter 1 to cover the present problem. We assume that the wavefunction in the z direction (perpendicular to the Si–SiO$_2$ interface) varies as $\sim z\exp(-\lambda z)$, where λ is to be a variational parameter in what follows, and λ^{-1} is a measure of the thickness of the electron gas layer. Clearly, electrons are free to move across the surface of area $\mathcal{A}$ in parabolic energy bands, and so the overall normalized wavefunction for electrons in the inversion layer is of the form

$$F(\boldsymbol{r}, \lambda, \boldsymbol{k}_{||}) = (2\lambda)^{3/2}z\exp(-\lambda z)\exp(i\boldsymbol{k}_{||} \cdot \boldsymbol{r})/\mathcal{A}.$$

Assuming that there are n electrons per unit area in the inversion layer, we solve Poisson's equation for the charge density implied by $|F|^2$ to obtain the electrostatic potential, which for $z > 0$ is given by

$$\Phi(z) = -(4\pi n e/\varepsilon_\text{s})\,4\lambda^3\exp(-2\lambda z)(z^2/4\lambda^2 + z/2\lambda^3 + 3/8\lambda^4),$$

where ε_s is the semiconductor dielectric constant, and the two constants of integration are both zero since both $\Phi(z)$ and $\text{d}\Phi/\text{d}z$ vanish as $z \to \infty$. This potential and its derivative is matched at $z = 0$ to a field in the dielectric of thickness δ and dielectric constant ε_0, which extends to a potential of V_g at the metal–oxide interface, given that $\Phi(\infty) = 0$ (Fig. 5.3(a)). This matching gives a relation between n, V_g and λ:

$$n = V_\text{g}/[\delta/\varepsilon_0 + 3/(2\lambda\varepsilon_\text{s})]$$

which is trivially the capacitance relation obtained if all the inversion layer electrons were at a distance $3/(2\lambda\varepsilon_\text{s})$ in from the semiconductor–oxide interface.

The analysis is taken further (Stern and Howard 1967; Kelly and Falicov 1977) by calculating the total energy of the inversion layer electrons in the Hartree approximation (i.e. including the effects of electron–electron Coulomb interactions, but not any of the more complex exchange interactions) and obtaining a relation between n and λ. In practice, the interface is usually oriented normal to the (001) crystallographic plane of Si. In setting up the Schrödinger equation for the electrons, it is convenient to note that

$$-\text{d}^2(z^2e^{-2\lambda z})/\text{d}z^2 = \lambda^2(z^2e^{-2\lambda z})$$

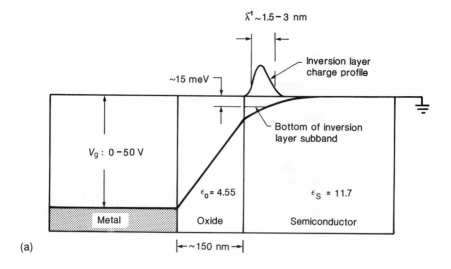

(a)

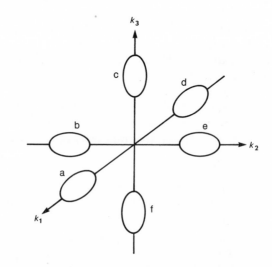

(b)

Fig. 5.3 (a) The potential in a MOS structure with a positive gate bias; (b) the sixfold conduction-band degeneracy in bulk Si. (After Kelly and Falicov 1977.)

and to consider λ as a type of k_z. The Schrödinger equation is obtained by taking the effective mass expansion about the conduction-band minimum and replacing each k component with $-i$ times the equivalent component of ∇ (Luttinger and Kohn 1955). With $F \equiv F(r, \lambda, k_{||})$, we have

$$-\hbar^2\left(\frac{\partial^2 F/\delta x^2}{m_{xx}} + \frac{\partial^2 F/\delta y^2}{m_{yy}} + \frac{\partial^2 F/\delta z^2}{m_{zz}}\right) + |e|\,\Phi(z)F = EF.$$

The sixfold degeneracy of the Si conduction bands (Fig. 5.3) means that we must consider kinetic energy terms of the form $\hbar^2 \lambda^2 / 2m_{zz}$ and here the larger longitudinal mass ($m_{zz} = m_1 = 0.92m_e$) of the two $\langle 001 \rangle$ valleys implies a lower kinetic energy compared with the need to use the transverse mass ($m_{zz} = m_t = 0.19m_e$) for the other four valleys. In the following, only the two heavy mass valleys are assumed to contain electrons, and each contains an equal number. The total one-electron energy (i.e. the sum of all the energies of the occupied states) from the solution of the Schrödinger equation is

$$E_1/\mathcal{A} = \pi\hbar^2 n^2 / 4m_t + \hbar^2 \lambda / 2m_{zz} + 15\pi n^2 e^2 / (8\varepsilon_s \lambda).$$

With a constant density of states in a 2D system, the average in-plane kinetic energy is half the Fermi energy, and for one valley occupied the Fermi energy is $E_f = n/N(E)$, and so the total kinetic energy is $n^2/2N(E)$. Here the density of states is determined by the effective masses in the two directions perpendicular to $\langle 001 \rangle$, namely m_t. The factor of 1/4 in the first term on the right-hand side comes from two valleys being occupied; the density of states is effectively doubled. The terms in the expression for $E_1/\mathcal{A}$ can be seen respectively as the kinetic energy in the plane of the electron gas, the kinetic energy at right angles (i.e. of confinement), and the electrostatic potential energy. The electron–electron contribution is more complicated to obtain so as not to double-count the interactions. The simplest way to proceed (Kelly and Falicov 1977) is to refer to the technical definitions of the Hartree term which can be written as either (i) the sum of the kinetic energies, the external potential, and half the total electron–electron interaction terms, or (ii) the sum of the kinetic energies and the self-consistent potential contribution minus half the electron–electron terms. Since in our case the external potential is merely the gate voltage, we arrive at the total Hartree energy per unit area as

$$E_{tot}/\mathcal{A} = \pi\hbar^2 n^2 / 4m_t + n\hbar^2 \lambda^2 / 2m_{zz} - neV_g/2 + 15\pi n^2 e^2 / (16\varepsilon_s \lambda).$$

This expression is minimized with respect to λ, subject to the earlier capacitance relation between n, V_g and λ. In the process one obtains the balance between kinetic and electrostatic energies. That earlier expression gives an almost linear relationship between n and V_g, so that we find here that λ varies as $(n/a^*)^{1/3}$; the dimensions in this relation are preserved by the effective Bohr radius factor a^*. For typical values of $\delta = 140$ nm, and $\varepsilon_s = 11.7$ and $\varepsilon_0 = 4.55$, λ^{-1} is of the order of 1.5–3 nm, approaching the limit of validity of the effective mass and envelope function theory. In practical research MOS-FETs, it takes about 0.5V bias on the gate to reach the threshold for forming the inversion layer, and a device with c.150 nm of oxide will break down with a gate voltage of c. 100V. One can achieve a variation in n of over two orders of magnitude. For this reason alone the MOSFET is a very flexible tool for studying electron–electron interactions, since the ratio of the typical interelectron spacing r_s to the effective Bohr radius a^* can be varied in one device over the range 0.7 to 7 simply by changing the gate bias. The results of more detailed and sophisticated calculations show that the above model contains the key physics ideas and that the results are quantitatively relatively accurate, particularly given the delicate balance between kinetic and electrostatic energies implied by the minimization to obtain λ.

The first excited state in the potential well has a normalized envelope variational wavefunction of the form

$$F_1(r, \lambda, k_{||}) = \sqrt{(12\lambda^3)}\, z\, (1 - 2\lambda z/3) \exp(-\lambda z)[\exp(ik_{||} \cdot r)/\mathcal{A}]$$

with a one-electron kinetic energy of $7\hbar^2\lambda^2/6m_{zz}$. This function has a node in the z direction and therefore has a p_z type symmetry in the z direction, allowing optical transitions from the ground state. Because the dispersion in the plane is parabolic in both cases, the optical transitions are concentrated at the same inter-sub-band energy and are easily observed. In addition, the separation in energy between ground and first sub-band varies as $\lambda^2 \sim (n)^{2/3}$, while the Fermi energy varies as n. At high enough Fermi energies, excited sub-bands should become occupied. In the Si MOSFET, the situation is complicated by the second ladder of sub-bands (derived from the $\langle 100 \rangle$ and $\langle 010 \rangle$ valleys with $m_{zz} = m_t$), and it is the occupation of these that is seen in transport experiments (Ando *et al.* 1982). Experiments, and indeed more detailed calculations, show that the variational approach to the excited states, while explaining and predicting the correct trends, is less accurate than results achieved with the ground states only. Furthermore, once an appreciable fraction of the electrons are optically excited, electric charge is redistributed, and this fact must be incorporated into the variational calculation.

5.3.2 The modulation-doped heterojunction

Just as with the Si MOSFET, there is a wide range of calculations of ever increasing sophistication of the electrostatics of a modulation-doped hetero-junction (Ando 1982). The physics within the GaAs layer is similar to that within Si in the MOSFET. The source of the electrons is a layer of doped AlGaAs. If we assume that all donor levels in the AlGaAs layer are activated and the carriers are transferred to the GaAs layer, we have a 2DEG of density $N_d d$ per unit area, where N_d is the doping density in the AlGaAs layer and d is the thickness of the doped layer. The gate beyond the AlGaAs layer acts to raise or lower the conduction band minimum throughout the structure, and the 2DEG can be depleted via external contacts. The situation, as far as the electrostatics is concerned, is shown in Fig. 5.4.

A layer of thickness d doped at N_d provides a quadratic solution of the Poisson equation for the potential, and therefore a curvature as shown. There are heterojunction steps at $z = 0$ and within the layer denoted c (for capping), otherwise within the layer c and s (for spacer) the potential is linear. The electron states in the 2DEG are such that the potential derivatives can be matched at $z = 0$. Given a fixed potential at the back of the substrate and the work function of the metal, the number of carriers in the 2DEG is fixed and there are no further degrees of freedom. By varying the parameters, one can have a normally on device (i.e. one containing a 2DEG) or a normally off device as shown in Fig. 5.5(a); the results of calculations of the bound states

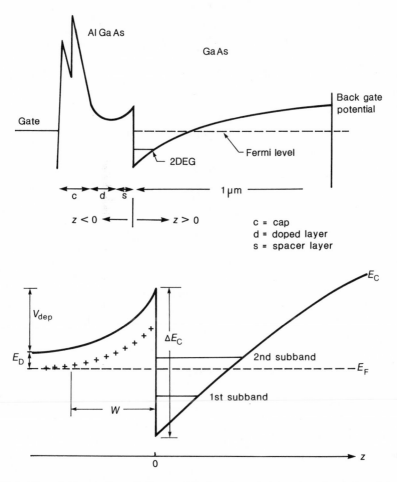

Fig. 5.4 The electrostatics of the modulation-doped heterojunction, showing the conduction band profile in a direction perpendicular to the wafer surface.

and their occupancy, including the effect of a gate voltage, are shown in greater detail in Fig. 5.5(b). The typical electron densities are in the range 10^{14}–10^{15} m^{-2} for physics studies and up to 10^{16} m^{-2} for practical devices.

5.3.3 Generic two-dimensional electron gas phenomena

There are several aspects of the physics of a 2DEG that are common to both the MOSFET and the HEMT. The first of these are the plasmon properties. In 3D systems, the Coulomb interaction between electrons is divided into two parts, of which one is effectively short range and leads to screening, and the other is long range and leads to the collective excitations of the electron gas—the plasma oscillations (Ashcroft and Mermin 1976; Madelung 1978). In 3D, these oscillations take place at a frequency given by

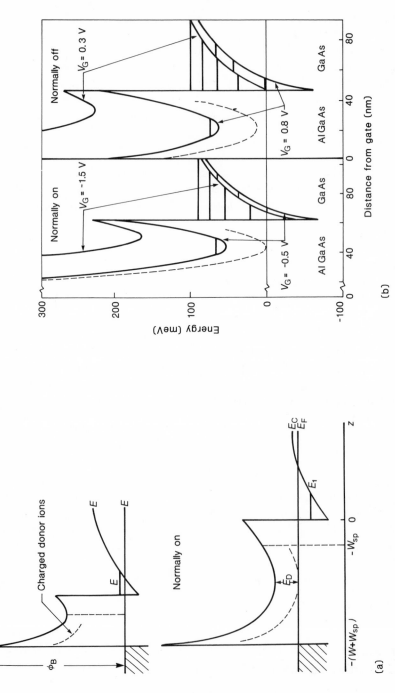

Fig. 5.5 Electron states in modulation-doped transistor structures with different biases between the gate and the 2DEG in both the normally-on and normally-off transistor configuration. (After Weisbuch and Vinter 1991.)

$$\omega_p = \sqrt{[4\pi n e^2/m^*]}$$

and have a weak dependence on wavelength in the case of finite-wavelength oscillations. In 2D, the derivation of the plasma frequency occurs via the same procedure: one seeks the zeros of the frequency- and wavevector-dependent dielectric function $\varepsilon(q,\omega)$, which imply that the electron system can support an internal electric field without any external applied electric field. The full details are outside the scope of the present text, but can be found in Stern (1967) and Ando *et al.* (1982). In a pure 2D system, the plasma frequency has a linear dispersion, namely

$$\omega \sim cq/\sqrt{\varepsilon_s}$$

In a Q2D system, i.e. where the electron gas in embedded in dielectric material and other conducting planes are present (as in a metallic gate and back contact to a semiconductor sample), the plasma frequency has a $\sqrt{q}$ dependence at shorter wavelengths

$$\omega_p \sim \sqrt{[4\pi n e^2 q/m^* \varepsilon_s]}$$

while at longer wavelengths the finite thicknesses d_{sc} and d_{ins} of the semiconductor and insulator play a role and

$$\omega \sim q\sqrt{\left(\frac{4\pi n e^2/m^*}{\varepsilon_{sc}/d_{sc} + \varepsilon_{ins}/d_{ins}}\right)}.$$

These plasma oscillations have been seen in far-infrared absorption experiments, and they are important in the operation of some devices.

Just as the collective modes of the electron gas are modified by the dimensionality, so too are the screening properties. In 3D, a fixed positive impurity charge modifies the local density of electrons so that the extra charge is partially screened out. In the Thomas–Fermi model, we solve

$$\nabla \cdot (\varepsilon_s \nabla_\phi) = -4\pi(\rho_{ext} + \rho_{ind})$$

where ρ_{ext} and ρ_{ind} are the external charge and the induced charge densities, the former introduced to the ideal system and the latter being the internal charges flowing in response to ρ_{ext}. The Thomas–Fermi approximation is to write the induced charge density as being proportional to the local change in potential, weighted by the density of states at the Fermi energy and the electronic charge:

$$\rho_{ind}(r) = -e[\phi(r) - \phi(r)]|_{\rho_{ind} = 0}(dN/dE_F).$$

This leads to an equation of the form

$$\nabla^2\phi - Q_s^2\phi = -4\pi\rho_{ext}/\varepsilon_s,$$

where Q_s is the 3D screening parameter, and the solution for an external point charge Ze is one of an exponentially screened Coulomb potential

$$\phi = (Ze/\varepsilon_s r)\exp(-Q_s r) \quad \text{and} \quad Q_s^2 = 6\pi n e^2/E_F.$$

In a Q2D MOSFET system (Ando *et al.* 1982), the induced change density is effectively in a plane, although the potential is 3D, with the resulting equation being of the form

$$\nabla \cdot (\varepsilon_s \nabla_\phi) - 2\varepsilon q_s \phi(r, z = 0)\delta(z) = -4\pi\rho_{ext}$$

where

$$q_s = (2\pi e^2/K)(dN/dE_F)$$

with $K = (\varepsilon_{si} + \varepsilon_{SiO_2})/2$, which is a constant in 2D since dN/dE_F is a constant. The asymptotic solution $\phi(r)$ for the potential (averaged over the z direction) from a charge Ze at $r = 0$, but $z = z_0 < 0$ within the oxide takes the form (r is the in-plane distance from $r = 0$):

$$\phi(r) \sim Ze(1 + q_s z_0)/(Kq_s^2 r^3)$$

which indicates rather less screening in 2D than in 3D (i.e. power law rather than exponential). This is because of the restricted possibilities of electrons confined to 2D to move to screen the effect of extra charge.

5.4 Electrical transport experiments and their interpretation

The first electrical transport data to be considered are the low-field mobility and its dependence on temperature, which we show in Fig. 5.6 for both the MOSFET and the HEMT. Starting with the Si device, we are able to plot the mobility as a function of voltage above threshold for inversion, i.e. as a function of the carrier density. We see in all cases a mobility that is dropping in value above $c.100$ K with a dependence that varies between T^{-1} and $T^{-1.5}$. In contrast with the similar variation in the AlGaAs–GaAs heterojunction case discussed below, a full theoretical explanation of this behaviour in a MOSFET is not available. A leading term is the scattering of the electrons by acoustic phonons in Si which gives an expression for the scattering time in the mobility ($\mu = e\tau/m^*$) that varies as (Ando *et al.* 1982)

$$1/\tau \sim n^{1/3}T.$$

The temperature variation is approximately correct, but note that the experimental results for carrier dependence are non-monotonic. At high temperatures, many sub-bands are occupied, and the scattering between them should also be incorporated. In addition, there are known to be residual charges in the oxide (as many as 2×10^{15} m^{-2}), and these give rise to strong impurity scattering. Attempts to obtain a qualitative fit to the data with theory are thwarted by the many unknown parameters: interface roughness, and the scattering it produces, the precise disposition of the oxide charges, how the phonons in bulk Si are modified at an interface with (amorphous) SiO$_2$, etc. The situation at low temperatures is only slightly more clear. The lack of dependence of mobility on temperature is characteristic of a metal. As the carrier density increases, so does the degree of confinement (cf. the λ of

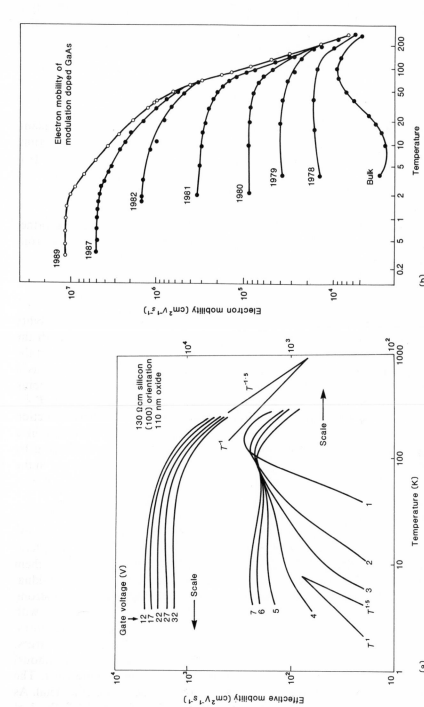

Fig. 5.6 The mobility of the 2DEG in (a) a MOSFET (Fang and Fowler 1968) and (b) a HEMT as a function of temperature (Pfeiffer *et al.* 1989). The latter shows the dramatic improvement in low-temperature mobility as the quality of the epitaxial layers has improved over 20 years.

the previous section). Therefore the carriers are closer to the interface with SiO_2, and so are increasingly susceptible to interface roughness and charges in the oxide—hence the reduction in mobility, even taking into account the extent that the carriers in the 2DEG partially screen them. At very low carrier densities, i.e. near the threshold voltage for the formation of the inversion layer, the few carriers present move with a quite different temperature dependence, and extensive research was carried out on this in the 1970s. Here most of the electrons are in localized states produced by the disorder in the oxide, and they move by hopping from one localized state to another; this is seen by confirming an $\exp(-a/T^{1/3})$ dependence for the conductivity predicted by Mott for this model (Pepper *et al.* 1974).

The data for the modulation-doped HEMT is simpler to interpret. In Fig. 5.6(b), we show data from heterojunctions with a fairly low 2DEG carrier concentration as the materials properties have improved over the years. In Fig. 5.7(a) we show data from pure GaAs, together with the temperature dependences that have been derived from calculations of the various scattering mechanisms in GaAs. Note the importance of optic phonon scattering at higher temperatures and ionized impurity scattering at low temperatures. The modulation-doping concept is used to remove the charged impurity centres from the GaAs and place them remotely in the AlGaAs. We can think of reducing the effective density of these charged centres to almost zero and so raising the limit on mobility as imposed by ionized impurity scattering as indicated. The low-temperature scattering from neutral impurities puts a limit on the mobility, but even this contribution has been reduced with improvements to the crystal growth technique, given the achievements of mobilities in excess of $10^3 \, m^2 \, V^{-1} \, s^{-1}$ at 1–2 K. In view of the exceptionally high mobilities, even the estimate for piezoelectric scattering has had to be revised. Indeed, the calculations shown in Fig. 5.7(a) were derived for bulk material, and the Boltzmann approximation was used; the Fermi level was near mid-gap, and the probability that the state into which an electron might scatter is occupied was negligible. In the 2DEG we have to use Fermi–Dirac statistics, and at low temperature the Pauli exclusion principle does play a role (Stormer *et al.* 1990). In the purest HEMT samples at the lowest temperatures, the mobility limitations imposed by these scattering processes can be rather less stringent than is indicated in Fig. 5.7(a) (cf. Fig. 5.7(b)). The mean free path of electrons in the purest samples is now a few millimeters. Modulation doping to produce 2D hole gases has resulted in mobilities of $5 \times 10^5 \, cm^2 \, V^{-1} \, s^{-1}$. Here, the more complex nature of the valence bands and the increased opportunity for hole scattering are responsible for the lower limits.

The mobility studied so far was achieved with very low electric fields. As the fields increase, so electrons are accelerated to higher energies and a wider range of final states are available for scattering. Some new processes occur, such as the emission of optic phonons, along the lines described in Chapter 1 for bulk materials, but there are also a number of processes in Si MOSFETs

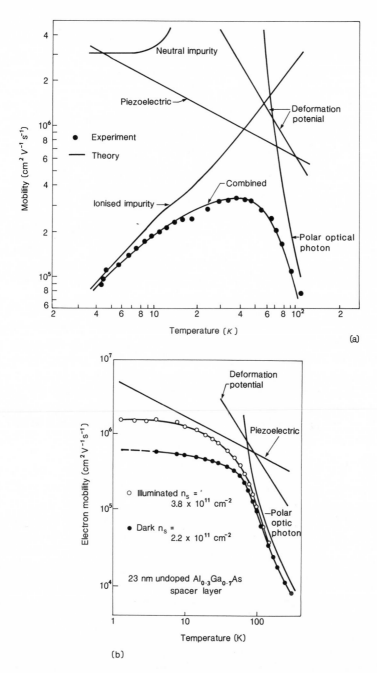

Fig. 5.7 Theoretical and experimental mobility: (a) high purity GaAs (Stillman and Wolfe 1976); (b) at a heterojunction (DiLorenzo *et al.* © 1982 IEEE). The theoretical lines give the limits placed on the mobility by the relevant scattering mechanism.

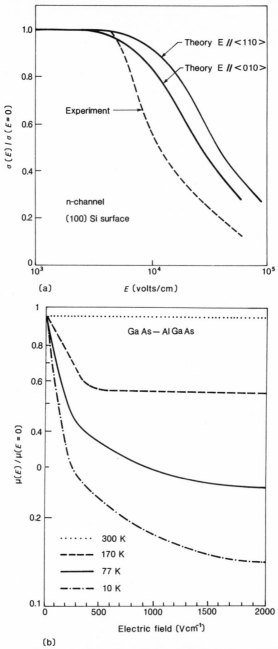

Fig. 5.8 Electron mobility as a function of electric field in (a) a Si MOSFET (theory (– –) and experiment (—)) (Hess and Sah 1974) and (b) a GaAs–AlGaAs heterojunction (Keever *et al.* 1982) as a function of temperature. Note in (b) that the sharp drop in mobility with electric field reflects the strongly enhanced low-field mobility as shown in Fig. 5.6(b).

that scatter electrons between different conduction band minima. The conductivity of the 2DEG falls with increasing electric field and eventually it saturates (Fig. 5.8(a)). The saturated drift velocity in a Si MOSFET is only about 60 per cent of the value in bulk Si, so that the extra methods of hot electron relaxation peculiar to the surface are important, as indeed is the non-uniform spatial density of the carriers themselves in the z-direction.

The mobility for electrons at the GaAs–AlGaAs interface shows a similar reduction as the electric field is increased (Fig. 5.8(b)). Here the emission of optic phonons is most important and efficient once the carriers accumulate an excess energy of about 36 meV. The greater relative importance of optic phonon emission in III–V materials comes from the fact that within each unit cell an optic phonon sets up a dipole with which an electron can interact directly through its charge (this is not the case with Si). The relative drop in mobility with electric field is much greater in the heterojunction. The high-field mobility is the one of most relevance to practical devices. The saturated drift velocity of electrons at a heterojunction has been widely investigated in device structures and it is thought not to differ greatly from that in bulk GaAs, i.e. $c.10^7$ cm s^{-1}. The velocity overshoot seen in the velocity–field characteristics of bulk GaAs (cf. Fig. 1.11) is not unambiguously seen in heterojunction structures.

The modifications to transport in the presence of a strong electric field generate warm or hot electrons, and a more detailed description of these phenomena is incorporated in Chapter 7. The discussion of the quantum corrections to the Boltzmann transport resulting from electron–electron interactions and disorder is deferred until after a description of transport in magnetic fields, as the results are used in the diagnostic experiments of quantum interference. Further details of the processes involved in setting the mobility of semiconductors are given by Ridley (1993).

5.5 Magnetotransport phenomena

The 2D density of states in a strong magnetic field changes from a constant value (in the absence of a field) to a set of broadened discrete levels, as described in Chapter 4. The implications of this for transport properties is dramatic. Even the changes in modest magnetic fields give further information about transport phenomena in the 2DEG. In Fig. 5.9(a) we show typical low-temperature data for the oscillations in conductivity of a Si MOSFET as a function of gate voltage, for a fixed magnetic field, together with the Hall conductivity. The conductivity is proportional to the density of states (via the Einstein relation (cf. Chapter 4)), and so we might expect a set of very sharp peaks in the longitudinal conductivity (as indeed was seen in Fig. 4.6), but in a Si MOSFET the degree of broadening of the Landau levels by interface roughness, oxide charge, etc. gives rise to the broadened structure shown. The Si MOSFET data are complicated by the twofold valley degeneracy that comes from the Si conduction band structure, and also the spin-splitting in

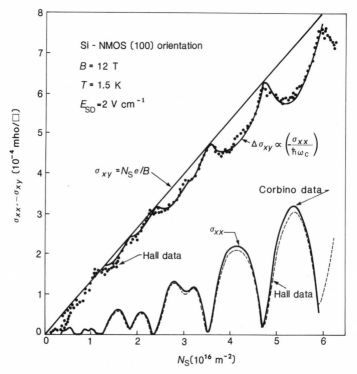

Fig. 5.9 The conductivity and Hall conductivity of a 2DEG in a Si MOSFET as a function of carrier density at fixed magnetic field (Wakabayashi and Kawaji 1980). Note that the deviation of the Hall conductivity from the straight line is proportional to the conductivity. The equivalent diagram for the GaAs–AlGaAs heterojunction is given in Fig. 4.6.

the higher magnetic fields. Thus each Landau level gives rise to four peaks in the conductivity, although broadening reduces the number resolved. During the 1970s, much experimental and theoretical work was devoted to interpreting the data in terms of the microscopic properties of the SiO_2 interface (Ando *et al.* 1982). Such work was halted in the 1980s with the advent of the GaAs–Al-GaAs heterojunction and the superior quality of the data (see Fig. 4.6).

In order to explain these new data, we must return to consider the effects of any remaining disorder. The fluctuations in potential from the discrete nature of the Coulomb charges in the AlGaAs layer are reflected in the nature of the electron states at the interface. The broadened Landau levels will have energy regimes of both localized and 'extended' electron states (see Fig. 4.5). In practice, a measure of the conductivity as a function of temperature and of magnetic field allows one to distinguish between metallic and insulating regimes. A conductivity that goes to zero as the temperature is reduced and has a finite activation energy is characteristic of localized state conduction, while conductivity that remains finite indicates metallic conduction with extended states. At lower magnetic fields there are more Landau levels, each with a reduced degeneracy, and the number of oscillations increases, although as the contributions from adjacent Landau levels overlap, the amplitude of the oscillations is reduced.

The Hall conductivity, which should have the value ne/B in a classical model, shows systematic deviations in Si and more prominently in GaAs. In high-quality material, the Hall conductivity is better described as a set of plateaux separated by steep rises. The quantized value of the Hall conductivity on each plateaux was explained in Chapter 4, but using an argument based on filled Landau levels. As the carrier density increases in Si, we see that, as $\sigma_{xx} \to 0$, σ_{xy} tends to the classical value just where it equals the quantum Hall value. As long as σ_{xx} remains zero, σ_{xy} maintains its quantum Hall value. We interpret this in terms of the extended/localized nature of the electron states at the Fermi level. In pure materials, the density of states is very high and not much broadened, so that the Fermi energy is in a region of extended states only over a small range of magnetic fields, giving finite σ_{xx} and non-quantized values of σ_{xy}.

The detailed theory of the quantum Hall effect has been revolutionized during the 1980s, with the advent of the Landauer–Buttiker formalism, and the results of a number of experiments, particularly in small structures, which indicate that most of the Hall current passes around the perimeter of the sample in so-called edge states (Buttiker 1992). This can be seen from Fig. 5.10(a) which shows that the cyclotron motion (in the limit of a vanishing longitudinal electric field) does not transport electrons except around the edges. If we return to section 4.3.3 and the argument leading to the quantum Hall effect, and consider the field lines in the limit of a weak longitudinal field, we obtain Fig. 5.10(b) which leads to the same conclusion about the importance of edge states. We also saw in Section 4.3.4 that 1D non-scattering channels are characterized by a quantized conductance, and the edge states here are just a special case.

From the definitions of the conductivity and resistivity tensors in a 2D system, we obtain the result that

$$\rho_{xx} = \sigma_{xx}/(\sigma_{xx}^2 + \sigma_{xy}^2)$$

which implies that $\rho_{xx} \to 0$ as $\sigma_{xx} \to 0$. This unusual result can be understood with reference to Fig. 5.10. The electrons move down the Hall bar under the Hall field because of the finite Hall conductivity! There is no longitudinal resistance.

The quantum Hall effect has a simple explanation in terms of single electron states. Once the magnetic field is sufficiently large that all electrons are in the lowest Landau level, there should be no further plateaux in σ_{xy} or zeros in σ_{xx}, and so it was surprising that plateaux and zeros were seen under the condition that the lowest level was one-third and two-thirds full, and as the quality of the material improved, at further fractions (1/5, 2/5, 2/7, 2/9, 2/11, . . .) (Fig. 5.11). The theory for these plateaux requires gaps to open up within

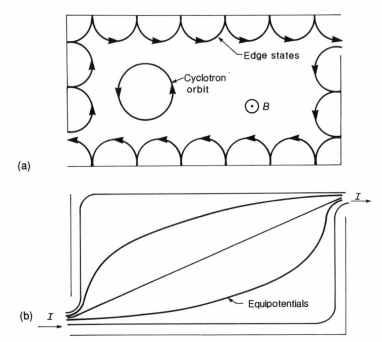

Fig. 5.10 Edge states viewed from (a) the cyclotron motion point of view and (b) the limit of the equipotential point of view. These are used to describe the quantum Hall effect and 1D motion (see Section 5.6).

the Landau levels (and indeed these gaps have been measured by activation energy studies of the conductivity near the zeros in σ_{xx}) and invokes many-electron effects—there are bound states containing three electrons whose excitations have an effective charge of 1/3. The details take us far beyond the scope of this text (see Prange and Girvin 1990; Chakraborty and Pietilainen 1988)).

5.6 Quantum interference and corrections to Boltzmann conduction

The Boltzmann conductivity is derived from single-electron theories; however, there are formal two-electron field theories for the conductivity, involving the destruction of electrons at one position and their recreation elsewhere (Fradkin 1991). At the end of the 1970s, the implications of quantum interference between these electrons was investigated; again the details are beyond the scope of this text (Lee and Ramakrishnan 1985). Whether the scattering of an electron is by disorder or by another electron, in 2D there are logarithmic corrections to the conductivity arising from the way that different electrons interfere in the transport process between two points in a system (as described in Section 4.3.2). Since the interference depends on electron phase, and magnetic fields introduce phase differences between electrons that take

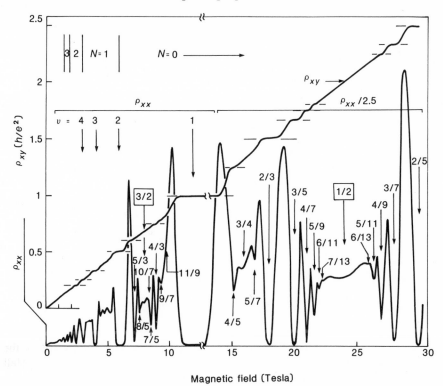

Fig. 5.11 The fractional quantum Hall effect, showing evidence of a rich variety of many-body phenomena in a 2DEG in a high magnetic field. (After Willett *et al.* 1987.)

different trajectories, the weak-field magnetoresistance of the 2DEG is a powerful technique for measuring the quantum interference (Pepper 1989; Ulloa *et al.* 1992). There is a wide body of data from electron gases in Si and GaAs and theoretical fits to both longitudinal and transverse magneto-resistance in the weak, intermediate, and high magnetic field regimes. We describe some of the details in Section 5.9 in the study of the transition from 2D to both 3D and 1D as the various predicted functional forms of the magnetoresistance have been verified. In addition, the values of free parameters (e.g. the power law of the energy dependence of principal inelastic scattering rates) have been determined.

5.7 Optical properties

An early set of experiments on *n*-channel Si MOSFETs shows the full power and the range of physics that can be investigated via the absorption of far-infrared radiation (Kneschaurek and Koch 1977, Ando *et al.* 1982.) The experimental set-up

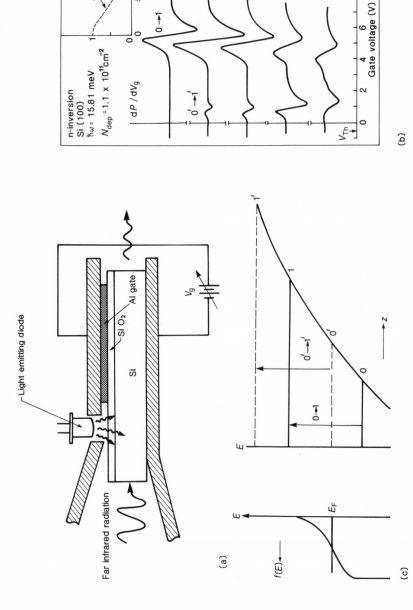

Fig. 5.12 (a) The experimental set-up, (b) the data, and (c) the transitions that are excited during a voltage sweep of a 2DEG in a MOSFET under illumination of a fixed-frequency far-infrared laser. The trends in the data are described in the text. (After Kneschaurek and Koch 1976.)

is shown in Fig. 5.12(a), the data in the form of the differential of the optical absorption with respect to gate voltage and temperature at a fixed excitation energy are shown in Fig. 5.12(b), and a diagram showing the relevant excitations is given in Fig. 5.12(c). As in earlier discussions (Chapters 1 and 4), optical absorption requires filled initial states, empty final states, and non-zero matrix elements. The optical absorption will be over a narrow band of energy, as both the ground and excited sub-bands have the same in-plane effective mass and hence kinetic energy. At low temperatures, only the lowest-energy sub-bands associated with the two $m_{zz} = 0.916m_e$ valleys are occupied, and at some bias the first excited states will have just the correct energy to absorb power from the exciting laser (the 0–1 transition in Fig. 5.12(b)). As the temperature is increased, there is a finite probability that the ground-state sub-band associated with the $m_{zz} = 0.19m_e$ valleys will become occupied by thermal excitation. In turn, these electrons can be excited into their excited sub-bands (0′–1′ in Fig. 5.12(b)). Because of the smaller mass in this second ladder of sub-bands, the energy separation between the sub-bands is greater. Since the separation also depends on λ^2 (see Section 5.3.1 above), any absorption at fixed energy will occur for a smaller value of λ, and hence a lower carrier density ($n \sim \lambda^3$) and hence gate voltage. At higher temperatures still, the finite probability of thermally occupying the 1 and 1′ levels further reduces the optical absorption strength. The data in Fig. 5.12(b) contain all these features.

The quantitative interpretation of the inter-sub-band absorption is complicated by an important set of many-electron correction terms, giving rise to what is known as the Moss–Burnstein shift (Chen *et al.* 1976; Ando *et al.* 1982). The final result is that an optically determined inter-sub-band energy is a combination of the actual one-electron inter-sub-band energy and an effective plasma frequency:

$$\omega^2(\text{exp}) = \omega^2_{(\text{bare})} + \omega^2_{(\text{plasma})} \quad \text{and} \quad \omega^2_{(\text{plasma})} = 4\pi n e^2 f / \varepsilon_s m^* d_{\text{eff}}$$

where f is the oscillator strength of the transition and d_{eff} is the thickness of the 2DEG ($\sim 3/2\lambda$). The origin of this second term is identical to local field corrections in the theory of dielectrics (Ashcroft and Mermin 1976).

The optical properties of a 2DEG at a GaAs–AlGaAs heterojunction are similar to those in a Si MOSFET, but simpler because of the absence of multiple conduction band valleys. The single heterojunction system has not been studied widely, as the modulation-doped quantum well has proved a more powerful physics tool and is of direct relevance to optical devices (see Chapter 10).

5.8 Recent advances

In this section we describe two recent advances that have been achieved using the flexibility of MBE or MOCVD growth in the design of 2DEG structures.

5.8.1 Back-gating

The relatively small conduction bandgap offset of about 0.3 eV between GaAs and AlGaAs is 10 times smaller than that between Si and SiO_2, and as a result

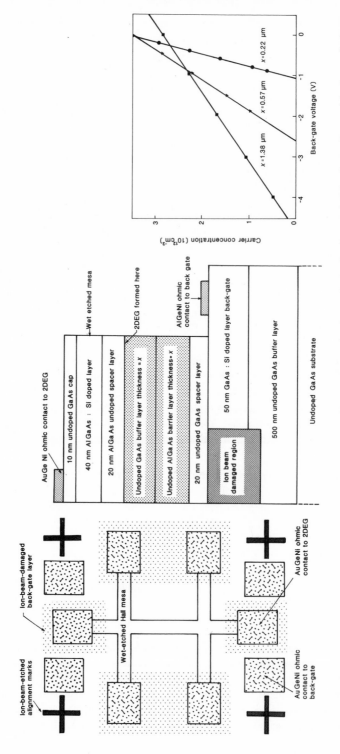

Fig. 5.13 A focused ion beam back-gated high-electron-mobility structure is shown in both plan and side view, along with the variations in carrier density observed as the back-gate voltage induces different 2D carrier densities. The slopes of the curves agree well with the distance between the 2DEG and the back-gate conducting layer, as expected from a simple capacitor relation. (After Linfield *et al.* 1993.)

large fields cannot be applied to surface gates to modulate the carrier density in GaAs, as is possible in Si. Over the last decade, a number of attempts have been made at back-gating, i.e. varying the potential profile beneath the 2DEG by having a further electrode there. As a bias is applied to this back-gate, the shape of the energy bands can be modified to some extent, with some form of gate or back-gate leakage setting the practical limits. With contacts to the source and drain, the electron density of the 2DEG can be varied. Early results were obtained on crude samples (e.g. direct application of $c.100$ V to the back of a wafer (Stormer *et al.* 1982)), but gradually, as the quality of the grown crystals improved and with the ability to achieve spatially precise low-resistance ohmic contacts, it became possible to achieve about 30 per cent modulation of the carrier density in a 2DEG with bias applied to a conducting layer set back over 1 µm (Hamilton *et al.* 1992). More recently, focused ion beams have been used to damage selected volumes of doped material in the back-gating layer, rendering it insulating at low temperatures, in just those parts beneath the position of the ohmic contacts to the 2DEG. The structure and results are shown in Fig. 5.13 (Linfield *et al.* 1993). This has enabled studies to be made of the 2DEG where the carrier density has been varied between zero and $3 \times 10^{16}\,\mathrm{m}^{-2}$. The mobility varies precisely as $n^{1.5}$ over the entire range of carrier density, as expected from the residual remote Coulomb scattering (Ando *et al.* 1982). Note that one can bring the back-gated layer to within 0.22 µm of the 2DEG, five time closer than ever achieved with other techniques (such as relying on shallow ohmic contacts to the 2DEG that do not penetrate through to the back-gated layer). This system is very useful for checking the theory of the electron states, as one can be reasonably precise about the conditions of back-gate bias and total carrier concentration at which a second sub-band becomes occupied. The simple variational theory developed for Si, in Section 5.3.1 proved to be inadequate when modified for GaAs. However, the introduction of a second variational parameter λ_2, the inverse thickness of the second sub-band wavefunction, led to more accurate predictions (Kelly and Hamilton 1991). In Fig. 5.14, the experimental data come from the Hall mobility, which reduces with a second occupied sub-band and the extra scattering channels available, and from Shubnikov–de Haas magnetoresistance oscillations, as a second period is established once a second sub-band is occupied. In thicker ($c.0.1$ µm) modulation-doped quantum-well-type layers it is possible to form a second 2DEG at the second interface of the quantum well with an appropriate bias.

5.8.2 Two 2DEGs in parallel

Although it has always been possible to grow multilayer structures with many 2DEGs in parallel (as in multiple quantum wells with doped barrier layers), the selective contact of two closely separated ($c.20$–50 nm) 2DEG layers has only been mastered recently by a variety of techniques which allow a number of simple experiments to be performed (Fig. 5.15(a)). A source–drain current in one 2DEG can set up and, by electron–electron interactions (a sort of

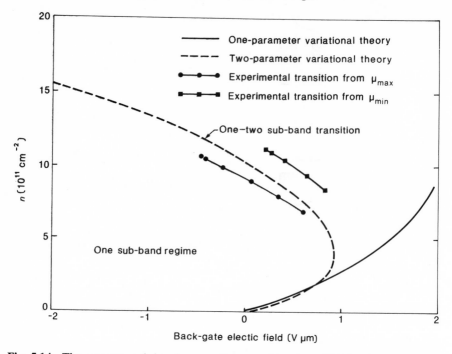

Fig. 5.14 The one–two sub-band occupation transition in a 2DEG with a back-gate varying the shape of the 2D electron states. Theory and experiment are in good agreement. (After A. Hamilton, Thesis, University of Cambridge, 1993.)

Coulomb drag), induce a current in the second 2DEG (Gramila *et al.* 1992). The experimental Coulomb drag resistance (voltage drop generated over the second 2DEG divided by current in the first 2DEG) varies as T^2 at low temperatures (Fig. 5.15(b)), and this is the form of the temperature dependence of electron–electron scattering (Ashcroft and Mermin 1976). Here, the electrons that scatter from each other are in the different 2DEGs. There have also been experiments on tunnelling between two 2DEGs induced by a bias between the two electron gases (Eisenstein *et al.* 1992).

5.9 The transition from three to two dimensions

The transport properties of electron gases in 3D and 2D have different functional dependences on temperature and magnetic field (see Chapter 4). Studies of the transition from 3D to Q2D systems have been investigated in a GaAs Schottky gate FET (also known as a MESFET, see Chapter 16). A layer of n-doped GaAs about 1 μm thick is grown on top of semi-insulating GaAs, and a reverse bias on a Schottky gate is used to vary the thickness of the depletion region underneath and hence the thickness of the remaining conducting channel (Fig. 5.16(a)). Ultimately the device can be pinched off, with no conduction between source and drain, and just prior to pinch-off the

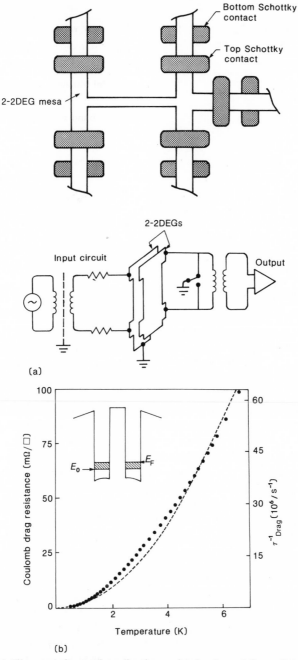

(a)

(b)

Fig. 5.15 (a) The experimental realization, with back and front Schottky gates, of selective contacts to two closely separated but parallel 2DEGs; (b) the resulting Coulomb drag resistance as a current in one 2DEG induces a field in the second. (After Gramila *et al.* 1992.)

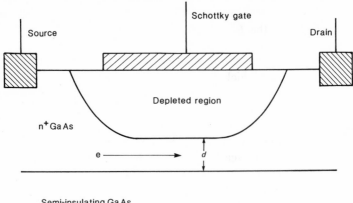

(a)

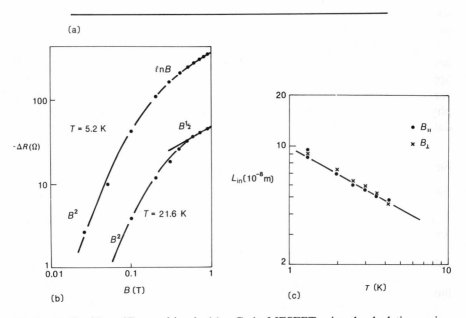

(b) B (T)   (c) T (K)

Fig. 5.16 The 3D-to-2D transition in (a) a GaAs MESFET using the depletion region of a reverse biased Schottky gate to define the thickness of the conducting layer, as seen in (b) the magnetoconductance data and (c) the inferred L_{in}. (After Newson *et al.* 1985.)

conducting layer is sufficiently thin for 2D transport effects to be observed. An extensive set of calculations have been made of the energy levels as a function of channel thickness in the regime where the layers are thin. Changes in the occupation of these energy levels, and hence the conductivity, in the presence of a magnetic field are used to monitor the thickness of the channel. (Such studies of magnetic depopulation are more widely used in studies of

Q1D systems, and so are discussed in the next chapter.) The theory for transport in such a system, and the electron–electron interaction and interference corrections to the Boltzmann conductivity in particular, indicate a negative magnetoresistance, with a B^2 variation at low fields in 3D and 2D, changing to $\sqrt{B}$ and $\ln B$ behaviour in 3D and 2D respectively at higher fields. With a layer only 44 nm thick (as inferred from calibration data), only one sub-band is occupied at low temperatures and 2D transport is expected. At higher temperatures, thermal excitations to many sub-bands are possible, and the inelastic length can become smaller than 44 nm so that the 3D conductivity is anticipated at higher magnetic fields. This is seen in Fig. 5.16(b). In Fig. 5.16(c) we plot the inelastic scattering length L_{in} as a function of temperature as inferred from the weak-field magnetoconductance (in the regime where the cyclotron length L_c satisfies the condition $L_c^2 > dL_{in}$, where d is the thickness of the channel):

$$\Delta\sigma(B) = (e^2/2\pi^2\hbar)\ln(d^2L_{in}B^2e^2/3\hbar^2 + 1)$$

which clearly shows L_{in} varying between 100 and 40 nm between 1 and 5 K, with a $1/\sqrt{T}$ dependence (Newson *et al.* 1985).

Studies of the 2D-to-1D transition are integral to the subjects raised in the next chapter, and so are discussed there. The ability to make structures of appropriate size, and to use gate voltages, magnetic fields, etc. to vary the various important length scales, has been one of the principal reasons why the physics of semiconductor microstructures has been at the forefront of elucidating the nature of quantum transport in solids.

References

Ando, T. (1982). Self-consistent results for a GaAs/Al$_x$Ga$_{1-x}$As heterojunction. I. Subband structure and light-scattering spectra. *Journal of the Physical Society of Japan*, **51**, 3893–99.

Ando, T., Fowler, A. B. and Stern, F. (1982). Electronic properties of two-dimensional systems. *Reviews of Modern Physics*, **54**, 437–672.

Ashcroft, N. W. and Mermin, N. D. (1976) *Solid state theory*. Holt, Rinehart, Winston, New York.

Buttiker, M. (1992). The quantum Hall effect in open conductors. In *Semiconductors and semimetals*, Vol. 35. (ed. M. Reed), pp. 191–277. Academic Press, New York.

Chakraborty, T. and Pietilainen, P. (1988). *The fractional quantum Hall effect: properties of an incompressible quantum fluid*. Springer Series in Solid State Physics, Vol. 58. Springer-Verlag, Berlin.

Chen, W. P., Chen, Y. J. and Burstein, E. (1976). The interface EM modes of a 'surface quantised' plasma layer on a semiconductor surface. *Surface Science*, **58**, 263–5.

DiLorenzo, J. V., Dingle, R., Feuer, M., Gossard, A. C., Hendel, R., Hwang, J., C. M., *et al.* (1982). Materials and device considerations for selectively doped heterojunction transistors. *Technical Digest, International Electron Devices Meeting*, pp. 578–81 IEEE, New York.

Dingle, R., Stormer, H. L., Gossard, A. C. and Wiegmann, W. (1978). Electron mobilities in modulation-doped heterojunction superlattices. *Applied Physics Letters*, **33**, 665–7.

Eisenstein, J. P., Gramila, T. J., Pfeiffer, L. N. and West, K. W. (1992). Resonant tunnelling in GaAs/AlGaAs quantum wells. *Surface Science*, **267**, 377–82.

Fang, F. F. and Fowler, A. B. (1968) Transport properties of electrons in inverted silicon interface. *Physical Review*, **169**, 619–31.

Fradkin, E. (1991). *Field theories of condensed matter systems*. Addison-Wesley, New York.

Gramila, T. J., Eisenstein, J. P., MacDonald, A. H., Pfeiffer, L. N. and West, K. W. (1992). Electron–electron scattering between parallel two-dimensional electron gases. *Surface Science*, **263**, 446–50.

Hamilton, A. R., Frost, J. E. F., Smith, C. G., Kelly, M. J., Linfield, E. H., Ford, C. J. B., *et al*, (1992). The back-gate split-gate transistor: a one-dimensional ballistic constriction with variable Fermi energy. *Applied Physics Letters*, **60**, 1881–3.

Hess, K. and Sah, C. T. (1974). Hot carriers in silicon surface inversion layers. *Journal of Applied Physics*, **45**, 1254–5.

Keever, M., Kopp, W., Drummond, T. J., Morkoc, H. and Hess, K. (1982). Current transport in modulation-doped (AlGaAs/GaAs) heterojunction structures at moderate field strengths. *Japanese Journal of Applied Physics*, **21**, 1489–95.

Kelly, M. J. and Falicov, L. M. (1977). Electronic ground state of inversion layers in many-valley semiconductors. *Physical Review B* **15**, 1974–82.

Kelly, M. J. and Hamilton, A. (1991). The electronic structure of the back-gated high electron mobility transistor. *Semiconductor Science and Technology*, **6**, 201–7.

Kneschaurek, P. and Koch, J. F. (1977). Temperature dependence of the electron intersubband resonance on (100) Si surfaces. *Physical Review B*, **16**, 1590–6.

Lee, P. and Ramakrishnan, T. (1985). Disordered electronic systems. *Reviews of Modern Physics*, **57**, 287–337.

Linfield, E. H., Jones, G. A. C., Ritchie, D. A. and Thompson, J. H. (1993) The fabrication of a back-gated high electron mobility transistor—a novel approach using MBE growth on an *in situ* ion beam patterned epilayer. *Semiconductor Science and Technology*, **8**, 415–22.

Luttinger, J. M. and Kohn, W. (1955). Motion of electrons and holes in perturbed periodic fields. *Physical Review*, **97**, 869–83.

Madelung, O. (1978). *Introduction to solid state theory*. Springer-Verlag, Berlin.

Newson, D. J., McFadden, C. M. and Pepper, M. (1985). Quantum corrections and the metal–insulator transition as a function of dimensionality in the GaAs impurity band. *Philosophical Magazine B*, **52**, 437–58.

Nicollian, E. H. and Brews, J. R. (1982). *MOS (metal oxide semiconductor) science and technology*. Wiley, New York.

Pepper, M. (1989). Quantum interference in semiconductor devices. In *Bandstructure engineering in semiconductor microstructures* (ed. R. A. Abram and M. Jaros). NATO ASI Series B, Physics, Vol. 189, pp. 137–47. Plenum, New York.

Pepper, M., Pollitt, S., Adkins, C. J. and Oakley, R. E. (1974). Variable-range hopping in a silicon inversion layer. *Physics Letters A*, **47**, 71–2.

Pfeiffer, L., West, K. W., Stormer, H. L. and Baldwin, K. W. (1989). Electron mobilities in excess of 10^7 cm^2/V sec in modulation doped GaAs. *Applied Physics Letters*, **55**, 1888–90.

Prange, R. E. and Girvin, S. M. (1990) *The quantum Hall effect* (2nd edn). Springer-Verlag, New York.

Ridley, B. K. (1993). *Quantum processes in semiconductors* (3rd edn). Oxford University Press.

Schrieffer, J. R. (1957). Mobility in inversion layers: theory and experiment. In *Semiconductor Surface Physics* (ed. R. H. Kingston), pp. 55–60. University of Pennsylvania Press, Philadelphia, PA.

Stern, F. (1967). Polarisability of a two-dimensional electron gas. *Physical Review Letters*, **18**, 546–8.

Stern, F. and Howard, W. E. (1967). Properties of semiconductor inversion layers in the electric quantum limit. *Physical Review*, **163**, 816–35.

Stillman, G. E. and Wolfe, C. M. (1976). Electrical characterisation of epitaxial layers. *Thin Solid Films*, **31**, 69–88.

Stormer, H. L. (1992). The many phases of 2D electrons in high magnetic fields. *Physica Scripta*, **T45**, 168–73.

Stormer, H. L., Gossard, A. C. and Wiegmann, W. (1982). Observation of intersubband scattering in a 2-dimensional electron system. *Solid State Communications*, **41**, 707–9.

Stormer, H. L., Pfeiffer, L. N., Baldwin, K. W. and West, K. W. (1990). Observation of a Bloch–Gruneisen regime in two-dimensional transport. *Physical Review B*, **41**, 1278–81.

Ulloa, S. E., MacKinnon, A., Castano, E. and Kirczenow, G. (1992). From Ballistic transport to localization. In *Handbook on Semiconductors* North Holland, Amsterdam Vol. 1 (revised edn) (ed. P. T. Landsberg), pp. 817–962.

Wakabayashi, J. and Kawaji, S. (1980). Hall conductivity in n-type silicon inversion layers under strong magnetic fields. *Surface Science*, **98**, 299–307.

Weisbuch, C. and Vinter, B. (1991). *Quantum semiconductor structures*. Academic Press, New York.

Willett, R., Eisenstein, J. P., Stormer, H. L., Tsui, D. C., Hwang, J. C. M. and Gossard, A. C. (1987). Observation of an even-denominator quantum number in the fractional quantum Hall effect. *Physical Review Letters*, **59**, 1776–9.

6 The one-dimensional electron gas

6.1 Introduction

The role of 2D systems in elucidating details of quantum transport in semiconductors was a key theme of the previous chapter, which concluded with an investigation of the transition between three and two dimensions. Until recently, 1D systems have been the playground of the theorist; some of the most intractable problems in many-body physics admit analytical or transparent numerical solutions (Lieb and Mattis 1966). It has long been appreciated that an infinitesimal amount of disorder is sufficient to localize all electron states in one dimension (Mott and Twose 1961). Thouless's ideas, described in Chapter 4, Section 4.3.2, were advanced on the basis of 1D arguments. Once the extent of the applications of Q2D systems had been appreciated (cf. Chapter 16), attempts were made to extend the technology to realize Q1D systems and to investigate the Q2D–Q1D transition. It has been important to find ways to confine the Q2D electron gas spatially in one of the two in-plane directions; an increasing number of methods have been found. In this chapter, we mirror the previous one, and describe the wealth of phenomena in Q1D systems, as investigated in the 1D electron gas (1DEG). In later chapters (11, 12, and 15) we consider other aspects of physics in Q1D systems not considered here.

6.2 Practical realizations

6.2.1 The Si MOSFET

In an n-channel MOSFET, conduction from source to drain takes place in an inversion layer of electrons that connects two heavily doped (n^+) contacts. The inversion layer is formed by the action of a bias applied to the gate. The earliest method of forming a 1D channel was the addition of heavily doped (p^+) contacts separated by a gap of 1–2 µm in a line running between source and drain (see Fig. 6.1(a)). These contacts can be formed by ion implantation or diffusion (alloying) before the gate metal is added (Fowler et al. 1982; Dean and Pepper 1984). A reverse-biased p–n junction can be formed between the n-channel and the p-contacts by a suitable bias on the latter, and the depletion layer thus formed has a bias-dependent thickness that can in turn be used to control the lateral extent of the n-channel. This method of fabrication relies on control over the diffusions or implants so that there is not too much lateral spread. In practice, it is difficult to control to better than the width of a 1 µm channel to better than about 10 per cent along a typical length of

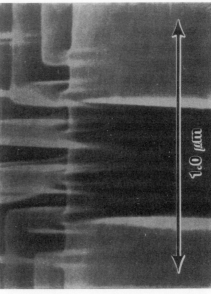

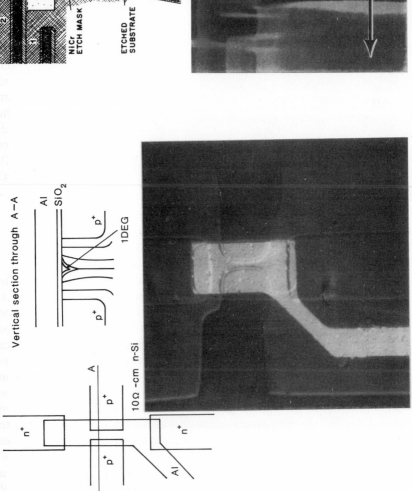

Fig. 6.1 1D electron gases in Si MOSFETs formed by (a) biasing p$^+$ side contacts (after Hartstein 1985), and (b) using a narrow gate defined by electron beam lithography as a mask to etch away the rest of the structure (after Skocpol 1986).

5–10 μm (Hartstein 1986). This technique has the advantage of making only modest (about 1 μm) demands on lithography. One of the persistent problems in this approach is the theoretical modelling of the electron states in a complex electrostatic pattern generated by the biases on the gate and the p^+ implants (see Section 6.3).

Alternative techniques for making narrow channels include the direct writing by electron beam lithography of a thin gate between the source and drain contacts whose bias generates a narrow channel. Gates as narrow as 60 nm have been formed, although the fringing fields in the oxide mean that a channel of typically c.100 nm is formed at the Si–SiO$_2$ interface. Yet another technique is to use reactive ion etching or wet chemical etching to remove physically much of the oxide not covered by the gate (see Fig. 6.1(b)). This inhibits lateral spread of the gating action, and surface depletion of the thinner oxide layers also prevents lateral spread of the inversion layer. These latter techniques have the added flexibility that the lithography can be used to define various side-arms to the one-dimensional channel from which voltage and/or currents can be measured (Skocpol 1986).

6.2.2 The modulation-doped heterojunction

By the mid-1980s, high electron mobilities in the 2DEG at modulation-doped heterojunctions were readily available. Electron beam lithography had already proved its worth in the study of Q1D systems in Si. Studies of Q1D physics in heterojunction systems have been carried out almost exclusively in conjunction with electron beam lithography forming patterned gates, of by now considerable sophistication, the negative bias on which now defines the 1DEG (Thornton *et al.* 1986). Multiple-gate structures (Fig. 6.2) have been made in recent structures, with eight or more independent contacts being available to control the electrostatics in the plane of the 2DEG (Field *et al.* 1993). The action of the gates here is straightforward; a sufficient negative bias on a gate depletes all the carriers in a region under the gate. In order to control the lateral spread of the gate fields, research has enabled high-mobility electron gases to be prepared ever closer to the top surface, i.e. at a depth of less than 20 nm instead of the typical 100 nm (Frost *et al.* 1991).

6.2.3 Generic 1D systems

Early work on Si devices involved channels that were more than 1 μm, and typically about 10 μm in length, this being within the limits set by the then available Si technology. The important transport length scales for these channels are all rather smaller, and so the 1D systems are long, i.e. of the type to which some of the Thouless arguments might be applied. Indeed, as we shall see below, the transport studies all dealt with multiple-scattering regimes for the electrons in the 1D channels. The real revolution came in the modulation-doped heterostructures, where, with high quality material, the phase-breaking length (the distance over which an electron could retain memory of its phase) could exceed several micrometres and entire structures

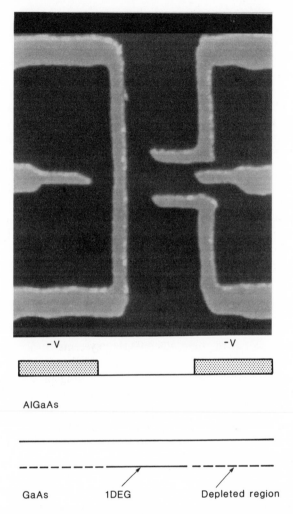

Fig. 6.2 (a) Multiple-gate patterns for studying Q1D transport (reported in Field *et al.* 1993) and (b) the depleting action of two adjacent gates to define a narrow channel.

could be made smaller than this length. The first observation of ballistic motion (van Wees *et al.* 1988; Wharam *et al.* 1988*a*), via the quantization of the 1D ballistic resistor, set off an intense and continuing burst of research activity that takes up most of this chapter.

6.3 The theoretical description for the electron states

From the outset, it has been appreciated that 3D modelling of the self-consistent electrostatic fields in these Q1D systems is complicated, requiring detailed numerical solutions. These have been performed for both Si and

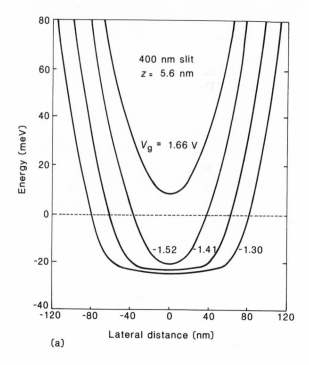

(a)

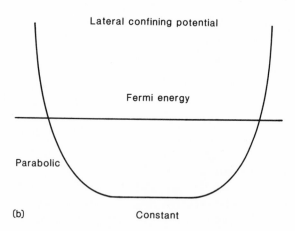

(b)

Fig. 6.3 (a) The self-consistent potential profile in a line ′at 5.6 nm below a GaAs–AlGaAs heterojunction across a 1D channel defined in a 2DEG by a gate separation of 400 nm with different biases applied to the gates. Note the distortion to the parabolic shape as more electrons actually populate the sub-band (Laux *et al.* 1988); (b) a practical approximation to the potential profile.

GaAs structures (Laux and Stern 1986; Laux *et al.* 1988), but in practice simpler analytical models have been invoked to explain and analyse the data. In Fig. 6.3 we show the electrostatic potential profile and the self-consistent sub-band energies in a cut across a long one-dimensional channel in a GaAs–AlGaAs system formed by two long gates split by about a micrometre. The earliest work on the 1D channels was based on a quadratic potential profile across the 1D channel, and so harmonic oscillator wavefunctions were used. In practice, one sees that the bottoms of the sub-bands are relatively flat, and some simplified, but numerical, solutions are available that include a small flat region between the two halves of the harmonic potential. This insertion has the effect of modifying the energy levels of successive sub-bands, and the thickness of the flat region (Fig. 6.3(b)) is used as a fitting parameter.

The parabolic potential, namely $V(y) = m^* \omega_0^2 y^2 / 2$ with energy levels at $\hbar \omega_0 (n + 1/2)$, is a convenient starting point for studies of magnetic depopulation that have proved invaluable in determining the properties of 1D channels. The ω_0 is a measure of the lateral confinement in a long 1D channel, and it is hoped to obtain a relationship between the gate bias and the equivalent ω_0. The application of a strong magnetic field along the channel imposes an extra confinement potential in the y direction given by $V'(y) = m^* \omega_c y^2 / 2$, where $\omega_c = eB/m^*$ is the cyclotron frequency. The magnetic field increases the sub-band energies to $\hbar \omega (n + 1/2)$, where $\omega^2 = \omega_0^2 + \omega_c^2$, and with sufficiently large fields $\hbar \omega$ can become an appreciable fraction of the Fermi energy for electrons in the 1D channel. The overall wavefunctions in the absence of a magnetic field are now of the form

$$\Psi(x,y,z) = (1/\sqrt{L})\exp(ik_x x)\phi_n(y)F_m(z),$$

where ϕ is an oscillator wavefunction and F is as given in the previous chapter. In the absence of a magnetic field the one-electron energy is

$$E(k_x, n, m) = \hbar^2 k_x^2 / 2m^* + \hbar \omega_0 (n + 1/2) + E_m$$

where E_m is the appropriate solution for the bound-state energy level in the z direction normal to the multilayers. In a magnetic field, the change from ω_0 to ω causes a further separation in the energy levels, and for a fixed carrier density, successive energy sub-bands have too great an energy to remain occupied. (The calculation of the Fermi energy is the same as that described in Chapter 2, but using the appropriate form of the Q1D density of states from Chapter 4.) The magnetic fields at which successive levels become depopulated can be measured experimentally, and the comparison with theory used to refine the models for the confining potential (Fig. 6.4), for example by using the flat bottom in Fig. 6.3(b) as a parameter (Berggren *et al.* 1986). The detailed shape of the potential profile within a short Q1D system has attracted less attention in the context of transport experiments as described below, where the important matter is whether a particular one-dimensional subband is either occupied or not. Finer details of the transport data contains

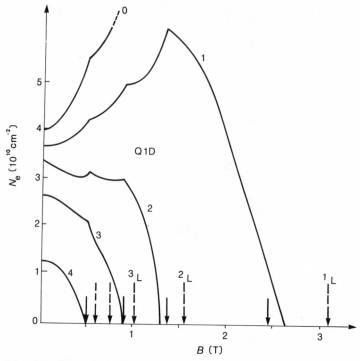

Fig. 6.4 Magnetic depopulation of Q1D energy levels. In the 0.15 μm channel formed from a 2DEG with concentration of 1.5×10^{11} cm^{-2}, the ground sub-band and four excited sub-bands are occupied at zero magnetic field. With increasing magnetic field, calculations show that the fourth excited state depletes at about 0.5 T as the magnetic contribution to its energy lifts it above the Fermi level, with the third, second, and first states following at 0.85 T, 1.3 T, and 2.65 T. The full arrows correspond to experimental magnetoconductance peaks associated with the depopulation, while the broken arrows show the depletion of the 2D Landau levels. (After Berggren *et al.* 1986.)

information relevant to the shape of the channel and this may one day become a renewed topic of research.

The relative smoothness of the potential profile in Fig. 6.3(a) belies one important factor, namely the microscopic potential fluctuations within the 1D channel caused by the detailed arrangement of the donor ions set back in the AlGaAs layer in the case of the heterojunction structures. In the case of the 2DEG, the exceptionally high mobilities imply modest scattering, from which one must infer a weak impurity scattering caused by charged impurities. Detailed calculations have been performed on the potential distribution, and the fluctuations have been shown to be large (i.e. several meV), and in space-dependent models of the potential profile the simple picture hardly seems plausible (see Fig. 6.5 from Nixon and Davies 1990). The situation is resolved, as far as transport is concerned, in that small-angle (Φ) scattering from such an impurity potential profile will have little effect on the mobility

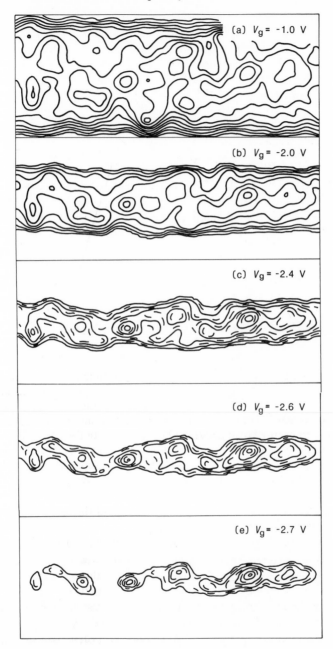

Fig. 6.5 The electron density contours in a 1 μm × 0.4 μm channel defined by Schottky gates above a 2DEG: (a) $V_g = -1.0\,\text{V}$; (b) $V_g = -2.0\,\text{V}$ (c) $V_g = -2.4\,\text{V}$; (d) $V_g = -2.6\,\text{V}$; (e) $V_g = -2.7\,\text{V}$. The calculations include a realistic impurity potential distribution with increasingly negative gate bias. Note the role of fluctuations in breaking up the wire between −2.6 and −2.7 V. (After Nixon and Davies 1990.)

in 2D systems, where a $(1 - \cos\Phi)$ term appears in the integral for the scattering rate, and at higher electron densities the fluctuations are screened out. The quantization of the ballistic resistance (cf. Chapter 4, Section 4.3.4) is only seen in short 1D channels, (0.5 µm) and it may be scattering from these same potential fluctuations that limit the length scale over which ballistic transport is observed.

6.4 Transport experiments and their interpretation

6.4.1 Diffusive transport

We begin by considering Q1D systems in which there are many scattering events as the carriers move along the channel. In Fig. 6.6(a), we show transport data from a Q1D channel in a Si MOSFET at three temperatures. In Fig. 6.6(b) we show the temperature dependence of the logarithm of the conductivity; this shows a transition (in Fig. 6.6(c)) from an exponent of 1/2 at low gate voltages to 1/3 at high gate voltages in a fit to the expression

$$\ln(G/G_0) = -(T_0/T)^{-n}.$$

All the electrodes (including the side-gate control electrodes) are held at constant potential, and the gate voltages increased (i.e. the carrier density is varied). This behaviour is consistent with transport involving electrons hopping between localized states for which the conductivity varies with an exponent $n = 1/(d + 1)$, where d is the dimensionality of the conducting system (Mott and Davis 1979). At lower gate biases the conduction is 1D changing to 2D conduction at higher gate biases where the 1D channel has been opened out to a width several times the characteristic transport length (i.e. the mean hopping distance of 30 nm in this system).

The striking feature of Fig. 6.6(a) is the very strong structure in the lowest-temperature data (see Fig. 6.6(d) for greater detail). Such structure is quite reproducible within a given sample, but there is no correlation between samples. Probes other than gate voltage (e.g. magnetic field) can be used, but in each case strong 'universal conductance fluctuations', on the scale of e^2/h, can be observed (Stone 1985, 1992). The fluctuations decrease in significance as the sample size increases in a well-defined statistical sense, as we now describe. We can extract the characteristic transport length scale, namely the inelastic diffusion length L_{in}, from the low-field magnetoconductance $G(B)$ as

$$[2G(B = 0)/\{G(B) - G(B = 0)\}]^2 = L^2[L_{in}^{-2} + (e/\hbar)^2 w^2 B^2/3],$$

where $L(w)$ is the 1D channel length width. If we regard the sample as an incoherent average of L/L_{in} samples of effective length L_{in}, then the scale of the conductance fluctuations should be

$$\delta G_{rms} \sim (L_{in}/L)^{3/2} e^2/h,$$

a relationship which has been verified in Q1D (and Q2D) systems. It is in this sense that the fluctuations are considered universal. (Hartstein 1986)

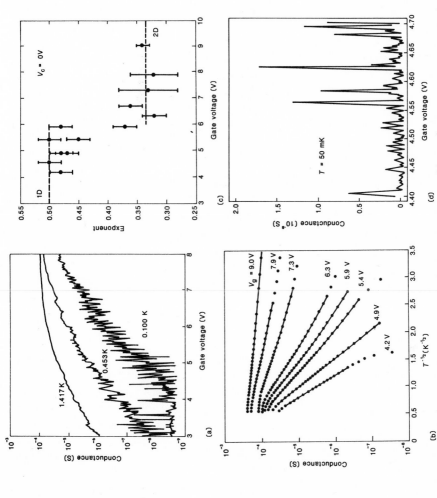

Fig. 6.6 (a) Measured conductance versus gate voltage at three temperatures; (b) conductance versus temperature at several voltages; (c) exponents for the temperature variation from (b); (d) conductance data at the lowest temperatures. (After Hartstein 1985.)

6.4.2 Ballistic transport

If the transport length scale L_{in} is rather larger than both the length and the breadth of the Q1D channel, carriers can pass through the channel without any scattering. This situation can easily arise with heterojunction material where split Schottky gates can be used to define channels of submicrometre width and length, when the inelastic length is several micrometres. The most dramatic consequence in this regime is the quantization of the conductance in units of $2e^2/h$ (van Wees *et al.* 1988; Wharam *et al.* 1988*a*). The data is shown in Fig. 6.7, and the simple derivation of the result was given in Chapter 4, Section 4.3.4. The key point is that in one dimension, there is a cancellation between the energy dependence of the density of states and of the velocity of

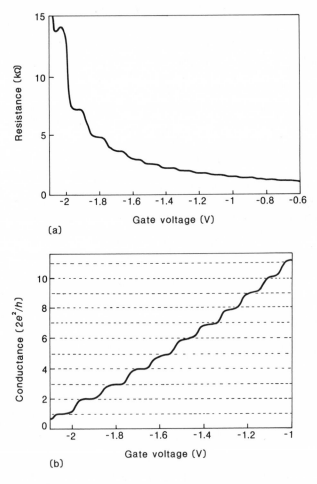

Fig. 6.7 The quantization of the Q1D ballistic resistance: (a) the raw data of resistance versus gate voltage; (b) the conductance versus gate voltage after making corrections for contact resistances. (After van Houten *et al.* 1992.)

carriers, resulting in a fixed conductance per channel, provided only that it is occupied. As a slight bias is applied along a short Q1D channel with N sub-bands occupied, the resulting conductance is $2Ne^2/h$. As a control parameter (such as the bias on the Schottky gates that define the electrostatic potential and hence the width of the channel) is varied, the bottoms of the sub-bands can be brought down below the Fermi level and the conductance rises in units of $2e^2/h$. Note that this type of result can also be obtained using the Landauer–Buttiker formalism of Chapter 4, Section 4.3.6, as we discuss further below. The accuracy of this quantization is only a few per cent, as opposed to the 10^{-6} per cent in the case of the quantum Hall effect.

There is also a classical formulation which produces the same result, and which led to some theoretical controversies in the early days because it drew attention to the coupling of electron wavefunctions in the contact areas to those in the channel, and the possibilities that reflections might alter the accuracy of the quantization. The argument can be seen with reference to Fig. 6.8(a) (van Houten *et al.* 1992): a small excess density δn of electrons on the left travelling at the Fermi velocity v_F are incident on the gap, giving an overall flux of $\delta n v_F \langle \cos\phi\theta(\cos\phi)\rangle$, where $\theta(x) = 1$ for $x > 0$ and $\theta(x) = 0$ for $x < 0$. If the mean free path is much greater than the width W, and the flux is fully transmitted, the current is

$$I = e W \delta n v_F \int \cos\phi\, d\phi/2\pi = (e/\pi) W v_F \delta n.$$

The δn is set up by a difference $\delta\mu = eV$ in chemical potential on either side, and since $\delta n/\delta\mu$ is just the density of states at the Fermi level ($= m^*/\pi\hbar^2$) in 2D, we obtain the result that the conductance $G = I/V$ is given by

$$G = (2e^2/h)(k_F W/\pi).$$

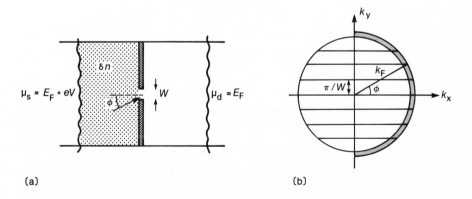

(a)   (b)

Fig. 6.8 (a) A classical picture of ballistic transport through a Q1D channel with net current flow from an imbalance of chemical potential, and (b) the quantum version using the displaced Fermi sea with the finite width of the channel picking out allowed wavevectors. These are real and reciprocal space versions. (After van Houten *et al.* 1992.)

In practice, the product $k_F W$ has to be evaluated half-way along the channel, as one can see from the lithographic definition of the gates. One can appreciate from the effects of the electrostatic fringing fields that the width W is not fixed along the length of the channel. Note that in the classical model $k_F W / \pi$ does not have to take integer values.

In the quantum argument (Fig. 6.8(b)), the sub-band energies are of the form

$$E_n(k) = E_n + \hbar^2 k^2 / 2m^*,$$

and the number of occupied sub-bands is the largest integer N such that $E_N < E_F$. The current per unit interval of energy per occupied sub-band is the product of the group velocity and the density of states (a product which, as we noted above and in Chapter 4, is independent of the energy and does not depend on the sub-band index). With a chemical potential imbalance $\delta\mu$, the current carried per sub-band is $ev_n(E)N_n(E)\delta\mu$. The number N of occupied sub-bands is given by $\mathrm{Int}(k_F W / \pi)$ and again, since $G = eI/\delta\mu$, we obtain

$$G = (2e^2/h)N,$$

which is precisely a special case of the Landauer–Buttiker formulation for two-terminal conductances, assuming 100 per cent transmission in each of the occupied sub-bands:

$$G = (2e^2/h) \sum_{m,n=1}^{N} |t_{mn}|^2$$

where $t_{mn} = \delta_{mn}$. The 100 per cent transmission implies no reflection, as generally occurs when electron waves impinge on potential steps and discontinuities. The classical and quantum results are identical for a channel of fixed width but with infinite confining potentials.

Deviations from the quantized condition can be anticipated from several sources. First, the above results are derived in a zero-temperature approximation. At finite temperatures the occupation of states may not be 100 per cent, and so a more general expression is required. This follows the standard derivations of temperature corrections to the zero-temperature conductivity of metals, leading to

$$G(E_F, T) = \int G(E, 0)(-\mathrm{d}f/\mathrm{d}E)\,\mathrm{d}E = (2e^2/h) \sum_{n=1} f(E_n - E_F)$$

where f is the finite-temperature Fermi occupation factor. This form for G leads to rounding of the steps between plateaux, and they are weak by 4.2 K (Fig. 6.9). Secondly, the coupling between the wide regions of the 2DEG and the narrow Q1D electron gas should allow some reflection of the incident electron waves. This should occur in particular at edges to the potential profile which are sharp on the length scale of the Fermi energy electrons, namely c.50 nm in the case of a typical $E_F \sim 10$ meV. Such reflections have been the subject of considerable calculation and debate; they should allow mixing between the different subband electron states (Ulloa *et al.* 1992). There

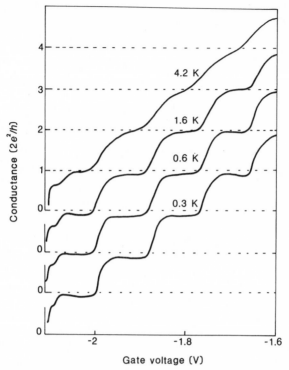

Fig. 6.9 Temperature dependence of quantized conductance. (After van Houten *et al.* 1992.)

may also be reflections when the electron exits the channel, and indeed resonant reflection effects have been predicted when the channel length is a multiple of half-wavelengths of the Fermi wavelength. In practice all these effects contribute to some extent to the plateaux as measured, but definitive experiments to isolate each factor are not possible. If the entry and exit of the channel are smooth on the scale of the Fermi wavelength, then the quantization is expected to be good. Even here there is a further possibility: electrons can tunnel through short barriers of modest height, leading to finite transmission though a nominally closed channel in short structures. This situation has been analysed as described below, in which we approximate the potential near the centre of the channel as having a saddle-point structure (see Fig. 6.13(a) below):

$$V(x, y) = V_0 - m^*\omega_x^2 x^2/2 + m^*\omega_y^2 y^2/2$$

where x is the direction of current flow and y is the transverse direction. Note that this is a valid assumption for the potential profile as the Schottky gates lift the energy of bottom of the conduction band in the channel. In the absence of tunnelling, the threshold for current flow in the nth channel is given by

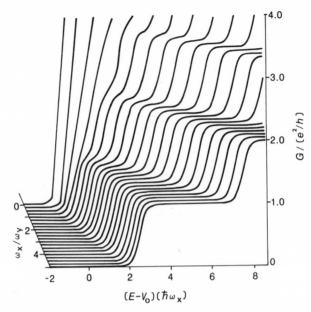

Fig. 6.10 The conductance of a saddle-point channel as a function of the longitudinal energy for various scales of transverse energies. (After Buttiker 1992.)

$$E_n = V_0 + \hbar \omega_y(n + 1/2) < E_F.$$

The tunnelling contribution can be calculated explicitly, and the transmission takes the form

$$T_{mn} = \delta_{mn}/[(1 + \exp(-\pi \varepsilon_n)],$$

where

$$\varepsilon_n = 2[E - \hbar \omega_y(n + 1/2) - V_0]/\hbar \omega_x,$$

as shown in Fig. 6.10.

Other factors leading to non-quantized values of the conductance include series resistances in two-terminal experiments, and non-ideal reservoirs in practical devices where the condition that the width of the 2DEG is much greater than that of the channel is not met.

6.5 Magnetotransport phenomena

6.5.1 Edge states

The concept of an edge state, which we now describe, has been of great importance in clarifying the nature of magnetotransport in Q1D systems, and the results have led to a revision of the quantum Hall effect (Buttiker 1992).

In Chapter 4, Section 4.3.3, we discussed the motion of an electron in a high magnetic field, in terms of cyclotron orbits. If an electric field is applied, the orbit drifts in a direction perpendicular to the electric and magnetic fields. The

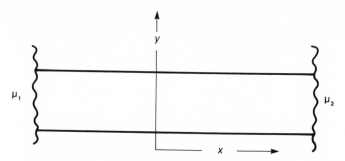

Fig. 6.11 A strip of 2DEG in a magnetic field.

situation is not the same in a small structure or for electrons that are within a cyclotron radius of confining potentials at the edges of the structure. To illustrate this we consider a strip of 2DEG, as shown in Fig. 6.11, and solve the Schrödinger equation for the electrons in the presence of a strong perpendicular magnetic field with

$$H = (\mathbf{p} - e\mathbf{A})^2/2m + V(y),$$

where $V(y)$ is the confining potential that defines the strip and we adopt the Landau gauge with $\mathbf{A} = (By, 0)$. In the absence of any explicit x dependence in the Hamiltonian, we can write the solutions in the form

$$\Phi_{n,k} = \exp(ikx)f_{n,k}(y),$$

where k is the momentum in the x direction, $f_{n,k}$ is a function of y only, and n is a discrete index. The equation for $f_{n,k}$ then becomes

$$Ef = -(\hbar^2/2m)(d^2f/dy^2) + V(y)f + m\omega_c^2(y - y_0)^2f/2,$$

where $\omega_c = eB/m$ is the cyclotron frequency, $y_0 = -kl_B^2$ and $l_B = \sqrt{[\hbar/(eB)]}$ is the cyclotron radius. In this form, we have a Schrödinger equation for an electron moving in a 1D confining potential $V(y)$ to which is added a magnetic confining potential $E_B = m\omega_c^2(y - y_0)^2/2$, which is a harmonic oscillator potential centred at y_0.

In the case where $V(y)$ is zero for $|y| < W/2$ and infinite elsewhere, we see that as long as y_0 is several times l_B from the boundaries, we have an undisturbed oscillator solution, giving

$$E_n = \hbar\omega_c(n + 1/2)$$

which is independent of both k and y_0. Once y_0 approaches the boundary at $\pm W/2$ the two potentials must be considered together. We can think of the infinite wall as adding to the confinement of the electrons, and the energy levels will rise the closer y_0 approaches the walls, having a complex functional form $E_{n,k} = E[n,\omega_c,y_0(k)]$. An important consequence of this form for the energy is that the semiclassical velocity of carriers has the form

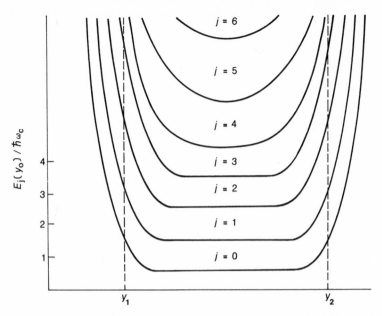

Fig. 6.12 The energy levels for the electrons in a strong magnetic field as a function of the centre coordinate y_0, with y_1, y_2 being the limits of the strip in Fig. 6.11. (After Buttiker 1992.)

$$v_{n,k} = (1/\hbar)(\mathrm{d}E_{n,k}/\mathrm{d}k) = (1/\hbar)(\mathrm{d}E_{n,k}/\mathrm{d}y_0)(\mathrm{d}y_0/\mathrm{d}k).$$

Whereas in the bulk of the sample, the energy levels are independent of y_0 and hence the velocity is zero (we think of the electrons making cyclotron orbits about a fixed position), near the walls $\mathrm{d}E_{n,k}/\mathrm{d}y_0$ is non-zero and the electrons move along the walls in skipping orbits (cf. Chapter 5, Fig. 5.10). For a given density of electrons it is possible for the Fermi level in the bulk of a material to be between Landau levels, but at the edge there are always edge states at the Fermi energy to carry current, even in the absence of an external electric field. Since each of these channels has a whole set of k-quantum numbers associated with it, the density of states can be obtained as

$$\mathrm{d}N/\mathrm{d}E = (\mathrm{d}N/\mathrm{d}k)(\mathrm{d}k/\mathrm{d}E) = (1/2\pi)(2\pi/\hbar v_{n,k}) = 1/\hbar v_{n,k}$$

and the inverse relationship with velocity means that the $N(E)v(E)$ product independent of energy. This independence was required above for the quantized conductance condition. The current that can be injected into an edge state by a difference in chemical potentials is then simply

$$I = v_n(\mathrm{d}N_n(E)/\mathrm{d}E)(\mu_1 - \mu_2) = (e/h)(\mu_1 - \mu_2).$$

The subsequent quantized conductance of these 1D edge states leads to the alternative model for the quantum Hall effect.

The derivation just given assumes the infinite wall potential. The treatment can be extended to treat a form of $V(y)$ that varies only slowly on the scale of l_B. A second-order Taylor expansion of $V(y)$ about some point near the edge of the sample can be made. The derivative term $\mathrm{d}V/\mathrm{d}y$ acts like a field applied in the y direction, and one obtains from it a further drift velocity, just as already encountered for crossed electric and magnetic fields. The motion is centred, not a y_0 as before, but at

$$Y = y_0 - (\mathrm{d}V/\mathrm{d}y)/(m\omega_c^2).$$

At the Fermi level, the separation between the centres y_0 of successive Landau levels Δy is given by

$$\Delta y = y_0(k_n) - y_0(k_{n+1}) = \frac{\hbar\omega_c}{(\mathrm{d}V/\mathrm{d}y)}.$$

Since we have harmonic oscillator wavefunctions in the y direction, for which the width of the state is $\sigma = \sqrt{(\langle n|(y-y_0)^2|n\rangle)} = l_B\sqrt{(n+1/2)}$, the ratio $\Delta y/\sigma$ can be very large if the change in potential energy over a distance l_B, namely $l_B\mathrm{d}V/\mathrm{d}y$, is small compared with $\hbar\omega_c$. In such a case, adjacent edge states may be very far apart spatially, and the scattering probability between different edge states is then exponentially small. The weak variation in potential on the scale of l_B is often encountered in structures where fringing fields from reverse bias Schottky gates are used to define conducting regions.

In the case of a narrow channel in a magnetic field, the calculations described above associated with a saddle-point potential can be repeated again with a harmonic oscillator potential added to the $V(x, y)$ term. In very high fields, the trajectories follow the equipotentials of the saddle point (Fig. 6.13(a)) and the two-terminal conductance as a function of magnetic field for a given shape of the channel gives rise to better-defined plateaux (Fig. 6.13(b)), because deviations from unity transmission from tunnelling between levels are reduced in high magnetic fields.

For transport in edge states, the mean free paths can be macroscopic at low temperatures in pure samples. The only mechanism for scattering (via impurities etc.) is scattering between edge states and the chance that the final edge state might be associated with a different physical edge (i.e. back-scattering, (cf. Fig. 5.10)). The Aharonov–Bohm effect, as described in the Chapter 4, may break down in high fields if there is no scope for interference between the electrons associated with edge states carrying current in the opposite direction.

6.5.2 Magnetotransport studies

The overriding importance of edge states in the understanding of high-field magnetotransport in the Q1D system, and the fact that transport in Q2D systems reduces to an edge-state description, has led to a wide range of further experiments. The first was to establish a spin-splitting in the quantization of the ballistic resistance in the presence of a magnetic field due to contributions of $\pm g\mu_B B$ to the electron energy—a doubling of the periodicity of the steps and a halving of their height were seen (Wharam *et al.* 1988*a*).

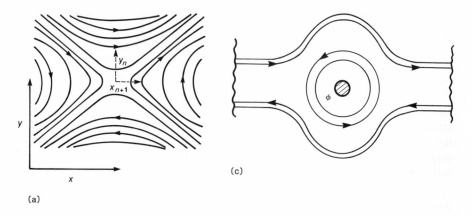

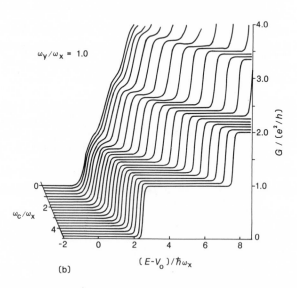

Fig. 6.13 (a) The equipotentials and edge states (arrowed) at a short narrow (saddle-point) channel, (b) the two-terminal conductance that results in high magnetic fields, and (c) the break-down of the Aharonov–Bohm effect in the absence of interference between edge states on opposite sides of the channel. (After Buttiker 1992.)

The resistance of two ballistic channels in series was seen to be close to the higher of the two channel resistances rather than the sum (Wharam *et al.* 1988*b*). Two channels in parallel give a resistance derived from the sum of two conductances. These results have been analysed in terms of transmission probabilities through pairs or one of a pair of channels (Beton *et al.* 1989).

The addition of a magnetic field to the Q1D electron gas (Q1DEG), and the form of skipping orbits generated, has been tested in a series of 'focusing' experiments, where electrons launched from the exit of one channel may or may not pass into and through another channel depending on the trajectory of the skipping orbit. Applications to complex structures, such as multiple channels in series where magnetic deflection can take place between the channels, has so far verified the predictions based on simple theory (van Houten *et al.* 1992). Perhaps the most elegant effects seen in this context are associated with modifications to the Hall voltage that can be achieved in small structures: it can be enhanced, quenched, or become negative at low magnetic fields. The classical theory of the Hall effect gives a Hall resistance proportional to the magnetic field. In small structures, in particular thin wires with thin leads attached, there is a region around $B = 0$ where the Hall resistance is zero, and beyond which it rises to a value above the classical value, which it maintains (the so-called 'last plateau') before joining the classical value at a higher magnetic field. Some of the typical data are given in Fig. 6.14, which shows Hall resistances for channels defined by the split-gate technique, with the pattern of the Schottky gates given in the insets. Different negative biases applied to the gates can be used to define the width of the conducting channel as well as the widths of the leads through which the voltage is measured. The detailed analysis in terms of the Landauer–Buttiker formalism predicts that the Hall resistance should take the value

$$R_\mathrm{H} = (h/e^2)(T_+ - T_-)/(2T_\mathrm{d}(T_\mathrm{d} + T_+ + T_-) + T_+^2 + T_-^2)$$

where the T terms are the transmission probabilities for electrons incident upon the cross at the probes going directly through or being deflected to the

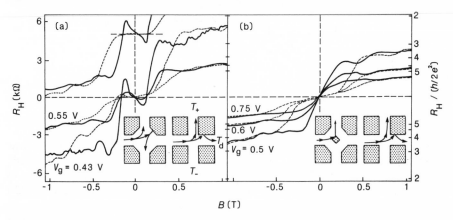

Fig. 6.14 The Hall resistance for different Schottky gate biases for two structures as a function of magnetic field, showing negative and enhanced Hall resistances at low magnetic fields, for different biases applied to the Schottky gates that define the channel. The dotted lines refer to the normal crosses, while the full lines refer to crosses with a widening at the intersection (which magnifies the effects being sought). The inset shows the types of trajectory that give enhanced and negative contributions to the Hall resistances. (After Ford *et al.* 1989).

left or right (T_d, T_+, T_- respectively). Depending on the magnetic field and the detailed geometry of the region where the leads meet the channel, the curvature of electron orbits and elastic reflections from the potential side-walls of the channel, an electron may exit from any one of the three possible channels, thereby contributing to a positive, zero, or negative Hall resistance. Possible trajectories are shown in the insets. In the case where a small dot remains at the centre of the cross (Fig. 6.14(b)), the reflections/trajectories assist the top exit (i.e. contributing to T_+) and an enhanced Hall resistance is the result. The last plateau phenomenon comes from the regime where $T_+ \gg T_d \gg T_-$ when $R_H \sim (h/2e^2)/(N + T_d)$ and N is the number of occupied edge states. If T_d were zero, we would have the quantum Hall effect, but a small value from the scattering of some carriers into other edge states ($0 < T_d < 2$, when $N \sim 15$ in practice) is sufficient to induce the plateau (see Buttiker (1992) for further details).

6.6 Applications of ballistic motion in a 1DEG

Many of the dramatic effects associated with ballistic motion in the 1DEG occur with small applied voltages, low temperatures, and high magnetic fields, where electron scattering is not present or is inhibited. Any applications are likely to be associated with problems in metrology or in clarifying further aspects of physics, about which there is further discussion in Chapter 12. There is always the hope that a subset of the physics might be robust enough to survive to 77 K or room temperature, and to the presence of high magnetic fields. Few preliminary experiments for which the results look encouraging have been carried out to date. The source–drain bias used to measure conductances, namely a few microvolts, is a small fraction of the Fermi energy in the 2DEG, typically about 10 meV. Practical transistor action involves source–drain biases of the order of 1 V. One can ask what happens to the Q1D ballistic resistor in the high longitudinal electric field regime. With respect to Fig. 6.15, a source–drain bias of 100 meV is easily applied. In this case all occupied electron states in the cathode have final states in the anode in which to propagate (Fig. 6.15(b)). We use a simple model for the current:

$$J = e \int N(E) v(E) T(E) \, dE,$$

where $N(E)$ is the density of states at the anode, $v(E)$ is the velocity of the electrons, and $T(E)$ is the transmission coefficient ($T(E) \equiv 1$ to date). Once the source–drain bias eV exceeds the Fermi energy, there are no more initial states to access, and the integral applied at the cathode predicts a current saturation (which for typical devices is about 0.1 μA per 1D sub-band). In addition, the explicit calculation of the transmission coefficient at a potential step shows an increasing reflectance (so that $T(E) < 1$) with increasing bias. These two factors applied to the equation for the current leads to a prediction of negative differential resistance, i.e. the current falls as the bias is increased

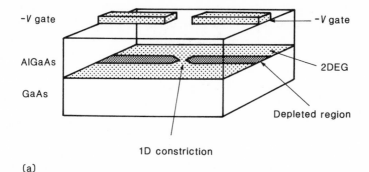

1D constriction

(a)

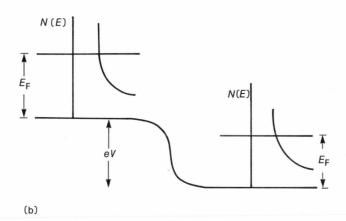

(b)

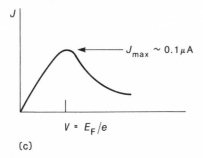

(c)

Fig. 6.15 The Q1D ballistic resistor in the high-field regime: (a) the structure, (b) the density of states, and (c) the predicted I–V characteristics. (After Kelly *et al.* 1990.)

(Fig. 6.15(c)). One consequence is that any circuit containing such an element should be unstable against oscillations, unless special precautions are applied to control them. Experiments show clear evidence of the negative differential resistance, in that once the source–drain bias exceeds *c*.20 meV, the current trace becomes very noisy if a 1D channel has been defined (Brown *et al.* 1989).

This effect can in principle apply at room temperature with some provisos. First, the source–drain bias must exceed the Fermi energy by a few kT, so that the mechanism of current saturation can be invoked. Secondly, to the limited extent that scattering can be induced while the electron is in the channel (which is $c.0.1$ μm long, and for which the transit time is of the order of 1 ps), the reduction of the transmission coefficient may be less than predicted by the zero-temperature no-scattering approximation. Some weak evidence of instabilities has been observed at room temperature, although the experimental programme has not been followed up (Kelly *et al.* 1990). The practical device implications are discussed in Chapter 17. There is scope for further investigation, as some of the effects, such as negative Hall resistance, or the use of the non-linearities in the *I–V* characteristics for frequency multiplication etc. (Gödel *et al.* 1994), might also survive to higher temperatures.

6.7 Optical properties

Most of the work on the optical properties of the Q1DEG have been performed on samples with many Q1D channels in parallel. This increases the

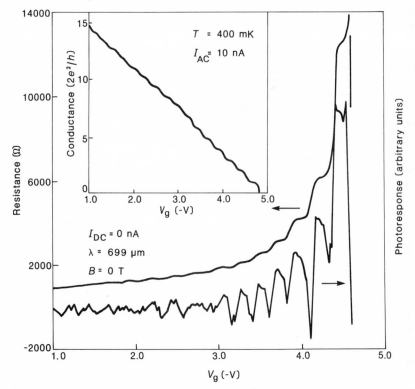

Fig. 6.16 Correlations in the photoresponse and resistance of a Q1D channel in the absence of a magnetic field. The inset shows the quantization of the conductance. (After Patel *et al.* 1990.)

strength of the optical signal and allows conventional optics experiments to be performed. As with the optical transitions associated with 2D systems, there are inter-sub-band transitions, and these are again in the far infrared (Fig. 6.16). The split-gate device has the advantage of tunability of the Q1D channel and its electric sub-bands. Thus it is possible to take a fixed laser wavelength and tune the Q1D inter-sub-band transitions through the energy difference. The specific dependence on tuning is a clear indication of the fact that inter-sub-band transitions are being monitored. The effects of many-body interactions and a self-consistent readjustment of the electrostatic potential are greater in the 1D than in the 2D system. In the experiments, the size of the photoresponse (i.e. the excess current due to the absorption of light) is largest when the conductance is between plateaux, i.e. just where a Q1D sub-band is being depopulated. This is also where the device is most affected by thermal broadening and tunnelling as described above. Indeed, data from the device at 0.3 K and subject to far-infrared excitation are in close agreement with data for the same device at 0.9 K but with no irradiation. If the experiments are repeated in a high magnetic field, the data can be interpreted in terms of transitions between edge states. Additional resonant structure can be obtained when the excitation energy equals the cyclotron energy (Patel *et al.* 1990).

6.8 Recent advances

In the last chapter, the role of back-gating in altering the carrier density in a 2DEG was explained. In the structures used to obtain the Q1DEG, the back-gate is a further control terminal with which to alter the properties. In Fig. 6.17, we show the layout of the device, with the multilayer structure that incorporates a back-gate to the 2DEG and the biasing arrangement when the split-top-gate is added (Hamilton *et al.* 1992). In Fig. 6.18 we show the carrier concentration and the gate leakage current that can be achieved with a technology that uses shallow ohmic contacts to the 2DEG. (The range of back-gate bias that can be sustained without significant leakage is being extended to about 5 V with the focused ion beam technology described in Chapter 3.) In Fig. 6.19 we show how the split-gate bias and the back-gate bias can be used to induce changes in the conductance by a combination of modifying the width and the carrier density in the Q1DEG. The two gates modify the electrostatic potential that forms the channel in a complex way, but a detailed set of calibrations gives a recipe for being able to measure the electronic properties as a function of either carrier density or channel width. Such freedom allows for close comparison of the results with theory (Hamilton *et al.* 1992).

A variation of the split-gate structure under high source–drain bias will be discussed in the next chapter in the context of generating hot electrons. A Q1D channel can be pinched off by a sufficiently negative bias on the Schottky gate, but in turn a source–drain bias can be used to inject hot electrons from

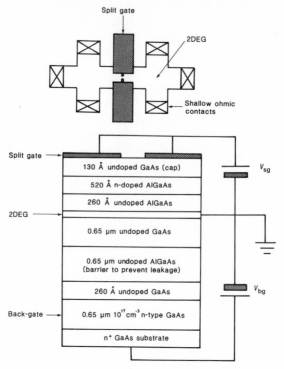

Fig. 6.17 A multilayer structure that contains a 2DEG subject to a back-gate bias from the substrate and a split-gate bias from surface Schottky gates. (After Hamilton *et al.* 1992.)

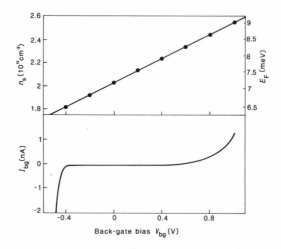

Fig. 6.18 The 2DEG carrier concentration and the back-gate leakage as a function of back-gate bias. (After Hamilton *et al.* 1992.)

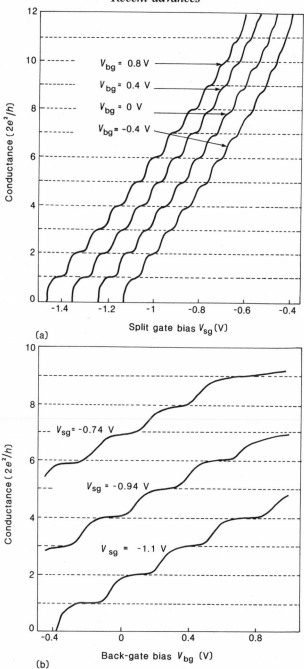

Fig. 6.19 The quantized conductance of a back-gated split-gate transistor structure as a function of (a) split-gate and (b) back-gate biases. With the two gates, it is possible to vary the biases in a correlated way to maintain channel width and vary the carrier density, or vice versa. (After Hamilton *et al.* 1992.)

a Q1D channel into a 2DEG, or indeed another Q1DEG. Studies can then be
made of the relaxation of the hot carriers. In Chapters 9–12 some of the
structures examined in this chapter will reappear in the context of superlattices, quantum dots, and counting individual electrons in and out of different
potential wells.

References

Berggren, K.-F., Thornton, T. J., Newson, D. J., and Pepper, M. (1986). Magnetic
depopulation of 1D subbands in a narrow 2D electron gas in a GaAs/AlGaAs
heterojunction. *Physical Review Letters*, **57**, 1769–72.

Beton, P. H., Snell, B. R., Main, P. C., Neves, A., Owens-Bradley, J. R., Eaves, L.,
et al. (1989). The resistance of two quantum point contacts in series. *Journal of
Physics: Condensed Matter*, **1**, 7505–11.

Brown, R. J., Kelly, M. J., Pepper, M., Ahmed, H., Hasko, D. G., Peacock, D. C.,
et al. (1989). Electronic instabilities in the hot-electron regime of the quasi-one-
dimensional ballistic resistor. *Journal of Physics: Condensed Matter*, **1**, 6285–90.

Buttiker, M. (1992). The quantum Hall effect in open conductors. In *Semimetals and
semiconductors*, Vol. 35, (ed. M. Reed), pp. 191–277. Academic Press, New York.

Dean, C. C. and Pepper, M. (1984). One-dimensional electron-localisation and
conduction by electron–electron scattering in narrow silicon MOSFETs. *Journal of
Physics C*, **17**, 5663–76.

Field, M., Smith, C. G., Pepper, M., Ritchie, D. A. R., Frost, J. E. F., Jones, G. A.
C., and Hasko, D. G. (1993). Measurements of Coulomb blockade with a
non-invasive voltage probe. *Physical Review Letters*, **70**, 1311–14.

Ford, C. J. B., Washburn, S., Buttiker, M., Knoedler, C. M., and Hong, J. M. (1989).
Influence of geometry on the Hall effect in ballistic wires. *Physical Review Letters*,
62, 2724–7.

Fowler, A. B., Hartstein, A., and Webb, R. A. (1982). Conduction in
restricted-dimensionality accumulation layers. *Physical Review Letters*, **48**, 196–9.

Frost, J. E. F., Ritchie, D. A., Ingram, S. G., and Jones, G. A. C. (1991). The growth
of shallow high mobility two-dimensional electron gases. *Journal of Crystal Growth*,
111, 305–8; **113**, 726.

Gödel, W., Manus, S., Wharam, D. A., Kotthaus, J. P., Böhm, G., Klein, W.,
et al. (1994). Ballistic point contacts as microwave mixers. Electronics Letters, **30**,
977–9.

Hamilton, A. R., Frost, J. E. F., Smith, C. G., Kelly, M. J., Linfield, E. H., Ford,
C. J. B., *et al.* (1992). The back-gate split-gate transistor: a one-dimensional ballistic
constriction with variable Fermi energy. *Applied Physics Letters*, **60**, 1881–3.

Hartstein, A. (1986). 1D structures—field confinement approach. In *The physics and
fabrication of microstructures and microdevices* (eds. M. J. Kelly and C. Weisbuch),
pp. 266–79. Springer–Verlag, Berlin.

Kelly, M. J., Brown, R. J., Pepper, M., Hasko, D. G., Ahmed, H., Peacock, D. C.,
et al. (1990). Room temperature negative differential resistance in a
quasi-one-dimensional ballistic resistor. *Electronics Letters*, **26**, 171–3.

Laux, S. E. and Stern, F. (1986). Electron states in narrow gate-induced channels in
Si. *Applied Physics Letters*, **49**, 91–3

Laux, S. E., Frank, D. J., and Stern, F. (1988). Quasi-one-dimensional electron states
in a split-gate GaAs/AlGaAs heterostructure. *Surface Science*, **196**, 101–6.

Lieb, E. H. and Mattis, D. C. (1966). *Mathematical physics in one dimension*. Academic Press, New York.

Mott, N. F. and Twose, W. D. (1961). The theory of impurity conduction. *Advances in Physics*, **10**, 107–63.

Mott, N. F. and Davis, E. A. (1979) *Electronic process in non-crystalline materials*, (2nd edn). Clarenden Press, Oxford.

Nixon, J. A. and Davies, J. H. (1990). Breakdown of quantised conductance in point contacts using realistic potentials. *Physical Review B*, **41**, 7929–32

Patel, N. K., Janssen, T. J. B. M., Singleton, J., Pepper, M., Jones, G. A. C., Frost, J. E. F., *et al.* (1990). Far-infrared photoconductive response of a one-dimensional GaAs/AlGaAs ballistic channel. In the *Proceedings of the 20th International Conference on The Physics of Semiconductors*. (eds. E. M. Anastassakis and J. D. Joannopoulos), pp. 2371–4. World Scientific, Singapore.

Skocpol, W. (1986). Transport physics of multicontact Si MOS nanostructures. In *The physics and fabrication of microstructures and microdevices* (eds. M. J. Kelly and C. Weisbuch), pp. 255–65. Springer-Verlag, Berlin.

Stone, A. D. (1985). Magnetoresistance fluctuations in mesoscopic rings and wires. *Physical Review Letters*, **54**, 2692–5.

Stone, A. D. (1992). Theory of coherent quantum transport. In *Physics of nanostructures*. (eds. J. H. Davies and A. R. Long), pp. 65–100. Institute of Physics, Bristol.

Thornton, T. J., Pepper, M., Ahmed, H., Andrews, D., and Davies, G. J. (1986). One-dimensional conduction in the 2-D electron gas. *Physical Review Letters*, **56**, 1198–201.

Ulloa, S. E., MacKinnon, A., Castano, E., and Kirczenow, G. (1992). From ballistic transport to localization. In *Handbook on semiconductors*, Vol. 1 (revised edn) (ed. P. T. Landsberg), pp. 817–962. North Holland, Amsterdam.

van Houten, H., Beenakker, C. W. J., and van Wees, B. J. (1992). Quantum point contacts. In *Semimetals and semiconductors*, Vol. 35, (Ed. M. Reed), pp. 9–112. Academic Press, New York.

van Wees, B. J., van Houten, H., Beenakker, C. W. J., Williamson, J. G., Kouvenhoven, L. P., van der Marel, D., and Foxon, C. T. (1988). Quantised conductance of point contacts in a two dimensional electron gas, *Physical Review Letters*, **60**, 848–50.

Wharam, D. A., Thornton, T. J., Newbury, R., Pepper, M., Ahmed, H., Frost, J. E. F., *et al.* (1988*a*). One-dimensional transport and the quantisation of the ballistic resistance. *Journal of Physics C*, **21**, L209–13.

Wharam, D. A., Pepper, M., Ahmed, H., Frost, J. E. F., Hasko, D. G., Peacock, D. C., *et al.* (1988*b*). Addition of the one-dimensional quantised ballistic resistance. *Journal of Physics C*, **21**, L887–91.

7 Hot electron phenomena

7.1 Introduction

Much interesting physics, and most device applications, are associated with carriers that have been excited in some manner to have an energy well in excess of their energy when in thermal equilibrium. In some cases the carriers behave in a manner that can be associated with an elevated temperature, giving rise to the name hot carrier. The name is used more widely to refer to any highly excited carrier. We discussed the velocity–field curves for bulk Si and GaAs in Chapter 1, Section 1.7. We described the transition from Ohm's law behaviour at low electric fields, where the carriers are little perturbed from equilibrium, through to the regime of saturated drift velocity at high electric fields. Here the carriers are in a steady state; they are highly excited but are giving up their excess energy to the lattice (via phonon emission) at the same rate that they acquire energy from the electric field. We also introduced the idea of avalanching in very high electric fields where very hot electrons excite further electron–hole pairs, and we related the uncontrolled form of this phenomenon to dielectric breakdown. In Chapter 5, Section 5.6, we described the reduction of mobility seen in a 2DEG as a function of increased applied field. In Chapter 6, Section 6.6, we briefly discussed the current–voltage characteristics of a Q1DEG in the presence of a strong electric field where eventually there is less current for a greater applied bias. In this chapter we gather together the main hot carrier effects in semiconductor multilayers, associated with the non-linear response to strong electric or optical fields. Using heterojunctions, one can inject carriers from wider-gap into narrower-gap semiconductors. The injected carriers are initially very hot, having an initial velocity as much as 10 times that of the saturated drift velocity, and we can study the detailed mechanisms by which they lose their excess energy and momenta and return to thermal equilibrium temperatures. The modification to the density of states in less than three dimensions slows down the rate of relaxation of hot carriers.

7.2 Methods for generating hot carriers: the role of heterojunctions

A carrier can be heated by several methods. In bulk semiconductors, the two most common methods are the application of electric or optical fields. In the former, electrons and holes are accelerated under the applied field and gain energy such that they occupy quantum states of the crystal for which the equilibrium Fermi function would be vanishingly small. In a semiclassical picture (Ashcroft and Mermin 1976), the carriers in an electric field E move along the band structure according to the rule that

$$\hbar \dot{k} = - eE$$

which persists until some scattering takes place on a time scale $\tau \sim 1$ ns for acoustic phonon emission or $\tau \sim 1$ ps for optic phonon emission. Some carriers avoid scattering and can be accelerated to higher energies, giving rise to phenomena such as intervalley transfer and avalanching (as described in Chapter 1). The mean free paths for such carriers depend sensitively on the carrier density, the electric field, and the lattice temperature, and they typically vary over the range 0.01 to 1 μm. Optical absorption generates electron–hole pairs, provided that the difference in energy between the electron in the valence band and that in the conduction band matches the optical absorption energy. For visible and ultraviolet light, the electrons may be excited high into the conduction bands, where they undergo energy relaxation processes on the time scale of $c.1$ ps. A steady state population of hot carriers is built up once the rate of optical absorption matches the cooling rate. Transient populations may be generated with pulsed optical excitation and their relaxation probed with other pulsed techniques, as we describe later. There are other methods for generating hot carrier distributions in bulk semiconductors. These include direct local heating (as with transient heat pulses), or using some injecting contact, such as a Schottky barrier at the surface of the semiconductor.

With multilayers, prepared by the techniques described in Chapter 2, there are new methods for generating hot electrons. Some relevant multilayer structures are shown in Fig. 7.1. We can form electrostatic barriers, such as the planar doped barriers, and the electrons that pass over such a barrier by thermionic emission are initially hot in the post-barrier region. The potential heights can vary from a small fraction of 1 eV to several eV, depending on the doping–thickness product of the p^+ layer (cf. Chapter 2, Section 2.6). The observation of hot electron effects requires a structure where the shorter arm of the planar doped barrier potential triangle is rather shorter in length than the mean free path for significant hot electron relaxation (see next section). Note that the electrostatic barrier so formed can be imitated over a relatively narrower range of energies by an appropriate Al profile (triangular in the case of a planar doped barrier).

Alternatively, heterojunctions involving two different chemical compositions can be used to generate hot electrons. Here, carriers created in a wide-gap semiconductor gain a pulse of energy as they cross into a narrower-gap semiconductor, and the instantaneous velocity that they acquire can be very large. Consider an electron in thermal equilibrium in, say, $Al_{0.3}Ga_{0.7}As$ with a kinetic energy of $kT = 0.025$ eV at room temperature. If it crosses the heterojunction and gains an energy of 0.25 eV (cf Chapter 2, Section 2.8), corresponding to a velocity of more than 10^6 ms^{-1}, i.e. 10 times the saturated drift velocity in GaAs. Note too that the electron velocity in the AlGaAs can

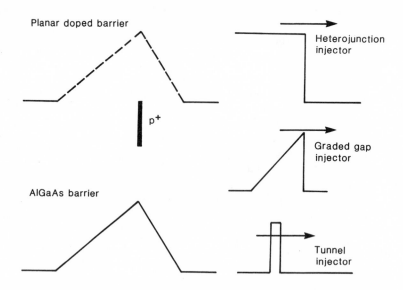

Fig. 7.1 The conduction band edge profiles used to generate hot electrons (the full lines represent Al profiles, and the broken lines represent electrostatic profiles from planar doped barriers (see Chapter 2, Section 2.6)).

be in any direction as long as it has a component that allows the electron to approach the heterojunction. Once over the heterojunction, the electron motion is in a relatively narrow cone ($c.10°$ across) since the velocity impulse is given in a direction normal to the interface. This collimation of the hot electron distribution has advantages in the analysis of hot electron relaxation, as described in this chapter, as well as in improving the speed and other aspects of practical devices described later in Chapters 16–19.

The hot electron injection mentioned so far involves electrons in the conduction band, or holes in the valence band, moving in extended states away from the band edge. If a (relatively) thick barrier of height ΔE is interposed between two regions of the same composition, carriers can pass over the barrier by thermionic emission, with a characteristic $T^2\exp(-\Delta E/kT)$ temperature dependence for the current density (cf. Chapter 2, Sections 2.5.4 and 2.6). We shall see in the chapters on devices that this strong temperature dependence is highly undesirable for electronic systems that have to operate in ambient conditions from the north pole to the equator, i.e. over the temperature range $-40°C$ to $+80°C$. If the barrier between the two layers is sufficiently thin, electrons can tunnel through it, and this process has several advantages (see Chapter 8). First, the tunnelling current has a much smaller dependence on temperature. It is possible to obtain hot electron distributions that are quasi-monochromatic, in that they have a spread in energy typified by a Fermi energy of a few meV rather than a wide spread of energies characteristic of thermionic emission over a barrier. Multiple thin barriers can be used to narrow the distribution even further. The tunnel current is generated by applying a bias across the tunnel barrier as shown in Fig. 7.1. We include these tunnelling forms of injection for completeness, although the detailed physics of tunnelling is deferred to the next chapter.

7.3 Hot electron spectroscopy (transport studies)

In the quest for a fast unipolar transistor (see Chapter 16 for further details on various types of transistor), several multilayer structures were investigated in the mid-1980s. In Fig. 7.2 we show the conduction band profiles of the mutilayers involved. They share the following characteristics: there are three conducting layers (emitter, base, and collector in the usual transistor nomenclature) separated by two barriers. In the early samples, the conducting layers were GaAs, while the barriers could be formed electrostatically using the principles of the planar doped barrier (Chapter 2) and/or compositionally using AlGaAs layers of varying shapes that allowed thermionic currents to be generated (or tunnel currents in special cases). The transistor operates with the emitter–base barrier under forward bias and the base–collector barrier under reverse bias. If the differential resistance of the former is less than the latter, the transistor action (transfer and amplification of resistance from the input to the output circuit) can take place. A successful transistor must be fast, and so all possible *RC* time constants associated with switching a transistor on and off must be minimized. In particular, the conducting regions must have low resistance, and in practice a low base resistance is the most difficult to achieve. This is because of the other requirement for a transistor—high gain. If a fraction α of the emitter carrier is actually collected, the ratio of the collector current to the base current, which is the current gain, has the value $\alpha/(1 - \alpha)$. The challenge is to collect as high a fraction of the emitter current as possible (certainly more than 95 per cent). The simple fact that has held back the development of the hot electron transistor is that a heavily doped base, as required for a low base resistance, is very efficient at taking away the excess energy of injected hot electrons. Only in recent years and in special cases has α exceeded 50 per cent, with 99 per cent now achieved.

The measurements of α are made as follows. The current–voltage characteristics of the base–collector barrier are measured for different values of the forward bias on the emitter and hence the injected emitter current. The ratio $\Delta I_c/\Delta I_e$ is determined numerically. Typical raw data are shown in Fig. 7.3(a), and these are transformed into a useful form in Fig. 7.3(b) by using a detailed analysis of the collector barrier height (obtained from $I–V–T$ studies of the base–collector barrier as a diode, just as in obtaining the height of the Schottky barrier or the planar doped barrier in Chapter 2). The ΔI_c data at a given emitter bias are plotted against the energy of the barrier with respect to the Fermi energy in the base region. For a fixed emitter current, the differential increase in collector current comes from a differential lowering of the collector barrier. We are performing energy spectroscopy of the hot

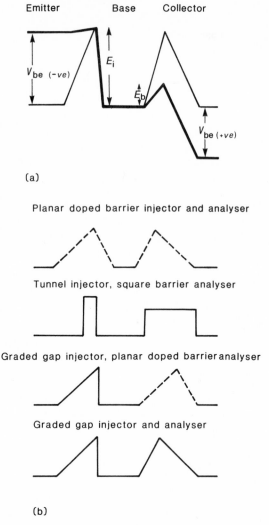

(a)

Planar doped barrier injector and analyser

Tunnel injector, square barrier analyser

Graded gap injector, planar doped barrier analyser

Graded gap injector and analyser

(b)

Fig. 7.2 (a) A hot electron transistor consisting of three heavily doped GaAs layers (emitter, base, and collector) separated by two energy barriers for electrons. During action, the emitter is forward biased to inject hot electrons, and the collector barrier is reversed bias to collect the hot electrons. (b) The conduction band profiles of various hot electron transistor structures that have been examined in the GaAs–AlGaAs materials system.

electrons. The important features of Fig. 7.3(b) are (i) the modest peak at an energy close to that of the electrons injected into the base from the emitter and (ii) the larger broad peak at rather low energies. The first peak contains those injected electrons that traverse the base without any energy loss, the so-called ballistic electrons, and those that lose up to a few tens of meV. In a

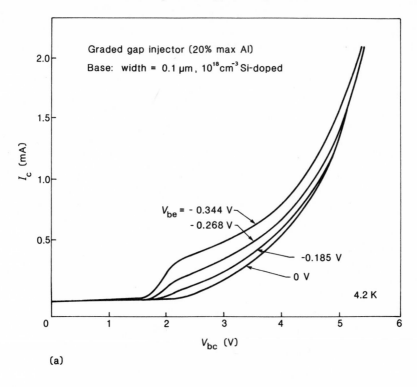

(a)

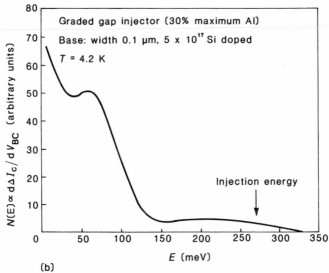

(b)

Fig. 7.3 (a) The raw data of the increase in collector current with increasing emitter current; (b) the distribution in energy of this extra collector current. (After Long *et al.* 1986.)

good transistor this peak should dominate the results, whereas in practice the large broad peak at lower energy contains those injected electrons that have undergone multiple energy loss processes. They are only collected when the reverse bias on the collector is sufficient to reduce the collector barrier height almost to zero. Indeed, simulations of these structures show that some electrons are excited out of the base (gaining energy from the injected electrons) and these also contribute to the collector current. From a series of experiments with different base widths, the mean free path for significant energy loss ($c.50$ meV) is of the order of 20–30 nm when the electron density is of order 10^{24}m^{-3}, a typical high value of doping. Over the base width that exceeds 0.1 μm, most injected hot electrons have lost most of their excess energy. Therefore a thin base region is essential if $\alpha > 50$ per cent is to be achieved. In order to keep a low base resistance, one must increase the doping in the base still further. As the Fermi energy in the base rises, so the barriers must be raised. The scattering of hot electrons increases as the base doping increases, generating a vicious circle. While high gain (and a high α) can be achieved by keeping the base doping low, the RC time constants increase, resulting in a slow transistor. Note that very thin bases provide a severe technology challenge: one must make an ohmic contact to the base layer, without shorting out either barrier and making undesired contact to the emitter or collector layers.

Detailed simulations have been set up to analyse results of the type shown in Fig. 7.3 obtained from many different experiments, with the emitter barrier height, the electron density in the base, and the base layer thickness being key variables. Several important processes can be identified by which injected hot electrons could loss their excess energy. First there is the emission of phonons, just as occurs when electrons in bulk semiconductors are heated by strong electric fields and reach a saturated drift velocity. Secondly, because the base is doped, there are two types of electron–electron interaction. A single cold electron in the Fermi sea in the base can extract energy from a single hot electron. Alternatively, the hot electron can excite one of the (collective) plasmon modes in the electron gas of the base. Formalisms are set up for calculating the scattering rate once the frequency and energy-dependent dielectric function have been obtained; the details are beyond the scope of this text (see Hayes and Levi (1986), and references cited therein, and Long *et al.* (1986)). It has taken very detailed simulations to establish the precise balance of the various energy loss processes. One important fact, appreciated early on, is that the electromagnetic nature of polar optic phonons (with a net dipole per unit cell as the Ga and As atoms move against each other) allows them to interact with the plasmon modes to form coupled hybrid excitations. The creation of one of these excitations with energy of $c.60$ meV is a particularly effective method for a hot electron to lose energy. Indeed, the simulations all show a satellite single-loss peak just below the ballistic peak, and there have been several unsuccessful attempts to isolate this feature. Both the finite width of the injected energy distribution and the single-electron energy loss pro-

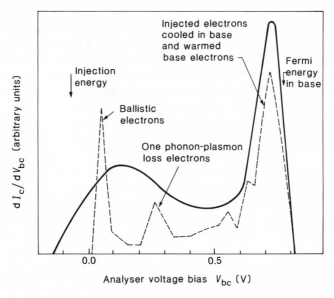

Fig. 7.4 Comparison of hot electron spectra derived from experiment on a narrow base device (Hayes and Levi © 1986 IEEE) and from Monte Carlo simulation (Long *et al.* 1986). The simulations indicate that some of the current peak near the Fermi energy comes from electrons excited out of the base with energy from the injected hot electrons. Note that with the base–collector forward biased (in the spectroscopy mode of operation) as the ordinate, the hot electrons appear at low bias and the cool electrons at higher bias.

cesses going on in parallel have masked any distinct feature. The overall result of the simulation exercises is that the energy losses to single-electron processes and to the collective plasmon–phonon modes are approximately equally effective. In Fig. 7.4 we show one of the better agreements obtained between experiment and simulation that supports these points.

In Fig. 7.5(a) we show the effect of a perpendicular magnetic field on the hot electron spectrum. The most noticeable fact is that, under the influence of the Lorentz force ($e\boldsymbol{v} \times \boldsymbol{B}$), the injected hot electron trajectory is bent away from the narrow forward cone described earlier. Two consequences follow (Fig. 7.5(b)): the electron trajectory in the base is lengthened, and so the opportunity for scattering increases, and some of the kinetic energy in the forward direction is changed to kinetic energy in the transverse direction. Thus the collector barrier looks higher to the electrons that impinge on it. Simulations of these data are useful in corroborating the earlier findings.

In recent years there has been one scientific advance which holds promise, although the technology challenge of making low-resistance ohmic contacts to the base remains. If the 3D bulk-doped base is replaced by a 2DEG, a sufficiently low resistance (*c.*300 $\Omega/\square$) can be achieved for fast transistor operation. The collector barrier is modulation doped to generate the 2DEG

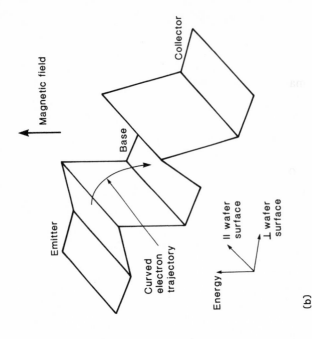

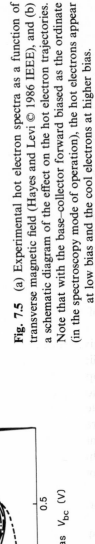

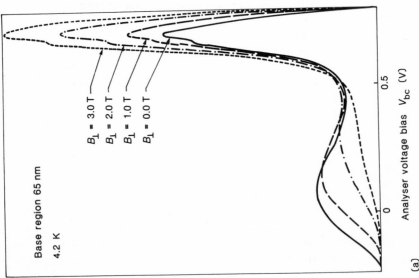

(b)

Fig. 7.5 (a) Experimental hot electron spectra as a function of transverse magnetic field (Hayes and Levi © 1986 IEEE), and (b) a schematic diagram of the effect on the hot electron trajectories. Note that with the base–collector forward biased as the ordinate (in the spectroscopy mode of operation), the hot electrons appear at low bias and the cool electrons at higher bias.

in the otherwise undoped base region. A schematic diagram of the transistor concept and typical hot electron spectra at low temperature are shown in Fig. 7.6(a) and 7.6(b) respectively. The transfer efficiency in this device exceeds 99.9 per cent at low temperatures, as the plasmon modes for a 2DEG are quite different from those in 3D (see Chapter 5, Section 5.3.3). However, conventional methods of making shallow ohmic contacts are inadequate; these devices break down under modest electric fields as the collector barrier is shorted out. Further details on this and other related transistors are discussed in Chapter 16.

7.4 Hot electron spectroscopy (optical studies)

Electrons that have been excited in optical interband absorption may re-emit light as they relax back to the valence band. If they lose some of their energy before that optical relaxation process, the emitted light will be of lower energy. The study of photoluminescence, and indeed a number of other optical phenomena (e.g. Raman spectroscopy, not discussed here), has several advantages which are not available in transport studies. Very-high-speed optical techniques are available, so that relaxation of the energy distributions of hot carriers can be measured in real times (from about 0.1 ps to about 1 ns). The energy distribution can be measured over a wide energy range at a given instant. There are optical techniques for producing a hot electron distribution with a very narrow initial spread in energies. In some cases optical techniques can be applied to structures where the hot carrier distribution is generated by high electric fields. In addition, there is a wide range of steady state optical experiments which provide data that complements the transport studies. Finally, there are many optical phenomena of instrinsic interest and of importance to the operation of high-speed optical devices; these are discussed in Chapter 18. The heavily doped contact layers required for fast electronic devices act as reabsorbers of luminescence, and this places a limitation of the use of optical techniques directly on many forms of practical device structures, and on the interpretation of experimental results.

The importance of luminescence studies can be appreciated easily (Shah 1986). If $\alpha_0(\omega)$ is the absorption coefficient in the absence of excited carriers, then at low densities of carriers the absorption is approximately

$$\alpha_0(\hbar\omega)(1 - f_e - f_h),$$

where f_e and f_h are the electron and hole distributions, and the luminescence intensity is given by

$$L(\hbar\omega) = A\,\alpha_0(\hbar\omega)f_e f_h.$$

In some circumstances the luminescence intensity can be deconvoluted to give the carrier distributions. Furthermore, the light that is given out in the interband relaxation might be reabsorbed, and in practice, if the electron and hole distributions have the Fermi–Dirac form, the luminescence intensity at

Hot electron phenomena

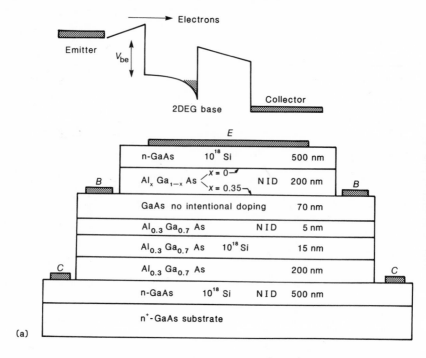

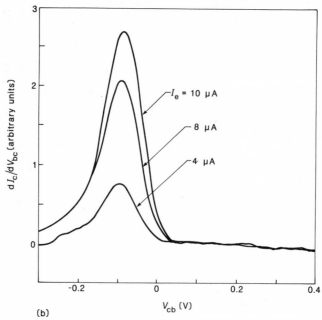

Fig. 7.6 (a) The 2DEG base vertical hot electron transistor (NID, no intentional doping), and (b) typical hot electron spectra. (After Matthews *et al.* 1992*a*.)

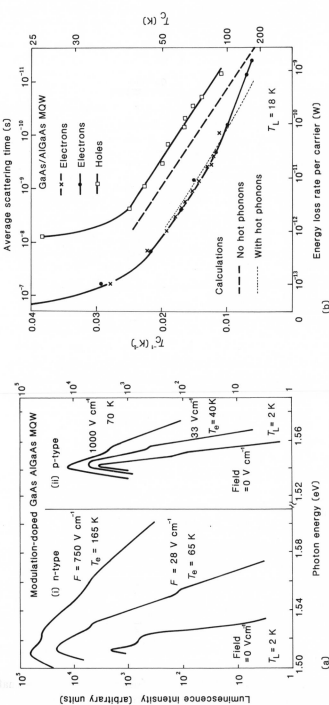

Fig. 7.7 (a) The energy dependence of luminescence from n- and p-type modulation-doped GaAs–AlGaAs multiple quantum wells, subject to varying electric fields in the plane of the wells. The slope on the high-energy side of the peak is a measure of the carrier temperature which is also listed. (b) In the steady state, the power input per carrier $e\mu F^2$, where F is the electric field, equals the energy loss rate to the lattice. When the inverse carrier temperature is plotted against this energy loss rate, the striking feature is the factor of 25 difference between the electron and hole energy loss rates. Further analysis indicates that the electron loss rate is very low, and a non-equilibrium optical phonon distribution (so-called 'hot phonons') has been invoked to explain the reduction in the electron energy loss rate through a decrease in the electron–phonon scattering matrix element. (After Shah © 1986 IEEE.)

high energy falls off at a rate varying as $\exp(-\hbar\omega/kT_{\text{eff}})$, where T_{eff} is the effective temperature in the Fermi–Dirac distribution, which is a suitable measure of the hot carrier distribution for modest electric fields. In Fig. 7.7(a) we show the steady state luminescence intensity for quantum wells (see Chapter 9) with n- and p-modulation doping when electric fields are applied along the wells. The data show how the carrier temperature evolves with electric field. Note that, in the steady state, the energy loss rate of the excited carriers must equal the (electrical) power input $e\mu F^2$, leading to the data in Fig. 7.7(b) showing a 25-fold greater energy loss rate for holes than for electrons at a given effective temperature.

A recent study has been made of the relaxation of hot carriers formed by injection over a heterojunction (Lyon and Petersen 1992). The multilayer semiconductor structure, which is an n–i–p structure with the optical process coming from hot electrons generated in wide-gap AlGaAs recombining in low-doped p-GaAs, is shown in Fig. 7.8(a). Incidentally, the inset shows how band-bending can lead to current injection by tunnelling through a barrier, and how the energy distribution can be quite narrow. The electroluminescence (i.e. the light given off by current-generated carriers) is analysed spectroscopically and shown in Fig. 7.8(b). Light generated in the AlGaAs is detected, together with a series of peaks at lower energy from electrons as they recombine before any phonon emission (the peak labelled as originating from ballistic electrons) and after a series of discrete (phonon) losses. The results are interpreted as indicating a mean free path for optic phonon emission of about 0.1 μm, a value inferred from other optical and transport data on bulk and low-doped (or undoped) GaAs. A typical electron will have emitted five phonons before entering the p^+-doped region where the luminescence takes place rapidly and where structure in the luminescence is lost.

In time-dependent studies, a hot carrier distribution is generated by pulsed optical excitation and the luminescence is measured as a function of time delay. In Fig. 7.9(a) we show typical raw data, and in Fig. 7.9(b) we show the 'cooling curves' which are a plot of T_{eff} as a function of time delay. The relatively slow times for hot carrier relaxation in quantum wells compared with bulk semiconductors have given rise to a strong and continuing debate on the effects of confinement into thin layers on the electron states, the phonon modes, the electron–phonon interaction, and the non-equilibrium occupation of the electron states and phonon modes. Some aspects of this debate are covered in Chapter 15.

7.5 High-field transport structures

7.5.1 *Minority transport in high fields*

Until now, reference to hot carrier transport has implied majority carrier transport. With focused laser excitation, it is possible to create electron–hole pairs in a bulk semiconductor and to use time-resolved luminescence and other techniques to obtain both the electron and hole velocities (as current pulses at appropriate

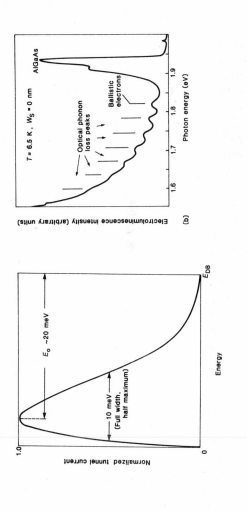

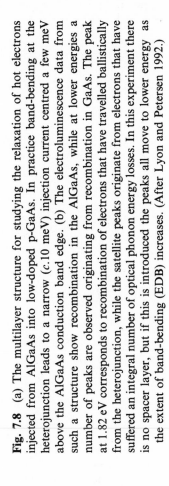

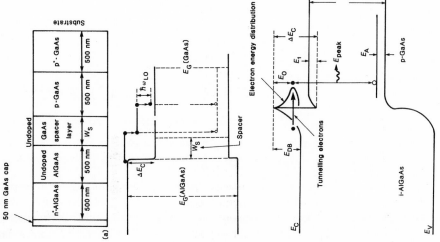

Fig. 7.8 (a) The multilayer structure for studying the relaxation of hot electrons injected from AlGaAs into low-doped p-GaAs. In practice band-bending at the heterojunction leads to a narrow (c.10 meV) injection current centred a few meV above the AlGaAs conduction band edge. (b) The electroluminescence data from such a structure show recombination in the AlGaAs, while at lower energies a number of peaks are observed originating from recombination in GaAs. The peak at 1.82 eV corresponds to recombination of electrons that have travelled ballistically from the heterojunction, while the satellite peaks originate from electrons that have suffered an integral number of optical phonon energy losses. In this experiment there is no spacer layer, but if this is introduced the peaks all move to lower energy as the extent of band-bending (EDB) increases. (After Lyon and Petersen 1992.)

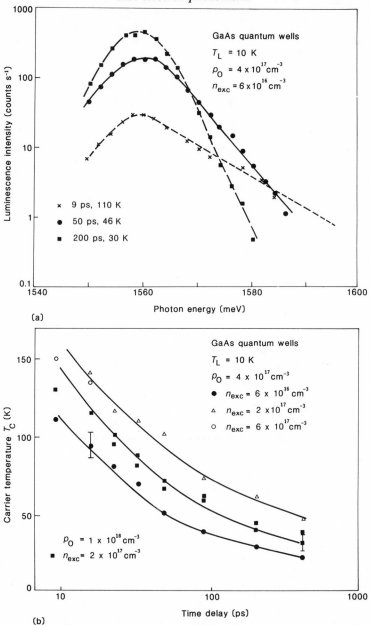

(a)

(b)

Fig. 7.9 (a) Luminescence spectra from p-doped quantum wells as a function of time delay, as carriers relax after a small optical excitation that produces an electron density nearly a factor of 10 smaller than the equilibrium hole density. Again, the slope of the luminescence on the high-energy side is a measure of the electron temperature at the different times. (b) The data are collected as 'cooling curves', a measure of the effective carrier temperature as a function of time and of the density of excited carriers, for two levels of modulation doping. (After Kash *et al.* 1984.)

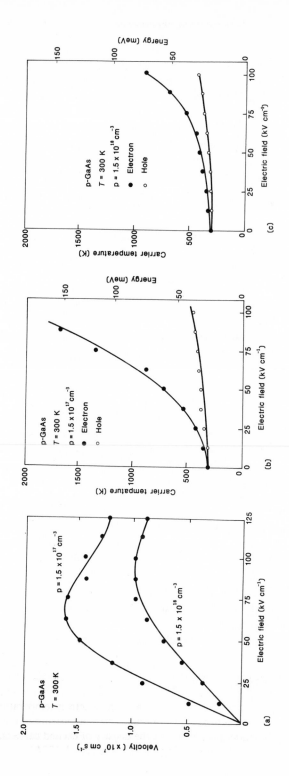

Fig. 7.10 (a) Velocity–field curves for electrons in p-doped GaAs; (b), (c) the field dependence of the effective electron and hole temperatures for two values of p-doping in GaAs. (After Furuta *et al.* 1992.)

electrodes) in applied electric fields. Typical results for electron transport in p-GaAs are shown in Fig. 7.10(a) for two levels of doping, and the effective electron and hole temperatures as a function of electric field are plotted in Fig. 7.10(b) for each case. The data in Fig. 7.10(a) are qualitatively similar to that shown earlier (Chapter 1, Fig. 1.7), but there is a trend of reducing peak electron velocity with reduced n-type and then increasing p-type doping. The heavier mass of the holes, and the higher density of final states for scattering, implies a much lower mobility for the holes than the electrons. It is these velocity–field curves that are appropriate for describing transport in the base region of bipolar transistors.

7.5.2 Parallel transport and real-space transfer

The effects of high fields on the mobility of electrons and holes in Q2D gases was described in Chapter 5. We describe here some further aspects of high-field transport within the plane of multilayers. In the first place, the velocity–field curves are normally obtained from samples that are much longer than any length scale associated with the transport. It is now possible to make FET structures with deep submicrometre spacings between the source and drain, and even shorter length gates (down to about 20 nm). It is possible that the electrons are accelerated under the gate without having a chance to scatter, and one would like to take full advantage of any high velocity that might be achieved before the degradation of carrier motion via scattering can set in. This so-called overshoot velocity occurs on short length and time scales, and has been predicted in simulations of electronic transport for many years without firm experimental confirmation. An effective increase in the source–drain transit velocity in small devices over the values that one extracts from large devices would be used to infer such a velocity overshoot. In Fig. 7.11 we show data from thin epitaxial layers of (a) light and (b) heavily n-doped GaAs used in the fabrication of MESFETs. At a gate length of 40 nm there is a 40 per cent increase over the transit velocity of $1 \times 10^7 \, \text{cm} \, \text{s}^{-1}$ typical of larger devices with low-doped ($2 \times 10^{17} \, \text{cm}^{-3}$) GaAs, with only a 15 per cent increase from the $8 \times 10^6 \, \text{cm} \, \text{s}^{-1}$ velocity in higher-doped ($1.5 \times 10^{18} \, \text{cm}^{-3}$) GaAs. Note that at gate lengths below about 40 nm the velocity increase is degraded because a minimum field length is needed to accelerate the electrons to the peak of the velocity–field curves in the first place.

The patterning of Schottky gates above a 2DEG featured strongly in the context of 1D transport in Chapter 6, and in particular it was shown that strong longitudinal electric fields could be applied to Q1D channels. Here we describe another application of the technique, this time to form a lateral hot electron transistor structure. In Fig. 7.12 we show a schematic diagram of a planar hot electron transistor structure, where the three conducting regions are different regions of a 2DEG. A strong negative gate bias applied to the Schottky gates creates the two barriers. At low temperatures this type of device could give more than 99 per cent base transfer efficiency, leading to gains in excess of 100. Furthermore, the effective mean free path is inferred to be $c.0.48 \, \mu\text{m}$ for low energy injection of the hot electrons (i.e. with less than

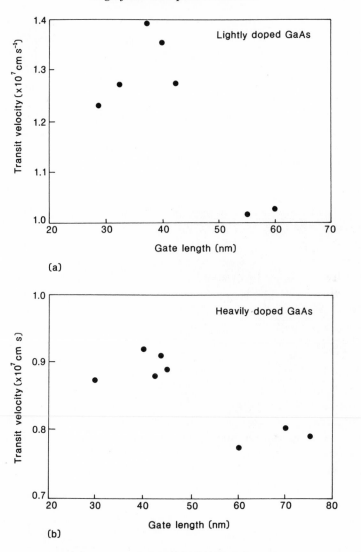

Fig. 7.11 Transit velocities in (a) lightly and (b) heavily doped GaAs channels as a function of the length of the gate (equal to the accelerating field length) in the limit of small gate lengths. (After Ryan *et al.* 1989.)

the 36 meV needed for optical phonon emission). This value drops sharply once the injected energy exceeds 36 meV. This system is very useful for studying the motion of ballistic electrons, but leakage currents at higher temperatures limit any wider application. We return to a development of this structure in Section 7.6.2, where a hot electron current is injected into a 2DEG from a Q1D channel, and a clear picture emerges of the energy loss processes of the hot electrons.

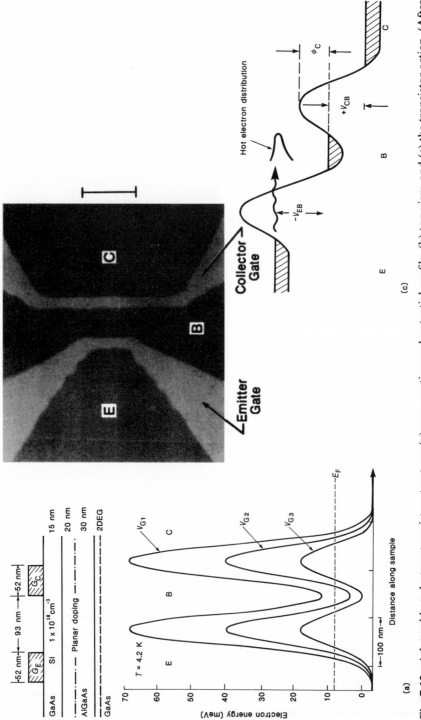

Fig. 7.12 A lateral hot electron transistor structure: (a) cross-section and potential profile, (b) top view, and (c) the transistor action. (After Palevski *et al.* 1987, 1989).

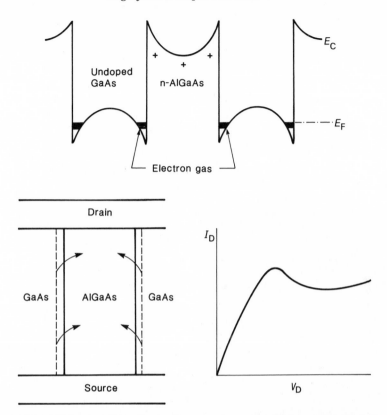

Fig. 7.13 The modulation-doped quantum-well structure, the configuration of source and drain, and the current–voltage characteristics of a real-space transfer diode. Heated electrons transfer back to the low mobility AlGaAs layer. Compare and contrast this with the *k*-space transfer equivalent in Chapter 1, Fig. 1.12. (After Luryi 1990; From S. M. Sze, ed © 1990. Reprinted with permission of John Wiley and Sons Inc.; Hess *et al.* 1979.)

If a sufficiently large source–drain voltage is applied to a HEMT structure, it is possible for some electrons to reach an energy that exceeds the height of the AlGaAs–GaAs potential step at the heterojunction. It is then possible for the electrons to scatter back into the AlGaAs layer (cf. Fig. 7.13). In the absence of any other sink for these electrons, they will reach the drain contact via transport in the lower-doped alloy. This is a real-space analogue of the high-field transport in GaAs where the transfer is in *k*-space (see Chapter 1, Fig. 1.12). A form of negative differential resistance (NDR) has been observed, as the mobility of the electrons is lower in the doped alloy layer, and the NDR has been used to form an oscillator that is an analogue of the Gunn diode (cf. Chapter 17, Section 17.3). A speed limitation on any device is set by the time taken for excited electrons to relax back into the GaAs channel. At about 10^{-11} s, this is several times slower than the relaxation of carriers in *k*-space in Gunn diodes.

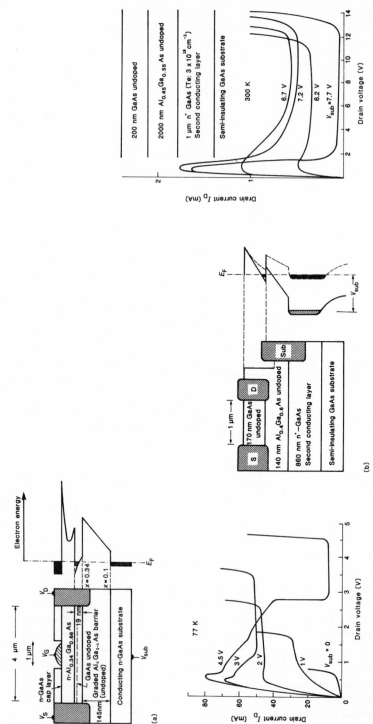

Fig. 7.14 Two forms of a real-space transistor structure and their related transistor characteristics. In (a) a 2DEG exists in the undoped GaAs layer by virtue of modulation doping in the AlGaAs layer, while in (b) the carriers are induced in the base by the bias applied to substrate. In both cases hot electrons generated in the channel by a source–drain bias can scatter into the substrate layer and be collected at the substrate contact. This diversion of electrons results in the strong NDR at 77 K and 300 K respectively for the two structures. (After Luryi 1990 from S. M. Sze (ed) © 1990. Reprinted with permission of John Wiley and Sons Inc.)

The geometry of real-space transfer in HEMT-type devices lends itself to a natural extension in a transistor structure. The hot electrons that scatter into the AlGaAs layer are then subject to another strong field in that layer, and are collected at another terminal (Fig. 7.14(a)). In this structure, the conventional gate is really only a means of generating a local high field to accelerate the electrons between source and drain. This can be eliminated by a variation as shown in Fig. 7.14(b) where the source–drain distance is reduced to about 1 μm and the second conducting layer acts as the third electrode. The characteristics of these two device types are shown in Fig. 7.14(c) and 7.14(d), and indicate strong negative differential resistance, controllable by the substrate voltage. We return to the device implications of these structures in Chapter 16.

7.5.3 Vertical transport

In subsequent chapters on tunnelling and other transport physics of multi-layers, hot electron phenomena are often present, even if not commented upon. We show here, with one clear example, how features in the current–voltage characteristics reflect the relaxation of hot carriers. The structure is a simple AlGaAs tunnel barrier within an n–i–n structure, but placed towards one end of the i layer, i.e. not unlike the p^+ layer in a planar doped barrier. Under bias, a diode-like characteristic is achieved, but on a fine scale there is an oscillation, periodic in the applied bias, which can be amplified greatly by taking the derivative. The structure and the experimental results are shown in Fig. 7.15. Clearly, the general shape of the current–voltage curve is determined by tunnelling through a barrier, and the exponential increase in current with bias is related to the associated linear decrease in the height of the potential barrier. What is the origin of the fine structure? The separation in energy of about 36 meV suggests that phonon emission is playing a role, and that somehow the total current is sensitive to this. The actual explanation turns out to be very subtle. After passing through the tunnel barrier, the electron traverses an extended depletion region before reaching the anode. During this transit, it has sufficient energy to emit one or more optic phonons. As the bias over the depletion region is increased to reach the threshold for the emission of another optical phonon, the probability for doing so increases as the final states for both the electrons and phonons are empty. This in effect opens up another channel to allow current to flow in the region downstream from the tunnel barrier. In turn this extra current is able to support an extra fraction of the bias across the whole device structure (i.e. the so-called space-charge effect). This reduces the bias over the tunnel barrier, and hence reduces the tunnel current itself. This effect, which was first described and explained by Hickmott *et al.* (1984), appears in a number of device contexts, and is due to hot electron relaxation in the anode region of multilayer diode structures. When double-barrier and other structures are present in the device, the Hickmott effect is reduced in significance. Phonon emission by hot electrons is present in all practical vertical transport devices, as the biases exceed the threshold of *c*.0.036 V for such emission. (Note that the strong

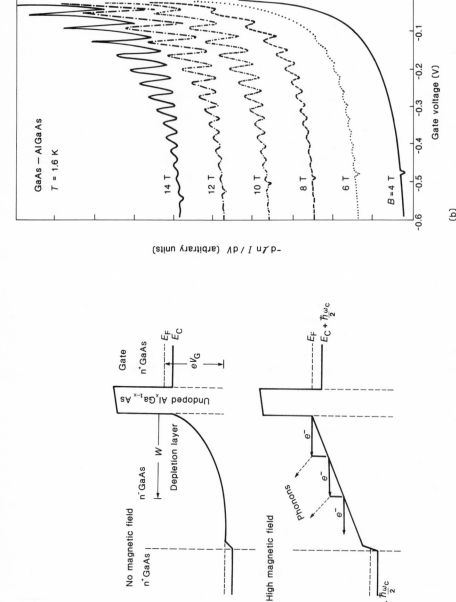

Fig. 7.15 (a) A single tunnel barrier with a low highly doped emitter and low-doped collector layer under bias. In a strong magnetic field the donor levels become deeper and retrap electrons, resulting in a uniform field over the low-doped layer after the tunnel barrier. (b) A plot of dI/dV against V for this structure with increasing magnetic field. The precise origin of the oscillations (separated in energy by the 36 meV associated with optic phonon emission by hot electrons) is described in the text. (After Hickmott *et al.* 1984.)

magnetic field deepens the donor-binding energies, and so neutralizes their charge in the depletion region. Electron scattering from their random positions would obscure effects associated solely with phonon emission at low magnetic fields.)

7.6 Recent progress

In this section, we describe two recent investigations of hot electrons.

7.6.1 Interaction of hot electrons with 2DEG

It has been possible to inject hot electrons into a two-dimensional electron gas using a large-area version of the 2DEG base hot electron transistor. This device is arranged so that only about 60 per cent of the injected carriers were collected; the other 40 per cent are trapped in the base region where they equilibrate with the electron gas, heating the latter in the process. In Fig. 7.16(a) we show the device layout, and note that there are enough contacts to the base to make Hall and Shubnikov–de Haas measurements; the layer sequence and the etching profile of different layers is the same as that in Fig. 7.6. The current–voltage characteristics between the emitter and base are shown in Fig. 7.16(b), spanning the range of no current through to several microamperes of injected current. The Shubnikov–de Haas and Hall measurements over the corresponding range of emitter biases are shown in Figs 7.16(c) and 7.16(d). The data reveal a rich variety of phenomena. First, there are changes in the position of the Shubnikov–de Haas features and the quantum Hall plateaux with emitter bias even in the regime of no injected current. This is a back-gating phenomenon (see Chapter 5, Section 5.8.1) that is reducing the carrier density in the base region and its mobility by 6 per cent and 13 per cent respectively between zero and 220 mV emitter bias. Such trends continue as the bias is increased, but are swamped and reversed by the effects of the injected current. The primary effect of increasing the bias to about 365 mV is to raise the temperature of the electron gas from the lattice temperature of 1.2 K to an estimated 4.2 K (as compared with raising the lattice temperature in the absence of an injected current). The injected current has sufficient energy to raise the temperature by 30 K, and so we deduce that optical phonon emission accounts for 90 per cent of the energy relaxation of the injected electrons. The exchange of electrons and energy between hot and cold electron distributions is an important area of contemporary research.

7.6.2 One-dimensional injectors of hot electrons

The patterned Schottky gate was used above to define a lateral hot electron transistor structure (cf. Fig. 7.12). The concept can be extended in several directions. In Fig. 7.17(a) we show a mesa structure etched from a high mobility 2DEG layer structure, with the heterojunction having been etched away in the lightly shaded area. The gates are shown as darkly shaded areas. The pair marker G_1 can form a Q1D channel, which can be pinched off with sufficient negative gate bias. However, a bias between the two 2DEG regions can allow current to be injected from the region I into the region B_1, where

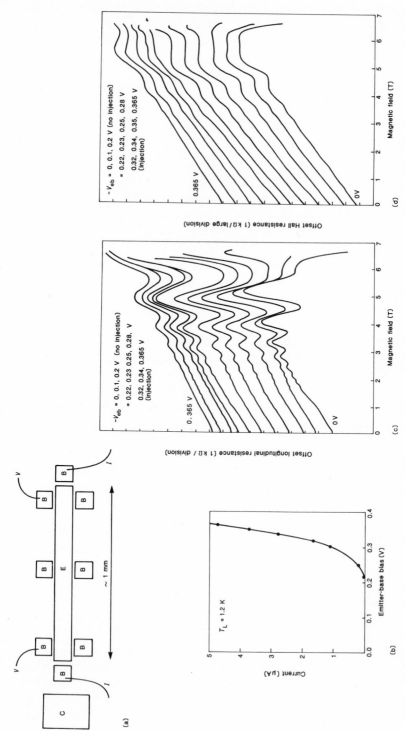

Fig. 7.16 (a) Top view of a hot electron transistor structure with a 2DEG base, (b) the emitter–base characteristics at 1.2 K, and (c) the Shubnikov–de Haas and (d) the Hall data of the base as a function of the emitter bias. The two regimes, with and without hot electron injection, are described in the text. (After Matthews *et al.* 1992*b*.)

four-terminal measurements can be made on the 2DEG there. The control gate G_c prevents hot electrons from reaching the region B_2 where control measurements on the cold 2DEG can be made. The potential profile is shown in Fig. 7.17(b). In the absence of any injected current, the resistance in the region B_2 can be measured as a function of temperature, and these data are used as a thermometer of the electron temperature in B_1 when current is injected. The resistance in B_1 is now monitored as a function of injected current and its injection energy, and the data are converted into an effective temperature shown as a function of the injection energy ($-eV_i$ as shown in Fig. 7.17(b)) for currents of 10, 20, . . ., 80 nA in Fig. 7.17(c). With the lattice held at 1.2 K, it is possible to reach electron temperatures in excess of 20 K with increased injection current. The oscillations show the efficiency of optical phonon emission to relax the energy of the hot electrons without much further heating of the 2DEG. Indeed, a regular sawtooth structure would be expected if the phonon mechanism were 100 per cent effective. The modest deviation from this, and the approximately constant slope of electron temperature with injection energy (modulo 36 meV) show both the dominance of the phonon mechanism and the efficiency of the electron mechanism for taking up the remaining excess energy.

7.7 Device implications

Typical biases for transistors and lasers are a few volts, while some microwave and optical devices (e.g. avalanche diode oscillators and photodetectors respectively) operate at much higher voltages. The basic performance and the estimates of speed, power consumption, noise levels, etc. all depend on the high-field properties of semiconductors. The physics at low fields has intrinsic interest, and some applications (as in metrology), but most devices are dominated by high-field effects, at least in the active regions. The access to these active regions, for example, under the gate of a FET or the quantum wells in optical devices, are designed to operate with lower electric fields, and they take advantage of some of the desirable low-field phenomena, such as low resistance, high mobility, etc. The detailed design and modelling of devices has an intimate mix of hot, warm, and cool electron phenomena to take into account. When the active device volume has electron states of reduced dimensionality, this represents an added complication. Where large volumes of device material are involved, it is often possible to extract averaged quantities from test structures and use these as inputs into device models, for example resistances as a function of field. Where the active device region is small, then ballistic motion of electrons, for instance, has a quite different physical basis than drift–diffusion motion, and a different design considerations apply. In practice, Monte Carlo simulations (Kizilyalli and Hess 1989) are often used to give qualitative, and semi-quantitative, guides to device performance. These simulations involving tracking thousands of typical electrons entering the active region of the device and undergoing various

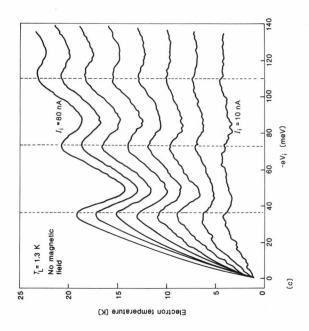

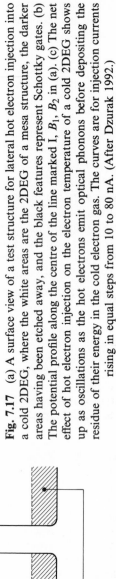

Fig. 7.17 (a) A surface view of a test structure for lateral hot electron injection into a cold 2DEG, where the white areas are the 2DEG of a mesa structure, the darker areas having been etched away, and the black features represent Schottky gates. (b) The potential profile along the centre of the line marked I, B_1, B_2 in (a). (c) The net effect of hot electron injection on the electron temperature of a cold 2DEG shows up as oscillations as the hot electrons emit optical phonons before depositing the residue of their energy in the cold electron gas. The curves are for injection currents rising in equal steps from 10 to 80 nA. (After Dzurak 1992.)

scattering processes at random. The inputs to such simulations (e.g. relative strengths of different scattering mechanisms) are extracted from the physics results presented in this and other chapters.

References

Ashcroft, N. W. and Mermin, N. D. (1976). *Solid state physics*. Holt, Rinehart, Winston, New York.

Dzurak, A. S., Ford, C. J. B., Kelly, M. J., Pepper, M., Frost, J. E. F., Ritchie, D. A. *et al.* (1992). Two-dimensional electron gas heating and phonon emission by hot ballistic electrons, *Physical Review B*, **45**, 6309–12.

Furuta, T., Taniyama, H., Tomizawa, M., and Yoshii, A. (1992). Hot-carrier transport in p-GaAs. *Semiconductor Science and Technology*, **7**, B346–50.

Hayes, J. R. and Levi, A. F. J. (1986). Dynamics of extreme non-equilibrium electron transport on GaAs. *IEEE Journal of Quantum Electronics*, **QE-22**, 1744–52.

Hess, K., Morkoc, H., Shichijo, H., and Streetman, B. G. (1979). Negative differential resistance through real space electron transfer. *Applied Physics Letters*, **35**, 469–71.

Hickmott, T. W., Solomon, P. M., Fang, F. F., Stern, F., Fischer, R., and Morkoc, H. (1984). Sequential single-phonon emission in $GaAs–Al_xGa_{1-x}As$ tunnel junctions. *Physical Review Letters*, **52**, 2053–7.

Kash, K., Shah, J., Block, D., Gossard, A. C. and Wiegmann, W. (1984). Picosecond luminescence measurements of hot carrier relaxation in III–V semiconductors using sum-frequency generation. *Physica B*, **134**, 189–98.

Kizilyalli, I. C. and Hess, K. (1989). Physics of real-space transfer transistors. *Journal of Applied Physics*, **65**, 2005–13.

Long, A. P., Beton, P. H., and Kelly, M. J. (1986). Hot electron transport in heavily doped GaAs. *Semiconductor Science and Technology*, **1**, 63–70.

Luryi, S. (1990). Hot-electron transistors. In *High-speed semiconductor devices* (ed. S. M. Sze), pp. 399–461. Wiley, New York.

Lyon, S. A. and Petersen, C. L. (1992). Ballistic electron luminescence spectroscopy. *Semiconductor Science and Technology*, **7**, B21–5.

Matthews, P., Kelly, M. J., Hasko, D. G., Law, V. J., Ahmed, H., Frost, J. E., F., Ritchie, D. A., and Jones, G. A. C. (1992*a*). The Physics of the Two-Dimensional Electron Gas Base Vertical Hot Electron Transistor. *Semiconductor Science and Technology*, **7**, B536–9.

Matthews, P., Kelly, M. J., Hasko, D. G., Law, V. J., Ahmed H., Frost, J. E., *et al.* (1992*b*). Interactions between hot injected electrons and cold or warm electrons in the two-dimensional electron gas base of a vertical hot electron transistor. *Surface Science*, **263**, 141–6.

Palevski, A., Heiblum, M., Umbach, C. P., Knoedler, C. M., Broers, A. N. and Koch, R. H. (1987). Lateral tunnelling, ballistic transport and spectroscopy in a two-dimensional electron gas. *Physical Review Letters*, **62**, 1776–9.

Palevski, A., Umbach, C. P., and Heiblum, M. (1989). A high gain lateral hot-electron device. *Applied Physics Letters*, **55**, 1421–4.

Ryan, J. M., Han, J., Kriman, A. M., Ferry, D. K., and Newman, P. (1989). Overshoot saturation in ultra-submicron FETs due to minimum acceleration lengths. *Solid State Electronics*, **32**, 1609–13.

Shah, J. (1986). Hot carriers in quasi-2D polar semiconductors. *IEEE Journal of Quantum Electronics*, **QE-22**, 1728–43.

8 Tunnelling phenomena

8.1 Introduction

This is the first of three chapters devoted to the new physics that has been made accessible with the advent of semiconductor multilayers. We shall be concerned with the quantum-mechanical phenomenon that an electron of total energy less than that of a thin potential barrier might not be reflected, as would always happen classically, but rather might pass through the barrier with an exponentially small but finite probability. Tunnelling has been an exploitable phenomenon in solid state physics for some time, and Josephson junction technology makes intimate use of superconducting tunnel junctions. In a wider sense, many of the leakage currents that plague high-performance devices have, in fact, come from carriers tunnelling through barriers that are undesirably thin. We are now able to control the tunnel barrier thickness precisely and to engineer the current–voltage characteristics. The subject has received an added boost through the discovery of resonant tunnelling involving two potential barriers: for certain well-defined energies the tunnelling probability is not exponentially small, but approaches unity. Application of a suitable bias to a double-barrier structure can establish the condition of resonant tunnelling with a high current density. A further increase in bias takes the structure off resonance, and the current falls. This gives rise to a new form of negative differential resistance with attractive attributes that we shall describe below. Although it has taken some time, the device spin-offs are now becoming apparent. In the mean time, many physicists have explored nearly every conceivable aspect of resonant tunnelling.

We start with a description of a single-tunnel barrier, the design and evaluation of the current–voltage characteristics, and the effects on these of temperature, pressure, magnetic fields, etc. The device potential, which is described further in Chapter 17, will become apparent. We proceed to the double-barrier diode and summarize the wealth of investigations, concentrating on those that might have device implications. We then describe a few of the more complex multiple-barrier structures that have been proposed for various reasons. This leads naturally into the subject matter of the following chapter, namely the physics of a periodically repeated set of tunnel barriers that constitute a superlattice. The highly non-linear current–voltage characteristics of tunnel structures have been proposed as new design elements for transistors, and they have the effect of reducing the number of these required in integrated circuits to perform given functions (Chapter 16, Section 16.6).

8.2 The single-tunnel barrier

8.2.1 Simple theory for tunnelling and current–voltage characteristics

Consider the elementary quantum mechanics of the problem posed in Fig. 8.1 of an electron with kinetic energy E, incident on a barrier of height $V(>E)$ and thickness W. The traditional approach is to consider the solutions to the Schrödinger equation on each side of (and within) the barrier, and to analyse the progress of a normalized plane-wave incident from the left. At each barrier interface (i.e. $z = 0, W$) we match the solutions of the Schrödinger equation and its derivative with respect to position. This is equivalent to conserving numbers of particles and the current that they carry (Landau and Lifshitz 1977). With definitions of the carrier momentum given by

$$k^2 = 2m^* E/\hbar^2$$

outside the barrier and

$$K^2 = 2m^*(V - E)/\hbar^2$$

inside the barrier, the wavefunctions in the three regions are

I: $\qquad\qquad\qquad \exp(ikz) + r\exp(-ikz),$

where r is the reflection amplitude,

II: $\qquad\qquad\qquad \alpha\exp(Kz) + \beta\exp(-Kz),$

and

III: $\qquad\qquad\qquad\qquad t\exp(ikz)$

where t is the transmission amplitude. The four matching conditions are sufficient to determine r, t, α and β. The most important quantity for us is the transmission probability $T(E) = |t|^2$, where

$$t(E) = [\exp(-ik\,W)(1 - \phi^2)]/[\exp(KW) - \phi^2\exp(-KW)]$$

and $\phi = (K + ik)/(K - ik)$. Note that for $E < V$, $|\phi| = 1$ so that the form of $t(E)$ is dominated by the $\exp(KW)$ term in the denominator, and the transmission amplitude is exponentially small for $E < V$ (see Fig. 8.1). Note that for $E > V$, we use

$$k'^2 = 2m^*(E - V)/\hbar^2$$

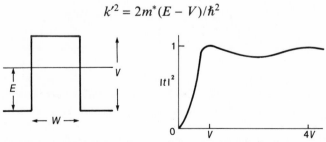

Fig. 8.1 The simplest (1D) analysis of tunnelling through a single barrier.

inside the barrier, and obtain

$$t = [\exp(-ikW)(1 - \theta^2)]/[\exp(ik'W) - \theta^2\exp(-ik'W)]$$

where $\theta = (k' + k)/(k' - k)$. There is now no exponential suppression, and for energies $E = V, 4V, 9V, \ldots$ the transmission is identically unity (Fig. 8.1).

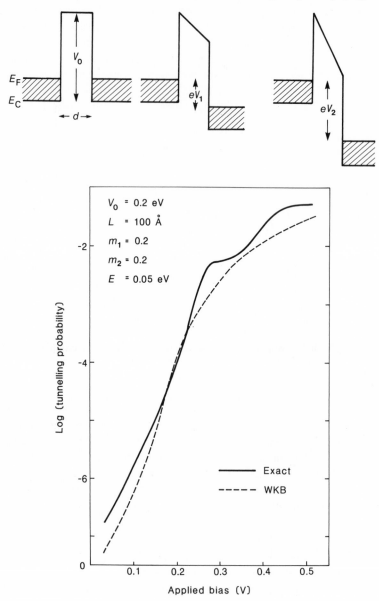

Fig. 8.2 The simple single-barrier tunnelling structure and its *I–V* characteristics showing (i) the differences between an exact solution and a Wenzel–Kramers–Brillouin

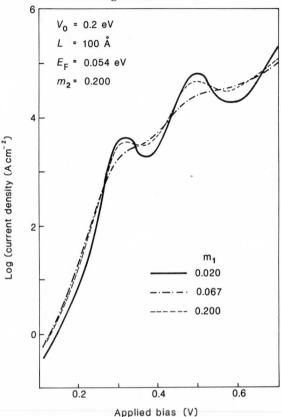

$V_0 = 0.2$ eV
$L = 100$ Å
$E_F = 0.054$ eV
$m_2 = 0.200$

m_1
——— 0.020
—·—·— 0.067
— — — 0.200

Fig. 8.2 (cont.) (WKB) approximation (which misses structure), and (ii) the effect of changing the effective mass of the carriers in the highly doped contact layers. (After Mendez 1987.)

This simple tunnelling problem can be realized in practice with a thin AlGaAs barrier, say about 3 nm thick with about 40 per cent Al, between layers of GaAs. We consider first a simplified structure, as in Fig. 8.2. For the moment, we consider the GaAs layers to be doped so as to produce a Fermi energy E_F (a few tens of meV for $c.10^{18}$ cm^{-3}) and that the AlGaAs layer is undoped, and furthermore that there are no space-charge effects. These simplifications will be removed in the following section. If a bias V is applied to the tunnel barrier, the current at a finite temperature is obtained as follows. We consider those electrons in the cathode side that are approaching the barrier ($k_z > 0$). These electrons have a total energy $E = \hbar^2 k^2/2m^*$, of which the longitudinal and transverse parts are given by

$$E_1 = \hbar^2 k_z^2/2m^* \qquad E_t = \hbar^2(k_x^2 + k_y^2)/2m^*$$

respectively. We now take all these incident electron states and sum over the products of their occupancy, their normal incident group velocity, and the probability that they are not reflected under bias. Note that under bias, the familiar conservation law for number $1 = |T|^2 + |R|^2$ must be supplemented by $k_i = k_t|T|^2 + k_r|R|^2$ for the conservation of current. Note that tunnelling depends only on E_1 in this model. This allows us to write the current density as

$$J \rightarrow = (e\hbar/4\pi^3 m^*) \int\int\int k_z(1 - |R(E_1,V)|^2) f(E)[1 - f(E + V)]$$

$$\times \, dk_x dk_y dk_z(:k_z > 0)$$

$$= (em^*/2\pi^2\hbar^3) \int\int (1 - |R(E_1,V)|^2) f(E)[1 - f(E + V)] dE_1 dE_t.$$

The prefactors come from the density of states in **k**-space, and all symbols have their usual meaning with $R = |r|^2$ the reflection intensity. The net current density is $J = J \rightarrow + J \leftarrow$, and so

$$J = (em^*/2\pi^2\hbar^3) \int\int (1 - |R(E_1,V)|^2) [f(E) - f(E + V)] dE_1 \, dE_t.$$

We recall from Chapter 1, Section 1.4, that the Fermi function $f(E)$ (strictly $f(E,T)$) must be solved self-consistently for E_F from the equations for (i) the Fermi–Dirac statistics, (ii) the doping density N_d in the contact layers, and (iii) the 3D density of states respectively:

$$\text{(i) } f(E) = \{\exp[(E - E_F)/kT] + 1\}^{-1}$$

$$\text{(ii) } N_D = \int_0^{E_F} n(E)f(E) \, dE$$

$$\text{(iii) } n(E) = \sqrt{(2/m^* E)} m^* / \pi^2 \hbar^3.$$

If we take a low-temperature limit, and assume $E_F = 20 \, \text{meV}$, then the current (density)–voltage characteristics are as shown in Fig. 8.2. There are several points to note.

1. At low bias the linear *I–V* dependence comes from assuming an approximately constant tunnelling probability, but the increasing bias has two effects, first to increase the density of occupied electron states capable of tunnelling, and second to reduce the tunnelling barrier height and so increase the tunnelling probability. As a result the overall current–voltage profile is superlinear.

2. In the presence of a linear potential profile, i.e. under a constant electric field, there are solutions to the Schrödinger equation known as Airy functions, and one can proceed to calculate this solution exactly as well (Landau and Lifshitz 1977).

3. There is no point taking the voltage to too high a level, as the approximations in this simple model become less reliable.

4. The only temperature dependence in the current–voltage characteristics comes from the Fermi occupation functions, and these vary over the range

0–400 K by a much smaller extent than do currents that have a thermionic component, which is an attractive feature in device applications.

5. The current–voltage characteristics are non-linear but antisymmetrical. If we extend the model by having equal-length undoped regions of GaAs on each side, the current–voltage characteristics are expanded on the voltage axis, as part of the bias would be dropped over these undoped regions, but otherwise they are anti-symmetrical in the absence of space-charge effects. If the undoped regions are of unequal length, the current–voltage characteristics become asymmetrical. There are important device implications for both types of generalized structure (subharmonic mixers and more general mixer–detector diodes for microwaves respectively, with the latter being an attractive alternative to Schottky diodes or planar doped barrier diodes as discussed in the next section), but the explicit model approximations made above must be relaxed and more detailed calculations are required in the design of practical devices.

6. We shall see that the agreement between theory and experiment is at best modest, and improvements are the subject of continuing research, as described in Section 8.2.4.

8.2.2 Practical structures: the design of current–voltage characteristics

A typical single-barrier structure for physics and device applications is more likely to have the form shown in Fig. 8.3. In practice one needs no intentional doping in the barrier, as the extra variations of a donor-level potential can give rise to an undesired potential profile. We return to the

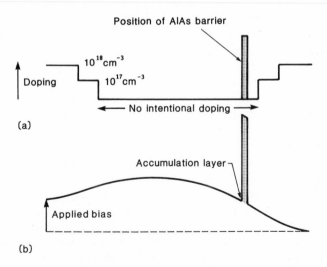

Fig. 8.3 (a) The doping and composition profile of a practical single tunnel barrier structure; (b) a schematic diagram of the conduction band minimum of that structure under an applied bias. (After Syme *et al.* 1992.)

question of the detailed relationship between a given semiconductor layer structure and its potential profile in Section 8.2.4. The thin layer of AlAs (typically about 3 nm thick) forms the tunnel barrier. The asymmetry in the undoped GaAs layers on each side will be reflected in the current–voltage characteristics of diodes made from such a multilayer. The heavily doped layers are used as sources and sinks for electrons, and it is to these layers that ohmic contacts are made. The layers (c.20–100 nm) of interim doping are optional, but they serve to smooth out the space-charge effects which we discuss now.

In Chapter 2, we described the physics of the n–i–n structure, where it was shown that a dipole layer is set up at an abrupt interface between two doping levels; some mobile carriers spill over from the heavily doped to the lightly doped or undoped region. One solves a self-consistent Poisson equation for the distribution of carriers, using the Thomas–Fermi approximation (that the local excess density of carriers is proportional to the net change of the local conduction band edge from its value in undoped material). In the presence of a bias, the conduction band is further distorted by an additional redistribution of spill-over charge at each end of the undoped region; again, the potential profile must be recalculated self-consistently, and this involves numerical rather than analytical solutions. One assumes that the extra charge densities from the injected current are modest. Having achieved a potential profile as in Fig. 8.3(b), the calculation of the current proceeds just as described in the previous section, namely with the calculation of $T(E)$, $R(E)$, and the tunnel current density (Syme *et al.* 1992). In this approach to calculating the current–voltage characteristics, it is assumed that the rate-determining step is the tunnelling of electrons from an accumulation layer adjacent to the tunnel barrier (see Fig. 8.3), but tunnelling electrons must first surmount the electrostatic barrier from the space-charge effects by thermionic emission. This last factor gives a temperature dependence to the current–voltage characteristics, with a barrier height of typically c.100 meV. The qualitative shape of the current–voltage characteristics are associated with the distortion of the conduction band under bias and the relative ease of current injection in either direction. The quantitative values are more problematical, as $T(E)$ and $1 - R(E)$ are exponentially dependent on the thickness (cf. the $\exp(KW)$ term).

In Fig. 8.4 we show two sets of results for the room temperature current–voltage characteristics for a design structure with 40:1 asymmetry in the undoped spacer layer thicknesses, corresponding to the growth using (a) MBE and (b) MOCVD. In Fig. 8.4(a), the thickness of the barrier has been used as a parameter to fit with experiment. In Fig. 8.4(b) the calculation is for the design structure and the data from the MOCVD-grown structure. In this latter case no simple adjustment of the barrier thickness would give satisfactory agreement, suggesting that the square barrier profile might not be a suitable approximation in this case. We return to this and other complications in Section 8.2.4.

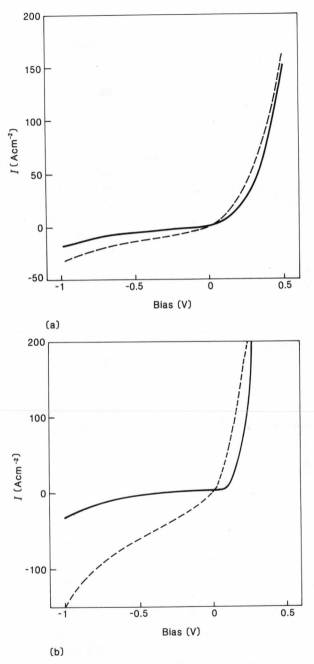

Fig. 8.4 (a) The current–voltage data (dashed line) from MBE-grown diode versus simulation (solid line) using the barrier thickness as a variable; (b) the current–voltage data for the same ideal structure compared with data from a MOCVD-grown diode. (After Syme *et al.* 1992.)

8.2.3 Effects of temperature, stress, magnetic fields, etc.

A range of further experiments is possible to verify the simple model for tunnelling. Diodes can be subject to variations in temperature, pressure, magnetic fields, etc. At lower temperatures, the diode characteristics sharpen up, with a reduced reverse bias current and a more rapid turn on for positive bias. In some of the MBE-grown diodes, further weak structure appears in the current–voltage data at low temperatures. In practice, the accumulation layer is qualitatively similar to an inversion layer in a MOSFET, and tunnelling should take place from a number of 2D sub-bands (i.e. effectively with discrete k_z), rather than from a continuum of 3D states, and the number of these sub-bands increases with bias, giving rise to the extra structure.

The relative position of energy bands, and particularly their edges, move under pressure, and the results of pressure-dependent measurements can be used to verify aspects of the model for tunnelling (Fig. 8.5). AlAs is an indirect-gap semiconductor, and tunnelling might take place via X states for which there is a lower energy barrier. In practice, the low-temperature data support the idea that, in barriers of less than 4 nm, the tunnelling takes place through the Γ states. In the case of AlAs barriers in GaAs, the Γ–Γ barrier height is about 1 eV, while the Γ–X barrier is only about 0.3 eV high. The Γ–X tunnelling mechanism has been clearly established for thicker barriers from the trends produced in studies of pressure and temperature dependence.

The detailed quantitative agreement between theory and experiment in the case of Γ–X tunnelling is more complicated. The simple wavefunction-matching arguments must be modified once it becomes possible for an electron wavefunction to couple to two or more energy bands. After all, simple wavefunction matching assumes a common Bloch function throughout the entire structure. Some form of process may also be required to take up the change in electron momentum in changing bands. For more details on these complications to the electronic structure, see Jaros (1986).

Magnetic fields applied parallel to the direction of current tend to have little effect on the current–voltage data. Electrons spiral forward around the trajectory that they would have in the absence of a magnetic field. When B and J are perpendicular, the situation is different, and yields further information on the barrier. The argument is as follows (see Eaves *et al.* (1986) for further details). In the heavily doped regions, the magnetic field has little effect on the transport as the scattering from ionized impurities etc. prevents Landau-level-type trajectories from forming. In the undoped region, this is not so, as shown schematically in Fig. 8.6(a). With the substitution of $(\mathbf{p} + e\mathbf{A})$ for $\mathbf{p}$, the Hamiltonian that describes the motion changes from

$$H = p^2/2m^* + V(z) - eEx$$

to

$$H = p^2/2m^* + V(z) + e^2 B^2 x^2/2m^* + p_y eBx/m^* - eEx$$

where E is the electric field over the barrier under bias. The two magnetic field terms have an effect similar to that encountered with the hot electron

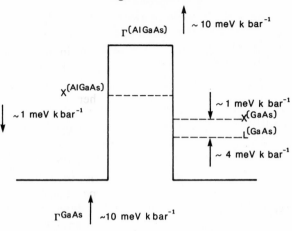

(a)

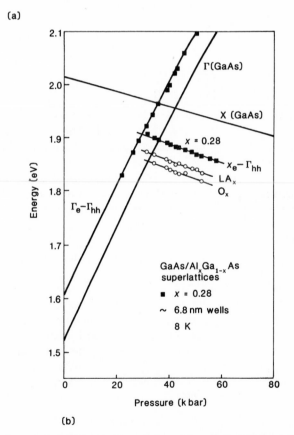

(b)

Fig. 8.5 (a) The shifts of the Γ, X, and L points of the GaAs and AlGaAs band structures under pressure. (b) the effect of pressure measured by optical techniques in superlattices: e, electron; hh, heavy hole, LA_x and O_x relate to phonon replica features associated with indirect-gap transitions. (After Wolford *et al.* 1986.)

transistor structure: they transfer energy and momentum from the original direction of motion (x) to the transverse direction (y). If one assumes a narrow distribution of energies in the tunnelling current, the transmission term can be calculated to second order in B, which takes the form

$$T = T_0\exp\{-[2KW/4(V - \varepsilon)](-eEW + e^2B^2W^2/3m^*)\}$$

where T_0 is the value calculated for energy ε using the expression in Section 8.2.1, with K and W having the same definitions as given there. Two facts can be deduced: (i) for a fixed bias over the tunnel barrier, the current density varies as $J = \exp(\beta B^2)$, where $\beta = e^2W^3K/6m^*(V - \varepsilon_F)$; (ii) to maintain a constant current, an extra voltage $\Delta V_B = -EW$ must be applied to the barrier to offset the effects of the applied magnetic field $\Delta V_B = e^2B^2W^2/3m^*$. Experimental verification of the former is shown in Fig. 8.6(b). The second

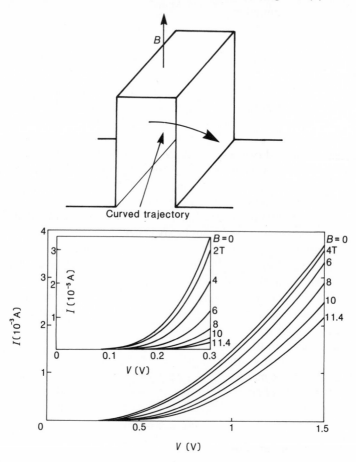

Fig. 8.6 Tunnelling through a single barrier with a transverse magnetic field which has the effect of increasing the distance that the electron tunnels in the barrier. The current at fixed bias drops with increasing magnetic field. (After Eaves *et al.* 1986.)

implication has also been tested to give the value of m^* in the barrier layer, but transport in the spacer layers complicates a simple analysis.

8.2.4 The single-tunnel barrier: reconciling theory and experiment

A constant theme in the previous sections has been the modest quantitative agreement between theory and experiment. Apart from any issues of numerical accuracy (and care is needed when handling exponentially varying functions), a number of theoretical issues are unresolved. We have not taken into account the different effective mass in the barrier layer, but that is straightforward via a modification of the definition of K and a generalized current continuity equation involving $1/m^*$ as a prefactor in each material. There is also the 2D nature of the initial tunnelling states in the accumulation layer, non-parabolic energy bands, multiple energy bands (Γ, X- and L-related) on each side of a heterojunction that are suitable for wavefunction matching, scattering of non-ideal interfaces, non-square potential profiles, etc. All these effects have been investigated. While individually they may be be significant only in special cases, their combined effects may be quite important in most practical examples.

The experimental side is also fraught. Band-gap offsets are not known with satisfactory precision, so that the barrier height has some uncertainty. All the studies of multilayers reveal less than ideal structures. The interfaces are not atomically flat, and steps at interfaces act as scattering centres. The interfaces are not often completed over one monolayer, and in some cases the potential profile may be more Gaussian or Lorenzian than square. There are variations across a wafer, even without specific localized defects. Considering the $\exp(-KW)$ term in the tunnelling expression, we are uncertain about both the K (i.e. the $\sqrt{(E - V)}$ term) and the W. These are just some of the difficulties encountered in reconciling theory and experiment.

The tunnel barriers are thin, so what role does the statistical distribution of discrete Al and Ga ions play in setting up a uniform potential profile? If fractions of a monolayer are present, what role is played by steps on either interface? If there are aggregations of ions, is the barrier one of spatially varying height, and so how does this barrier relate to the effective barrier height extracted from experimental results?

It is the combination of complications in both theory and experiment that makes nonsense of many attempts to tighten up exclusively on one or the other. Progress will be made with multiple and correlated studies of the type that have been more commonly applied to double and multiple barriers, to which we now turn.

8.3 The double-barrier diode

8.3.1 Simple theory of resonant tunnelling

We begin this section by extending the theory of Section 8.2.1. Consider the elementary quantum mechanics of the problem posed in Fig. 8.7, namely an

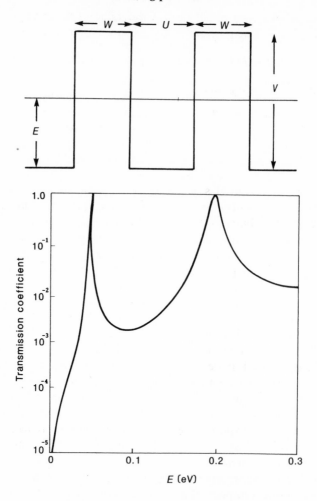

Fig. 8.7 The simplest (1D) analysis of resonant tunnelling through a double-barrier structure. The transmission verses energy calculation is performed for two δ-function barriers of strength 0.3 eV times 5 nm, separated by 10 nm. (J. H. Davies, University of Glasgow, Private Communication.)

electron with kinetic energy E, incident on a double barrier structure, each barrier of height $V(>E)$ and thickness W, and separated by a well of thickness U. The traditional method of solution is to extend the treatment of the single barrier by setting up forward and backward travelling wavefunctions in three of the five regions of the structure, and exponentially growing and decaying wavefunctions in the two barriers. A normalized wave is incident from the left, some of which is reflected and some transmitted to the right of the entire structure. The same definitions of k, K, and k' are used outside, under, and above the barriers. The result of this analysis leads to a transmission coefficient which is a more complicated generalization of that obtained from the single barrier:

$$t(E) = 8\exp\{-i[k(U + 2W)]/[D_1\exp(-2KW) + D_2\exp(2KW)]\}$$

where, for $E < V$ with $\theta = ik/K$,

$$D_1 = 2\cos(kU)(2 + 1/\theta + \theta) - i\sin(kU)(\theta^2 + 2\theta + 2 + 2/\theta + 1/\theta^2)$$

and

$$D_2 = 2\cos(kU)(2 - 1/\theta - \theta) - i\sin(kU)(\theta^2 - 2\theta + 2 - 2/\theta + 1/\theta^2).$$

Note again the exponential suppression of the transmission amplitude for $E < V$, from the $\exp(2KW)$ factor within D_2; the factor of 2 in the exponent comes from the two barriers. However, there is a real difference: regarding D_2 as a function of energy, then if we can obtain the condition $D_2(E) = 0$, electrons would be transmitted with a probability of order unity. It is as if there were no barrier at such an energy. This is the condition called resonant tunnelling, which will concern us for most of the rest of this chapter. In Chapter 10 on quantum wells, we discover that there are discrete-energy bound states for electrons within quantum wells (thin barriers of GaAs between thicker layers of AlGaAs provide a practical example). The condition $D_2(E) = 0$ for resonant tunnelling is the same as that for bound states in quantum wells—indeed, if we let V and W both increase, we obtain states of energy $E_n = \hbar^2(n\pi)^2/2m^*$, where n is an integer. For finite V and W the solution must be obtained numerically. A plot of $T(E) = |t(E)|^2$ is shown in Fig. 8.7. When $|t| \sim 1$, the amplitude of the forward and backward waves in the central well grow exponentially large, with multiple reflections forming waves that interfere constructively to overcome the exponential decay due to tunnelling in either barrier. Note that this very simple picture will cause problems below when we introduce the electronic charge and electrostatics into the problem. It is important to note that the $T(E) = 1$ condition depends on the two barriers being identical and the absence of applied bias. In practice, there are bound states in wells subject to electric fields and with unequal barrier widths. If there is such a bound state at an energy E_i, at which the barriers have transmission T_1 and T_2, then near the resonance energy we can approximate the transmission $T(E)$ as (Ricco and Azbel 1984)

$$T(E) = \frac{4T_1T_2}{(T_1 + T_2)^2} \frac{\gamma^2}{[(E - E_i)^2 + \gamma^2]}$$

where $\gamma = h/\tau$ is the lifetime width of the resonant state (see below) and in a quasi-classical approximation $\gamma \sim E_i(T_1 + T_2)$. It is only when $T_1 = T_2$ that there is complete transmission. More generally, the peak transmission is *not* exponentially small.

The derivation of the current–voltage characteristics for the double-barrier structure from the form of $T(E)$, $R(E)$ is exactly the same as that for the single

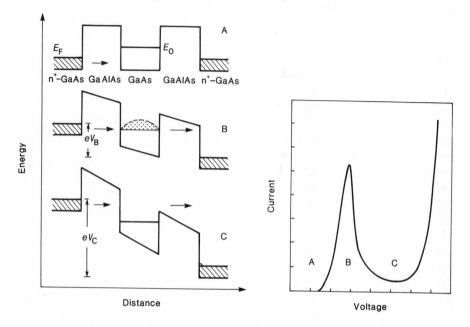

Fig. 8.8 A simple model double-barrier diode and its current–voltage characteristics (as a generalization of Fig. 8.2). Note the strong drop in current with increasing bias once the bound level in the quantum well falls below the conduction band minimum in the cathode layer. (After Mendez 1987.)

barrier. In practice, where numerical methods are used, great care must be taken with the algorithms used in integration, so that the weight under quite narrow resonant tunnelling structures is represented accurately. One of the first features to emerge from a typical calculation is that current peaks are associated with resonant tunnelling. The qualitative picture, equivalent to that in Fig. 8.2 above, is shown in Fig. 8.8. As the bias is applied across the two outer contact layers (heavily doped), the voltage is dropped over the (un-doped) double-barrier structure, lowering the energy of the bound-state in the central well, until it comes into resonance with the electrons in the cathode and the current rises with increasing tunnelling probability. As the bias is increased further, the energy of the resonant level falls below the bottom of the conduction band edge in the cathode and resonant tunnelling can no longer occur. The current drops, rising again as another bound state comes into resonance or as the second barrier becomes small enough to permit electrons to pass over it. The region of negative differential resistance has attracted much interest from device inventors. As we shall see below, the resonant tunnelling process is very fast, and the fastest purely electronic device is a resonant tunnelling diode operating at 712 GHz (Brown *et al.* 1991).

8.3.2 *Practical structures: the design of current–voltage characteristics*

The practical precaution of having no doping in the proximity of the tunnel structure applies equally well to double and multiple barriers as it did to the single barrier. Furthermore, there is often an extended region of undoped GaAs on the anode or cathode side of the double barrier, just as in the case of the single barrier; this helps to expand the voltage axis, as only a part of the applied bias is dropped over the double-barrier region to bring the structure into and out of tunnelling resonance. Figure 8.9(a) shows a typical good experimental result for a structure that is nominally symmetrical. The peak current densities in forward and reverse bias are approximately equal (which does not often occur), but the voltages at which the peaks appear are not so similar. This is probably because the doping profile of the contact layers is not perfectly symmetrical with respect to the double-barrier tunnel structure. The jagged form of the current–voltage characteristics in the region of negative differential resistance is commonly observed; this is because the device is spontaneously oscillating in the measuring circuit, and all that is being measured there is some form of average over the oscillation cycle, which itself also depends explicitly on circuit, as opposed to solely device, parameters. Indeed, one can obtain bistability in the current–voltage data (Fig. 8.9(b)). A parallel capacitance can be applied to eliminate such oscillations and to obtain a result qualitatively similar to the simulation. The large slope resistance near the origin reflects the low tunnel current before the resonance condition is reached.

In Fig. 8.9(b) we are also able to show the typical results of simulations and experiments when applied to a given design structure. The quantitative agreement is less satisfactory than in the case of the single-barrier diode; the reasons for this include those mentioned in Section 8.2.3, and we have more to add here. The exact voltage profile is complicated in a resonant tunnel diode, as the charge that builds up in the quasi-bound state in the central well (cf. the exponentially large amplitudes there in the simple model) is sufficient to alter the actual potential profile. Indeed, this effect is partially responsible for the bistability commonly observed. Here we see that the peak in the simulation current comes at too low a voltage. This means that depletion at the anode and 2D accumulation at the cathode have not been accounted for adequately. Although it is not directly apparent, the peak current density achieved in theory and experiment is in quite good agreement; this is also a common feature of simulations. The most striking disagreement is that simulations always grossly underestimate the valley current after the resonant tunnelling peak. Many reasons have been put forward, most of which represent another channel for current not included in the simple models, such as scattering by phonons, impurities, extra tunnelling via impurity states in the barriers, tunnelling via X- and L-related electron states rather than just the Γ states considered so far, disorder effects in the alloy barriers, and interface steps and roughness more generally. Even the representation of the

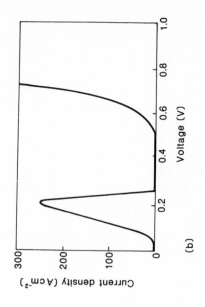

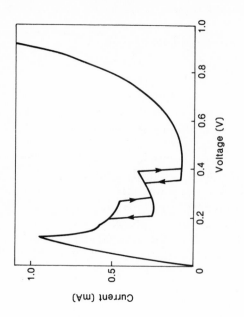

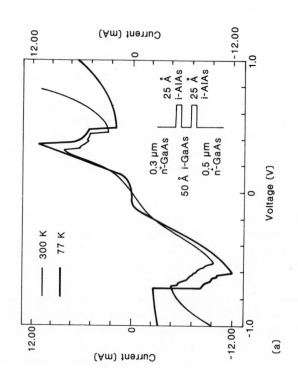

Fig. 8.9 (a) The current–voltage characteristics of a symmetrical double-barrier diode at 77 K and 300 K, showing a drop in peak current and a rise in valley current (after Morkoc *et al.* 1986); (b) the state of agreement between simulation and experiment (after Hodson *et al.* 1988). (Note the bistability in the current–voltage data).

Coulomb effects associated with charge building up in the well can broaden out the voltage range over which tunnelling occurs and narrow the voltage range of the valley.

One figure of merit widely quoted for the double-barrier diode is the peak-to-valley current ratio. This, together with the valley-to-peak voltage ratio, is an important factor in microwave device applications, as well as in suggestions for multistable devices for digital applications (cf. Chapters 16–17). In general, one requires high current densities for high-speed devices: a fixed number of electrons is required to charge or discharge some part of a device or circuit, and these must be delivered as quickly as possible. One needs thin barriers so that the tunnel current is high, and a heavily doped cathode so that there is a good reservoir of filled initial states for tunnelling. If one is interested in digital circuits, the features in the current–voltage characteristics need to appear on a voltage scale comparable with the bias supply (a few volts). This means that a thin well is required so that the quasi-bound state is high in energy and takes about 0.3 V to be brought into resonance, with a high-doped cathode and a low-doped depletion region at the anode to increase the voltage swing to about 3 V at the device terminals. Peak-to-valley ratios of 4:1 have been achieved at room temperature with the GaAs–AlGaAs materials system, and even higher values (50:1) with other combinations of materials that extend the range of parameters of the tunnel structure further.

8.3.3 Effects of temperature, stress, magnetic fields, etc.

The double-barrier diode structure has been rather more widely investigated than the single-barrier structure. A temperature dependence as shown in Fig. 8.9(a) is typical. Thermionic emission over the tunnel barriers and scattering within the tunnelling process (as by phonon emission/absorption) can occur at high temperatures. The valley currents rise with temperature. The peak tunnelling current falls as the partial occupancy of both initial and final states cancel each other in the $f(E) - f(E + V)$ factor in the tunnelling integral. The combination degrades the peak-to-valley ratio from typically more than 10 : 1 at 77 K to 2–4:1 at room temperature, to $c.1:1$ at 150–200 °C. Any smaller tunnelling features due to resonant tunnelling using higher-energy bound states vanish at lower temperatures.

Pressure studies of tunnelling in double-barrier structures complement those on single barriers. The diagrams in Fig. 8.10 show the effect of pressure on the tunnel barriers (Fig. 8.10(a)), the current at constant bias (Fig. 8.10(b)), and the voltage of the greatest negative differential conductance (approximately half-way between peak and valley) as pressure is applied (Fig. 8.10(c)). For biases above 0.4 V in this device, we are beyond the first resonant tunnelling peak, and in fact only the first tunnel barrier is effective. The clear break at $c.4$ kbar is an indication that the X states are playing a role in contributing to the (non-resonant) tunnelling current; the density of tunnelling states is higher for X states, and the change of effective barrier height with

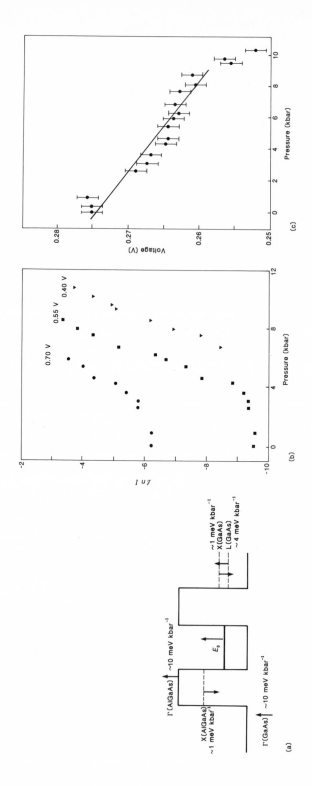

Fig. 8.10 (a) The double-barrier tunnel structures under pressure, and (b) the variation of current with pressure at fixed bias, and (c) variation of voltage of resonant tunnelling feature with pressure in a structure that consists of two 10 nm $Ga_{0.6}Al_{0.4}$ barriers separated by a 4 nm GaAs layer. (After Mendez *et al.* 1986*a*).

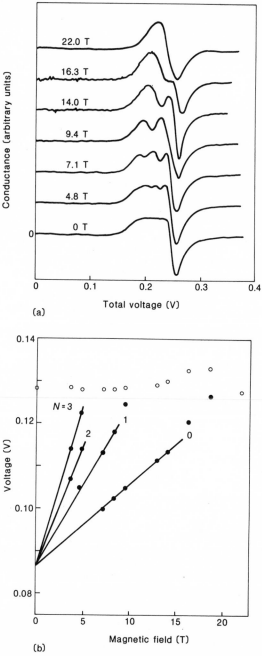

Fig. 8.11 (a) Conductance of a double-barrier diode (the same as shown in Fig. 8.10 (a)) as a function of magnetic field and (b) a fan diagram of the principal features showing that Landau levels in the accumulation layer at the cathode are the origin of the field-induced features in the conductance (After Mendez *et al.* 1986*b*.)

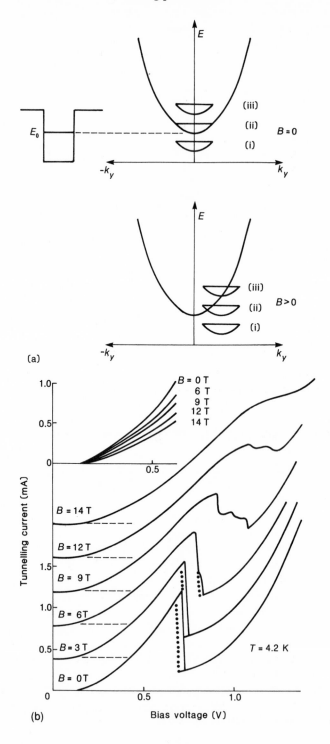

(a)

(b)

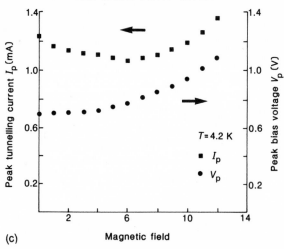

Fig. 8.12 (a) The alignment of the occupied initial states (the partial parabolas cut off at the Fermi energy in the cathode layer) against the final state dispersion (full parabola) for different biases with and without magnetic field. The structure is viewed from the anode with (i) no bias, (ii) the bias for the peak current, and (iii) the bias for the valley current. A finite magnetic field perpendicular to the current direction shifts the initial states by Δk_y with respect to the final states as shown. (b) The current–voltage data from a resonant tunnel diode as a function of magnetic field perpendicular to the current. At a sufficiently strong magnetic field Δk_y is so large that there is no overlap between initial and final states for resonant tunnelling. (c) The data are used to map the in-plane dispersion of electron states in the quantum well. (After Ben Amor *et al.* 1988.)

pressure from 1 meV kbar^{-1} for Γ states to 12 meV kbar^{-1} for X states is apparent. In contrast, the unchanging gradient of -1.76 mV kbar^{-1} in the voltage of the maximum negative differential conductance is indicative of Γ-state resonant tunnelling out to the highest pressures, even with 10 nm barriers; note that only half the applied voltage is acting to reduce the energy of the quasi-bound state in the quantum well so that this result is consistent with the value of 1 meV kbar^{-1} quoted above.

Magnetic field studies at low temperature have proved useful in elucidating the physics of resonant tunnelling. When the field is parallel to the current, the Landau levels (cf. Chapter 2) will be formed in the plane of the accumulation layer. The same thing happens in the plane of quasi-bound states (cf. Chapter 4) whose parabolic energy bands in the $k_x - k_y$ plane are converted into discrete Landau levels with an in-plane energy separation of $\hbar\omega_c$. The filling of these Landau levels changes with magnetic field, and thus so do the conditions for resonant tunnelling. The effects are shown in Fig. 8.11, where the conductance dI/dV is plotted against total bias voltage at different fields. The structure is plotted (solid circles) in conventional fan diagrams (using only half the total applied voltage as a measure of energy changes in the central well), confirming a 2D effective mass in the well close to that of bulk GaAs. The main negative differential conductance peak (open circles) barely moves with magnetic field, suggesting that the ground-state Landau levels in the 3D terminals and the 2D quantum well do not change relative to each other, and so are determined by the same effective mass.

A rich variety of new transport conditions have been investigated with the magnetic field perpendicular to the current. The physics derives from the trajectory of a ballistic electron subject to a Lorentz force, as encountered previously in the hot electron transistor and the single-tunnel barrier. An electron injected into a tunnel structure will transfer some momentum and kinetic energy from the forward to the transverse direction (Fig. 8.12(a)). In 'conventional' tunnel structures, where the well and barrier thicknesses are in the 3–10 nm range, the changes in the current–voltage characteristics can be interpreted in terms of the in-plane momentum $\Delta k_y = e B \Delta s / \hbar$ acquired by the electron in going a distance Δs from the heavily doped electrode to the centre of the well of the tunnel diode. Resonant tunnelling will only occur if an extra bias is applied such that $e\Delta V/2 = \hbar^2 (\Delta k_y)^2 / 2m^*$. Thus the variations in features with magnetic field can be used to map out the 2D dispersion of the electron states in the central quantum well (Fig. 8.12(b, c)).

If structures are grown with thick central wells, complete cyclotron orbits can be accommodated within the wells for moderate magnetic fields. In this case fine structure in the current–voltage characteristics is observed, whose detailed origin is in the nature of the orbits of ballistic hot electrons in the central well.

These studies confirm the basic picture that has been developed above on the nature of ballistic motion and tunnelling.

8.3.4 *The double-barrier diode as a physics laboratory*

The precision with which double-barrier diode structures can be designed and fabricated has meant that they have been useful in testing a wide variety of physics ideas, some of device significance. In this section we discuss some examples, starting with a continuation of the previous section.

8.3.4.1 Cyclotron orbits of ballistic hot electrons   If, as shown in Fig. 8.13(a), the well of a double-barrier diode is expanded to $c.100$ nm, there are many more quasi-bound states, and while the resonant tunnelling features of any one state might be small, a rich structure is retained in the conductance as shown in Fig. 8.13(b). At high biases the electrons pass over the second barrier, but the resonance features are still retained. The beating observed in the amplitude of the dI/dV features has been correlated with interference between electrons that are reflected by the potential steps at the first and second interfaces of the second barrier. In the presence of magnetic fields perpendicular to the field and current in the well, various types of orbit are possible for the electron motion. Figure 8.13(c) shows examples of orbits that skip along one interface and orbits that traverse the quantum well several

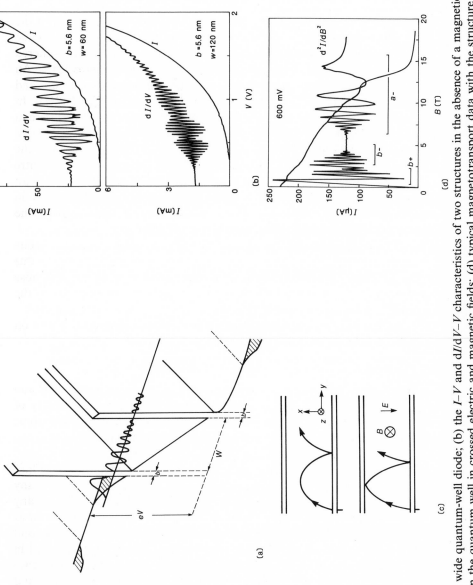

Fig. 8.13 (a) A wide quantum-well diode; (b) the I–V and dI/dV–V characteristics of two structures in the absence of a magnetic field; (c) possible orbits in the quantum well in crossed electric and magnetic fields; (d) typical magnetotransport data with the structure identified with the orbits just given. (After Eaves *et al.* 1991.)

times. A complex pattern emerges in the current–voltage data as a function of magnetic field (Fig. 8.13(d)), but the structure has all been assigned successfully to particular orbits. These striking results are possible because the mean free path of the ballistic electrons must exceed $c.100$ nm at low temperature, and the geometry and fields under which the electrons move are well defined.

8.3.4.2 Traversal times of tunnelling structures

In semiclassical transport, we associate a velocity with a crystal momentum via $v = \hbar k/m^*$. What is the velocity of a tunnelling electron? How fast is the resonant tunnelling process? Theorists have conducted a spirited debate on these issues. For an electron above the barrier in Fig. 8.1, the velocity is reduced from $\surd(2m^*E)$ to $\surd(2m^*(E - V))$, and so to zero if $E = V$. What if $E<0$? The simple semiclassical extension would have us use $v = \surd 2m^*(V - E))$, but do we have any justification for this? Some early calculations used Gaussian wavepackets with a narrow distribution of k-vectors. If $\Delta\phi$ is the change of phase across the barrier, then the tunnelling time is taken to be $\tau_\phi = h\partial\Delta\phi/\partial E$, and for an almost opaque barrier ($KW>>1$ in the terminology of Section 8.2.1), we obtain $\tau_\phi = 2m^*/hK$, which is independent of the thickness of the barrier (a counter-intuitive result) and diverges as the incident kinetic energy goes to zero. All the d.c. transport and magnetotransport data are consistent with the semiclassical formulation above, and some sophisticated theory has been used to rationalize the results (see Landauer and Martin (1994), and references cited therein, for further details).

The problem in resonant tunnelling is even more complex. In Fig. 8.7, the width in energy of the transmission resonance can be interpreted as a lifetime for an electron placed in the well state: $\tau \sim h/\Delta E$, and so $|\psi|^2 \sim \exp(-t/\tau)$. If the barriers are very thin, there may be corrections to this form. In the simple model, the exponential build-up of charge in the well comes from multiple reflections within the well, suggesting a long net traversal time for a given electron. There are no definitive real-time experiments on double-barrier structures, but the fact that devices operate at 420 GHz in GaAs–AlGaAs multilayers and at higher frequencies in other materials systems (Brown *et al.* 1991) suggests that the electron velocities cannot be much less than the semiclassical values: $c.1$ ps to traverse 100 nm at 10^5 ms^{-1} corresponds to 1 THz. We shall return to this problem in Chapter 11.

8.3.4.3 Hole versus electron tunnelling

So far in this chapter we have been concerned with electron tunnelling. There is an equivalent, but smaller, body of work on hole tunnelling using p-doped contacts. The tunnelling phenomena are less prominent for holes than for electrons. This is due to several factors. The hole mass is larger than the electron mass, meaning that barriers have to be thinner to achieve comparable tunnelling probabilities. The hole band structure is much more complex in a quantum-well structure (see Appendix 3). It originates from triply degenerate valence bands at $k = 0$ in a bulk semiconductor, with the degeneracy split by spin–orbit effects and the reduced symmetry imposed by the tunnel barriers. The valence band offsets are also smaller than the conduction band offsets in GaAs–AlGaAs. Scattering rates for holes are much greater than for electrons as the number and type of final states are greater. Nevertheless, all the effects described above for electrons have been followed by comparable experiments on holes. The results of hole tunnelling in crossed electric and magnetic fields are shown in Fig. 8.14 (see Fig. 8.12 for the equivalent electron tunnelling data).

8.3.4.4 Sequential versus coherent resonant tunnelling The traversal time associated with tunnelling is comparable with the inverse of some scattering rates, for example the emission of phonons by hot electrons. If there are multiple reflections of electron waves in a quantum well, dephasing because of scattering might well destroy the coherence of the electron wavefunctions that has been implicitly assumed. An alternative picture for resonant tunnelling is that electrons tunnel into the well, thermalize and equilibrate there, and tunnel out of the well through the second barrier via an independent process which has no phase correlation with the tunnelling-in. In practice, incoherent sequential tunnelling is more readily observed with thicker barriers, when the electron state lifetime in the well exceeds transport scattering times. In high-speed devices the barriers are thin and coherent tunnelling is possible.

A detailed treatment of sequential tunnelling is not given here (see Luryi 1985), but the current is related to the lesser of the tunnelling probabilities of the two barriers. In the case of coherent tunnelling, if the two barriers are sufficiently different (as is nearly always the case under usual biases), the expression for $T(E)$ in terms of T_1 and T_2 (Section 8.3.1), in the limit of both a bias and a Fermi energy for the incoming electrons much greater than the resonance width, is also proportional to the lesser of the two probabilities. Analyses show that the sequential and coherent pictures give identical results provided the resonance width does not exceed the range of incoming energies (Weil and Vinter 1987). In Fig. 8.15, we give the sequential tunnelling picture that leads to negative differential resistance. The top diagram shows the double-barrier structure, with 3D electron gases in the contacts to an energy E_F and a quasi-bound level in the quantum well at energy E_0. The middle diagram shows the structure under bias so that there is an overlap in energy between E_0 and occupied states in the emitter contact layer. The bottom diagram shows the Fermi sphere (radius p_F) of occupied states in the emitter contact. It is assumed that tunnelling into the well is only possible if the momenta p_x, p_y and the total energy are conserved. If the energy E_0 (above the bottom of the conduction band edge in the quantum well) corresponds to $p_0^2/2m^*$, conservation of momentum in the tunnelling process implies that only those electrons on the (shaded) disc $p_z = p_0$ can participate in tunnelling. At low temperature, tunnelling into the quantum well only occurs over a voltage range in which the shaded disc moves from the equatorial plane to the pole.

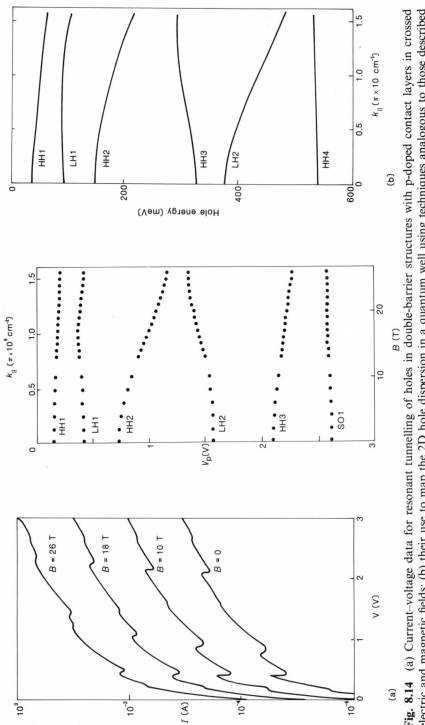

Fig. 8.14 (a) Current–voltage data for resonant tunnelling of holes in double-barrier structures with p-doped contact layers in crossed electric and magnetic fields; (b) their use to map the 2D hole dispersion in a quantum well using techniques analogous to those described in Section 8.3.3. (After Hayden *et al.* 1992.)

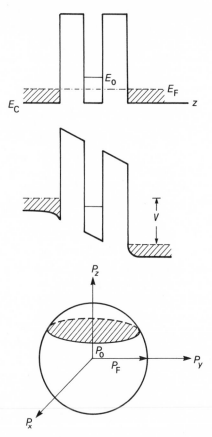

Fig. 8.15 The sequential tunnelling picture for negative differential resistance as described in the text. (After Luryi 1985.)

At higher biases $p_0^2 < 0$ and no tunnelling takes place. This is sufficient to produce a current peak, and hence a negative differential resistance beyond the peak. Tunnelling out of the quantum well follows the same reasoning, but there are always final states available. The rate-determining step is tunnelling through the thicker barrier.

8.3.4.5 Tunnelling Versus Thermionic Emission It is noted here that the only temperature dependence in tunnelling is in the Fermi occupation factors in the formula for the tunnelling current. As long as tunnelling through barriers is the rate-determining step, this temperature dependence is set by the scale of the Fermi energy in the contacts, and this can exceed $c.0.1$ eV ~ 1000 K. In practice the barrier heights for thermionic emission are somewhat higher, typically $c.0.6$ eV. However, the tunnel barriers themselves are often only $c.0.3$ eV high, or 0.2 eV above the Fermi energy. The thermionic emission current over such barriers means that most tunnel structures have an appreciable temperature dependence. Conversely, the current in thermionic emission structures (Schottky barriers or planar doped barriers) is dominated by tunnelling at sufficiently low temperatures. For the design of practical devices which must operate between -40 and $+80$ °C on the earth, any means of reducing the temperature dependence of operation is highly desirable; this means that complex circuitry to compensate for temperature variations can be dispensed with. The barrier heights obtained in the AlGaAs–GaAs system seem slightly too small to guarantee a sufficient temperature independence, but the broader range of III–V material combinations that are available may provide the solution, leading to the much wider use of devices designed around tunnelling currents (see Chapter 17).

8.3.4.6 Light Emission from Tunnelling Structures   The spectroscopic advantages of optical techniques can be exploited in the context of tunnelling structures. The principle of photoluminescence spectroscopy is explained in Fig. 8.16(a). Light from a He–Ne laser (at 632.8 nm) is absorbed in a top contact 1 μm thick. The holes thus created drift and diffuse towards the tunnel barrier, where they accumulate and tunnel into the quantum well where their density is $c.10^7$–$10^8\,\mathrm{cm}^{-2}$. They recombine with the tunnelling electrons present at a much higher density $(c.10^{11}\,\mathrm{cm}^{-2})$. The intensity and linewidth of the main luminescence feature can be monitored as a function of applied bias; typical results are shown in Fig. 8.16(b). The main luminescence energy drops with increasing bias (see Chapter 10 for further details of the cause, namely the quantum-confined Stark effect). The linewidth correlates with the current, giving an indication of charge building up in the quantum well, and indirectly an indication of the speed of charge transfer. If a strong magnetic field is applied, sharp Landau levels are formed and the carriers can occupy the lowest Landau level, so that the effects of linewidth broadening from a finite energy distribution of carriers are lost. The integrated luminescence intensity is a measure of the extent of electron charge build-up in the well. At higher biases, the higher resonant tunnel features do not show up in luminescence as strongly as they do in the current, indicating that the electrons are spending less time in the well—they can escape more easily. These results show the corroborative power of optical spectroscopy (even with very weak signals and low efficiency optical processes) in explaining the operation of tunnelling.

8.4 Multiple-tunnel barriers

The wealth of novel phenomena encountered in the move from a single-barrier to a double-barrier diode has prompted many studies of multiple-barrier structures. Here we mention a few results from triple-barrier structures that shed light on aspects of sequential and coherent tunnelling. There are two key regimes in which the central barrier is either (i) thinner or (ii) thicker than the outer barriers. In the former (Fig. 8.17(a)), the quasi-bound states in the

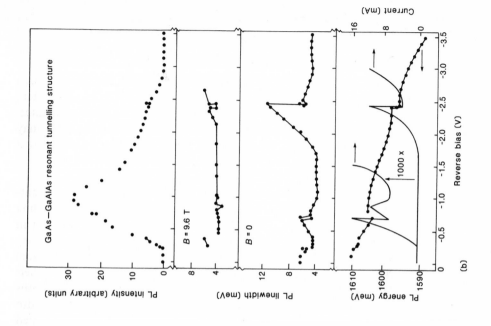

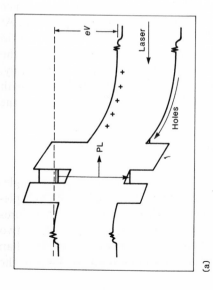

Fig. 8.16 (a) The principle of photoluminescence (PL) spectroscopy on double-barrier diodes: light absorption in the surface layer creates holes that drift to the double-barrier structure, greatly increasing the possibility of radiant recombination in the double barrier structure. (b) The data from a typical structure correlating the energy of the quantum-well photoluminescence, and its intensity and linewidth, with features in the current–voltage characteristics as the bias is increased. The data give an indication of the build-up of charge in the well during resonant tunnelling. (After Skolnick *et al.* 1992.)

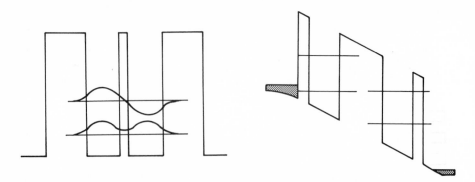

Fig. 8.17 Typical triple-barrier structures with (a) a thin and (b) a thick central barrier in which the physics of quantum-well energy levels and tunnelling have been investigated.

two wells interact (cf. Chapter 10), and under bias the tunnelling is via one of the two energy levels. In the latter (Fig. 8.17(b)), the rate-determining step is tunnelling through the thick barrier, and this will occur when the quasi-bound states in each well are brought into alignment with each other and with electrons in the emitter layer.

Two examples of the wide range of multiple-barrier structures under investigation are discussed now. A series of barriers and wells with a gradual change in the relative thickness of well and barrier can be considered as a 'digital' version of a graded gap (Fig. 8.18(a)). The quasi-bound states are higher in energy in the successively narrower wells. A bias in one direction will bring them into alignment; a bias in the reverse direction, will take them further out of alignment. The diode properties have been established, but the residual features associated with resonant tunnelling (peaks and shoulders in the current–voltage data) are undesirable in device applications. The problem of obtaining perfectly antisymmetric current–voltage characteristics was alluded to with reference to Fig. 8.9(a). This has been overcome to some extent by using multiple rather than single barriers to define a quantum well (Fig. 8.18(b)), where the effects of interface roughness are averaged out and any migration of dopant seems less significant.

8.5 Multiple-barrier diodes: reconciling theory and experiment

The problems of obtaining close agreement between the current–voltage data simulated for a particular design of a tunnel structure and that obtained experimentally for a particular realization have already been covered in Section 8.2.4. The situation is further complicated with multiple barriers and resonant tunnelling. In one study (Rimmer *et al.* 1991) many of the known characterization techniques were brought to bear on a triple-barrier structure to determine the structure parameters as precisely as possible; in this example

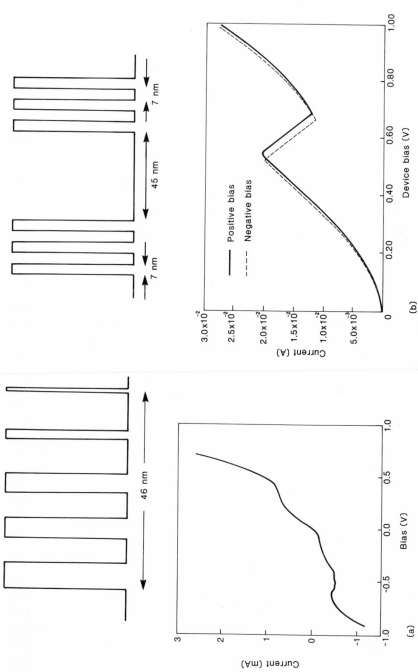

Fig. 8.18 Multiple-barrier structures and their current–voltage characteristics: (a) a tunnel structure with graded barrier/well thicknesses showing some asymmetry in the current–voltage characteristics near the origin (Syme *et al.* 1991); (b) a double-barrier diode in which each barrier is itself a triple barrier, showing a much greater degree of symmetry than is achieved with two single barriers. (After Reed *et al.* 1986.)

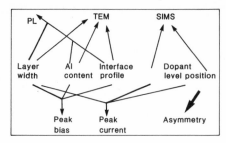

Structure-property relationships

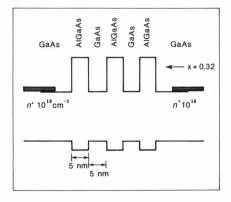

Design specification

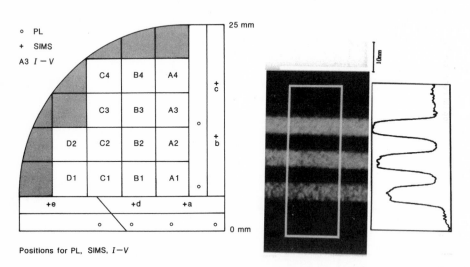

Positions for PL, SIMS, *I*−*V*

TEM values for layer widths and compositions and SIMS data for dopant asymmetry

Distance to centre (mm)	Layer thickness (nm)					Interface (nm)	Al (x) content	Dopant asymmetry
	Barrier 1	Well 1	Barrier 2	Well 2	Barrier 3			
2	6.00	5.86	5.89	5.80	5.93	0.87	0.34	12 nm
10	5.85	5.55	5.79	5.51	5.77	0.76	0.32	17 nm
21	5.01	4.97	5.12	5.08	5.20	0.91	0.31	13 nm

Fig. 8.19 The structure–property relationships of multilayer structures, exemplified by a collage of analytical data (including transmission electron microscopy and secondary ion mass spectroscopy) from a quarter wafer of a triple-barrier structure. The 2 inch wafer is shown to have a radial variation in layer thicknesses of about 10 per cent from

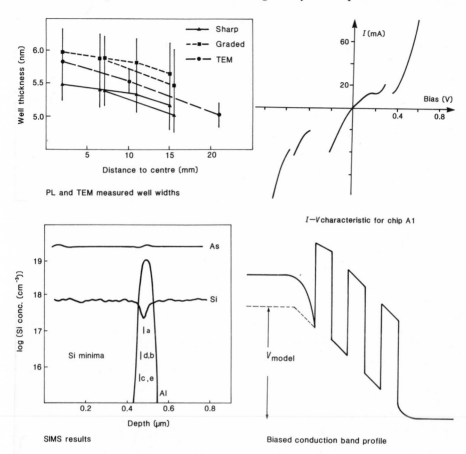

PL and TEM measured well widths

I—V characteristic for chip A1

SIMS results

Biased conduction band profile

Measured and simulated bias voltage for first $I-V$ peaks with inferred Si asymmetry

| Chip | Bias at first | | Simulated peaks | | Fitted |
	Substrate+ve	Substrate−ve	Substrate +ve	Substrate −ve	asymmetry
A1	0.28 (0.25)*	0.22 (0.14)*	0.25	0.14	3.8 nm
C1	0.37 (0.29)*	0.30 (0.18)*	0.26	0.16	3.0 nm
D1	0.45 (0.36)*	0.41 (0.29)*	0.30	0.16	

*Corrected for series resistance

the centre to edge, and the asymmetry of the current–voltage characteristics from a nominally symmetrical structure is related in part to a misalignment of the Al profile with a gap in the Si dopant profile. Simulations of the current–voltage data suggest that the full explanation may be even more complicated. The absence of such analytical detail can render uncertain many inferences on the physics from the transport or optical properties alone. (After Rimmer *et al.* 1991.)

there was a systematic radial variation in the thickness of the MBE-grown structure. Figure 8.19 shows a collage of the data obtained from transmission electron microscopy (TEM), photoluminescence (PL), secondary ion mass spectroscopy (SIMS), and the prediction of the position of the first peak in the current–voltage characteristic for forward and reverse bias compared with the actual data. The asymmetry in the current–voltage data is probably due to the imperfect alignment of the centres of the AlGaAs tunnelling barrier structure with the dip in the doping profile, although there is also a slight difference in the well thicknesses. The asymmetry in the experimental current–voltage data is used in the simulations to infer the offset of the centre of the doping profile, but this is in poor agreement with that obtained experimentally. To date, the simulation package has the effects of depletion included in an approximate manner, and may yet be improved to help produce a better agreement. In terms of detailed characterization, the techniques have all been pushed to the state of their art. There is clearly a need for further research across all fronts to improve the situation.

8.6 Device implications

In Chapters 16–19 we consider the new devices that have been made using semiconductor multilayers. The main contributions of tunnelling structure have been (i) the tailoring of the current–voltage characteristic to some desired end by using a combination of composition and dopant profiling, and (ii) the achievement of very high speed by using thin layers, so that an active region of less than 0.1 µm is involved. In Chapter 10, the optical properties of barrier–well structures, from which a whole new set of devices follow, are discussed further.

References

Ben Amor, S., Martin, K. P., Rascol, J. J. L., Higgins, R. J., Torabi, A., Harris, H. M., and Cummers, C. J. (1988). Transverse magnetic field dependence of the current–voltage characteristics of double-barrier quantum well tunnelling structure. *Applied Physics Letters*, **53**, 2540–2.

Brown, E. R., Soderstrom, J. R., Parker, R. D., Mahoney, L. J., Molvar, K. M., and McGill, T. C. (1991). Oscillations up to 712 GHz in InAs/AlSb resonant-tunnelling diodes. *Applied Physics Letters*, **58**, 2291–3.

Eaves, L., Stevens, K. W. H., and Sheard, F. W. (1986). Tunnel currents and presence of an applied magnetic field. In *The physics and fabrication of microstructures and microdevices* (eds. M. J. Kelly and C. Weisbuch), pp. 343–51. Springer-Verlag, Berlin.

Eaves, L., Leadbeater, M. L., Alves, E. S., Sheard, F. W., and Toombs, G. A. (1991). Resonant tunnelling effects in semiconductor heterostructures. In *Physics and technology of heterojunction devices.* (eds. D. V. Morgan and R. H. Williams), Chapter 2, pp. 33–52. Peter Peregrinus, London.

Hayden, R. K., Takamasu, T., Maude, D. K., Valadares, E. C., Eaves, L., Ekenberg, U., *et al.* (1992). High-magnetic-field studies of hole energy dispersion, cubic anisotropy and space-charge build-up in the quantum well of p-type resonant tunnelling devices. *Semiconductor Science and Technology*, **7**, B413–17.

Hodson, P. D., Robbins, D. J., Wallis, R. H., Davies, J. I., and Marshall, A. C. (1988). Resonant tunnelling in AlInAs/GaInAs double barrier diodes grown by MOCVD. *Electronics Letters*, **24**, 187–8.

Jaros, M. (1986). Electron states in semiconductor microstructures. In *The physics and fabrication of microstructures and microdevices* (eds. M. J. Kelly and C. Weisbuch), pp. 197–209. Springer-Verlag, Berlin.

Landau, L. D. and Lifshitz, E. M. (1977). *Quantum mechanics* (3rd edn), Pergamon, Oxford.

Landauer, R. and Martin, Th. (1994) Barrier interaction time in tunnelling. *Reviews of Modern Physics*, **66**, 217–28.

Luryi, S. (1985). Frequency limit of double-barrier resonant-tunnelling oscillators. *Applied Physics Letters*, **47**, 490–3.

Mendez, E. E. (1987). Physics of resonant tunnelling in semiconductors. In *Physics and applications of quantum wells and superlattices* (eds. E. E. Mendez and K. von Klitzing), NATO ASI Series B, Physics, Vol. 170, pp. 159–88. Plenum, New York.

Mendez, E. E., Calleja, E., and Cunningham, J. E. (1986a). Resonant tunnelling through indirect-gap semiconductor barriers. *Physical Review B*, **34**, 6026–9.

Mendez, E. E., Esaki, L., and Wang, W. I. (1986b). Resonant magnetotunnelling in GaAlAs/GaAs–GaAlAs heterostructures. *Physical Review B*, **33**, 2893–6.

Morkoc, H., Chen, J., Reddy, U. K., Henderson, T., and Luryi, S. (1986). Observation of a negative differential resistance due to tunnelling through a single barrier into a quantum well. *Applied Physics Letters*, **49**, 70–2.

Reed, M. A., Lee, J. W., and Tsai, H. L. (1986) Resonant tunnelling through a double GaAs/AlAs superlattice barrier, single quantum well heterostructure. *Applied Physics Letters*, **49**, 158–60.

Ricco, B. and Azbel, M. Y. (1984). Physics of resonant tunnelling. The one-dimensional double-barrier case. *Physical Review B*, **29**, 1970–81.

Rimmer, N., Syme, R. T., Frost, J. E. F., Jones, G. A. C., Kelly, M. J., and Stobbs, W. M. (1991). Wafer uniformity of an MBE-grown $Al_x Ga_{1-x}As/GaAs$ tunnelling structure. *Proceedings of the Semiconductor Materials Conference, Oxford, 1991*, Institute of Physics Conference Series, 117, no. 8, pp. 577–80. Institute of Physics, Bristol.

Skolnick, M. S., Simmonds, P. E., Hayes, D. G., White, C. R. H., Eaves, L., Higgs, A. W., *et al.* (1992). Electron transport in double-barrier resonant tunnelling structures studied by optical spectroscopy. *Semiconductor Science and Technology*, **7**, B401–8.

Syme, R. T., Kelly, M. J., Condie, A., and Dale, I. (1991). A resonant tunnelling microwave detector for power levelling and limiting applications. *Electronics Letters*, **27**, 2025–6.

Syme, R. T., Kelly, M. J., Robinson, M. F., Smith, R. S., and Dale, I. (1992). Novel GaAs/AlAs tunnel structures as microwave detectors. In *Quantum Well and Superlattices IV*, (eds. G. H. Dohler and E. S. Kotalas), SPIE 1675, pp. 46–56.

Weil, T. and Vinter, B. (1987). Equivalence between resonant tunnelling and sequential tunnelling in double-barrier diodes. *Applied Physics Letters*, **50**, 1281–3.

Wolford, D. J., Keuch, T. F., Bradley, J. A., Gell, M. A., Ninno, D., and Jaros, M. (1986). Pressure dependence of $GaAs/Al_x Ga_{1-x}As$ quantum-well bound states: the determination of valence-band offsets. *Journal of Vacuum Science and Technology B*, **4**, 1043–6.

9   Superlattices and minibands

9.1 Introduction

The repetition of a pair or more of multilayers (say repeats of m monolayers of GaAs and n monolayers of AlAs) introduces a new periodicity into a crystal structure. The new underlying lattice is called a superlattice; it retains the lattice constant of GaAs in the plane of the layers, but has a new lattice constant in the perpendicular direction set by $md_{GaAs} + nd_{AlAs}$, where the d_i are the interlayer spacings of the component materials in the direction of the crystal growth. There is an interaction between the discrete bound states found in 1D wells when they are placed adjacent to each other. A narrow band, called a miniband, is formed in the direction of the crystal growth. Its width in energy is determined by the thicknesses and heights of the AlGaAs or AlAs barriers, while its mean energy is determined by the thickness of the GaAs layers. The attraction of superlattices is that new crystals are being prepared which are not available in nature. We can engineer the electronic and optical properties of these crystals at the stage of preparing the growth specification for the multilayers. In this chapter we discuss the theory and experimental results relating to superlattices and minibands.

The original proposal for preparing superlattices contained an idea for a new and potentially very fast form of negative differential resistance based on electron transport in minibands with applied electric fields (Esaki and Tsu 1970). This phenomenon (the so-called Bloch oscillation described in Section 9.3) is yet to be observed directly in transport studies, although some subtle optical experiments with very modest fields, certainly too small to be of any direct device significance, have given evidence of such oscillations (Feldmann et al. 1992; Leo et al. 1992; Roskos et al. 1992). Indeed, transport in minibands has turned out to be more complex than originally thought. Nevertheless, the superlattice idea has been behind many of the materials, physics, and technology advances over the last two decades that are discussed throughout this text.

For the purposes of this and the next chapter, we distinguish between superlattices and multiple-quantum-well structures on the basis that in the former the barrier layers are sufficiently thin to allow strong wavefunction overlap between adjacent wells, while in the latter the barriers are sufficiently thick that the different wells are independent of each other.

9.2 Electronic structure of an ideal superlattice

There are many methods for introducing the electronic structure of a superlattice. We use two that are complementary in terms of the insights that they offer. As with the tunnel structures in the previous chapter, we use a 1D analysis in the z direction of the successive multilayers, but retain the effective mass energy bands in the $x-y$ plane of the individual layers. In this section we use much nomenclature of Weisbuch and Vinter (1991).

9.2.1 The tight-binding approach

Consider the coupled quantum-well structure in Fig. 9.1. It is shown as two identical but adjacent quantum wells with potential profiles V_1 and V_2. Anticipating the results of more detailed calculations in the next chapter, there are bound states in each well, and each state has a (real as opposed to complex) wavefunction of the qualitative form shown as ψ_1, ψ_2, which extends into the barrier layers since the barrier heights are finite. In semiconductors these ψ_1 are to be regarded as envelope wavefunctions in the sense of Chapter 1, Section 1.4 (see Chapter 10, Section 10.2.1, for details). We regard electrons in ψ_1, ψ_2 as being tightly bound in their parent well V_1, V_2 respectively. As the two wavefunctions are brought together, the exponentially small tail of the wavefunction in one well penetrates into the other well. This is precisely the problem encountered when the electron wavefunctions of two hydrogen atoms overlap, leading to the formation of a hydrogen molecule (Harrison 1980). To a first approximation, we ignore the modifications that this overlap causes to either wavefunction, which can be justified by the results of a more rigorous calculation. In practice, we use ψ_1, ψ_2 as the basis functions for a degenerate perturbation theory calculation. We assume that the eigenfunctions for the combined system are some linear combination of these, say $\Psi = \alpha\psi_1 + \beta\psi_2$, and we obtain their eigenenergies. If the Hamiltonian for one well, giving a bound state of energy E_0, is

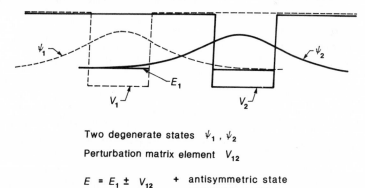

Two degenerate states ψ_1, ψ_2

Perturbation matrix element V_{12}

$$E = E_1 \pm V_{12} \quad \begin{array}{l} + \text{ antisymmetric state} \\ - \text{ symmetric state} \end{array}$$

Fig. 9.1 The overlap of electron states ψ_1, ψ_2 in two adjacent quantum wells V_1 and V_2 which interact (by analogy with the picture of the hydrogen molecule) to form bonding and antibonding energy levels as described in the text. This is a 1D model for the direction perpendicular to the wells. Electron states in the planes of the wells are unaffected to a first approximation, and remain parabolic bands of appropriate effective mass and band edge in each layer. (After Weisbuch and Vinter 1991.)

$$H_0 = -(\hbar^2/2m^*)\partial^2/\partial z^2 + V_1,$$

then we must add the V_2 of the other well to obtain the Hamiltonian H of the combined system. The Schrödinger equation can now be written as

$$H\Psi = E\Psi$$

and, if expanded in the basis of ψ_1, ψ_2, then multiplication of both sides by ψ_1^*, ψ_2^* and integration over all space leads to a matrix form

$$H\Psi = E\Psi \Rightarrow \begin{pmatrix} E_0 + \mathcal{V}_1 - E & \mathcal{V}_{12} \\ \mathcal{V}_{12} & E_0 + \mathcal{V}_1 - E \end{pmatrix} \begin{pmatrix} \alpha \\ \beta \end{pmatrix} = \begin{pmatrix} 0 \\ 0 \end{pmatrix}$$

where the potential terms are calculated as matrix elements:

$$\mathcal{V}_1 = \langle \psi_1 | V_2(z) | \psi_1 \rangle = \langle \psi_2 | V_1(z) | \psi_2 \rangle$$
$$\mathcal{V}_{12} = \langle \psi_1 | V_2(z) | \psi_2 \rangle = \langle \psi_2 | V_1(z) | \psi_1 \rangle.$$

Note that $\mathcal{V}_1$ is a measure of the ability of a single well to alter the energy of an electron state, whereas $\mathcal{V}_{12}$ is a measure of how one well can effect the energy of an electron state in the adjacent well. The secular equation, formed from the determinant of the matrix representation of $|H - E| = 0$, can be solved, resulting in energy levels $E = E_0 + \mathcal{V}_1 \pm |\mathcal{V}_{12}|$, i.e. split by an amount $2|\mathcal{V}_{12}|$. The lower eigenvalue is the symmetric combination and the upper the antisymmetric combination of the basis wavefunctions. We exploit this analysis further in Chapter 10.

The superlattice idea can be obtained by using the tight-binding model to extend the treatment just given to a repeated well structure. We consider N quantum wells in the topology of a ring (introducing periodic boundary conditions to eliminate the end wells and to simulate an infinite system (cf. Chapter 4, Section 4.2)). Part of this system is shown in Fig. 9.2. The equivalent $N \times N$ secular equation can be solved by the standard Bloch theorem technique of introducing a linear combination of envelope wavefunctions (Ashcroft and Mermin 1976). We use as a trial solution

$$\Psi_k(z) = (1/\sqrt{N}) \sum_n \exp(iknd)\, \psi_{\text{loc}}(z - nd),$$

where $\psi_{\text{loc}}(z - nd)$ is the (real) envelope function in the nth well centred at $z = nd$ and k is a Bloch wavevector. If we define overlap and self-energy integrals by direct analogy with the $\mathcal{V}_{12}, \mathcal{V}_1$ above, i.e. as

$$T = \int \psi_{\text{loc}}(z)\, V(z)\psi_{\text{loc}}(z - d)\, dz$$

$$S = \int \psi_{\text{loc}}(z - d)\, V(z)\, \psi_{\text{loc}}(z - d)\, dz$$

it can be shown by inspection that the $\Psi_k(z)$ are eigensolutions for $k = n\pi/Nd$, where n an integer, such that $-N/2 < n \leqslant N/2$, and that the energies are given by

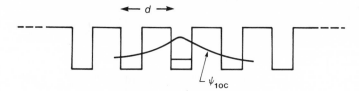

N wells

N-degenerate ground state

Tight-binding approximation

$\psi_k(z) = \frac{1}{\sqrt{N}} \sum_n e^{iknd} \psi_{\ell oc}(z - nd)$

$E = E_1 + S + 2T \cos(kd)$

$S = \int \psi_{1oc}(z-d) \, V(z) \, \psi_{\ell oc}(z-d) \, dz$

$T = \int \psi_{1oc}(z-d) \, V(z) \, \psi_{\ell oc}(z) \, dz$

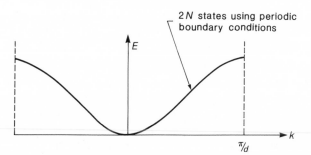

Fig. 9.2 The one-dimensional tight-binding model of a superlattice, where a narrow band is formed from the overlap of wavefunctions between adjacent wells in an infinite series of wells. The electron states in the other two directions are unaffected. (After Weisbuch and Vinter 1991.)

$$E(k) = E_0 + S + 2T\cos(kd).$$

For large N, these energies form a band of width $4|T|$ compared with the separation of $2|V_{12}|$ for the double well; the extra factor of two follows simply because each well communicates with two other wells in the superlattice.

In a real superlattice, the narrow bands in the z direction are known as minibands. They contain electron states that are spread over the entire superlattice in the z direction. In the x–y plane we still have the effective-mass motion of free electrons. It is a straightforward exercise to show that the density of states for this miniband can be written as

$$n(E) = N(m^*/\pi^2\hbar^2)\cos^{-1}[(E - E_0 - S)/2T] \quad \text{for } |E - E_0 - S| < 2|T|.$$

$$= Nm^*/2\hbar^2 \text{ for } E > E_0 + S + 2|T| \quad \text{(i.e. a 2D-like DOS).}$$

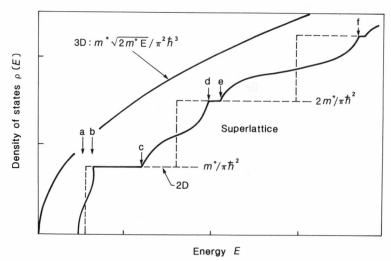

Fig. 9.3   The generation of the DOS of a superlattice by adding to the 2D step-like DOS from discrete sub-bands (cf. Fig. 4.2) the dispersion in the direction of the superlattice from Fig. 9.3. This latter shows up in the rounding of the 2D steps in the regions a–b, c–d, e–f, etc. (After Esaki 1983).

In practice, there may be many bound states, of individual energy E_n, in a quantum well, and each set contributes another miniband with energy and width determined by the S_n, T_n (which increase with n) and with its own DOS. The DOS of a superlattice is shown in Fig. 9.3, together with that of a uniform 3D solid and an isolated 2D system from one quantum well. We defer a further discussion of the widths and energies until the complementary approach to describing superlattices has been introduced.

9.2.2 The tunnelling approach

We return to generalize the calculation of tunnelling probabilities of the previous chapter. We consider a typical tunnel barrier in Fig. 9.4. We can write the solution outside the barrier quite generally in the form of a linear combination of forward and backward-going waves, and the only difference

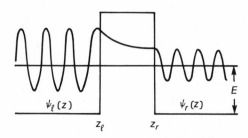

Fig. 9.4   A typical tunnel barrier in a superlattice showing the wavefunctions to the left and right of the barrier.

between the left and right sides of the barrier (under no bias) is the coefficients of the waves:

$$\psi_1 = a_1 \exp(ikz) + b_1 \exp(-ikz) \qquad \psi_r = a_r \exp(ikz) + b_r \exp(-ikz)$$

and $E = \hbar^2 k^2 / 2m^*$. From our previous experience, and more generally because the Schrödinger equation is linear and second order, its solution through the barrier can be used to establish two linear relations between the four coefficients above, expressed in terms of a transfer matrix S or a transmission matrix T written as

$$\begin{pmatrix} a_r \\ b_r \end{pmatrix} = \begin{pmatrix} s_{11} & s_{12} \\ s_{21} & s_{22} \end{pmatrix} \begin{pmatrix} a_1 \\ b_1 \end{pmatrix} \quad \text{or} \quad \begin{pmatrix} a_r \\ b_1 \end{pmatrix} = \begin{pmatrix} t \to & r_r \\ r_1 & t \leftarrow \end{pmatrix} \begin{pmatrix} a_1 \\ b_r \end{pmatrix}$$
$$\qquad (S) \qquad\qquad\qquad\qquad (T)$$

where $T = |t \to|^2$ is the probability of transmission of an electron incident from the left through the barrier, and $R = |r_1|^2$ is the probability of reflecting the same electron. Since, if ψ is a solution of the Schrödinger equation, then so is ψ^*, it is easy to show that $s_{11} = s_{22}$ and $s_{12} = s_{21}^*$, and four further relations can be established among terms in the S and T matrices:

$$t \to t \leftarrow^* = 1 - |r_1|^2, \quad |r_1|^2 = |r_r|^2, \quad t \to r_r^* = -r_1 t \leftarrow^*, \quad t \leftarrow r_1^* = -r_r t \to^*$$

so that we can write

$$S = \begin{pmatrix} 1/t \leftarrow^* & r_r / t \leftarrow \\ r_r^* / t \leftarrow^* & 1/t \leftarrow \end{pmatrix} \quad \text{with } |S| \equiv \det S = t \to / t \leftarrow = t \to^* / t \leftarrow^*$$

and finally conservation of probability current (with no applied bias) gives

$$|t \to|^2 + |r|^2 = 1 = |t \leftarrow|^2 + |r|^2 \quad \text{so that} \quad t \to = t \leftarrow$$

These are quite general relations and, before proceeding to the superlattice, we note that the single-tunnel barrier of Chapter 8, Section 8.1, can be recast in this formalism by calculating the S and T matrices explicitly to derive, for $E < V$,

$$T = |t|^2 = \frac{1}{1 + (1/4)(k/K + K/k)^2 \sinh^2(kW)}$$

and, if $t = |t| \exp(i\phi)$, then $\phi = \psi - kW$ and $\psi = \frac{1}{2}(k/K - K/k)\tanh(KW)$, with all other terms having their previously defined meaning.

If we translate the origin of the wavefunctions ψ_1, ψ_r through a distance D and recalculate the S matrix, we obtain the result

$$S_D = \begin{pmatrix} \exp(-ikD) & 0 \\ 0 & \exp(ikD) \end{pmatrix} \begin{pmatrix} S \end{pmatrix} \begin{pmatrix} \exp(ikD) & 0 \\ 0 & \exp(-ikD) \end{pmatrix}.$$

In addition, for two barriers in succession (barrier 2 to the right of barrier 1), the total transfer matrix is just

$$S_{\text{tot}} = S_2 S_1.$$

Note in this formalism how resonant tunnelling in a double barrier comes about. We consider one barrier extending from $-W/2$ to $W/2$ and the second barrier from $L - W/2$ to $L + W/2$ (so that the U in Fig. 8.7 of Chapter 8 has the value $L - W$). If we have calculated the r and t elements for one barrier, then the total transfer matrix for the double barrier system is $S_{tot} = S_L S$ where the elements of S_{tot} are

$$(s_{tot})_{11} = (1/t^*)^2 + \exp(-2ikL)|r/t|^2$$

$$(s_{tot})_{12} = (r/t)[\exp(-2ikL)/t + 1/t^*]$$

leading to

$$t_{tot} = 1/(s_{tot})_{22} = \frac{t^2}{1 + (t^2/|t|^2)|r|^2\exp(2ikL)}$$

$$= \frac{t^2}{1 + |r|^2\exp(2i(kL + \phi))}$$

and

$$T_{tot} = |t_{tot}|^2 = \frac{T^2}{|\{1 + |r|^2\exp[2i(kL + \phi)]\}|^2}.$$

$$= \frac{(1 - |r|^2)^2}{|\{1 + |r|^2\exp[2i(kL + \phi)]\}|^2}$$

The transmission is usually exponentially small, but when when the condition $2(kL + \phi) = (2n + 1)\pi$ is met, the transmission is identically unity. The condition can be rewritten as

$$\cotan[(k(L - W))] = \cotan(kU) = 1/2(k/K - K/k),$$

which is the condition for a bound state in a well of thickness U. We establish this result in Chapter 10, Section 10.2.1, and it has already been promised in Chapter 8, Section 8.3.1. Note that this form of the result is quite general, and is easily extended to more complex forms of barrier.

We continue this process of tunnelling through barriers and translating through wells many times in order to generate the conditions for a superlattice. We regard **S** as the transfer matrix for a barrier centred around $z = 0$. If a particular k state is to be allowed in a superlattice, the **S** matrix for all barriers should, up to a phase factor, be equivalent. The **S** matrix of the first barrier must be such that a wavefunction described by (a_r, b_r) seen from the second barrier has the same amplitude and phase relation as the wavefunction on the left of the first barrier (cf Fig. 9.5), i.e. we need

$$a_r\exp(ikL) = a_1\exp(i\phi) \quad \text{and} \quad b_r\exp(-ikL) = b_1\exp(i\phi)$$

and ϕ is a common phase. If this were not the case, and we repeated the process many times down the chain, the wavefunctions would grow indefinitely for z large in either the positive or negative direction, and we could not

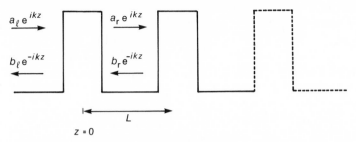

$a_\ell e^{ikz}$

$a_r e^{ikz}$

$b_\ell e^{-ikz}$

$b_r e^{-ikz}$

L

$z = 0$

Fig. 9.5 Transmission and reflection of waves in a superlattice as described in the text.

form normalizable wavefunctions for the electrons. By inserting these phase relations into S_{tot} for one period of the superlattice, we can regard this as an eigenvalue problem:

$$\begin{pmatrix} 1/t^* - \exp[i(\phi - kL)] & r/t \\ r^*/t^* & 1/t - \exp[i(\phi + kL)] \end{pmatrix} \begin{pmatrix} a_1 \\ b_1 \end{pmatrix} = \begin{pmatrix} 0 \\ 0 \end{pmatrix}$$

which has a solution if and only if

$$\exp(2i\phi) - \exp(i\phi)\{2\,\mathrm{Re}[t\exp(ikL)]/|t|^2\} + 1 = 0$$

or

$$\exp(i\phi) = \mathrm{Re}(t\exp(ikL))/|t|^2 \pm \sqrt{(\{\mathrm{Re}[t\exp ikL)]/|t|^2\}^2 - 1)}.$$

Real solutions for ϕ are only possible if

$$\mathrm{Re}[t\exp(ikL)] \leq |t|^2 \quad \text{or} \quad \cos(\phi + kT) \leq |t|^2.$$

In Fig. 9.6(a) we show how this condition can be met. As a function of energy, we plot $\pm|t|$, ϕ, and $\cos(\phi + kL)$ for a single barrier of thickness 5 nm ($c.0.5a^*$ for GaAs) as part of a superlattice of barriers and wells of equal thickness. The barrier height is 0.24 eV. The allowed solutions as a function of energy fall into bands shown as hatched intervals on the energy axis. The increasing width of minibands and decreasing width of minibandgaps cited in the previous section are clear in this diagram. In Fig. 9.6(b), we plot the width and position of the energy minibands for the same type of superlattice (i.e. of equal barrier and well widths), but as a function of the width of the barrier or well for a fixed barrier height of 0.4 eV. Also shown are the energies for discrete levels in a single quantum well (see Chapter 10). The increasing widths of the minibands of increasing energy are apparent, and the explanation comes from the previous section in terms of increasing overlap of wavefunctions from adjacent wells. Note also that the energy levels above the barriers continue to show the miniband structure, although eventually the different minibands overlap in energy and the minibandgaps disappear.

This completes the introduction to superlattice electronic structure. There are many other techniques for introducing the electron states, but the two presented contain the most relevant information. The descriptions just given,

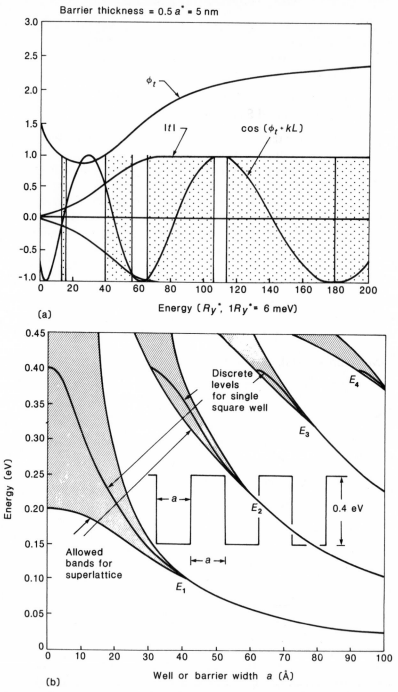

Fig. 9.6 (a) The construction of minibands for a superlattice (after Weisbuch and Vinter 1991); (b) the miniband energies and widths as a function of well/barrier thickness in a superlattice where the well and barrier thicknesses are kept equal (after Esaki 1983).

when translated into the GaAs–AlGaAs materials system, provide a range of new materials with electronic properties that can be tailored within the limits imposed by control over the materials growth. Square barriers are not essential, but they lend themselves to the simplest analysis.

9.3 Bloch oscillations and Stark ladders

The analytic results for the band structure of superlattices are insufficient, of themselves, to warrant the attention paid to them. We now show some of the interesting transport phenomena that follow from a cosine band structure of the form generated by the tight-binding approach. We assume that the parameter T is negative in the dispersion relation, so that the lowest-energy band is of the form shown in Fig. 9.7. We concern ourselves with this single sub-band in the first instance, and consider scattering phenomenologically at first. We later expand our investigations on these two points.

The semiclassical model for electron transport in an energy band $E(k)$ is contained in the two equations that relate the group velocity of a wavepacket of electrons with a well-defined k-vector, and the relationship between this k-vector and the applied electric field F:

$$v(k) = (1/\hbar)\nabla_k E(k) \quad \text{and} \quad \hbar \, dk/dt = -eF.$$

In a steady applied field, and in the absence of scattering, we have

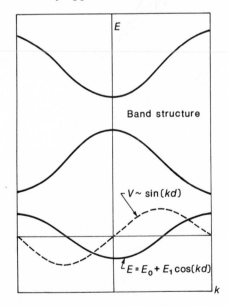

Band structure

$V \sim \sin(kd)$

$E = E_0 + E_1 \cos(kd)$

Bloch oscillations

$E = E_0 + E_1 \cos(kd)$

$v = (1/\hbar)(\partial E/\partial k) = (dE_1/\hbar)\sin(kd)$

$k(t) = k(0) - eFt/\hbar$

$v = (dE_1/\hbar)\sin\left[(eFdt/\hbar)\right]$

Fig. 9.7 A 1D analysis of the superlattice band structure and the semiclassical dynamics of an electron under a constant applied field. In the absence of any scattering, an electron velocity (and hence its trajectory in real space) is sinusoidal, giving rise to the so-called Bloch oscillations. (After Weisbuch and Vinter 1991.)

$$k(t) = k(0) - eFt/\hbar$$

Electrons in a miniband execute motion that is equivalent to their movement along the band structure at a constant rate. With the 1D analysis in the direction of growth, the $E(k)$ has a cosinusoidal dispersion relationship. Thus, for a constant field in one direction, the electron eventually reaches the upper half of the dispersion curve, where the velocity is in the opposite direction to k. This is equivalent to these electrons having a negative effective mass (cf. $\hbar^2/m^* = \partial^2 E/\partial k^2 \sim -T\cos(kd) \sim -E$). Eventually the electrons slow down as they reach the zone boundary, and there they reverse their direction as they jump to the opposite side of the dispersion curve (or continue on the same curve in the extended zone scheme); this is called Bragg reflection. The electron velocity under a constant applied field, and in the absence of scattering, is oscillatory. The period is given by the time taken for the electron to move from $-k_{ZB}$ to k_{ZB}, namely $\mathcal{T} = (2\pi/d)(eF/\hbar)^{-1}$, where d is the superlattice spatial periodicity. Note that this result is independent of the width of the miniband $4|T|$. Integrating the velocity with respect to time shows that the electron excursion is also oscillatory with time. This is the Bloch oscillation of electrons in a periodic potential.

This oscillation has never been seen in pure bulk solids: $\mathcal{T}$ takes a value of about 10^{-11} s for $F = 10\,\text{kV cm}^{-1}$ and $d = 0.35$ nm. However, this is a long time compared with transport scattering times. If d could be increased by a factor of 10–100, the chances of seeing this effect would be greater. This prospect gave the initial impetus to the studies of superlattices (Esaki and Tsu 1970). Before we give a critique of this simplest model, we consider the effect of collisions, using the standard approach to calculating the drift velocity of electrons in an infinite superlattice. A phenomenological scattering time τ is introduced which implies that, because of collisions, any given electron loses memory of its earlier dynamics on the scale of time τ. An electron with velocity v_z has an incremental velocity dv_z in a time interval dt given by

$$dv_z = (eF/\hbar^2)(\partial^2 E/\partial k_z^2)dt,$$

and so the average drift velocity imposed by collisions is

$$v_d = \int \exp(-t/\tau)dv_z = \int (eF/\hbar^2)(\partial^2 E/\partial k_z^2)\exp(-t/\tau)\,dt.$$

Using the sinusoidal form of the energy dispersion ($\sim 2T\cos(kd)$) allows us to solve this integral in the semiclassical model, with the result that

$$v_d = \frac{(\pi\hbar/m_{SL}d)\xi}{(1 + \pi^2\zeta^2)}$$

where $\zeta = eF\tau d/\pi\hbar$ and $1/m_{SL} = (1/\hbar^2)(\partial^2 E/\partial k_z^2) = (1/\hbar^2)(-2Td^2) > 0$. Note that v_d has a maximum value for $\pi\xi = 1$ and exhibits a negative differential mobility for higher fields. The condition for achieving this form of negative differential resistance (NDR), which does depend on the miniband width through $m_{SL} \sim -T$, is 2π times easier to fulfil than that required to achieve Bloch oscillations ($\tau > \mathcal{T}/2\pi$ rather than $\tau > \mathcal{T}$) even though they both reflect the same phenomenon, namely access to the regions of NDR in the upper half of the miniband.

The search for these phenomena has largely been unsuccessful (see Section 9.4), and the experimental reasons have been very instructive. However, the simple theory has been under attack for making too sweeping assumptions about the nature of miniband electron states in the presence of an electric field. The miniband widths of typical superlattices are a few tens of meV, while modest fields in practical devices are about 10^6 V m^{-1}. An electron starting out at the bottom of a miniband will reach the top of the miniband in a distance of a few tens of nanometres, i.e. only a few periods of the superlattice itself. Far from having transport in extended miniband states, the field strongly localizes the electron. There comes a point when the barrier presented by the structure for tunnelling into a higher miniband becomes sufficiently small that the probability for tunnelling becomes appreciable (Fig. 9.8(a)). This mechanism is known as Zener tunnelling or electric breakdown, and its absence in solids with unit cells of dimensions 1 nm or less is an important precondition for the validity of semiclassical models of transport. In fact, to the extent that the same electron states in adjacent wells differ in energy by $\Delta E = eFd$ in the presence of a field, the degree of overlap of the wavefunctions decreases and the concept of a miniband ceases to have any meaning. Indeed, as shown in Fig. 9.8(b), we should consider transport by some mechanism, i.e. one that includes some inelastic process that loses ΔE of energy, as an electron hops down a chain of localized states which is known as a Stark ladder. We shall show optical evidence for Stark ladders in Section

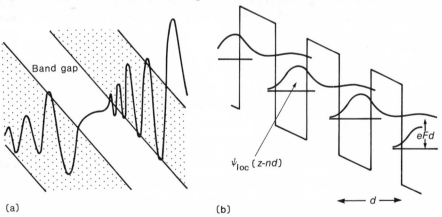

(a) (b)

Fig. 9.8 (a) Zener tunnelling in a superlattice as an electron in one band tunnels into the next-highest band; (b) the formation of Stark ladders in a superlattice under an applied electric field as the miniband concept breaks down once the electron states in adjacent wells are separated in energy by an amount that exceeds the miniband width at zero field.

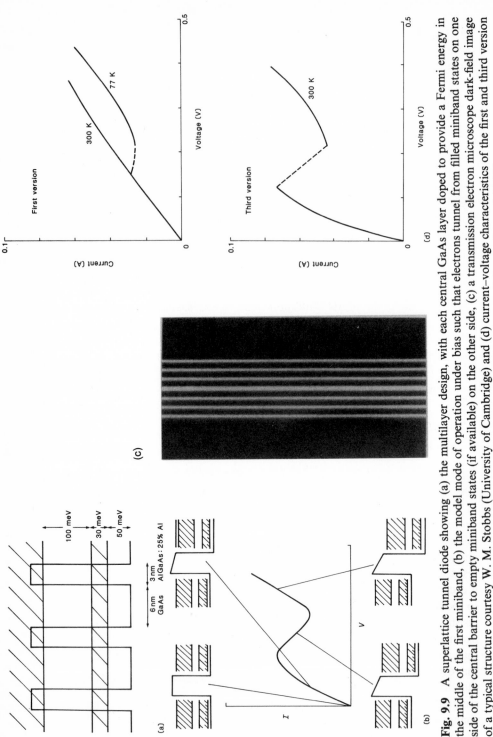

Fig. 9.9 A superlattice tunnel diode showing (a) the multilayer design, with each central GaAs layer doped to provide a Fermi energy in the middle of the first miniband, (b) the model mode of operation under bias such that electrons tunnel from filled miniband states on one side of the central barrier to empty miniband states (if available) on the other side, (c) a transmission electron microscope dark-field image of a typical structure courtesy W. M. Stobbs (University of Cambridge) and (d) current–voltage characteristics of the first and third version of this structure as the design principles became clearer. (After Kelly *et al.* 1987.)

9.5. Thus, while the miniband concept has been the spur for the experiments to be described, the theoretical basis for expecting to observe novel electronic transport is limited.

9.4 Measurements of transport in superlattices

9.4.1 Non-ideal superlattices

Most of the simple transport experiments on superlattices have been performed on non-ideal samples, either designed so deliberately, or becoming so because of imperfections in the growth. The first results indeed showed multiple negative differential resistance features, but the original successful interpretation of these data relied on imperfections in the superlattice, for example an unintentionally thick barrier which has been the focus of more recent studies. In Fig. 9.9 we show the design of a superlattice structure with an intentionally thick central barrier. Note that there are only three periods on each side. Theoretical calculations of tunnelling through multiple-barrier structures show that three periods are sufficient to give an adequate difference between transmission and reflection for the miniband concept to hold (Fig. 9.10). As described in the previous sections, superlattices are metals if they are doped: there is no true gap in their density of states once the energy levels in the direction perpendicular to the superlattice axis is taken into account (Fig. 9.3). The central third of each GaAs layer in Fig. 9.9(a) was (lightly) doped to a level that put the Fermi energy in the middle of the lowest miniband; the layers outside the superlattice structure were heavily doped.

The behaviour of this structure under bias is shown in Fig. 9.9(b), with most of the applied bias, to a first approximation, being dropped over the thicker undoped barrier in the centre. In fact, the structure is called a superlattice tunnel diode, as it acts just like a single-barrier tunnel diode (cf. Chapter 8, Section 8.2), but the superlattice segments act like filters of the kinetic energy in the direction normal to the tunnel barrier. In the absence of any scattering processes, elastic tunnelling can occur only when the filled part of the lowest miniband on the cathode side lines up with an empty part of a miniband on the anode side. Otherwise no current flows, and regions of negative differential resistance should be obtained. To the extent that this model of the voltage profile is accurate, such superlattice tunnel diodes can be, and have been, used successfully to perform energy spectroscopy on the minibands. The effects of the series resistance of a comparable structure without the superlattice were obtained, and against these results, the onset of negative differential resistance should occur when the extra bias across the superlattice equals the miniband width, and the width of the region before the 'normal' current resumes equals the width of the gap between the first and second minibands. Data from several structure corroborate this picture. The transport data taken with a transverse magnetic field show all the changes that confirm the above interpretation of the superlattice tunnel diode (see Fig. 9.11 and compare with Fig. 8.12 in Chapter 8).

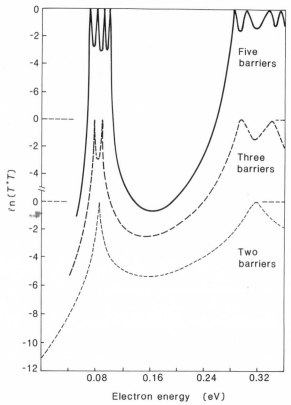

Fig. 9.10 Calculations of the transmission through two-, three- and five-barrier structures showing the evolution of minibands of unit transmission separated by energy ranges of almost unit reflection. (After Esaki © 1986 IEEE.)

9.4.2 Ideal superlattices

In Fig. 9.12 we reproduce the original conductance data on a 50-period superlattice, showing multiple negative differential conductance regions, and a picture of the first and second minibands behaving just as in Fig. 9.9(b). Here there was no intentional variation in the barrier thicknesses, but lessons from the superlattice tunnel diode show that an increase in thickness of less than 20 per cent is quite sufficient to nucleate a high-field domain over a single barrier, as opposed to letting the same field penetrate the full superlattice. While a theoretical analysis of this has not been pursued, the detailed electrostatics suggests that this model for transport in a less than ideal superlattice is valid; even quite modest fluctuations in layer profile might suffice. To date, no direct transport measurements on uniform superlattice structures have given direct evidence for conduction in minibands, beyond that inferred from structures as in Figs 9.9 and 9.12.

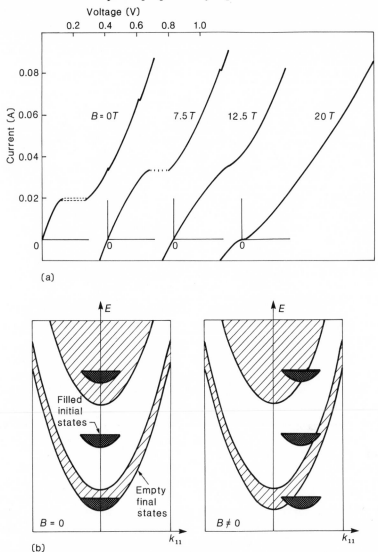

Fig. 9.11 (a) Current–voltage data for a superlattice tunnel diode as a function of transverse magnetic field; (b) a model of the filled initial states at different biases, at zero and finite magnetic field, showing a reduced overlap in the latter case leading to a higher onset voltage for negative differential resistance, and eventually its disappearance (Davies *et al.* 1987). This is a precursor of the work on resonant tunnelling (Chapter 8, Fig. 8.12, and 8.14).

9.5 Optical properties of superlattices

Optical absorption and photoluminescence probe particular combinations of filled initial and empty final states, and so they are ideally suited to investiga-

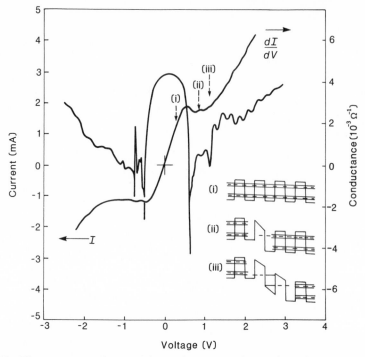

Fig. 9.12 The current–voltage and conductance–voltage data from a nominally uniform superlattice showing negative differential conductivity, along with an explanation in terms of high-field regions set up once the applied bias ceases to be dropped uniformly but, rather, concentrated at individual barriers. (After Esaki and Chang 1974.)

tions of the electronic structure of superlattices. Furthermore, some of the advantages of optical techniques described in Chapter 7, Section 7.4, are applicable here (e.g. spectroscopic and time-resolved studies). We concentrate here on typical studies of the optical properties of miniband transport and of recent claims to see Bloch oscillations optically. A more detailed and wider description of optical properties of quantum wells is contained in the next chapter. All experiments were carried out with the samples held at low temperature and subject to weak electric fields.

In Fig. 9.13(a) we show two types of superlattice that have been used in optical studies. The first is a regular superlattice, with the lowest-energy electron and hole minibands shown schematically. In the experiments below the ratio of barrier to well thicknesses remains unity, and the layers are 3 nm, 4 nm, and 7 nm thick. An expanded well (EW) is incorporated every 140 nm. Calculations (Fig. 9.13(b)) show that there are localized electron and hole states associated with the wide well, and the electron–hole energy difference is lower than that of the superlattice minibands.

In Fig. 9.14, we show the steady state luminescence spectra associated with three superlattices. The structures are excited with light sufficient to create

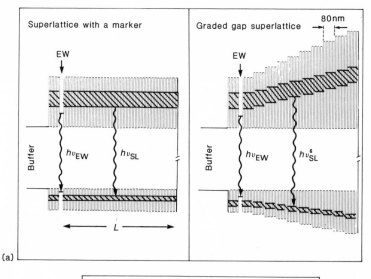

Superlattice with a marker

Graded gap superlattice

80 nm

EW

EW

Buffer

$h\nu_{EW}$ $h\nu_{SL}$

Buffer

$h\nu_{EW}$ $h\nu_{SL}^6$

$\longleftarrow L \longrightarrow$

(a)

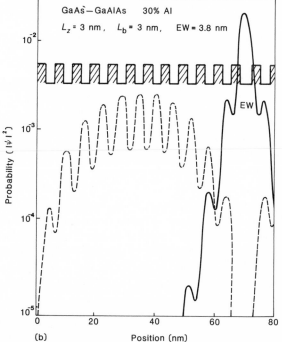

GaAs – GaAlAs   30% Al

$L_z = 3$ nm, $L_b = 3$ nm,   EW = 3.8 nm

10^{-2}

EW

Probability ($|\psi|^2$)

10^{-3}

10^{-4}

10^{-5}

0   20   40   60   80

(b)   Position (nm)

Fig. 9.13 (a) A uniform superlattice and a graded-gap superlattice as used for optical studies of miniband transport, each with wider or expanded well (EW) used to collect carriers which recombine radiatively and so act as a monitor of the carrier transport. The graded-gap superlattice consists of several periods of a uniform superlattice followed by another uniform superlattice, but with a 2 per cent change in Al concentration in the barriers. (b) Calculations of wavefunction intensity in a uniform superlattice with a wide well showing the localized state at the wide well used to collect carriers. (After Deveaud *et al.* © 1988 IEEE.)

electron–hole pairs in the superlattice. In each case there is a signal from the recombination of carriers in the EW, reflecting the ability of such wells to trap electrons and holes from the neighbourhood of the wells. The trend to note is the rapid decrease in the superlattice emission as the barriers are reduced from 7 nm to 3 nm. This is interpreted in terms of miniband transport, and is corroborated by further results to be discussed below. If there is miniband transport, as in narrow-barrier superlattices with wide minibands, carriers will

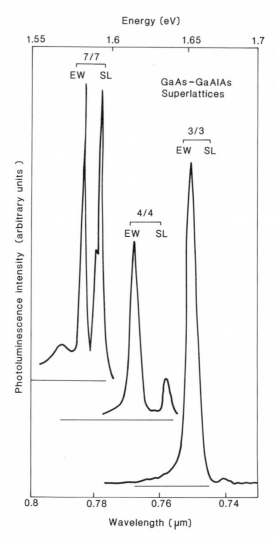

Fig. 9.14 Luminescence spectra from various superlattices (SL) containing an expanded well (EW) for optical calibration of the transport. The 70/70 refers to 70Å barriers and 70Å wells. The absence of SL luminescence in the 30/30 SL indicates efficient carrier transport to the extended well. (After Deveaud *et al.* © 1988 IEEE.)

find it easy to migrate to the EW and recombine there. If there is no miniband transport, as in wide-barrier superlattices, the photoexcited carriers are more likely to recombine while still in the superlattice. This is just what happens, with the 4 nm/4 nm structure being at the borderline between the presence and absence of miniband transport. In the absence of applied electric fields, we are relying on the diffusive motion of the carriers. A detailed analysis suggests that electron motion is dominant (over hole motion) in these structures with no applied field.

The second type of structure in Fig. 9.13(a) is a stepwise graded-gap superlattice: every 80 nm, the Al concentration changes by 2 per cent, so that the structure ranges from 35 per cent Al to 15 per cent Al in the barriers over approximately 1 μm. In such structures, the diffusion of photoexcited carriers is assisted by a kind of quasi-electric field as carriers approach the EW in the region of lower Al concentration. Both electrons and holes move towards the EW. In Fig. 9.15, the continuous-wave luminescence from the EW (note the log scale) compared with that from the various steps in the superlattice shows that photoexcited carriers are moving quickly through the superlattice. Note that magnetic fields have been applied parallel and perpendicular to the current, and the changes in luminescence are precisely those anticipated by the earlier discussion of magnetotransport in superlattices.

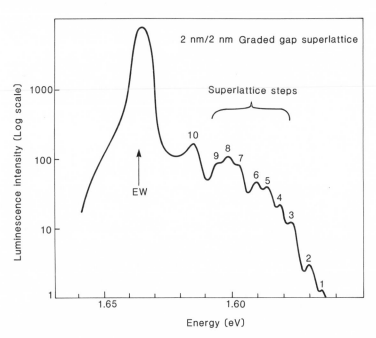

Fig. 9.15 Luminescence spectra from a graded-gap superlattice where the strong EW peak is a sign of strong miniband transport, and the weaker structure appears from luminescence from different steps in the graded-gap superlattice. (After Deveaud *et al.* © 1988 IEEE.)

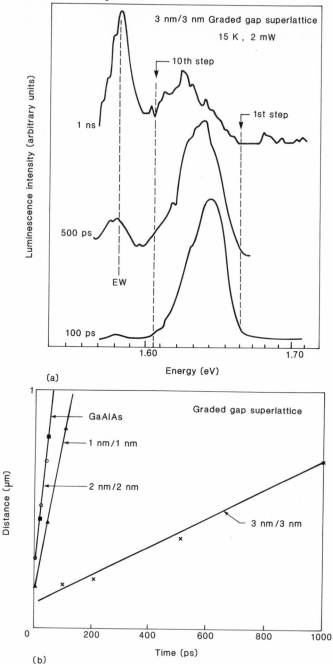

3 nm/3 nm Graded gap superlattice

Fig. 9.16 (a) Evolution with time of the luminescence from a graded-gap superlattice as the carriers congregate in the EW on the scale of 1 ns. (b) An interpretation of a collection of such data in terms of the drift of the peak position of the carrier wavepacket in different superlattices. (After Deveaud *et al.* © 1988 IEEE.)

If the luminescence is monitored in the time domain after a pulsed excitation, the evolution of the luminescence spectra (cf. Fig. 9.16(a)) can be used to track the drift–diffusion of the photoexcited electrons to the EW, and data on several superlattice structures of this type have been analysed to produce the results given in Fig. 9.16(b)). The fact that the diffusion is linear in time (rather than the more usual $\sqrt{t}$ dependence) suggests that the graded-gap steps are acting like a quasi-electric field. In this case, the data in Fig. 9.16(b) can be interpreted in terms of a low temperature hole mobility of 0.1 m^2 V^{-1} s^{-1}, about 10 times lower than bulk GaAs. The lower values reflect the imperfections introduced with each interface.

We conclude this section with a brief discussion of a recent optical detection of Bloch oscillations (Leo *et al.* 1992). The detailed and complex optical techniques do not need elaboration here, but the principle of the experiment is simple. Light in a 100 fs pulse is incident on a transparent sample from two directions. The response of carriers in the sample to the electric fields of the pulse sets up a transient diffraction grating, while a delayed second pulse from one direction (k_2 in Fig. 9.17(a)) can then be diffracted from the grating, and the diffracted signal k_3 can be detected. The intensity of this signal should fall off as a function of time as the transient grating disappears. The sample in this case is a 40-period superlattice (10 nmGaAs/1.7 nmAl$_{0.3}$Ga$_{0.7}$As) grown on an n-type substrate with a transparent Schottky gate on top to apply the bias. The substrate is removed near where the experiments are performed to allow the transmitted signal to be detected. The data in Fig. 9.17(b) show the decay of intensity of the diffracted beam with time for different applied fields across the superlattice. The oscillatory intensity modulation that is superposed on top of the decaying signal has the Bloch frequency ($1/\mathcal{T}$ from Section 9.3, using the parameters of the superlattice sample). The argument is that carriers in the superlattice are responding to the applied field by Bloch motion in the miniband, and these are modifying the relaxation of the transient grating in some manner. Furthermore, this signal is detected over the range $F = 2.5$–10 kV cm^{-1}, representing a 400 per cent tuning range of frequency $\omega \propto F$. Note that at most four full cycles are detected in this experiment at the highest fields. More recently, weak coherent radiation in the 0.3–5 THz range has been detected (Waschke *et al.* 1994). On taking the experiment from 4 K to 77 K, the intensity of this emitted radiation falls by a factor of 5, reflecting the effect of electron scattering in dephasing the electron motion; the extrapolation of this result to room temperature is not at all clear. Bloch oscillations have proved to be a very delicate phenomenon.

9.6 Applications of superlattices

The principal application for superlattices to date is not related directly to the topics discussed in this chapter so far. Many studies of epitaxial growth have noted that multilayers tend to smooth out any short-length scale irregularities

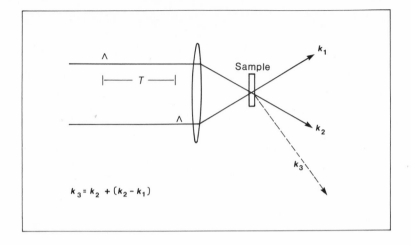

(a)

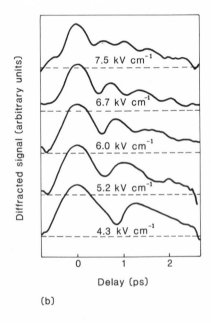

(b)

Fig. 9.17 (a) The principle of one form of four-wave mixing, where two light pulses with wave vectors k_1 and k_2 are incident simultaneously on a sample where their interference and absorption establish a transient grating, and a third beam (in this case another k_1 pulse, delayed by a time T) is diffracted from this grating. (b) A semilogarithmic plot of the diffracted signal against delay for various applied electric fields gives a measure of the time scale over which the grating disappears, and hence the carrier processes responsible for its disappearance. The structure in the decaying signal is evidence of Bloch oscillations. (After Leo *et al.* 1992.)

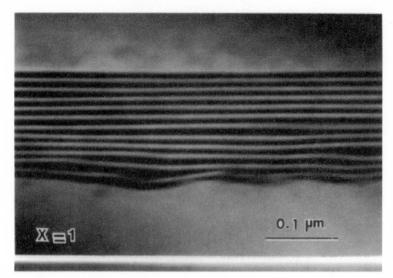

Fig. 9.18 The smoothing action of a superlattice buffer layer. (Courtesy of P. Petroff.)

on the surface of a substrate. An example is given in Fig. 9.18, where a few quantum wells smooth out features of $c.10$ nm height and $c.0.1\mu$m lateral extent. For many electronic and optical devices, it is essential to obtain wafer-scale uniformity, and a superlattice buffer layer is often the first feature to be grown below any active device volumes.

We regard the possible applications of multiple-barrier tunnel structures to devices via the design of current–voltage characteristics as having been covered in the last chapter. We anticipate the application in the next chapter of multiple quantum-well structures in optical devices. With these exclusions, specific applications of superlattices to date have been few. In most cases, the interest has been the basic materials science of superlattices as solids fully engineered at the atomic≡quantum scale. We have not had the opportunity here to discuss the results of research on short-period superlattices (where each component layer is only a few atomic monolayers thick), but these structures have been analysed by Lu and Sham (1989). The superlattice period may be as short as $c.2$ nm, and the interwell coupling is very strong. Conventional techniques for bulk band-structure calculations have been applied to them to confirm trends in the bandgaps, effective masses, and the nature of the upper conduction bands (the X and L valleys). In terms of new transport and optical phenomena, the results are disappointing; many properties can be predicted with reasonably accuracy on the basis of bulk solid concepts. Indeed, the differences between an $Al_{0.5}Ga_{0.5}As$ alloy and a short-period superlattice consisting of two monolayers of each of GaAs and AlAs seem rather modest. There is some interest in just where the semiconductor becomes indirect, but the various energy levels (Γ, X, L) are all close in energy in both systems.

The main device applications are likely to be in the future of high-speed millimetrewave and far-infrared applications, as discussed in Chapters 17 and 19. The 0.5–5 THz range is not well served by electronic or optical devices, but as the quality of superlattices improves, prototype devices operating in that range way well exploit some of the physics described in this chapter.

References

Ashcroft, N. W. and Mermin, N. D. (1976). *Solid State Physics.* Holt, Rinehart, Winston, New York.

Davies, R. A., Newson, D. J., Powell, T. G., Kelly, M. J., and Myron, H. W. (1987). Magnetotransport in a semiconductor superlattice. *Semiconductor Science and Technology*, **2**, 61–4.

Deveaud, B., Shah, J., Damen, T. C., Lambert, B., Chomette, A., and Regreny, A. (1988). Optical studies of perpendicular transport in superlattices. *IEEE Journal of Quantum Electronics*, **QE-24**, 1641–51.

Esaki, L. (1983). A perspective in superlattice development. In *Recent Topics in Semiconductor Physics* (eds. H. Kamimura and Y. Toyozawa), pp. 1–71. World Scientific, Singapore.

Esaki, L. (1986). A bird's-eye view on the evolution of semiconductor superlattices and quantum wells. *IEEE Journal of Quantum Electronics*, **QE-22**, 1611–24.

Esaki, L. and Chang, L. L. (1974). New transport phenomenon in a semiconductor 'superlattice'. *Physical Review Letters*, **33**, 495–8.

Esaki, L. and Tsu, R. (1970). Superlattice and negative differential conductivity in semiconductors. *IBM Journal of Research and Development*, **14**, 61–5.

Feldmann, J., Leo, K., Shah, J., Miller, D. A. B., Cunningham, J. E., Meier, T., *et al.* (1992). Optical investigation of Bloch oscillations in a semiconductor superlattice. *Physical Review B*, **46**, 7252–5.

Harrison, W. A. (1980). *Electronic structure and the properties of solids.* Freeman, San Francisco, CA.

Kelly, M. J., Davies, R. A., Couch, N. R., Movaghar, B., and Kerr, T. M. (1987). Novel tunnelling structures: physics and device implications. In *Physics and Applications of Quantum Wells and Superlattices* (eds. E. E. Mendez and K. von Klitzing), NATO ASI Series B: Physics, Vol. 170, pp. 403–21. Plenum, New York.

Leo, K., Haring Bolivar, H., Bruggemann, F., Schwedler, R., and Kohler, K. (1992). Observation of Bloch oscillations in a semiconductor superlattice. *Solid State Communications*, **84**, 943–6.

Lu, Y-T, and Sham, L. J. (1989). Valley-mixing effects in short-period superlattices. *Physical Review B*, **40**, 5567–78.

Roskos, H. G., Nuss, M. C., Shah, J., Leo, K., Miller, D. A. B., Fox, A. M., *et al.* (1992). Coherent submillimetre wave emission from charge oscillation in a double well potential. *Physical Review Letters*, **68**, 2216–19.

Waschke, C., Leisching, P., Haring Bolivar, P., Schwedler, R., Bruggemann, F., Roskos, H. G., *et al.* (1994). Detection of Bloch oscillations in a semiconductor superlattice by time-resolved terahertz spectroscopy and degenerate four-wave mixing. *Solid State Electronics*, **37**, 1321–6.

Weisbuch, C. and Vinter, B. (1991). *Quantum semiconductor structures.* Academic Press, New York.

10 Quantum wells and their optical properties

10.1 Introduction

It is arguable that the most widespread and successful applications of semiconductor multilayers to date exploit the optical properties of quantum wells, i.e. thin layers of a narrow-gap semiconductor sandwiched between layers of wider-gap material. This chapter contains a discussion of much of the relevant physics, with the device applications being described in more detail in Chapters 18 and 19. This chapter also completes the trio of chapters on novel physics in multilayers.

In Chapter 1, Section 1.5.2, we described the basic optical properties of bulk semiconductors in terms of the absorption of a photon to excite an electron from the valence band into an empty level in the conduction band. In Chapter 4, Section 4.4, this was taken further in the context of low-dimensional systems. In addition, the excitonic effect, originating in the residual Coulomb attraction between electron and hole was described, and both the linear and non-linear optical properties were outlined. In this chapter we describe the electron states and optical properties of a quantum well in the GaAs–AlGaAs materials system in greater detail. We discuss the effects (both linear and non-linear) of electric and magnetic fields. There are novel optoelectronic effects associated with coupled quantum wells. A quantum-well structure is able to capture carriers from an electric current, and we discuss the physics of this process. We shall discover that tailoring optical properties in quantum wells, and tuning these properties via electric fields or strong optical fields, is possible to a much greater extent than is the tailoring of electrical properties. The possibility of operating an optoelectronic device at a single frequency allows us to exploit the tunability to good effect, to be seen in Chapter 18.

10.2 The single quantum well: electronic and optical properties

10.2.1 Simple theory for electron and hole states

Consider the 1D potential profile of a quantum well structure in Fig. 10.1(a). It has infinitely high walls, and the energy levels and wavefunctions come from elementary quantum mechanics (cf. Chapter 4, Section 4.2, and Landau and Lifshitz (1977)). The more realistic profile for modelling the conduction band profile of a layer of (say) GaAs between layers of $Al_xGa_{1-x}As$ is a potential well of finite depth, as shown in Fig. 10.1(b). It is an inversion of the single-tunnel barrier of Fig. 8.1 in Chapter 8. At first, we treat only the

Particles in a box

1D case $V_0 = \infty$

$$-(\hbar^2/2m)(d^2\psi/dz^2) = E\psi$$

$$E_n = (\hbar^2/2m)(n\pi/L_z)^2 \quad n = 1, 2, 3, \dots.$$

$$\psi_n = A\sin(n\pi z/L_z)$$

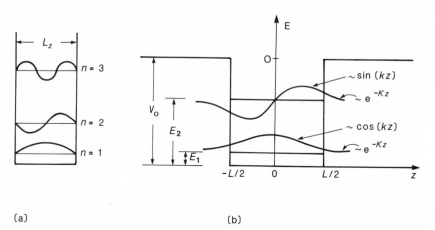

(a)   (b)

Fig. 10.1 (a) Infinite and (b) finite potential well in one dimension showing the low-lying electron states and their wavefunctions. (After Weisbuch and Vinter 1991.)

envelope function (cf. Chapter 4, Section 4.2, again), but we consider the details of the full Bloch function in the next section. We solve the one-electron Schrödinger equation by setting up trial solutions in the three regions, and match the wavefunctions and their spatial derivative at the boundaries ($z = \pm L/2$). Because the well is symmetrical about $z = 0$, the solutions will have either even or odd parity, and so we can be guided in our trial wavefunctions by the results in Fig. 10.1(a).

The trial wavefunctions for $V_0 < E < 0$, as shown, are

I $B\exp(K(z + L/2)]$ $z < -L/2$ where $k^2 = -(2m^*(V_0 - E)/\hbar^2)$
II $A\cos(kz)$ or $A\sin(kz)$ $|z| < L/2$ and $K^2 = -(2m^*E/\hbar^2)$
III $B\exp[-K(z - L/2)]$ $z > L/2$

and the matching conditions at $z = \pm L/2$ give rise, for the sin and cos form respectively, to implicit equations

$$k\tan(kL/2) = K \quad \text{or} \quad k\cot(kL/2) = -K$$

which can be recast, with $(k_0)^2 = 2m^*|V_0|/\hbar^2$, as

$$\cos(kL/2) = k/k_0 \quad \text{for } \tan(kL/2) > 0$$

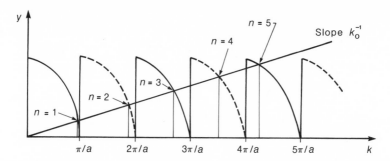

Fig. 10.2 Graphical solution for k values for a finite 1D quantum well as described in the text. (After Weisbuch and Vinter 1991.)

$$\sin(kL/2) = k/k_0 \qquad \text{for } \tan(kL/2) < 0.$$

These expressions are shown graphically in Fig. 10.2. There is always at least one bound state, and the number of bound states quite generally is

$$1 + \text{Int}\,[\sqrt{(-2m^* V_0 L^2/\pi^2 h^2)}].$$

The situation for hole states is much more complicated. This is because of the degeneracy of the hole states in the bulk valence bands before any reductions in symmetry take place. The full justification of the final results requires quite sophisticated group theory that we shall not introduce here (but see the discussion in Appendix 3). Indeed, most treatments of bulk valence band structure begin with simple models to which ever greater complications are added. The same is true here. We cite the results given by Kane (1957) (see the discussion in Seeger (1991)) for the bulk valence band structure. We take as a basis set the lowest conduction band which has s-type symmetry associated with orbitals at each atomic site, and the three uppermost valence bands which have p-type symmetry at each atom (all at $k = 0$). From these we form a fourfold degenerate set of orbitals in the absence of interatomic interactions, but once these are added in, the sharp energy levels broaden into bands. The two most important interaction terms are described by two parameters P and Δ; the former is an effective-mass-type term that involves interband interactions, while the latter is a spin–orbit parameter. The resulting energy bands are of the following form: the conduction band is given by

$$E = E_g + \hbar^2 k^2/2m^* + [\sqrt{(E_G^2 + 8P^2k^2/3)} - E_G]/2$$

and the valence band by

$$E = -\hbar^2 k^2/2m^* \text{ (heavy holes)}$$
$$E = -\hbar^2 k^2/2m^* - [\sqrt{(E_G^2 + 8P^2k^2/3)} - E_G]/2 \text{ (light holes)}$$
$$E = -\Delta -\hbar^2 k^2/2m^* - P^2k^2/\,(3E_G + 3\Delta) \text{ (split-off)}.$$

To these we have to add the perturbation represented by the reduction of the dimensionality in a quantum well. This is exceedingly complicated, and requires numerical calculations giving results of the type shown below in Fig. 10.4. Before proceeding to examine these results, we simplify the above expressions (noting that the conduction and light-hole valence bands are non-parabolic). Taking the heavy- and light-hole bands, and a parabolic approximation to the latter, and the doubly (spin) degenerate states at $k = 0$, we have four states that behave as a set with angular momentum $J = 3/2$, i.e. $J_z = \pm 3/2, \pm 1/2$. Near $k = 0$ in the direction of k_z, the bulk energy bands can be expanded as (Luttinger 1956)

$$E = -(\hbar^2 k_z^2/2m_0)(\gamma_1 - 2\gamma_2)$$

for $J_z = \pm 3/2$ and heavy-hole mass $m_0/(\gamma_1 - 2\gamma_2)$, and

$$E = -(\hbar^2 k_z^2/2m_0)(\gamma_1 + 2\gamma_2)$$

for $J_z = \pm 1/2$ and light-hole mass $m_0/(\gamma_1 + 2\gamma_2)$. For GaAs, we have $\gamma_1 = 6.790$ and $\gamma_2 = 1.924$. The appropriate expressions for the in-plane dispersions in a quantum well (i.e. $k_z = 0$, $k_y \neq 0$ say) are

$$E = (\hbar^2 k_y^2/2m_0)(\gamma_1 + \gamma_2) \text{ for } J_z = \pm 3/2$$
$$E = (\hbar^2 k_y^2/2m_0)(\gamma_1 - \gamma_2) \text{ for } J_z = \pm 1/2$$

These results are obtained by analogy with results in bulk semiconductors under uniaxial compressive stress. Note that heavy-hole bands in the z direction have light-hole masses in the transverse directions, and vice versa. The situation is shown schematically in Fig. 10.3, as we start with degenerate functions at $k = 0$, (Fig. 10.3(a)), follow the perpendicular and in-plane dispersions (Figs. 10.3(b) and 10.3(c), and note that higher-order terms produce an anticrossing of the energy bands in the transverse direction (Fig. 10.3(d)).

In fact, the effects of the quantum-well perturbation to the Hamiltonian and the $k \neq 0$ terms should be treated on the same footing. This is a numerical exercise with simple solutions when the well is infinitely deep, but otherwise it involves a complex interaction of the two phenomena. Typical results for the in-plane dispersion are shown in Fig. 10.4 for a multi-GaAs–AlGaAs quantum (68/71 atomic layers) well. Note the change of curvature of some bands (i.e. negative hole effective masses) and the strong non-parabolicity. There are no short cuts in this exercise (see Appendix 3). Furthermore, the interpretation of experimental results that depend on details of the valence band electronic structure can be very complicated.

10.2.2 More detailed theories of the electronic structure

So far we have used the most simple (square potential well) model to obtain electron states in quantum wells. In effect, we have used the envelope function approach and neglected all the subtleties associated with the periodic part of the Bloch function. The simple models have proved adequate for the design and interpretation of many of the optical properties of quantum wells, and

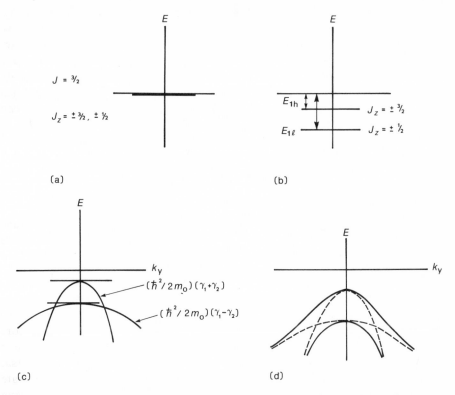

Fig. 10.3 Evolution of hole dispersion in a simple model of the valence bands in a quantum well. (a) The degenerate levels at $k = 0$ in the bulk are (b) split into heavy-and light-hole levels by the quantum-well potential. The Luttinger form of the $\mathbf{k} \cdot \mathbf{p}$ expansion of the energy bands (see Appendix 3) give the in-plane dispersions (c) that are light and heavy hole like for states that are heavy and light in the direction perpendicular to the quantum well. A full calculation of the hole bands gives rise to anticrossings as shown in (d). (After Weisbuch and Vinter 1991.)

this fact needs further examination. At first the good results for electron states in quantum wells is surprising. The original Luttinger–Kohn theory for impurity states in semiconductors (from which the envelope function model has been derived) included the assumption that the perturbing potential does not change appreciably on the scale of the unit cell (Luttinger and Kohn 1955). This is a good approximation for dopants, except for the region within $c.0.5$ nm of the impurity atom where central (unit) cell correction terms have been worked out in greater detail. Here the heterojunctions are abrupt on the scale of a unit cell and involve large energy shifts. This problem has been re-examined, with the conclusion that the effects on the energy levels are modest provided that the envelope function itself is smooth on the scale of more than four unit cells, a condition that is satisfied in most structures that we encounter (Burt 1989). The same conditions apply to the valence band

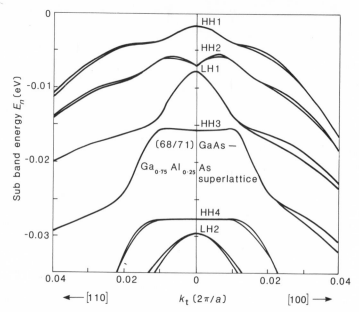

Fig. 10.4 The dispersion curves for hole bands in a multiple-quantum-well structure, showing the full complexity including negative hole masses in HH2 and highly non-parabolic bands from the anticrossings. (After Chang and Schulman 1985.)

wavefunctions but, as described above, these are already very complex. The relationship between microscopic calculations that involve the underlying Bloch functions and those that use only the simplified envelope functions have been examined for more complex quantum-well shapes (e.g. triangular wells and strained layers) and again the envelope function can account for most of the important phenomena (Jaros 1986).

10.2.3 Optical properties including excitonic effects and selection rules

In Chapter 4, Section 4.4, we gave a brief introduction to the optical properties in low-dimensional systems. We now develop that treatment in greater detail. We begin with a diagram (Fig. 10.5(a)) of a quantum well, showing the energy levels for the electrons and holes at $k = 0$ (referring to the in-plane momentum). In two dimensions the density of states in both the valence and conduction bands is a constant, and so if the optical matrix element were independent of k, the optical absorption should also be step-like, reflecting the joint density of states $G(E)$. For the lowest pair of sub-bands, and with $1/\mu = 1/m_e + 1/m_h$,

$$E_c(k) = E_0 + \hbar^2(k_{2D})^2/m_e,$$
$$E_v(k) = -\hbar^2/(k_{2D})^2/2m_h \Rightarrow G(E) = \mu/(\pi\hbar^2)\Theta(E - E_0)$$

where $\Theta(x) = 0$ if $x < 0$ and $\Theta(x)$ if $x > 0$. In addition, and just as in the 3D case, there is a strong absorption feature just below the threshold energy. This

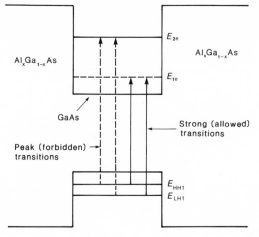

(a)

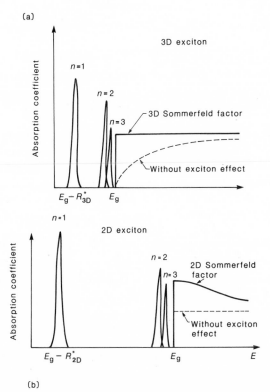

(b)

Fig. 10.5 (a) The simple interband optical transitions in a quantum well, and (b) the absorption spectra expected for 3D and 2D electron systems including both excitonic features and the many-body (Sommerfeld) factor modifying the density of states. The $n = 2, 3$ features are analogous to the 2s, 3s states in the hydrogen atom, and the Sommerfeld factor shows the effect of residual Coulomb attraction between continuum electrons and holes. (After Weisbuch and Vinter 1991.)

originates from the formation of excitons, i.e. electron–hole pairs that have a hydrogen-atom-like structure and are bound with a modified Rydberg energy of a few meV. There is a residual Coulomb interaction between electron and

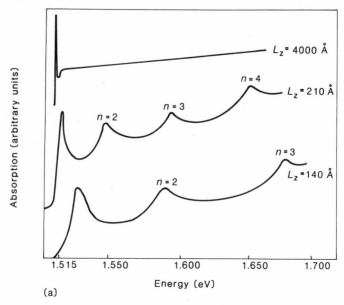

(a)

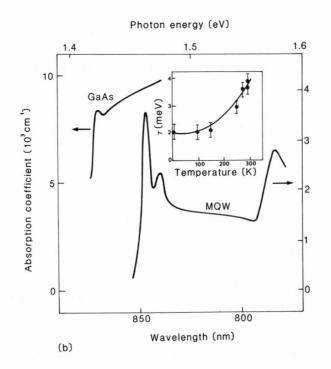

(b)

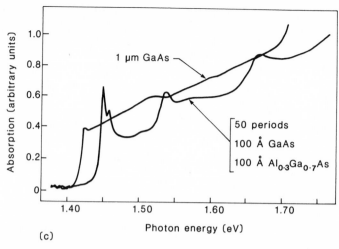

Fig. 10.6 (a) Early low-temperature optical absorption data from bulk GaAs and quantum wells of different thickness (After Dingle 1975). Compare with Fig. 2.8 in Chapter 2 for more recent data. (b) A comparison of bulk GaAs and multiple-quantum-well (MQW) absorption at room temperature showing the energy separation of the excitons associated with heavy and light holes (the inset gives the linewidth of the exciton peak as a function of temperature) (after Miller *et al.* 1982). (c) An 'equal-volume' comparison of the optical absorption of bulk GaAs and a multiple-quantum-well structure at room temperature (after Schmitt-Rink *et al.* 1989).

hole for excitation just above the bandgap, and this modifies the effective density of states from the simple one-electron picture used to date (described by a Sommerfeld factor, derived by Bastard *et al.* (1982)). Fig. 10.5(b) shows the modifications that these make to a 2D optical absorption spectrum.

In Fig. 10.6(a), we show the optical absorption at low temperature obtained from quantum-well structures, and compare this with the data for bulk material. Comparable data for room temperature are shown in Fig. 10.6(b), and an 'equal-volume' comparison of absorption in GaAs and a multiple-quantum-well structure is shown in Fig. 10.6(c). There are several important features to note.

1. In Fig. 10.6(a), the optical absorption features of the quantum well shift to higher energies, clearly exhibiting the quantum size effects (i.e. the extra kinetic energy of confinement for both the electron and hole states). The narrower the well, the higher is the energy. The excitations to higher sub-bands are also clearly seen (i.e. from the second hole state to the second electron state as allowed by the selection rules). These appear as extra steps on the staircase-like optical absorpton feature. The rounded shape of the steps indicates some of the residual electron–hole interactions referred to above.

2. There is an excitonic feature at the onset of each new sub-band structure, which in pure and regular samples is twin-peaked, revealing excitons that

incorporate heavy or light holes. The width of these excitonic features is a measure of their lifetime before recombining, either by giving out sub-band-gap radiation, or via some other non-radiative processes associated with impurity levels, phonons, etc.

3. In Fig. 10.6(b), we see that by room temperature the excitonic feature in bulk GaAs is much reduced in intensity and is broadened. The thermal energy kT is now greater than the exciton binding energy. In contrast, both the step-like features and the excitonic features in the optical absorption of the multiple quantum well remain at room temperature. The strength of the excitonic feature is surprising, giving an approximately threefold increase in binding energy, but in quantum wells the range of final states into which scattering can occur is also reduced. The inset in Fig. 10.6(b) shows the heavy-hole exciton linewidth as a function of temperature. The physics behind the fit includes a constant term describing intrinsic inhomo-geneities (e.g. steps at the interfaces at the edge of the quantum well) and a term for the possible emission of optical phonons of 36 meV as their probability ($\sim \exp(-\hbar\omega_{LO}/kT)$) becomes appreciable above about 77 K. The precise magnitude of the rise, depending on the cross-section for this process, is not well understood.

4. Note that the absolute scales of the absorption are quite different in Fig. 10.6(b). Fig. 10.6(c) is an attempt to be more quantitative. The optical absorption of a 1 μm thick layer of GaAs is compared with that of a multiple quantum well that is 1 μm thick and has equal well and barrier thickness. Here one sees more clearly the prominence of the exciton features and the step-like increases of the optical absorption, together with the many-body effects from the multiple-quantum-well structure. The overall optical absorption strength is comparable.

Although we treated selection rules in Chapter 4, Section 4.4.1, we comment on them here as they apply to GaAs/AlGaAs quantum wells. The simple derivation given earlier was based on parity considerations of the conduction and valence band Bloch functions as they applied to interband absorption, and the parity of the envelope wavefunctions as they applied to intra-sub-band absorption. In practice these derivations assume an infinitely deep potential. For finite potential wells, the envelope functions are not exactly orthogonal (the overall orthogonality involves differences in the Bloch func-tions in the cladding layers). In most cases, the devices are under bias (see below), or are within the depletion regions of doped layers, or are parts of modulation-doped structures. In all these cases the spatial variation of the confining potential means that the simple concept of parity (even–odd symmetry about some point) is no longer valid. One can observe a number of weak optical transitions that would be forbidden in an ideal structure. The full symmetry of the valence band states gives rise to the complex pattern of optical absorption (at $k = 0$) shown in Fig. 10.7 for light travelling in the x

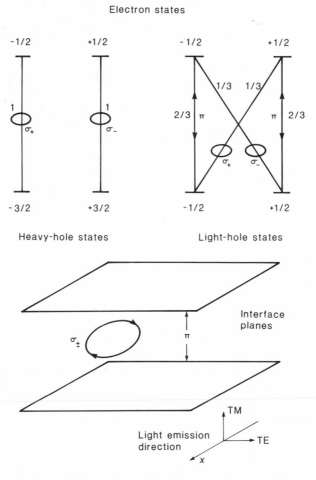

Fig. 10.7 Polarization of absorption and luminescence between valence and conduction bands of a bulk semiconductor and the particular geometry for a quantum well: σ, circular polarization; π linear polarization; the fractions indicate the relative dipole strengths. (After Weisbuch and Vinter 1991.)

direction and with electrons and photons polarized with respect to the z axis. The σ transitions are circularly polarized and the π transitions are linearly polarized, and the numbers correspond to the relative strengths of the transitions. When one adds the reduction in symmetry introduced by the quantum well and the fact that the light might be polarized in the plane of, or perpendicular to, the quantum wells, then a rich variety of polarization studies are possible. These are discussed by Weisbuch and Vinter (1991, Section 11, and references cited therein).

An early application of optical absorption was to estimate the ratio of the conduction to valence band offsets in the GaAs–AlGaAs system. Clearly, if this ratio is used as a parameter, one can plot the optical absorption peaks as a function of well width for different values of this parameter and then fit the optical data. The early result was a of 85 per cent to 15 per cent for the conduction to valence band discontinuity (Dingle 1975). Unfortunately, many corrections must be applied to the raw optical data before extracting the inter-sub-band energies, and the overall process is of less accuracy than was initially hoped. In addition to the excitonic corrections, there are other many-body effects and changes to the optical absorption energies from residual electric fields. Furthermore, the variation of the one-electron energies with the offset ratio is not very strong compared with these sources of possible error. The latest consensus is for a conduction band to valence band offset ratio of 60 ± 5 per cent to 40 ± 5 per cent obtained by a best fit to the various optical transitions in a series of quantum wells varying in thickness from about 5 nm to about 50 nm (cf. Chapter, Section 2.8, and Miller *et al.* (1984)).

10.2.4 Application of an electric and/or magnetic field

One of the most important applications of quantum wells comes from the modifications to their optical properties that follow the application of an electric field. The effects of a perpendicular electric field were shown in Chapter 4. Electron and hole states in the quantum well are moved closer together in energy—this effect can be calculated quite simply using Airy function solutions for carriers in a uniform constant field. We commence with the effects of a field on the one-electron properties, as shown schematically in Chapter 4, Fig. 4.12. A second-order perturbation theory calculation is straightforward, and leads to a quadratic reduction in the interband absorption energy with electric field (Miller *et al.* 1985). In practice, since the valence band discontinuity is relatively small, the holes are less confined in the potential, particularly the light holes, and the shift of their energy levels under an electric field is greater than those for electrons. In fact, the approximation in Fig. 4.12 is remarkably good: the electron levels hardly move, although the wavefunctions polarize, while the hole energies shift more rapidly. The applied electric field has an effect on the excitonic properties in addition to these one-electron energy shifts.

In bulk semiconductors, quite modest fields are sufficient to polarize an exciton and so cause dissociation. The addition of a linear potential to a Coulomb potential allows an electron to tunnel from its bound state (as shown schematically in Fig. 10.8(a)). In particular, an electron will easily tunnel if the electric field corresponds to a potential drop over an exciton diameter equal to the exciton binding energy (say 5 meV over 30 nm or 0.2 V/μm, a modestly high field often achieved in practical devices). The exciton in a quantum well behaves very differently for an electric field perpendicular to the well. The bandgap offsets in both the conduction and valence bands prevent the electrons and holes from escaping the quantum well until very large electric fields are applied (typically 50 times that required to cause the

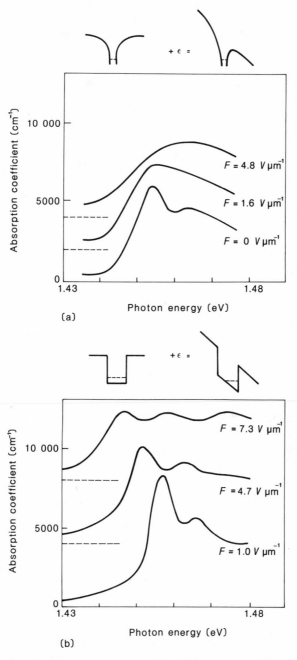

Fig. 10.8 Optical absorption in quantum wells subject to an applied electric field (a) in the plane of the quantum well and (b) perpendicular to the quantum well. The relative stability of the excitonic features is discussed in the text. (After Miller *et al.* 1985.)

same effect in a bulk semiconductor). The heterojunctions are far more effective than the Coulomb interaction in keeping electrons and holes in close proximity. Note that, as an electric field is applied in the plane of the quantum well, the dissociation of the exciton is only modestly more difficult, given the increase in binding energy of excitons in reduced dimensions. The situation is summarized in the schematic diagrams above the experimental data in Fig. 10.8.

The points to be noted in the data of Fig. 10.8 are as follows.

1. When the electric field is applied parallel to the layers, the exciton feature is quickly reduced in intensity, with very little shift in overall energy. The fields applied are modest, with the maximum field in Fig. 10.8(a) being a little over half the maximum field in Fig. 10.8(b). In contrast, the excitonic feature is still present at the highest field (7.3 V/μm) when applied perpendicular to the quantum well.

2. The shift in the peak absorption energy at over 10 meV is appreciable, and shifts of up to 30 meV have been measured before the excitonic feature broadens and weakens.

3. By taking a fixed energy (say $c.1.45$ eV), the application of an electric field can result in a strong increase in optical absorption. This is the basis of optical modulation, as described in Chapter 18. Multiple-quantum-well structures are designed to maximize this change in absorption. In other applications, a decrease in optical absorption with field is required, and this could be achieved by working at 1.46 eV. In Fig. 10.9, we show the peak energy shifts for heavy- and light- hole excitons, compared with theory, for a 9.5 nm well; there are no adjustable parameters. Most of the energy shift is associated with the one-electron energy-level shifts.

In contrast with electric field investigations, the effects of magnetic fields on interband absorption have been less widely studied, partly because they are smaller. The principal effect is an increase in the binding energy of the exciton, just as magnetic fields increase the binding energy of the hydrogen atom.

10.2.5 Non-linear and saturation phenomena

The intrinsically small number of electrons in quantum wells means that non-linear optical properties and saturation of optical transitions are readily observed. Excitons that form at room temperature have a short life before recombination or the formation of plasmas of unbound electrons and holes. These latter carriers occupy final states required for further optical absorption (known as phase-space filling). In addition, the Coulomb interaction between a given electron and hole is screened by the electron and hole plasmas created by the optical absorption. The scattering between electrons, holes, and excitons also increases. The two sets of results in Fig. 10.10 illustrate the principal non-linear and saturation phenomena. Fig. 10.10(a) shows the energy dependence of the low-power optical absorp-

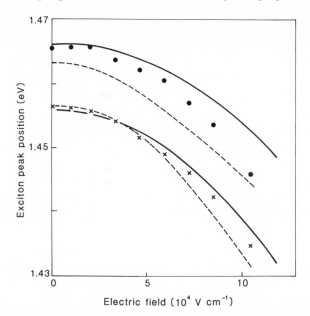

Fig. 10.9 Experimental exciton peak energy shift with electric field for light-hole (°) and heavy-hole (×) excitons compared with theory. The solid (broken) curves correspond to theoretical predictions for bandgap offset ratios of 57 per cent to 43 per cent (85 per cent to 15 per cent); there are no other free parameters in the theory. (After Miller *et al.* 1985.)

tion as a full curve, with the broken and dashed curves showing the absorption at different times after a strong optical excitation which completely bleaches out the excitonic features for a time of about 100 ps. Fig. 10.10(b) shows the intensity dependence of the absorption at the energy of the first exciton peak, for both bulk GaAs and a multiple-quantum-well structure; a 50 per cent reduction in intensity occurs for a rather lower incident power in the case of the multiple quantum wells (note the logarthmic power scale).

There have been more detailed studies of both steady state and transient changes of refractive index as a function of light intensity and energy. The results are important for the design of improved devices for optical signal processing. The details take us beyond the scope of this text but are reviewed by Schmitt-Rink *et al.* (1989).

10.2.6 Inter-sub-band absorption in the far infrared

The possibility of optical excitation between different bound levels in a doped quantum well was introduced in Chapter 4. Absorption is possible because of the parity of the bound-state wavefunctions and a non-zero value of the optical matrix integral $M \sim \int F_f(r)\boldsymbol{\varepsilon}\cdot\boldsymbol{r}F_i(r)\mathrm{d}r$. Note that with normalized envelope functions $F(r)$, this integral can take a value of order L (the thickness

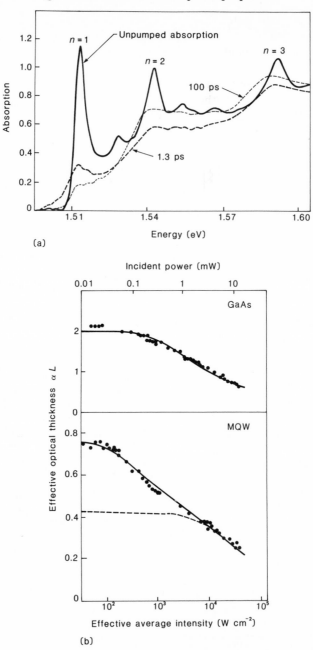

Fig. 10.10 (a) The absorption spectra of a multiple-quantum-well structure (20.5 nm wide) before and after strong excitation with light of energy 130 meV above the bulk bandgap (after Shank *et al.* 1983). The density of excited carriers is about 5×10^{11} cm^{-2}. (b) The dependence on incident light intensity of the optical absorption of the first exciton peak in bulk GaAs and in a multiple quantum well (MQW) (after Miller *et al.* 1982).

of the quantum well) instead of the interatomic dimension encountered with interband absorption. Thus there is a large dipole effect associated with the transition from the ground state to the first excited state in a quantum well. Note also that this absorption survives the application of quite strong fields perpendicular to the layers (cf. the situation in the conduction bands in Fig. 4.11 of Chapter 4). Given the nature of the envelope functions, only the component of the polarization vector of the light that is perpendicular to the

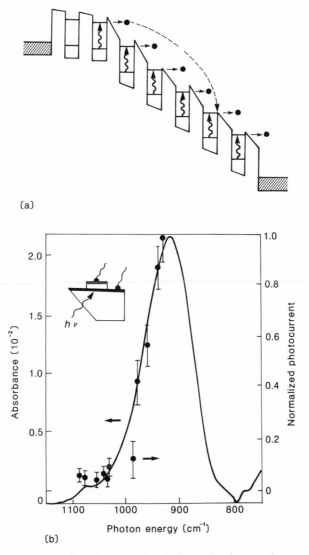

(a)

(b)

Fig. 10.11 (a) The principle of inter-sub-band absorption in an n–i–n structure as used for infrared photodetection; (b) the experimental arrangement and typical absorption data correlated with simple theory (After Levine *et al.* 1987).

quantum well is responsible for inter-sub-band absorption in the conduction band. This means that normally incident light is ineffective, and even oblique incidence is inefficient as the high refractive index bends this light towards the normal once it is inside the semiconductor. In practice, the back of the substrate can be beveled at $c.45°$ to allow light to reach the quantum wells (see inset to Fig. 10.11(b)). The optical absorption is essentially narrow-band, as the in-plane band structure in both sub-bands is the same free-electron motion with an effective mass m^*. In practice, the quantum wells are not perfect and fluctuations in well width result in a finite ($c.20$ per cent) bandwidth.

The application of this concept to infrared detection is described in Chapter 19, but we concentrate here on physics issues. A photoexcited carrier can relax back to the ground state emitting a far-infrared photon. For detector applications, one wants to sweep this excited carrier out of the device and detect the original photoexcitation process as a current. In the design of quantum wells, the well thickness and the height of the barriers are adjustable quantities. A detector designed to operate at a given wavelength (equivalent to the excitation energy between sub-bands) uses up one of the degrees of design freedom, but the second degree can be used to arrange for the excited level to be right at, or slightly above, the barrier energy. In this way, a modest vertical electric field will be able to sweep any photoexcited carriers away from the well either by tunnelling through a thin barrier or directly if the excited level is above the barrier. The former situation is shown in Fig. 10.11(a), while in Fig. 10.11(b) we show a typical structure and the absorption spectrum derived from it. Most work has focused on detection of radiation at about 10 μm. For this one typically uses about 36–40 per cent Al in the AlGaAs alloy and a well thickness of 7 nm or less. More recent studies have included the examination of inter-sub-band excitation of holes, where the complexity of the valence bands reduces the need to have light polarized in the plane of the layers (People *et al.* 1992). Fuller details of the device design of infrared detectors and the impressive performance figures now achieved are left until Chapter 19.

10.2.7 Luminescence as an analytical tool

The optical properties described so far have concentrated on absorption. The emission of light from quantum wells is also of interest. To date, luminescence from quantum wells has been a widely exploited analytical technique used to confirm the size and quality of quantum wells, as hinted in Chapter 2, Section 2.4.2. The light emitted by quantum wells at low temperatures conveys much information. Note that carriers created in the barrier layers are efficiently captured in the wells, increasing the efficiency of the quantum-well luminescence. Exciton formation is more likely in reduced dimensions than in the bulk, and the concentration of carriers in a small phase space further assists in exciton formation. However, luminescence from excitons is a forbidden process in first order; only if the electrons and holes have exactly the same momentum as the photon can a simple radiation process occur. In bulk materials, one relies on impurities to mediate in the luminescence process, giving emission referred to as donor-bound exciton and having an energy which reflects the donor binding energy. The same is true to a lesser extent in quantum wells. The reduced phase space makes it more likely that excitons can be formed under just the conditions to luminesce. In addition, a number of impurities in quantum wells can have their spatial location determined via their precise associated luminescence energy. Further, the fluctuations associated with the quantum-well interfaces (steps, kinks, and islands of material in a given atomic plane at the interface) mediate in the luminescence process. To the extent that these latter processes dominate in pure materials, the linewidth of a luminescence signal is a measure of the quality of the interfaces. As an example, one can determine that in some materials the quantum well is made up of regions of a fixed number of atomic spacings, with other regions having one less or more atomic plane. In the case where the lateral scale of these regions is large compared with the exciton diameter, a twin-peak luminescence signal is detected. Indeed, photoluminescence has been an important tool since the early days of epitaxial growth research (Weisbuch and Vinter 1991, and references cited therein).

So far, the luminescence has been described without reference to the excitation. In practice, a sample is flooded with light of an energy above that of the gap of the cladding layers, and one relies on capture of carriers in the quantum well. An alternative technique is to detect free or impurity-associated exciton luminescence at a particular energy while scanning the wavelength of a monochromatic light source. This process, known as photoluminescence excitation spectroscopy, tends to give a measure of the density of states at the excitation energy (Weisbuch and Vinter 1991, and references cited therein).

10.3 Multiple quantum wells: electronic and optical properties

In Chapter 9 we considered the electronic and optical properties of superlattices where bound states associated with a given layer of GaAs interacted strongly through thin AlGaAs barriers with the bound states in adjacent quantum wells. In this chapter, we have considered the optical properties of a single quantum well, and have assumed that, if the barriers are thick enough, we can treat the individual wells in a multiple-quantum-well structure as independent. We shall see that this is justified. However, we begin by considering wells that are separated by thin barriers.

10.3.1 A coupled pair of quantum wells

In Fig. 10.12, we show the optical absorption spectra of a single well, and of two, three, and ten closely coupled quantum wells. The markings associated with each figure are the results of calculations of the energy levels for the coupled quantum wells, determined by perturbation theory from the values of a single quantum well (cf. Chapter 9, Fig. 9.1). The good agreement is easily seen, as the very thin barriers (< 2 nm) allow significant interaction between adjacent wells. In the case of two wells, we obtain the familiar bonding–antibonding combination of envelope functions. The situation for three and more wells follows from the analogous calculation to a row of hydrogen atoms. Once the barriers are an appreciable fraction of an exciton radius (say more than 5 nm) the interaction through them falls off sharply (cf. Chapter 9, Fig. 9.6(b)), and once the barriers exceed 10 nm there is in effect no interaction between adjacent wells. At this stage, the optical effects associated with adjacent wells all add up in parallel.

The application of an electric field to coupled quantum wells can result in strong modifications to the splitting between bonding and antibonding levels, and so the excitonic features associated with them can be made to move more rapidly for smaller electric fields than the corresponding energies in the individual wells (Andrews *et al.* 1988). Multiple pairs of such closely coupled quantum wells have been used to tailor the properties of infrared detectors that rely on the inter-sub-band absorption (Vinter *et al.* 1992). Indeed, structure within one well, such as a linear grading of the Al concentration or a small amount of Al introduced into one half of the well, leads to modifications of the electron wavefunctions of subsequent excited levels. As will be seen in the chapters devoted to devices (Chapters 16–19), the ability to tailor the composition and thicknesses of many layers can be used to tailor the electrical and optical properties for a specific device function.

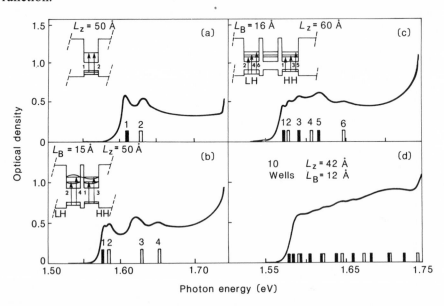

Fig. 10.12 Optical absorption spectra of (a) a single well, and (b) two, (c) three, and (d) ten strongly coupled quantum wells. The data are compared with simple calculations of where the features are anticipated. (After Dingle *et al.* 1975).

As a final point, we note the splitting between bonding and antibonding wavefunctions in a coupled-well structure (Fig. 10.12(b)). If the electrons can be excited within one well, a time-dependent perturbation theory analysis of this problem (Landau and Lifshitz 1977) implies that the electron will oscillate between the wells with an angular frequency $\Delta E/\hbar$ determined by the splitting of the energy levels. This effect has been seen recently as 'quantum beats' at terahertz frequencies (Roskos *et al.* 1992). This is the two-well analogue of Bloch oscillations (cf. Chapter 9).

10.3.2 Multiple quantum wells

Multiple quantum wells have become of widespread use in optical studies. The typical depth over which light is absorbed by interband processes in a semiconductor is comparable to the wavelength of the light, i.e. 1 μm. The quantum wells are designed for a particular optical transition, and vary in thickness between 3 and 25 nm. Once the barriers exceed about 10 nm, the wells act independently. Therefore one typically has a 50-quantum-well structure for optical studies. The optical response should be a simple multiple of the single well response, but this does not occur for several reasons as the wells are never identical. They fluctuate in thickness, no matter which growth technique is used; it may even be monolayer steps imposed by the condition of the surface the instant that the shutters close in an MBE machine. More systematic variations in temperature of parts of the growth machine may take place over the period of about 1 h that is required to prepare such a multiple-quantum-well structure. The situation is shown schematically in Fig. 10.13,

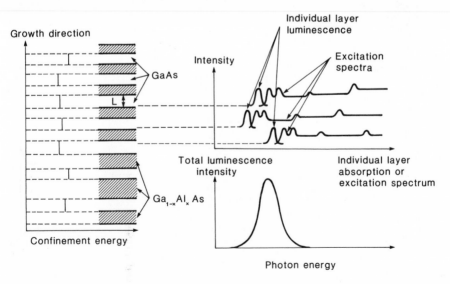

Fig. 10.13 A schematic diagram of the luminescence from a multiple-quantum-well structure as the sum of contributions from individual wells whose luminescence intensity, energy, and linewidth might vary. (After Weisbuch and Vinter 1991.)

where both the photoluminescence excitation spectra and the luminescence spectra are depicted for each layer. The spread then results in a broader overall luminescence spectrum, which is indicative of the overall uniformity of the multiple-quantum-well system.

A further complication associated with multiple-quantum-well structures can be inferred from Fig. 10.11(a), namely the effects of strong but non-uniform electric fields over different wells. In the intra-sub-band absorption of far-infrared radiation, the field is needed to sweep out any photoexcited carriers before they are retrapped in the same or an adjacent quantum well. In inter-sub-band absorption for optical modulation, the field is used to move the exciton absorption about in energy. In the two cases cited, the multiple

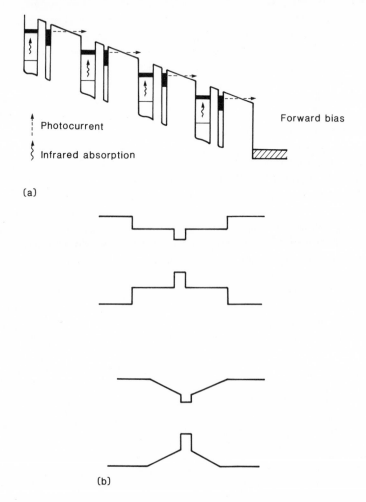

Fig. 10.14 Carrier capture in quantum wells is (a) hindered in infrared detectors (Choi *et al.* 1987(b)) and (b) helped in quantum-well lasers by the nature of the surrounding layers.

quantum wells appear the intrinsic region of an n–i–n structure or a reverse-biased p–i–n diode respectively. It is difficult to arrange for the field in the i-region to be a constant value, as there are usually depletion fields that act over the first and last few quantum wells. In the infrared detector, this broadens the absorption peak, which for many applications in an attractive side-effect. In the absorption modulators, this also broadens the exciton peak and reduces the contrast between high and low absorption at a given wavelength, which is an unwanted side-effect.

10.4 Quantum wells in a current

In the luminescence studies described above, and in lasers (see Chapter 18), we want quantum wells to capture carriers from the cladding layer. These carriers may be there because of optical excitation or they may form part of a current (as in lasers where radiative recombination is sought in the device function). In other contexts, such as the infrared photodetector, we do not want carriers to be trapped in the wells, but to pass through to the contact and the external circuit. The efficiency and the time scale of trapping of carriers by quantum wells are important phenomena. We have already encountered the quantum reflection of carriers at sharp potential boundaries (Chapter 9). We find that carriers with energy above the barrier still have an appreciable probability of reflection. If we do not want carrier capture in a quantum well, we should smooth out the potential barriers on the scale of several nanometres by grading the composition. An alternative scheme is to use coupled quantum wells, where the first excited level is bound in one well where the absorption takes place, but is free or almost free in the adjacent well under bias (Choi *et al.* 1987). If we do want capture, wider wells with higher Al contents can be placed on each side. The situation for these two cases is shown schematically in Fig. 10.14. In practical cases, the time for capture is comparable with the picosecond transit time (thickness divided by drift velocity) in typical multiple-quantum-well structures. Carrier capture in quantum wells has been widely investigated, and is discussed in the context of quantum-well lasers by Weisbuch and Vinter (1991).

References

Andrews, S. R., Murray, C. M., Davies, R. A., and Kerr, T. M. (1988). Quantum confined Stark effect in strongly coupled quantum wells. *Physical Review B*, **37**, 8189–204.

Bastard, G., Mendez, E. E., Chang, L. L., and Esaki, L. (1982). Exciton binding energies in quantum wells. *Physical Review B*, **26**, 1974–9.

Burt, M. (1989). Exact envelope function equations for microstructures and the particle in a box model. In *Band structure engineering in semiconductor microstructures* (eds. R. A. Abram and M. Jaros), NATO ASI Series B: Physics, Vol. 189, pp. 99–109. Plenum, New York.

Chang, Y. C. and Schulman, J. N. (1985). Interband optical transitions in GaAs–Ga$_{1-x}$Al$_x$As and InAs–GaSb superlattices. *Physical Review B*, **31**, 2069–79.

Choi, K. K., Levine, B. F., Bethea, C. G., Walker, J., and Malik, R. J. (1987). Photoexcited coherent tunnelling in a double-barrier superlattice. *Physical Review Letters*, **59**, 2459–62.

Dingle, R. (1975). Confined carrier quantum states in ultrathin semiconductor heterostructures. In *Festkorperprobleme XV* (ed. H. J. Queisser), pp. 21–48. Pergamon/Vieweg, Braunschweig.

Dingle, R., Gossard, A. C., and Wiegmann, W. (1975). Direct observation of superlattice formation in a semiconductor heterostructure. *Physical Review Letters*, **34**, 1327–30.

Jaros, M. (1986). Electron states in semiconductor microstructures. In *The physics and fabrication of microstructures and microdevices* (eds. M. J. Kelly and C. Weisbuch), pp. 197–209. Springer-Verlag, Berlin.

Kane, E. O. (1957). Band structure of indium antimonide. *Journal of the Physics and Chemistry of Solids*, **1**, 249–61.

Landau, L. D. and Lifshitz, E. M. (1977). *Quantum mechanics* (3rd edn). Pergamon, Oxford.

Levine, B. L., Choi, K. K., Bethea, C. G., Walker, J., and Malik, R. J. (1987). New 10 μm infrared photodetector using intersubband absorption in resonant tunnelling GaAlAs superlattices. *Applied Physics Letters*, **50**, 1092–4.

Luttinger, J. M. (1956). Quantum theory of cyclotron resonance in semiconductors: general theory. *Physical Review*, **102**, 1030–41.

Luttinger, J. M. and Kohn, W. (1955). Motion of electrons and holes in perturbed periodic fields. *Physical Review*, **97**, 869–83.

Miller, D. A. B., Chemla, D. S., Eilenberger, D. J., Smith, P. W., Gossard, A. C., and Tsang, W. T. (1982). Large room-temperature optical nonlinearity in GaAs/Ga$_{1-x}$Al$_x$As multiple quantum well structures. *Applied Physics Letters*, **41**, 679–81.

Miller, D. A. B., Chemla, D. S., Damen, T. C., Gossard, A. C., Wiegmann, W., Wood, T. H., and Burrus, C. A. (1985). Electric field dependence of optical absorption near the band gap of quantum well structures. *Physical Review B*, **32**, 1043–60.

Miller, R. C., Kleinmann, D. A., and Gossard, A. C. (1984). Energy-gap discontinuity and effective masses for GaAs–Ga$_{1-x}$Al$_x$As quantum wells. *Physical Review B*, **29**, 7085–7

People, R., Bean, J. C., Bethea, C. C., Sputz, S. K., and Perticolas, L. J. (1992). Broadband (8–14 μm), normal incidence, pseudomorphic Ge$_x$Si$_{1-x}$/Si strained layer infrared photodetector operating between 20 K and 77 K. *Applied Physics Letters*, **61**, 1122–4.

Roskos, H. G., Nuss, M. C., Shah, J., Leo, K., Miller, D. A. B., Fox, A. M., *et al.* (1992). Coherent submillimetre wave emission from charge oscillation in a double well potential. *Physical Review Letters*, **68**, 2216–19.

Schmitt-Rink, S., Chemla, D. S., and Miller, D. A. B. (1989). Linear and nonlinear optical properties of semiconductor quantum wells. *Advances in Physics*, **38**, 89–188.

Seeger, K. (1991). *Semiconductor physics: an introduction* (5th edn). Springer-Verlag, Berlin.

Shank, C. V., Fork, R. L., Yen, R., Shah, J., Greene, B. I., Gossard, A. C., and Weisbuch, C. (1983). Picosecond dynamics of hot carrier relaxation in highly excited multi-quantum well structures. *Solid State Communications*, **47**, 981–3.

Vinter, B., Vodjdani, N., Berger, E., Bockenhoff, E., and Costard, E. (1992). Switching between two mid-infrared intersubband absorption energies in double quantum wells. *Surface Science*, **267**, 601–4.

Weisbuch, C. and Vinter, B. (1991). *Quantum semiconductor structures*. Academic Press, New York.

11 Quantum pillars and boxes: electronic and optical properties

11.1 Introduction

New physics associated with hot electron injection, tunnelling, and quantum confinement is made possible with heterojunctions in multilayer semi-conductor structures. The new physics has been successfully applied to devices, where the heterojunctions ensure that the active volume is Q2D (as in FETs and quantum-well lasers). Even the tunnel structures are very thin in one of the three spatial dimensions. Given the proven advantages in going from bulk to Q2D systems, a further reduction in dimensionality might again lead to new and exploitable physics in semiconductors. We describe the new physics in this chapter and comment on its exploitability. We concentrate on the fabrication of structures that are on the quantum-length scale of $c.20$ nm in two, or all three, spatial dimensions, giving quantum wires (pillars) and quantum dots (boxes) respectively. We concentrate on the electrical and optical properties of the former and the optical properties of the latter. In practice, we encounter some new effects associated with the intrinsically small size of the structures (i.e. containing only a few carriers), but we leave the general physics of such mesoscopic systems to the following chapter. In addition, the finite charge of an electron plays a role in the energetics of small structures, and effects associated with this are also treated in Chapter 12.

11.2 Fabrication of quantum wires and dots

Quantum wires and dots are made using a combination of the epitaxy, lithography, and etching techniques described in Chapter 3. The availability of different doping levels in multilayers is very useful when fabricating semiconductor wires. In this section, we describe the fabrication of a range of quantum wires and pillars. There is a semantic point that wires run parallel to the surface of a substrate, while pillars rise vertically from it.

The simplest structure is the wire that is etched in a thin epitaxial layer of doped material deposited on an undoped substrate (Fig. 11.1). The epitaxial layer is doped to the extent that it is conducting at low temperatures (i.e. above about 2×10^{17} cm^{-3}). A pattern in resist can be made by lithographic techniques, incorporating both the wire and the multiple side-arms that allow four-terminal electrical measurements to be made. The remainder of the epitaxial layer can be removed by etching (the most common process) or rendered non-conducting by hydrogen implantation (proton isolation). This

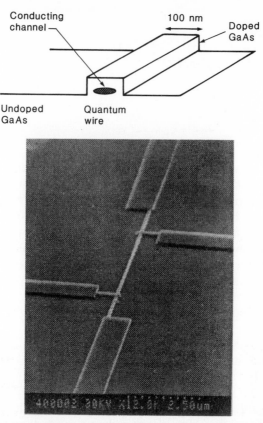

Fig. 11.1 The simplest quantum wire formed by etching away all but a thin strip of doped semiconductor on an undoped substrate: (a) schematic diagram; (b) practical example. (After Beaumont 1992.)

simple structure introduces a problem that has not yet been overcome or circumvented. The free surfaces contain many electron states that pin the Fermi energy in the middle of the gap. There is a depletion depth in from any free surface before the semiconductor can conduct at low temperatures (see Fig. 11.1). This depth is inversely proportional to the level of doping, and has a value of about 100 nm for about 10^{17}cm^{-3} doping. In practice we prefer to use as low a doping as possible, so that the Coulomb scattering between charged donor/acceptor ions and the mobile carriers is kept to a minimum. In addition, many of the chemical etching techniques are very sensitive to the level of doping. It is difficult to obtain very smooth features when etching heavily doped materials, as clusters of donor/acceptor ions can alter the electrostatic potentials and the various chemical etch rates locally (Potts *et al.* 1991). However, low doping limits the smallest diameter of conducting channel that can be achieved. In GaAs wires, the minimum diameter of the channels is about 0.08 μm. In addition, reactive ion etching in any environ-

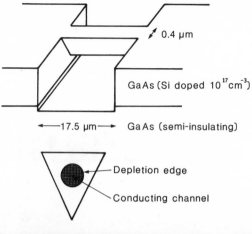

Fig. 11.2 (a) Schematic diagram of a free-standing quantum wire etched out of a layer of doped GaAs on an insulating substrate, and (b) an image of the end of one such region. (After Hasko *et al.* 1988.)

ment containing even trace amounts of hydrogen leaves any semiconductor material electrically dead to a depth of about 70 nm from the surface by hydrogen passivation, and this is before adding on any depletion depth.

A variation on this quantum wire involves making the conducting channel free-standing (Fig. 11.2). This is achieved by using wet chemical etching that undercuts the crystal from the surface, resulting in structures with a triangular cross-section (see also Chapter 3, Fig. 3.5). The physics specific to the free-standing aspect of this structure is described in greater detail in Chapter 15.

The fabrication of quantum pillars (Fig. 11.3(a)) in doped (multilayer) semiconductors is more complicated when electrical measurements are to be

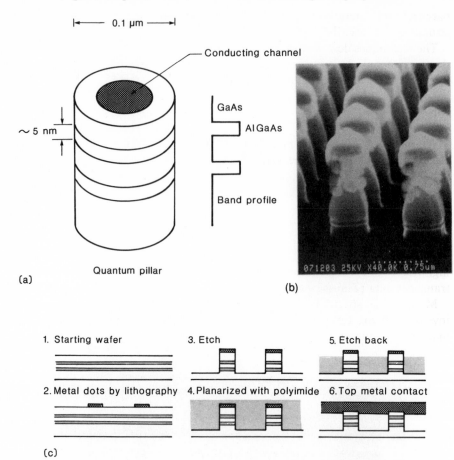

Fig. 11.3 A quantum pillar formed from resonant tunnelling semiconductor multilayers showing (a) a schematic diagram of the pillar, (b) the partially processed structure after the first etch, and (c) the full processing route. (M. Tewordt, private communication.)

made. The process steps are shown in Fig. 11.3(c). A metal dot of sub-micrometre diameter, which is laid down by lithographic steps, is used as a mask for reactive ion etching, resulting in the structure shown in Fig. 11.3(b). In order to make an electrical contact to the top of the wire, this structure is filled in with polyimide (a polymeric material) which is etched back to expose the metal dot. The whole surface can be coated with metal, making contact to the metal dots. There may need to be an annealing stage to ensure that the metal overlayer makes a good ohmic contact to the semiconductor layers through the original metal dots. The methods of fabrication have been refined so that single pillars can be contacted.

This same structure can also be used as a quantum *dot* if the two barrier layers are sufficiently thick that an electron can be confined vertically by the barriers and laterally by the side-walls. Optical experiments using such structures are described below.

The techniques described so far make inefficient use of the original semiconductor material. At best only a few per cent of the surface area is used, even when efforts are made to have different structures as close as possible. This has led to various attempts to use growth *per se* in the fabrication of quantum wires, employing a variation of Fig. 2.3 with more that 50 per cent AlAs in each layer.

We have already encountered a 1DEG in Chapter 6, where fringing electrostatic fields cause the extra confinement. These fields originate from Schottky gates on the surface or from lateral depletion layers formed by selected-area ion implantation. This latter technique is being extended to produce completely new materials systems. A sufficiently heavy Co implantation can be annealed to form buried $CoSi_2$ wires surrounded by Si. In recent experiments, the surrounding Si has been removed by etching to leave free-standing (metallic) wires (Fig. 11.4) of high quality, as estimated from the high degree of specular reflection of electrons from the surfaces inferred from transport data (Zimmerman *et al.* 1993).

Many other specialist techniques for making small structures are being invented all the time. Some of the smallest (subnanometre) structures are made using an intense high-energy electron beam to sublime certain inorganic

Fig. 11.4 $CoSi_2$ wires are formed by ion implantation through a mask into Si, followed by (i) an annealing stage when the lattice recrystallizes and buried disilicide wires are formed and (ii) an etch to reveal the wires. (After Zimmerman *et al.* 1993 courtesy of AT&T Bell Laboratories.)

materials, such as AlF_3, where the F is driven off, leaving small Al structures (Muray *et al.* 1984). The same technique, when applied to other materials (e.g. MgO), can sublime the whole material, leaving nanometre diameter holes (Turner *et al.* 1990). There has been long-term research on the intercalation of alkali metals between layers of graphite; this principle has been extended to fill microtubes based on the recently discovered C_{60} (Rosseinsky *et al.* 1992). While these small structures can easily be made and imaged, it is more difficult to perform optical or electrical measurements.

11.3 Electronic and optical properties

In this section we describe the results of some experiments on 1D systems and their theoretical interpretation. The topic is less mature and less developed than the 2D equivalents described in Chapters 5 and 10.

11.3.1 Lateral quantization: optical and electrical evidence

In Fig. 11.5(a) we show the conductance of 50 nm deep n^+-GaAs wires of differing lithographic thicknesses formed by dry etching for different times and a wet-etched wire for comparison. We note that there is an offset on the horizontal axis, indicating that the width of wet-etched wires must exceed about 80 nm before any conduction is possible. Furthermore, when dry etching is used, the offset increases by as much as a factor of 2. In this case, hydrogen in the plasma environment is passivating the subsurface volumes of the wire. In Fig. 11.5(b) we show a compilation of published data on the luminescence efficiency of quantum wires and dots, again as a function of the minimum lateral feature size. Below a diameter of about 80 nm, no light is emitted (and above that diameter there is a large spread in intensity). Thus, electrically and optically, there is a dead region adjacent to the surfaces.

The minimum inferred diameter of the conducting channel in quantum wires is also about 80 nm. On this scale, typically more than 10 laterally quantized 1D sub-bands are occupied, but with the scattering from the donor ions, no clear evidence of the lateral quantization has ever been seen in transport measurements. There is no analogue of the quantization of the Q1D ballistic resistance in doped quantum-wire structures. There are similar negative results from the optical studies. Unambiguous quantum size effects from lateral quantization are masked by fluctuations in the feature sizes and the intrinsically small effect being sought from pillars of diameter about 10 nm.

11.3.2 Analysis of tunnelling

Quantum pillars made from material containing resonant tunnelling double-barrier diodes have been used more successfully to analyse Q1D transport. Although we cannot see direct evidence of the lateral quantum size effects, we can see the influence of a discrete number of lateral sub-bands in electronic transport. Here the tunnel barriers and the immediately adjacent spacer layers are undoped, and effects associated with ballistic motion over short distances

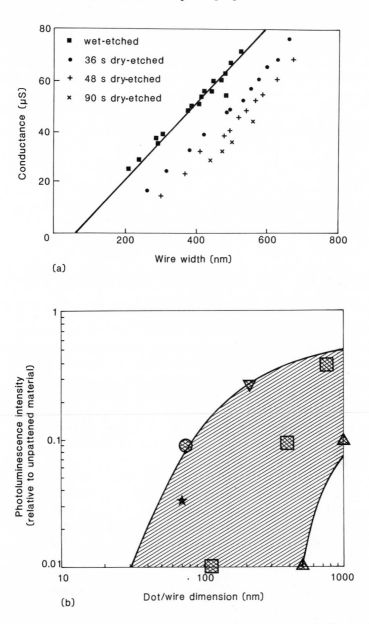

Fig. 11.5 (a) Plot of the conductance of quantum wires (such as in Fig. 11.1) as a function of wire width showing that wet-etched wires have a surface depletion depth of about 70 nm, while dry-etching leaves a wider depletion depth that increases further with exposure to the plasma, reaching about 230 nm after 90 s; (b) the luminescence efficiency of quantum wires and dots as a function of minimum feature size, with all data falling within the shaded region. The symbols are representative data points. (After Beaumont 1992.)

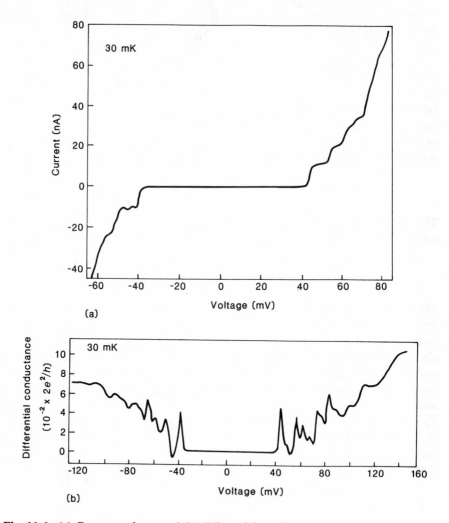

Fig. 11.6 (a) Current–voltage and (b) differential conductance–voltage characteristics of a double-barrier quantum pillar measured at low temperature. (After Tewordt *et al.* 1990*a*.)

can be seen to good effect. In turn, the analysis of transport in narrow quantum pillar resonant tunnel diodes has added to our understanding of tunnelling. Fig. 11.6 shows the relevant data (the low-temperature current–voltage characteristics and differential conductance–voltage curves) for a mesa diode about 0.1 μm in diameter. The qualitative form is quite different in several respects from data taken using large-area diodes (cf. Chapter 8, Fig. 8.9(a)). Here the current remains zero until a considerable bias is applied and then rises in a series of steps, as opposed to a smooth rise in current from zero bias in large-area diodes.

The behaviour of the small diodes can be explained using a model of a finite number of 1D channels containing ballistic electron motion, at least over the distance between the two heavily doped regions that act as contact layers. The important difference from the previous description of ballistic motion in Chapters 4 and 6 is that here the transmission of electrons through the double-barrier structure is a sharply varying function of energy and this must be incorporated explicitly. Taking the analysis in Chapter 4, Section 4.3.4, we can write the current for a single Q1D channel, by assuming that the energy of the bound state in the quantum well is determined from a bias being applied symmetrically over the double-barrier structure as (Tewordt *et al.* 1990b)

$$J = e \int_{E_F - eV}^{E_F} N(E) \, v(E) \, T(E + eV/2) \, dE.$$

The evaluation of this integral is dominated by the form of the transmission intensity $T(E)$. This is generally approximated by a Lorentzian:

$$T(E) = \frac{T_0}{1 + [2(E - E_0)/\Gamma]^2}$$

where $T_0 = \Gamma_e/\Gamma$, $\Gamma_e = h/\tau_e$, $\Gamma_i = h/\tau_i$ and $\Gamma = \Gamma_e + \Gamma_i$. Here τ_e is the characteristic tunnelling time of the electron through the barrier (related to the intrinsic linewidth of the quasi-bound state in the quantum well). This time is also an elastic scattering time. The time τ_i is an inelastic scattering time associated with phonon and other lattice or defect scattering processes that involve energy exchange. In the limit of no scattering (the resonant tunnelling limit), $\Gamma = \Gamma_e$, $\Gamma_i = 0$, and $T_0 = 1$. At the other extreme, if $\Gamma_i \gg \Gamma_e$, then $T_0 = \Gamma_e/(\Gamma_e + \Gamma_i) \sim \tau_i/\tau_e$ which is small. The $N(E)v(E)$ product is a constant in the above integral (cf. Chapter 4, Section 4.3.4), and the value of the current is dominated by whether the resonant peak in $T(E)$ falls within the range $E_F - eV$ to E_F

$$J = (2e/h) \int T(E + eV/2) \, dE$$

$$= (e/h) \, T_0 \Gamma \, \{\tan^{-1} 2[E + (eV/2) - E_0]/\Gamma\}|_{E_F - eV}^{E_F}$$

leading, for sufficiently small Γ, to current steps of height

$$\Delta I = e\pi\Gamma_e/h = e/2\tau_e.$$

The current steps in Fig. 11.6(a), obtained using $\Delta I = 10$ nA, imply that $\tau_e = 8$ ps a result comparable with a simple calculation of the linewidth of the double-barrier structure under the appropriate bias. The picture that emerges is that each time one of the bound states in the quantum well drops below the Fermi level of the emitter contact, a current step is seen.

While lateral quantization serves to highlight the steps, its precise origin is far from clear. In larger-area diodes, the much greater number of laterally quantized sub-bands would mean that the effects are blurred out. In practice, the number of current steps does decrease as one increases the diode area, but even in diodes of diameter greater than 1 μm two or three steps can be seen at the current threshold. There is evidence for an interpretation that the lateral confinement may be provided by the Coulomb potential of one strategically placed donor level in the structure. Dellow *et al.* (1992) have used reverse bias, applied to a lateral Schottky gate on top of the substrate (i.e. acting as a collar around the quantum pillar) to narrow the conducting diameter of the pillar continuously. The sudden disappearance of particular current steps with bias is consistent with the depletion region sweeping the edge of the conducting channel through the location of the defect so as to exclude its contribution to the current.

The expected exponential variation of τ_e with barrier thickness has been confirmed in a series of experiments using different structures. The results are plotted in Fig. 11.7, where samples with nominal barriers of 4.5, 5.0, and 7.1 nm (and actual barriers of 4.3, 5.0, and 7.1 nm, as checked by transmission electron microscopy) exhibit initial steps that vary from $c.100$ nA to $c.0.5$ nA. The same data are used to extract the total linewidth Γ from the voltage range over which the step is formed, via $\Gamma = \Gamma_e + \Gamma_i = e\Delta V/2$, assuming symmetric structures. This value is plotted in Fig. 11.7(b), and its independence of barrier thickness suggests that some other mechanism (phonons, interface roughness, etc.) is the dominant scattering mechanism in these diodes. Therefore the results are consistent with the sequential as opposed to the coherent picture of resonant tunnelling (cf. Chapter 3, Section 8.3.4.4).

The data in Fig. 11.6 can be taken further. From the expression leading to the current steps above, the differential conductance is a measure of the energy dependence of the transmission coefficient:

$$G(E) = \partial I/\partial V = (2e^2/h)(\partial E/\partial eV)\,T(E) = (e^2/h)T(E)$$

The final expression follows from a factor of 0.5 arising from the middle factor, again assuming a symmetrical device, and the energy level in the centre of the well shifting by half the applied voltage. The resulting peak value T_0 of $T(E)$ is 8–10 per cent, a value rather less than unity, and much smaller than could be explained by asymmetry of the barriers (a reduction to not less than about 60 per cent) or the effect of the bias. It is further evidence of the strong inelastic scattering, since if $\Gamma_i > \Gamma_e$, then $T_0 = \Gamma_e/(\Gamma + \Gamma_i) \sim \tau_i/\tau_e$ which is small.

A wide range of comparable and complementary data have been obtained by several groups throughout the world (starting with the pioneering experiments of Reed *et al.* (1988)). Extra structure in the current–voltage characteristics was observed, and some lateral quantization was invoked to separate out the different electron modes and give rise to resonant tunnelling at appreciably different voltages. With the further investigation of the current near threshold, the precise origin of lateral quantization (from the physical edge of the structure, defect levels, fluctuations in the widths of the tunnel barriers, etc.) has become a keen issue of continuing debate.

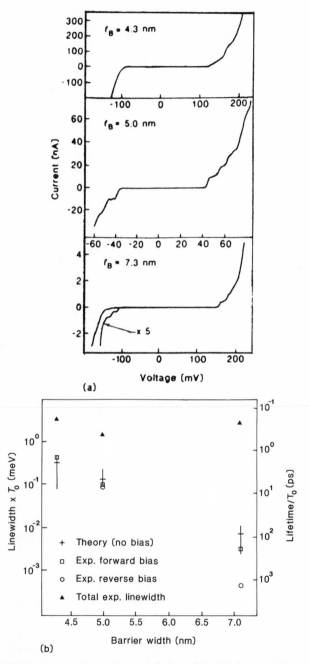

Fig. 11.7 (a) The current–voltage characteristics of double-barrier diodes with different barrier thicknesses (4.3, 5, and 7.1 nm), and (b) the linewidths and lifetimes inferred from the data. The scaling parameter T_0 is discussed in the text. (After Tewordt *et al.* 1991.)

11.3.3 Studies of arrays of quantum dots and wires

A fertile area of investigation has been the response of arrays of quantum wires or dots to various magnetotransport, far-infrared absorption, and other studies. In these experiments, the small signal associated with intrinsically small structures is multiplied. The results have important implications for the subject matter (few-electron systems) of the next chapter, but we concentrate here on the more conventional aspects.

In Fig. 11.8, we show a typical set of results. The surface of a wafer of modulation-doped structure containing a 2DEG is etched to form a series of 1DEGs in parallel. Resistance measurements are made as a function of magnetic field, and the transmission of microwaves or far-infrared light is also measured (from which one obtains the frequency-dependent conductivity $\sigma(\omega)$). Peaks occur in the conductivity just as 1D sub-bands are being

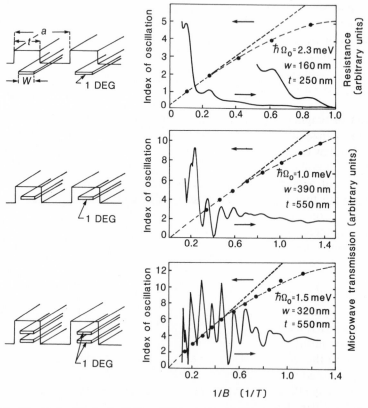

Fig. 11.8 The magnetoconductivity of the arrays of 1D systems shown on the left is measured directly or extracted from the microwave transmission data shown on the right. The peaks are analysed in terms of the 1D channel profile shown in Chapter 6, Fig. 6.3(b), leading to a flat bottom of width w and a parabolic side-wall potential characterized by Ω_0. In each case the etched width is wider because of surface depletion. (After Demel *et al.* 1988.)

depopulated (the density of states is very high). The transport data are plotted as $1/B$, as in 2D systems; the conductivity peaks should lie on a straight line (the fan chart from Chapter 4) as successive Landau levels are depopulated. Once the confining potential starts to interfere with the magnetic potential in the Schrödinger equation (as a low field, cf. Chapter 6, Section 6.3), we anticipate deviations from the 2D behaviour that can be analysed to extract the width of the 1D channel and the number of occupied sub-bands. The microwave transmission is also predicted to scale with conductivity, and the results agree well. This analysis of the data in terms of the inferred width w of the 1D structures and the effective confining potential $\hbar\Omega_0$ is included in the figure. This analysis assumes the conventional quantum confinement to form a 2DEG and an added parabolic potential to perform the lateral quantum confinement for each wire, leaving electrons free to move in one spatial dimension.

The analysis of the confined electron states is even easier if the lateral confinement is 2D, as in arrays of quantum dots (Fig. 11.9). The energy levels and wavefunctions of a 2D parabolic confining potential $V_{con}(r) = m^*\omega_0^2 r^2/2$ can be calculated explicitly, and the effect of a magnetic field (modelling by an extra confining potential $V_{mag}(r) = m^*\omega_c^2 r^2/2$, where ω_c is the cyclotron frequency) is easily added on. This implies that the energy levels and the excitation energies can be calculated in the one-electron approximation as a function of magnetic field. Damage associated with the etching of quantum dots immobilizes all the remaining 2DEG electrons on 200–400 nm diameter dots, but a few electrons in each dot are remobilized by exposure to light. Fig. 11.9 shows a collage of the structure, the transmission of far-infrared light as a function of a magnetic field, and the excitation spectrum compared with theory. Note that the $N = 210$ and $N = 25$ refer to the average number of mobile electrons on each dot, inferred from the strength of the absorption. The good agreement at this level of analysis sets the scene for the next chapter, where new phenomena associated with an intrinsically small number of electrons are investigated.

11.3.4 Lateral superlattices

There is a system that is intermediate between the arrays of 1DEGs and the uniform 2DEG, namely the density-modulated 2DEG. If an array of metallic lines on the surface is used as a gate, it is possible to imprint a weak periodic potential on the electron gas underneath. At present the pitch of the metallic grating is limited to distances greater than about 0.1 μm, but in future a range of one-electron effects (new bandgaps, novel Fermi lines, etc.) await discovery if the pitch can be reduced to c.0.03 μm. An alternative approach to imprinting the periodic potential has come from recent work using the MOCVD growth technique (cf. Chapter 2, Fig. 2.3) to grow 'fractional-layer' superlattices. Vertical layers of AlAs and GaAs are grown above an undoped GaAs layer with a vicinal surface (a few degrees off the $\langle 100 \rangle$). According to the vicinality, the period along the GaAs surface could be 16, 12, or 8 nm.

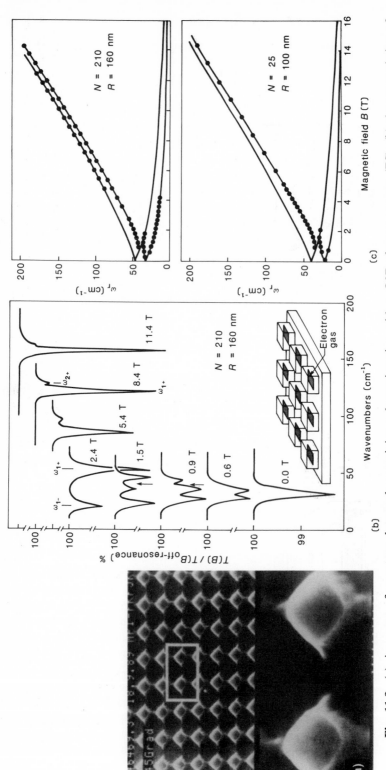

Fig. 11.9 (a) An array of quantum dots each containing a heterojunction with a Q0D electron system (ES), (b) the transmission of far-infrared radiation as a function of magnetic field, and (c) the experimental energy levels compared with theory. The values N and R are respectively the number of electrons in each dot and the radius of the electron system as best fits the data. The splitting of the energy levels at low field is caused by Coulomb interaction between adjacent electron systems. (After Heitmann *et al.* 1992.)

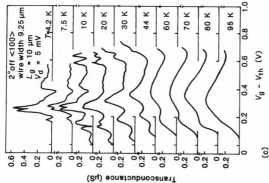

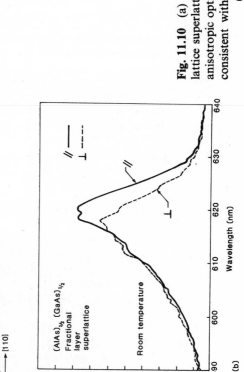

Fig. 11.10 (a) A 2DEG subject to a 1D periodic potential from a fractional lattice superlattice (of the kind shown in Chapter 2, Fig. 2.3(a)) showing (b) anisotropic optical properties and (c) extra structure in the transconductance consistent with the periodic potential imposed by the lateral superlattice. (After Tsubaki *et al.* 1992; Fukui *et al.* 1992.)

After a growth of these superlattices, the normal growth of a high mobility 2DEG structure is continued. The structure clearly shows up the effects of a periodic potential in one dimension when examined optically or electrically after patterning to give a wire that is 0.25 μm wide. In Fig. 11.10 we show a schematic diagram of the test structure, together with the anisotropy of the polarization-dependent photoluminescence spectra and the evolution with reducing temperature of structure in the transconductance of the device. From a preliminary analysis it has been inferred from the data that the amplitude of the periodic potential on the 2DEG is about 7 meV and the Fermi energy is about 18.5 meV. Other experiments have been performed on modulated 2DEG structures, relying on lithography and etching to imprint the periodic features. In the main, low-field magnetotransport measurements have been performed and extra structure is seen when the magnetic length is comparable to the artificial periodicity. This whole field was recently reviewed in detail by Hansen *et al.* (1992).

References

Beaumont, S. P. (1992). Quantum wires and dots: defect related effects. *Physica Scripta*, **T45**, 196–9.

Dellow, M. W., Beton, P. H., Langerak, C. J. G. M., Foster, T. J., Main, P. C., Eaves, L., *et al.* (1992). Resonant tunnelling through the bound states of a single donor atom in a quantum well. *Physical Review Letters*, **68**, 1754–7.

Demel, T., Heitmann, D., Grambow, P., and Ploog, K. (1988). One-dimensional electronic systems in ultrafine mesa-etched single and multiple quantum well wires. *Applied Physics Letters*, **53**, 2176–8.

Fukui, T., Tsubaki, K., Saito, H., Kasu, M., and Honda, S. (1992). Fractional superlattices grown by MOCVD and their device applications. *Surface Science*, **267**, 588–92.

Hansen, W., Kotthaus, J. P., and Merkt, U. (1992). Electrons in laterally periodic nanostructures. In *Semiconductors and semimetals*, Vol. 35, *Nanostructured systems* (ed. M. Reed), pp. 279–380. Academic Press, New York.

Hasko, D. G., Potts, A., Cleaver, J. R. A., Smith, C. G. and Ahmed H. (1988) Fabrication of sub-micrometre free-standing single-crystal gallium arsenide and silicon structures for quantum transport studies. *Journal of Vacuum Science and Technology B*, **6**, 1849–51.

Heitmann, D., Kern, K., Demel, T., Grambow, P., Ploog, K., and Zhang, Y. H. (1992). Spectroscopy of quantum dots and antidots. *Surface Science*, **267**, 245–52.

Muray, A., Isaacson, M., and Adesida, I. (1984). AlF₃—a new very high resolution electron beam resist. *Applied Physics Letters*, **45**, 589–91.

Potts, A., Kelly, M. J., Hasko, D. G., Smith, C. G., Cleaver, J. R. A., Frost, J. E. F., *et al.* (1991). Thermal transport in free standing semiconductor fine wires. *Superlattices and Microstructures*, **9**, 315–18.

Potts, A., Kelly, M. J., Hasko, D. G., Cleaver, J. R. A., Ahmed, H., Ritchie, D. A., *et al.* (1992). Lattice heating of free-standing ultra-fine GaAs wires by hot electrons. *Semiconductor Science and Technology*, **7**, B231–4.

Reed, M. A., Randall, J. N., Aggarwal, R. J., Matyi, R. J., Moore, T. M., and Wetsel, A. E. (1988). Observation of discrete electronic states in a zero-dimensional nanostructure. *Physical Review Letters*, **60**, 535–8.

Rosseinsky, M. J., Murphy, D. W., Fleming, R. M., Tycko, R., Ramirez, A. P., Siegrist, T., *et al.* (1992). Structural and electronic properties of sodium-intercalated C_{60}. *Nature, London*, **356**, 411–18.

Tewordt, M., Law, V. J., Syme, R. T., Kelly, M. J., Newbury, R., Pepper, M., *et al.* (1990*a*). Transmission probability of electrons through laterally confined resonant tunnelling structures. *Proceedings of the 20th International Conference on the Physics of Semiconductors, Thessaloniki, 1990* (eds. E. M. Anastassakis and J. D. Joannopoulos), pp. 2455–8. World Scientific, Singapore.

Tewordt, M., Law, V. J., Kelly, M. J., Newbury, R., Pepper, M., Peacock, D. C., *et al.* (1990*b*). Direct experimental determination of the tunnelling time and transmission probability of electrons through a resonant tunnelling structure. *Journal of Physics: Condensed Matter*, **2**, 8969–75.

Tewordt, M., Law, V. J., Kelly, M. J., Syme, R. T., Newbury, R., Pepper, M., *et al.* (1991). Electron-state lifetimes in submicron diameter resonant tunnelling diodes. *Applied Physics Letters*, **59**, 1966–8.

Tsubaki, K., Honda, T., Saito, H., and Fukui, T. (1992). Density of states of AlAs/GaAs fractional layer superlattice quantum wire in modulation doped structure. *Surface Science*, **267**, 270–3.

Turner, P. S., Bullough, T. J., Devenish, R. W., Maher, D. M., and Humphreys, C. J. (1990). Nanometre hole formation in MgO using electron beams. *Philosophical Magazine Letters*, **61**, 181–93.

Zimmerman, N. M., Liddle, J. A., White, A. E., and Short, K. T. (1993). Transport in submicrometer buried mesotaxial cobalt silicide wires. *Applied Physics Letters*, **62**, 387–9.

12 Mesoscopic phenomena and Coulomb blockade

12.1 Introduction

In Chapters 6 and 11 we encountered structures in which there were very few free and mobile carriers. Certainly, the usual condition for statistical averaging, namely that the number N of particles satisfies $1 << \sqrt{N} << N$, is violated. In this chapter, we concentrate on the physics of few-carrier systems, assuming that the fabrication and measurement techniques are those already described. The physics contains a number of new aspects, and their investigation is far from complete. Many new effects appear undesirable in terms of eventual device applications, while others are being proposed as the basis for radically new device technologies. The term mesoscopic (meso=middle) has been introduced to describe those systems that are neither microscopic (one or a few atoms) nor macroscopic. In practice, we shall always be dealing with structures whose volume might be several tens of nanometres on each side, but which contain only a few carriers taking part in the relevant transport or optical processes. Until now, our discussion has concentrated on the wave nature of electron transport behaviour and not on its charge nature. The capacitances encountered in these small structures are so small that the energy to charge a structure with even a single electron is such that $e^2/2C > kT$, and charge transfer into and out of the structure can be inhibited by this inequality, which is known as the Coulomb blockade.

12.2 Small structures

We re-examine some of the structures previously introduced to establish the sense in which they can be considered mesoscopic. In Chapter 6, the split-gate transistor was used to generate a Q1D electron gas. If the areal electron density in the Q1D channel is not less than that in the 2DEG at either end (say about 10^{15} m^{-2}), a channel that is 0.1 μm wide and 0.5 μm long contains a total of about 100 electrons. With a Fermi energy E_F of 2–3 meV, the number of those electrons within kT of E_F at 1 K is about five. Consider an experiment being performed in which a bias of $c.1$ meV is applied and the current is of order 10 nA. If the electrons move ballistically with the Fermi velocity, they are typically at least 1 μm apart, and the Q1D channel is empty of transport electrons for much of the time! We shall return to modified split-gate structures below to investigate single-electron phenomena and correlations in transport induced by the Coulomb charge of each electron.

The quantum dots used for far-infrared studies in Chapter 11, Section 11.3.3, contained typically 100 electrons per dot, but this number could be reduced to zero by the etching process used to make the dots, and then increased from zero by the application of light to detrap some electrons. With sufficient care the average number of electrons per dot can be controlled down to the level of single electrons.

The resonant tunnelling experiments in quantum pillars involve currents ranging from 1 to 100 pA, and with a low-field drift velocity of order 10^4 m s^{-1} the number of transport electrons in the dot is no more than 0.1 on average!

Having established that the structures are small in the sense of the number of carriers, we can also show another inherent property: the structures are highly individual and are not multiply reproducible. Consider the modulation-doped heterojunction that provides the 2DEG from which the Q1D system is formed by electrostatic squeezing. The Si atoms that form the donors in the AlGaAs are typically 30 nm apart in the plane of the AlGaAs layer. This means that there may be only about 15 dopant atoms above the Q1D channel. Given the statistical distribution of the x–y positions of the Si dopant atoms, and their distribution in distance from the 2DEG if the doped AlGaAs layer is of finite thickness, it is clear that one cannot expect to make many identical Q1D channels on a single wafer. There will be a significant statistical distribution of parameters associated with each channel; this is exemplified in Chapter 6, Fig. 6.5, where the potential profile seen by electrons in a 1DEG is a strong function of the disposition of dopant ions in the AlGaAs layer. Within these statistical fluctuations, there is also the rare possibility of a single defect or impurity atom adjacent to or even within the volume occupied by the Q1D channel. One distinguishing characteristic of small structures is their inherent irreproducibility.

Much of the interesting physics to be described below was first investigated in ultrasmall single-crystal, normal, and superconducting metal wires. The particular advantage that semiconductors have brought is the relatively low electron density in a very clean materials system, so that the Fermi wavelength is 50 nm rather than the 0.5–5 nm in metals. This makes interference and other phenomena that much easier to investigate.

12.3 Wavefunction coherence

In modern optics the concept of spatial and temporal coherence of light waves plays an important role in diffraction, interferometry, lasing, and many other phenomena which depend on precise differences in phase between interacting waves. The coherence of the electron wavefunction dominates our understanding of the electronic and optical properties of mesoscopic systems.

An electron passing through a semiconductor may undergo many different types of scattering process via its interaction with other electrons, ions, lattice vibrations, etc. In those scattering processes where the energy of the given electron is conserved (elastic collisions), its phase memory is preserved. Even though the electron motion may involve a change in direction, the relative phase of the wavefunction of the incoming and outgoing electron can be determined from the details of the scattering process. The implication is that if an electron returns to a particular point in a semiconductor having suffered only elastic collisions, it can interfere with itself. We shall see evidence for this below. In inelastic collisions, where energy is transferred to or from the electron, there is no fixed relation of the phase of any electron before and after the collision. At low temperatures and in pure materials, the inelastic scattering length can be macroscopic, while the mean free path between (elastic) scattering events may remain modest or also increase with reducing temperature. This will allow for the manifestation of electron coherence phenomena. In contrast, in disordered systems, particularly those which are non-crystalline, it is possible to have an inelastic length shorter than any elastic scattering length, precluding the observation of any coherence effects.

The measurement of the phase coherence length L_ϕ is made in reduced dimensionality systems by following the magnetoresistance as described in Section 12.4.3.

12.4 Manifestations of electron wavefunction coherence

In this section we describe experiments that provide evidence of electron coherence in semiconductors at low temperatures. Even though we have high-quality material, we still refer to the semiconductor samples as dis- ordered. By this we mean that they are imperfect, with dopant or other impurity atoms, non-ideal interfaces, etc. These imperfections play a role in elastic scattering which makes some effects of coherence more prominent and easier to measure.

12.4.1 Aharonov—Bohm effect

In Fig. 12.1(a) we show a structure that consists of a narrow ring with leads attached. It is formed by the definition of gates at two levels above a 2DEG, such that when a bias is applied to the gates, carriers deplete out from all regions other than that of the ring and its leads. In Fig. 12.1(b) we show the magnetoresistance of this loop, and in Fig. 12.1(c) we reproduce a Fourier analysis, showing that the variations of resistance have a periodicity of h/e (the magnetic flux quantum). The explanation was given in Chapter 4, namely that electrons approaching the loop have the choice of taking either arm and will interfere (if they retain their phase coherence) when they recombine to exit the loop. The effect of the magnetic field is to introduce a variable phase difference between the electrons which take either path. The total difference is $\delta = \pm \int A \cdot \mathbf{ds}$ according to which loop is taken (A is the vector potential and $\mathbf{ds}$ is the arc along the path taken). By Stokes' theorem, this is equivalent to $\int B \cdot \mathbf{d}A$, which is the integral of the magnetic flux density over the area enclosed by the loop. Every time an extra flux quantum enters the loop the

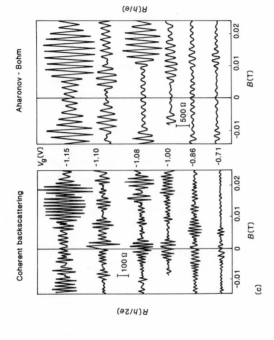

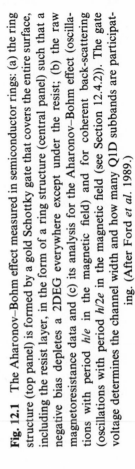

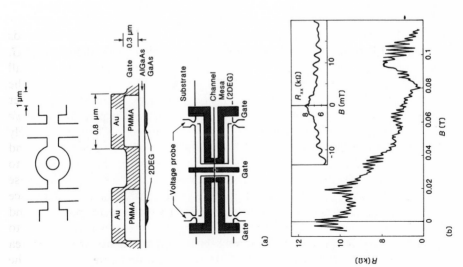

Fig. 12.1 The Aharonov–Bohm effect measured in semiconductor rings: (a) the ring structure (top panel) is formed by a gold Schottky gate that covers the entire surface, including the resist layer, in the form of a ring structure (central panel) such that a negative bias depletes a 2DEG everywhere except under the resist; (b) the raw magnetoresistance data and (c) its analysis for the Aharonov–Bohm effect (oscillations with period h/e in the magnetic field) and for coherent back-scattering (oscillations with period $h/2e$ in the magnetic field (see Section 12.4.2)). The gate voltage determines the channel width and how many Q1D subbands are participating. (After Ford *et al.* 1989.)

relative phase change is an extra 2π. Ideally, if the conducting channel that makes up the loop is monomode and there is no scattering, the resistance modulation should be 100 per cent. In practice, there are several lateral modes for which the cross-interference is modest, and there is some inelastic scattering with loss of phase coherence. Modulation of about 20–40 per cent has been achieved.

12.4.2 The h/2e oscillations and coherent backscattering

The filtered data shown in Fig. 12.1(c) also reveal a further set of oscillations with a period of $h/2e$. This corresponds to effectively twice as large an area being swept out in terms of extra flux being introduced. The explanation is that electrons go right around the ring in opposite directions and interfere at the entrance to the ring (Fig. 12.2). In doing so they enclose twice as much flux as in the Aharonov–Bohm effect. This new effect is a manifestation of coherent back-scattering. Since its clarification in (metallic and semiconducting) solids, the same effect due to the same physics has been sought and observed in optical systems: there is a narrow cone of reverse reflected light from a disordered set of transparent spheres (Kaveh *et al.* 1986).

12.4.3 Negative magnetoresistance and weak localization

In many instances, the magnetoresistance of metals is positive and is quadratic with applied magnetic field; this may be caused by the curved trajectories induced by a magnetic field and hence the extra distances travelled by the electrons, permitting more scattering. In our present regime of no inelastic scattering, the conductivity rises as a weak magnetic field is applied. The

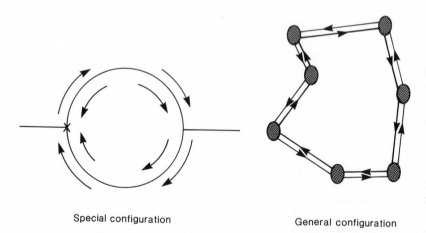

Special configuration   General configuration

Fig. 12.2 The phenomenon of coherent back-scattering whereby electrons return to a given point in space having taken a particular path (involving multiple elastic scatterings) or its time-reversed path around the loop. The electron wavefunctions are coherent and can interfere. A magnetic field threading the loop can impart a phase difference between electrons going in opposite directions.

reason is that all the electrons that contribute to the $h/2e$ oscillations undergo coherent back-scattering, and have exactly the same phase when they traverse any of their paths in the forward and reverse directions. The application of a small magnetic field (order of millitesla) removes the perfect phase coherence and the constructive interference of the electrons at the entry point to the loop. For small magnetic fields, there is a reduced probability of the electrons returning to their starting point and constructively interfering there. Therefore the conductance increases and negative magnetoresistance follows; this can be seen in Fig. 12.1(b). This negative magnetoresistance is a widespread phenomenon in disordered solids, particularly in two dimensions, as the explanation just given relies only on the loop trajectory of some electrons and not on the particular topology in which it is unambiguously seen and described. Any closed-loop trajectories that are traversed in a time-reversed manner can give rise to the same increasing conductance with the application of a weak magnetic field.

This negative magnetoresistance is used to measure the value of the phase coherence length L_ϕ and its dependence on temperature. There are a number of detailed formulations appropriate to particular geometry samples. We gave one example in Chapter 5, Section 5.9: a doped GaAs channel within a MESFET was biased to near pinch-off (cf. Fig. 5.15), and the magnetoconductance showed a logarithmic variation with magnetic field, the scale of which was determined by L_ϕ (called L_{in} for inelastic there) which varies as $1/\sqrt{T}$. In a Q1D channel in the regime of diffusive transport (cf. Chapter 6, Section 6.4.1), when the elastic scattering length l, the width W, of the Q1D channel the magnetic cyclotron length L_c, and the phase coherence length L_ϕ satisfy the conditions $l < W$ and $L_c < L_\phi$, the magnetoconductance varies as (Pepper 1988)

$$\delta G(B) = (2e^2/\hbar)(1/L_\phi^2 + 4W^2e^2B^2/3\hbar^2).$$

In Fig. 12.3(a) we show the raw data, along with a fit to this formula, and in Fig. 12.3(b) we show the value and temperature dependence ($T^{-1/3}$) extracted for $L_\phi = L_{in}$. An alternative method for obtaining L_ϕ is to analyse the scale of the universal conductance fluctuations described in the next section.

The constructive interference associated with carriers scattering from closed-loop trajectories and returning to an origin is considered a weak form of localization, and indeed many of the phenomena discussed in this chapter can be viewed from this starting point. The weak form is to be distinguished from the strong form of localization encountered in strongly disordered systems, where electrons can be confined to particular regions with wavefunctions that fall off exponentially with increasing distance. The localization here is rather weaker, being power-law-like in two dimensions. This refines the meaning of 'extended' in Chapter 4, Fig. 4.5(b), as it applies in two dimensions.

12.4.4 Universality of conductance fluctuations

The theory of macroscopic disordered systems has included as an important element a means of 'configuration averaging' when calculating the conductiv-

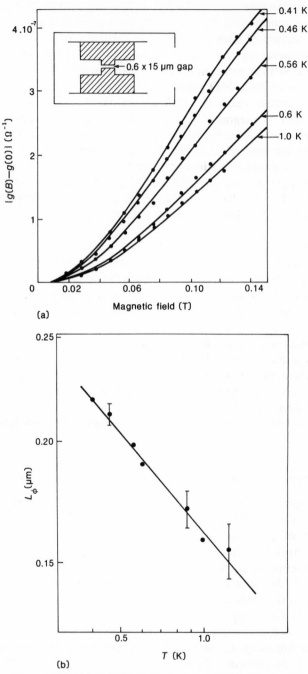

(a)

(b)

Fig. 12.3 (a) The magnetoconductance data of a Q1D channel and the fit to the theory as described in the text; (b) the value of the phase coherence length L_ϕ extracted from the fit. (After Thornton *et al.* 1986.)

ity of a particular realization of scatters. It is argued that a real disordered system is not represented by any given realization within some simplified model. One feature of such averaging is to remove any sharp structure (say in the conductance as a function of energy) associated with a particular configuration. Within semiconductor science, the measurements of the transport properties of fine Q1D wires (such as those described in Chapter 11, Section 11.2) reveal sharp structure in the conductance as some external parameter (such as magnetic field strength or even direction) is varied (cf. Chapter 6, Fig. 6.5 and 6.6(d)). Furthermore, such conductance fluctuations are always on the scale of e^2/h. It was quickly appreciated that configuration averaging is not appropriate in small structures, where the scattering centres precisely represent a particular configuration. This fact is clear in very small structures, but becomes less so as the structures become larger and contain more impurities or other scattering centres. The key issue is the phase coherence length L_ϕ and the ratio of the sample size L to L_ϕ. If we break up a sample into cubic volumes (square areas or equal lengths in reduced dimensions), each of side L_ϕ, and assume both perfect phase coherence within and no phase coherence between different volumes, we obtain an appropriate scaling of the fluctuations as the samples become macroscopic. The size of L_ϕ can be obtained from data on magnetoconductance as described earlier. It is also possible to use statistical correlation methods on the conductance fluctuation data itself to extract the value of L_ϕ.

Within volumes of samples defined on a scale L_ϕ, it is not possible to undertake averaging of scattering events for transport calculations. On this scale, a structure is said to be non-self-averaging. Self-averaging is possible on the scale of many L_ϕ, as has been amply demonstrated in the optical and electronic properties of, say, macroscopic specimens of disordered semiconductors.

12.5 Theoretical concepts for mesoscopic systems

The description of transport phenomena in mesoscopic systems involves the partial transmission of electrons in a coherent manner, with some scattering causing loss of phase coherence. In this context, the Landauer–Buttiker formalism comes into its own. Indeed, although used more widely in low-dimensional semiconductor physics, this formalism, which was introduced in Chapter 4, was developed to treat mesoscopic systems but is now used much more widely in the context of the Hall effect and Q1D transport (cf. Chapters 5, 6, and 11). The computation of the transmission coefficients depends on being able to define the potential seen by carriers to a very precise degree, which is not normally accessible, and one often resorts to extracting them from data for the particular device under consideration. When each sample is very individual, the precise location of impurity atoms can strongly influence the transport coefficients. The Landauer–Buttiker formalism is a linear response theory, and its extension into the non-linear regime from where any practical applications are likely to emerge is not complete. The correlations between the motion of the few electrons in a mesoscopic system are important and are considered in the next section. The theory of transport in mesoscopic systems is not a mature subject, but rather is under continuing investigation.

12.6 Coulomb blockade: theory

Until now, we have examined small structures with a wavefunction treatment of the electrons. If we consider the capacitance C of our small structures, we find that the energy required to put even one electron on them $(e^2/2C)$ can exceed kT at low temperatures. For the example in Chapter 11, Fig. 11.3, of a (square) quantum pillar of side, say, 0.1 μm, the capacitance between the heavily doped contact layers if they are, say, 20 nm apart is $C = 5 \times 10^{-17}$ F. The charging energy in this case is 3 meV, or kT for $T \sim 30$ K. In a small structure at low temperatures, we are able to see effects that depend on the discrete charged nature of the electron. This topic of research started with ultrasmall tunnel junctions in metals, but the kinds of structures discussed in Chapters 6 and 11, and earlier in this chapter, are ideal for further investigations. In this section we give an introduction to a theory for this effect in double-barrier tunnel structures. Experimental results for in-plane and vertical tunnel structures are given in the next section, with more general phenomena considered in later sections.

In Fig. 12.4 we show a schematic cross-section of a structure suitable for theoretical analysis (Fig. 12.4(a)) and a potential profile through such a structure (Fig. 12.4(b)). Practical realizations are described in the next section. Figure 12.4(a) shows a small quantum-dot structure connected to two large reservoirs by tunnel barriers. The dot is assumed to be small enough that the one-electron energy levels are well separated in energy (compared with kT at low temperatures). In the absence of a magnetic field, each level is empty or doubly occupied. The reservoirs are assumed to be metallic at thermal equilibrium at temperature T with a Fermi energy E_F. As the number of electrons localized in the quantum dot is assumed to be an integer N, an electrostatic potential difference $\Delta\phi(Q = -Ne)$ can be built up between the dot and the surrounding reservoirs. This potential difference $\Delta\phi$ can be considered as arising from two sources, one associated with the total capacitance C between the dot and the surrounding reservoirs, and the second being due to the effect of charges external to the dot, so that

$$\Delta\phi = Q/C + \Delta\phi_{\text{ext}}.$$

The electrostatic energy takes the form

$$U(N) = \int_0^{-Ne} \Delta\phi(Q')\,\mathrm{d}Q' = (Ne)^2/2C - Ne\Delta\phi_{\text{ext}}.$$

In the in-plane structure (a small structure defined by electrostatic barriers generated by Schottky gates above a 2DEG, as described in Chapter 6), the

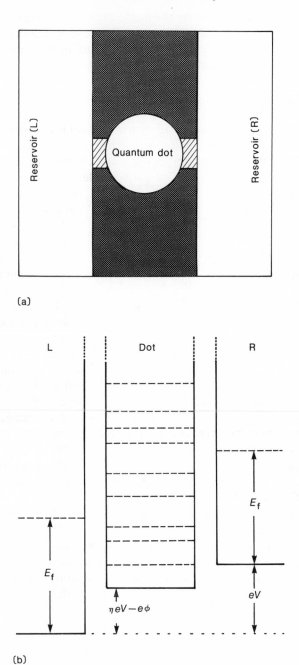

(a)

(b)

Fig. 12.4 The simplest Coulomb blockade structure: (a) the theoretical model, and (b) the potential profile for a double-barrier diode structure as a realization. (After Beenakker 1991.)

external charges are supplied by the modulation-doped layer and by the gate electrode. In the vertical device, donor levels in the depletion regions perform the same function, as do charges in surface states at the physical boundaries. Where the vertical structures have a Schottky gate around the pillar or on the substrate at the bottom of the pillar, an extra degree of control is possible. If we write $Q_{ext} = C\Delta\phi_{ext}$ as a definition of Q_{ext}, then the latter can be regarded as continuously variable so that the electrostatic energy term above can be rewritten as

$$U(N) = [(Ne - Q_{ext})^2 - Q_{ext}^2]/2C = \tilde{U}(N) - Q_{ext}^2/2C.$$

The $\tilde{U}(N)$ part is plotted as a function of Q_{ext} for different N in Fig. 12.5, and it can be seen that, if $e^2/2C >> kT$, the tunnelling of an electron from one reservoir through the quantum dot into the other reservoir is blocked unless we are at the degeneracy points, where the imbalance $Ne - Q_{ext}$ jumps from $-e/2$ to $+e/2$. Here Q_{ext} is the control variable which is varied via some gate voltage applied to one or other reservoir.

If a bias voltage V is applied between the two reservoirs, the potential profile is as shown schematically in Fig. 12.4(b) with a fraction η being considered dropped over one tunnel barrier and a fraction $1 - \eta$ over the other—the energy levels within the quantum dot are shown as broken lines. Current will flow by virtue of tunnelling from the right electrode into the dot and out to the left electrode. We assume that the energy separation ΔE between one-electron levels and kT are both much greater than $h(\Gamma_1 + \Gamma_r)$ where the Γ are the scattering rates determining the width of the tunnelling transmission resonance (cf. Chapter 8). This allows us to distinguish between the different levels on the dot. We also assume that the tunnelling process is elastic, although there are generalizations to include scattering. The current is obtained by analysing electrons tunnelling into or out of the quantum dot into the left or right reservoir. The initial or final energy in the reservoir is related to the energy level in the dot—the electrostatic energy $U(N + 1) - U(N)$ for

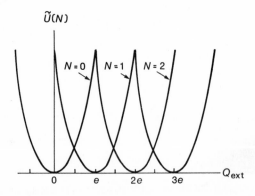

Fig. 12.5 Diagram of part of the total electrostatic energy of a quantum dot with N electrons, as described in the text.

electrons that end up on the dot or $U(N) - U(N - 1)$ for electrons that end up on the electrode. The probability of tunnelling is proportional to the Γ, and in the steady state the current through the left barrier equals that through the right. The full analysis is complex, but analytical, with the final expression for current being similar to that derived for Q1D tunnelling in Chapter 11, but adding the electrostatic terms to the energy conservation terms. In the limit where $h\Gamma << kT << \Delta E$, the current is dominated by one of the energy levels on the dot. A simplified version of this problem is as follows (Beenakker *et al.* 1991). The probability of finding N electrons on the quantum dot in equilibrium with the reservoirs is given by

$$P(N) \sim \exp[-(F(N) - NE_F)/kT]$$

where $F(N)$ is the Gibbs free energy, which approaches the ground-state energy $\mathcal{E}(N)$ of the dot as $T \to 0$. From the above discussion, we can estimate $\mathcal{E}(N)$ as

$$\mathcal{E}(N) = \sum_{p=1,N} E_p + (N)^2/2C - Ne\Delta\phi_{ext},$$

where the E_p are the one-electron energy levels in ascending order, which depend on the size of the quantum dot, the gate voltages, and any magnetic fields, but are assumed not to depend on N. $P(N)$ is zero at very low temperature unless the condition $\mathcal{E}(N) - NE_F = 0$ is satisfied for some N. For current to flow we need a finite probability for electrons entering and leaving the dot, i.e. a finite probability of there being either N or $N + 1$ electrons on the dot which implies

$$F(N + 1) - (N + 1)E_F = F(N) - NE_F \Rightarrow \mathcal{E}(N + 1) - \mathcal{E}(N) = E_F.$$

Note that these last two equations can be combined to give a renormalized energy for the N-electron quantum dot, namely

$$E_N^* = E_N + (N - \tfrac{1}{2})e^2/C = E_F + e\Delta\phi_{ext}.$$

There will be peaks in the current as the bias is applied such that this condition is satisfied for successive N. The magnitude of the current depends on the tunnelling rates and the detailed overlap of wavefunctions. Given the irregular boundary of most practical quantum dots and the typical value of N, the overlap of wavefunctions for different N can vary widely.

12.7 Coulomb blockade: experiments

In practical realizations, such as a 2DEG subject to a potential established by multiple gates, the tunnel barriers are obtained by biasing gates so that 1D channels are *just* pinched off (cf. Chapter 6 Fig. 6.2(b)). Under a small d.c. bias, electrons can tunnel through such barriers. In the inset to Fig. 12.6, we show a typical gate structure. A negative bias is applied to G_1, G_2, and G_4 such that a tunnel barrier is formed electrostatically in the 2DEG gas below the

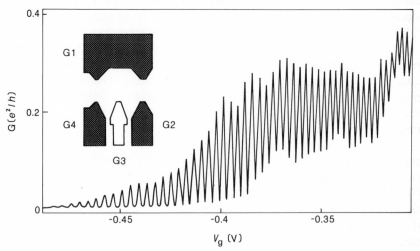

Fig. 12.6 The conductance through a quantum dot formed in a 2DEG by the pattern of gates shown in the inset, with the gate bias being applied to G_3. The separation between G_1 and G_4 is 0.4 μm, and the rest of the structure is approximately to scale. The peak in the conductance oscillations correspond to the removal (for increasingly negative gate bias) of one electron from the quantum dot, and there are about 100 electrons in the dot at −0.4 V gate bias. (After Ford *et al.* 1993.)

surface spanning the points of closest approach of G_1 and G_2 and of G_1 and G_4 (compare Figs 12.4 and 12.6). Experiments are performed to measure the conductance through the whole structure, which is held at a low temperature (typically *c.*0.1 K). This means that a very small bias is applied across the 2DEG subject to fixed gate biases and the current is measured. The bias on gate G_3 is used to modify the size of the quantum dot, and as the dot is squeezed smaller with increasing negative bias, a series of peaks occur each time the condition

$$E_N^* = E_N + (N - \tfrac{1}{2})(e^2/C) = E_F + e\Delta\,\phi_{\text{ext}}$$

is satisfied. For a sufficiently large dot ($N \geqslant 100$), this equality is satisfied for successive values of N. If $E_N \gg e^2/C$, then the peaks in conductance will occur roughly periodically, with the period determined by $\Delta\,(e\Delta\,\phi_{\text{ext}}) = e^2/C$, as seen in the conductance data in Fig. 12.6 where successive peaks going to the left arise from a quantum dot containing one less electron. The period of 7 meV can be used to infer the capacitance of the quantum dot, which scales approximately as the number of gates, and the charging energy is estimated to be $e^2/C \sim 1$ meV. Note the strong variations in the amplitudes of adjacent peaks in the conductance, which arise from the irregularity of the dot and the shape of the successive quantum-dot wavefunctions into and out of which the electrons tunnel. With the addition of a magnetic field, one can anticipate the energy levels on the quantum dot as arising from a combination of electrostatic and magnetic confinement, and some spin-splitting of the one-electron levels. Data are available giving an indication of both effects, but a full explanation cannot be provided as the data are very complex.

A second structure in which Coulomb blockade phenomena are seen is the resonant tunnelling quantum pillar of Chapter 11, Fig. 11.3. Note the qualitative similarity of the conductance data in Fig. 11.6 and 12.6. Here, in practical structures, the small capacitance, the large Fermi energy in the emitter, and the width of the tunnelling resonances make the data less easy to interpret other than in terms of a renormalized E_N^* (Tewordt *et al.* 1994).

The study of Coulomb blockade systems is in its infancy, and our understanding is being refined daily. A newer way of detecting the Coulomb effects is illustrated in the inset to Fig. 12.7, where the quantum dot is formed by G_1, G_3, G_4, and G_5, with G_4 as the gate that controls the dot size. The gates G_1 and G_2 form an adjacent measuring circuit. The latter is pinched off into the tunnelling regime, where the conductance is very sensitive to the local electrostatic fields including those caused by the quantum dot. The upper panel shows the detector circuit resistance. Its variation on a smoothly rising background is used to calculate the changes $\Delta\phi_{dot}$ dot in the electrostatic potential on the quantum dot. The conductance G of the quantum dot is also shown in the upper panel. Peaks and minima in G occur at the average values of $\Delta\phi_{dot}$. The minima in G occur when the Fermi level is midway between the levels E_N and E_{N+1}, while the maxima occur when the dot is occupied 50 per cent of its time with the extra $(N+1)$th electron, and $\Delta\phi_{dot} = 0$ for both. Calibrations and cross-checks are available by varying the different gate voltages, the temperature, and the magnetic fields.

The integration of quantum-scale structures in the manner suggested in Figure 12.7 is likely to yield more understanding of the electrostatic profile in semiconductor microstructures, and be used to probe the one-electron states in a quantum dot with greater finesse than can be achieved with, say, scattering experiments in atomic and nuclear physics and other areas of physics.

12.8 Random telegraph signals

The electrostatic profile of some mesoscopic systems described in this and earlier chapters is strongly modified if there is an impurity site, whose charge state is able to change (e.g. if it contains an energy level within kT of the local Fermi level), in close proximity. Such a change (from charged to neutral) can be sufficient to change the number of occupied sub-bands in the vicinity, and in turn modify the conductance or other transport coefficients of the structure. One of the oldest manifestations of a single-electron effect is the random telegraph signal noise in the current or resistance of a structure, which is associated with such charging and discharging of a strategically placed impurity in relatively large devices (Kirton and Uren 1989). The signal is pronounced in mesoscopic systems, as the fractional change in resistance can be quite large. In Fig. 12.8(a) we show the change in source–drain resistance

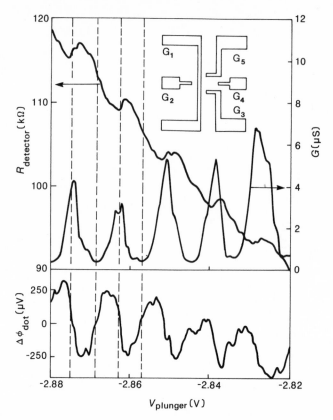

Fig. 12.7 A more complex Coulomb blockade circuit consisting of a variable quantum dot with a detector circuit in close proximity (see Figure 6.2a). The detector resistance is measured in the left-hand circuit (i.e. currents flowing in the gap between G_1 and G_2) for fixed bias, while the plunger G_4 voltage is varied. The raw resistance data are converted into conductance in the top panel. The charging energy ($\Delta\phi_{dot}$) of the quantum dot can be obtained from a detailed set of calibration measurements on the dot, and the relative temperature dependences of the quantum dot and detector circuit resistances (as described in the original paper), but the magnitude of the charging energy is in good agreement with that estimated by the capacitance of the structure.
(After Field *et al.* 1993.)

of a 0.15 μm × 1 μm Si MOSFET, and in Fig. 12.8(b) we show the resistance of a segment of a small MOSFET qualitatively similar to that shown in Fig. 6.1(b) of Chapter 6. The resistance change of a fraction of a per cent is shown over a time interval of the scale of 1 min. The resistance is seen to switch between two values. The variables are the gate voltage (and hence the Fermi level in the inversion layer) and the temperature. The rate of switching between the two states increases with temperature. As the Fermi level is swept by the gate voltage, the resistance changes from being predominantly in one state (say low resistance) to being predominantely in the other

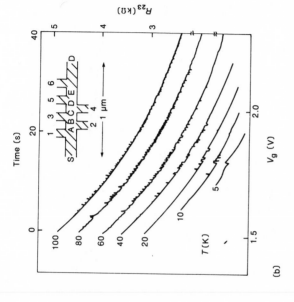

(a)

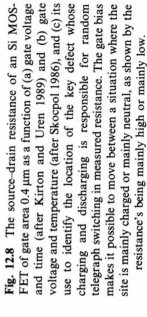

(b)

(c)

Fig. 12.8 The source–drain resistance of an Si MOS-FET of gate area 0.4 μm as a function of (a) gate voltage and time (after Kirton and Uren 1989) and (b) gate voltage and temperature (after Skocpol 1986), and (c) its use to identify the location of the key defect whose charging and discharging is responsible for random telegraph switching in measured resistance. The gate bias makes it possible to move between a situation where the site is mainly charged or mainly neutral, as shown by the resistance's being mainly high or mainly low.

state as the defect changes from being predominantly charged to neutral or vice versa.

The theory for random telegraph signals is relatively simple (Kirton and Uren 1989) and shares some superficial similarity with that introduced for the Coulomb blockade. We define E_c as the value of the Fermi level where a defect state changes its occupancy from n to $n + 1$ at $T = 0$. The zero of energy is chosen to be that of the defect occupied by n electrons. We consider the grand partition function of the defect state, which is assumed to have only two occupancies, of degeneracy $\gamma(n)$ and $\gamma(n + 1)$, as

$$Z_G = \gamma(n)\exp(-nE_F/kT) + \gamma(n + 1)\exp\{-[E_c - (n + 1)E_F]/kT\}.$$

The probability of finding the defect in the $(n + 1)$th electron state is

$$f = p(n + 1) = \frac{1}{1 + g\exp(E_c - E_F)/kT}$$

where $g = \gamma(n)/\gamma(n + 1)$. This probability can be recast in terms of the mean capture and emission times τ_c and τ_e, which can be read off the data, as

$$f = \tau_c/(\tau_c + \tau_e)$$

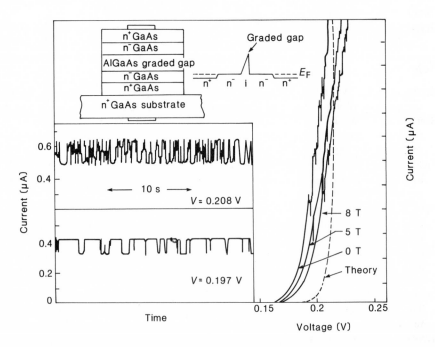

Fig. 12.9 Random telegraph signals measured in a large-area (*c.*180 μm diameter) diode, where the charging and discharging of a strategically placed defect near the maximum Al composition in the graded AlGaAs layer can lower the potential barrier in its vicinity and so introduce a significant current filament. (After Judd *et al.* 1986.)

In practice, the data can be used to extract the value of E_c and hence, if the nature of the defect is guessed at (a particular charge state of an oxide defect in SiO_2), its location can be established given the gate voltage and the electric fields in the device, as shown schematically in Fig. 12.8(c). The statistical distributions of large numbers of such traps have been used to explain aspects of low-frequency $1/f$ noise in devices (Kirton and Uren 1989).

The structures do not have to be intrinsically small for the effects to be seen. In graded-gap diodes (cf. Chapter 7, Fig. 7.1) a Si donor placed in the AlGaAs layer near the heterojunction could raise and lower the conduction band edge in the vicinity. This then introduces a filament of extra current, which can be quite large (because of the thermionic nature of emission over the barrier) particularly near the turn-on of the diode (Fig. 12.9). The defect has to be close to the interface, and once its energy is known, one can even use this data to infer the local Al concentration as Si donor levels go deep above $x = 25$ per cent Al.

12.9 Conclusion

The field of mesoscopic research is advancing very rapidly, with a greatly increasing number of publications. While the ground rules in this chapter are likely to survive, the number and sophistication of mesoscopic phenomena that can be analysed is likely to increase.

References

Beenakker, C. W. J. (1991). Theory of Coulomb-blockade oscillation in the conductance of a quantum dot. *Physical Review B*, **44**, 1646–56.

Beenakker, C. W. J., van Houten, H., and Staring, A. A. M. (1991). Influence of Coulomb repulsion on the Aharonov-Bohm effect in a quantum dot. *Physical Review B*, **44**, 1657–62.

Field, M., Smith, C. G., Pepper, M., Ritchie, D. A., Frost, J. E. F., Jones, G. A. C., and Hasko, D. G. (1993). Measurements of Coulomb blockade with a noninvasive probe. *Physical Review Letters*, **70**, 1311–14.

Ford, C. J. B., Thornton, T. J., Newbury, R., Pepper, M., Ahmed, H., Peacock, D. C., *et al.* (1989) Electrostatically defined heterojunction rings and the Aharonov–Bohm effect. *Applied Physics Letters*, **54**, 21–4.

Ford, C. J. B., Simpson, P. J., Pepper, M., Kern, D., Frost, J. E. F., Ritchie, D. A., and Jones, G. A. C. (1993). Coulomb blockade in small quantum dots. *Nanostructured Materials*, **3**, 283–91.

Judd, T., Couch, N. R., Beton, P. H., Kelly, M. J., Kerr, T. M., and Pepper, M. (1986). The observation of discrete resistance levels in large area graded-gap diodes at low temperature. *Applied Physics Letters*, **49**, 1652–3.

Kaveh, M., Rosenbluh, M., Edrei, I., and Freund, I. (1986). Weak localisation and light scattering from disordered solids. *Physical Review Letters*, **57**, 2049–52.

Kirton, M. J. and Uren, M. J. (1989). Noise in solid state microstructures: a new perspective on individual defects, interface states and low-frequency ($1/f$) noise. *Advances in Physics*, **38**, 367–468.

Pepper, M. (1988). Quantum interference in semiconductor devices. In *Band structure engineering in semiconductor microstructures.* (eds R. A. Abram and M. Jaros), NATO ASI Series B: Physics, Vol. 189, pp. 137–47. Plenum, New York.

Skocpol, W. J. (1986). Transport physics of multicontact Si MOS nanostructures. In *The physics and fabrication of microstructures and microdevices.* (eds M J Kelly and C. Weisbuch), pp. 255–65. Springer-Verlag, Berlin.

Tewordt, M., Law, V. J., Nicholls, J. T., Martin-Moreno, L., Ritchie, D. A., Kelly, M. J., *et al.* (1994). Single-electron tunnelling and Coulomb charging effects in double barrier resonant tunnelling diodes. *Solid State Electronics*, **37**, 793–9.

Thornton, T. J., Pepper, M., Ahmed, H., Andrews, D., and Davies, G. J., (1986). One-dimensional conduction in the 2-d electron gas. *Physical Review Letters*, **56**, 1198–201.

13 Exotic materials combinations: new physics including strain

13.1 Introduction

So far, we have considered the physics of the GaAs–$Al_xGa_{1-x}As$ system. The alloy semiconductor is direct gap for Al contents below 45 per cent, and the effective mass approximation gives a satisfactory (semi-quantitative) account of all the important optical and electronic properties of multilayers and microstructures. For more than 45 per cent Al, the wider-gap material is indirect gap and some further care must be taken with the detailed matching of wavefunctions at the interfaces if quantitative results are sought. The GaAs–AlGaAs materials system is the most mature, as the lattice constant varies by less than 0.2 per cent across the entire composition range. Furthermore, the thermal expansion of the alloy and GaAs are the same, so that no further stresses are introduced as the material is cooled to room temperature and further to cryogenic temperatures from typical multilayer growth temperatures of about 580–600 °C. Even within this fairly straightforward system, the precise offset of the conduction and valence bands at the abrupt heterojunction has been a matter of continuing debate. It is now thought that a conduction band to valence band offset ratio of 65 ±5 per cent to 35 ±5 per cent is appropriate for describing the physics of the heterojunctions. However, abrupt doping profiles similar those of the planar doped barrier described in Chapter 2, Section 2.6, can introduce (deliberately or unintentionally) a dipole layer at a heterojunction, and so move the two band structures with respect to each other in energy (Shen *et al.* 1992).

It is already clear from Chapter 1, Fig. 1.8, that a much greater variety of III–V semiconductors is available, with different bandgaps and different possible bandgap alignments, even if we constrain ourselves to materials combinations with a similar lattice constant. For example, the GaSb–AlGaSb system spans an overall lower range of energy gaps. There is also the possibility of heterojunctions involving II–VI materials, with the CdTe–HgCdTe system spanning all bandgaps from CdTe at *c.*1.6 eV through to zero. HgTe is a semimetal with a negative bandgap (i.e. the energy band with conduction band symmetry in other semiconductors is lower in energy than the bands with valence band symmetry). The alloy with about 14 per cent Hg has a zero bandgap at 77 K. Staying with the same lattice constant also allows more exotic combinations, such as Ge–GaAs or ZnSe–GaAs, where we mix further materials from different groups of the periodic table. The ability to grow these heterojunctions and their stability if and when formed will concern us in this chapter, along with the extra range of electronic and optical properties that they may possess.

The increasing confidence of crystal growers has led to the availability of heterojunctions between materials of very different lattice constants, incorporating strains as great as 7 per cent between the two materials. One common example is the growth of $In_xGa_{1-x}As$ on GaAs, where replacement of the smaller Ga atom by the larger In atom implies an alloy with a larger lattice constant. An alloy will grow epitaxially on GaAs, i.e. continuing the GaAs lattice, by compressing itself in the plane of the heterojunction and relieving the strain by expanding the unit cell in the direction of the crystal growth. In practice, only thin layers of strained semiconductor can be grown on a given substrate–a thicker layer breaks up to incorporate a dense network of dislocations. It is possible to overgrow material of the original lattice constant (i.e. GaAs grown on $In_xGa_{1-x}As$) and so enhance the stability of a strained layer. The physics of that strain will be discussed in this chapter, and has been exploited in electronic and optical devices discussed further in Chapters 16 and 18. A comprehensive bibliography of epitaxially grown III–V heterojunctions has been compiled by Ploog and Graf (1984).

13.2 The physics of energy band alignment

Figure 13.1 shows the three types of energy band discontinuity that can be achieved at an ideal heterojunction. Type I is the straddling configuration, which we have considered so far in the GaAs–AlGaAs system, type II is a staggered configuration, which technically occurs in the GaAs–AlAs system where the conduction band edge of the AlAs material is determined by the X-point minimum in AlAs layers thicker than 3 nm, and type III is the broken-gap or non-overlapping-gap situation, represented by the widely studied GaSb–InAs system. Given a pair of semiconductors, and assuming

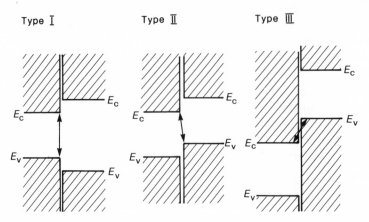

Fig. 13.1 The three types of energy band line-up that are possible at a heterojunction with different pairs of semiconductors: (a) type I; (b) type II; (c) type III.

that an ideal interface can be prepared, what determines the relative band line-up? There are two ingredients: the macroscopic properties of the two semiconductors, and microscopic interactions at the interface. It is fair to say that no theories have solved the problem of the relative line-up; the reasons are similar to those of the detailed theory of the Schottky barrier, namely the need to consider both the materials properties and the detailed micro-structure, both ideal and non-ideal in practice, at the interface. The experimental aspects of band alignment have been reviewed Margaritondo (1986), the theory has been reviewed by Tersoff (1987), and both have been reviewed by Matthai and Williams (1991), on which the next section is based.

The macroscopic models include the following.

1. An electron affinity rule, suggesting that the conduction band offset should be obtained from the difference in electron affinity χ between the two materials, as each χ is a measure of the energy required to remove an electron from the bulk of either material and take it to a distant position in the vacuum. This method ignores interface effects associated with dipole layers, interface reconstructions, etc.

2. A common anion rule, suggesting that in heterojunctions with common anions (as in GaAs–AlGaAs) the valence band offset should be small compared with the conduction band, but this is not well satisfied in practice either, as it takes no account of the different polarity of the bonds in the two materials.

3. An empirical deep-level model, suggesting a correction to model 1, in that a deep impurity level (such as is associated with a transition metal) is a better reference than the conduction band edge in terms of obtaining the relative energies in the two materials.

4. Linear models, in which all properties are linearly interpolated between known end-point data and band line-ups are transitive, i.e. for three materials (A, B, and C) we assume

$$\Delta E_v(A,B) + \Delta E_v(B,C) = \Delta E_v(A,C).$$

Many practical materials combinations appear to satisfy such rules, which are used in device modelling, but a theoretical analysis of examples such as $CaBr_2$–Ge raises doubts about the validity of the model.

The microscopic models take the opposite viewpoint, namely that atomic-like properties determine offsets. Three examples are given below.

1. Reference potentials are determined for the two materials as the mean interstitial potential in the relevant diamond or zinc-blende structures as determined by pseudopotential calculations. These reference levels are aligned and then offset by dipole corrections set up by the difference in electronegativity of the two materials.

2. Tight-binding models for solids give modifications to purely atomic energies associated with the near-neighbour environment in a solid, and differences in these modifications give the offsets.

3. The complex energy band structure defines a near-mid-gap energy that determines the change from a bonding (or valence band) to an antibonding (or conduction band) character for the wavefunctions, and this characteristic energy is aligned at a heterojunction.

Microscopic calculations have been performed on single interfaces in an attempt to incorporate local reconstructions or distortions to the atomic positions. These tend to have too many variables, so that agreement with experiment only narrows down some of the possible combinations of such variables. Just as the determination of Schottky barrier heights has been a fraught theoretical exercise (Chapter 2), so too has the determination of band line-ups.

In practice, energy band line-ups are determined experimentally, despite the difficulties already outlined in Chapter 2.

13.3 Unstrained heterojunctions of other materials

The subject of heteroepitaxy is worthy of a text in its own right (Kroemer 1983). Over the last decade scores of pairs of materials have been grown epitaxially. In many cases the heterojunctions have been difficult to grow, and the structural, electronic, and optical data even more difficult to interpret and/or reproduce. Rather than catalogue these problems, some of which are due to strain, we consider here the GaAs–Ge system, which is approximately lattice-matched, to reveal other difficulties.

In Fig. 13.2 we summarize the results of a study of the GaAs–Ge interface during MBE growth. The energy levels associated with core electrons respond to changes in the electronic structure of the valence bands. The ejection of optically excited electrons (i.e. photoemission) from the 3d core levels into the vacuum of both As and Ge during the deposition of Ge can be used to determine the evolution of the electronic structure at the interface. Photoemission (at binding energies of about 30–100 eV) is very surface sensitive, and the data can be used to determine the position of the Fermi energy as a function of Ge layer thickness. The details of the data analysis need not concern us here (Grant *et al.* 1987), but the results in Fig. 13.2(a) show that the Fermi energy is a sensitive function of the orientation of the GaAs surface, even though both the (100) and (110) surfaces are non-polar, i.e. they have the same number of Ga and As atoms. Furthermore, the Fermi energy is a function of the reconstruction of the surface layers on a (100) surface: the $C(4 \times 4)$, $C(8 \times 2)$, and $((4 \times 6))$ notation refers to different reconstructions of the surface and immediate subsurface layers of GaAs. The Fermi energy depends on the coverage of Ge, reaching a limiting value only when about 0.6 nm of Ge has been deposited. All these results indicate that

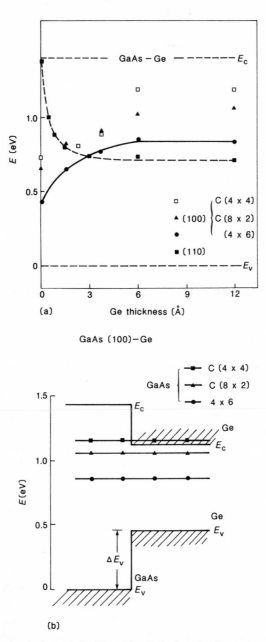

Fig. 13.2   (a) The evolution of the Fermi level during the formation of the GaAs–Ge interface, and its dependence on the Ge layer thickness and the detailed surface structure of GaAs. The data are inferred from core-level soft X-ray spectroscopy, as these levels are sensitive to the electronic structure in the valence bands. (b) A summary of the data for thick Ge overlayers as a function of the crystal orientation of GaAs surface. (After Chiaradia *et al.* 1984.)

the electronic structure of the interface depends sensitively on the detailed local chemical bonding, which in turn depends on the detailed atomic coordinates of the surface and the overlayer atoms, including any defects etc. Small charge imbalances and the polarization of bonds produce a large effect on the electronic structure of the heterojunction. The data in Fig. 13.2(b) summarize the results for thicker Ge overlayers; the valence band offset has been determined by techniques described in Chapter 2, Section 2.8, including the photoemission data, but the energy bands need to be distorted by space-charge effects (as is described in Fig. 2.13) to ensure that the local Fermi level at the heterojunction as shown aligns with the bulk Fermi energies in both GaAs and Ge.

In recent years, the crystalline integrity of the substrate and the quality of the overlayer has improved, so that a consistent picture is starting to emerge. However, the data do show that reproducible heteroepitaxy suitable for exploitation is some way in the future.

13.4 Quaternary alloy systems

The lines on the lattice constant–energy gap diagram in Fig. 1.9 of Chapter 1 allow one to follow ternary alloy systems. In the context of optical devices (Chapter 18), where a fixed wavelength of operation is required (e.g. 1.55 μm for long-haul communications in silica optical fibres), much research has been undertaken on the quality of the bulk material and on both thick and thin epitaxial layers of quaternary alloy systems such as $In_xGa_{1-x}As_yP_{1-y}$. This system covers areas on the lattice constant–energy gap diagram bounded by the ternary alloys. The precise values of x and y can be chosen to satisfy two independent constraints, for example lattice matching to a given substrate (InP, GaAs, or ZnSe) and a bandgap equivalent to a wavelength of 1.55 μm or some other desired value. An approximate interpolation scheme can be used to obtain the material parameter $P(x, y)$ for a general material $A_xB_{1-x}C_yD_{1-y}$ (Madelung 1991):

$$P(x,y) = (1 - x)yP(BC) + (1 - x)(1 - y)P(BD)$$

$$+ xyP(AC) + x(1 - y)P(AD).$$

For lattice matching to a substrate BD $((x = y = 0))$, the condition $a(x, y) = a(BD)$ leads to an x/y ratio given by

$$x/y = [a(BC) - a(BD)]/\{[a(BD) - a(AD)]$$

$$-[a(BC) + a(AD) - a(BD) - a(AC)]y\}.$$

For example $In_xGa_{1-x}As_yP_{1-y}$, with $x = 0.59$ and $y = 0.87$, has a bandgap equivalent to 1.55 μm and is lattice-matched to InP. The crystal growth, materials properties, and devices made in the GaInAsP materials system are described by Pearsall (1982).

The subject of quaternary alloys and their heterojunctions with binary compounds and/or ternary alloys is a matter of detailed materials science, with particular attention paid to the overpressures of the constituents during growth of bulk material and near heterojunctions during epitaxial growth. However, there is little new device-related physics which is not available in the binary and ternary alloy systems. If one departs from the conditions for lattice matching (either accidentally or deliberately), then the new physics associated with strain becomes important. The band structures of some of the quaternary alloys have been studied, but the interpolation scheme above has been used for energy gaps, satellite minima, and other electronic properties, in addition to the structural lattice parameter.

An alternative type of quaternary alloy is of the type typified by $In_{1-x-y}Al_xGa_yP$ and $In_{1-x-y}Al_xGa_yAs$ which allows yet other parts of the lattice constant–energy gap diagram (Chapter 1, Fig. 1.9) to be accessed. The latter system is often simpler to implement in MBE chambers, and is used to fine tune the energy band profiles of many devices (Chapters 16–19).

13.5 The stability of strained layers

The epitaxial growth of an overlayer on a substrate of given surface crystal structure will involve strain if the bulk lattice constant of the overlayer differs from that of the substrate. For modest differences in the lattice constants, the overlayer crystal will distort to maintain a registry with the substrate. In this section, we describe the physics associated with that strain and the effects on the electronic structure. We confine ourselves to growth of III–V materials in the [001] crystal direction, which has dominated the considerations of strained overlayers. There are further phenomena associated with piezoelectric effects that can occur in the [111] direction if the normal sequence of alternating planes of group III and group V atoms are interrupted in any way with a resulting lack of inversion symmetry (Lakrimi *et al.* 1992; Woolf *et al.* 1992).

A thin epitaxial overlayer will be under biaxial stress so that its in-plane lattice parameter a_e equals that of the substrate a_s (Fig. 13.3(a)). The strain $\varepsilon_\parallel$ in the overlayer is given by

$$\varepsilon_\parallel = \varepsilon_{xx} = \varepsilon_{yy} = (a_s - a_e)/a_e.$$

With standard linear elasticity theory, the layer will relax in the direction of crystal growth, with a strain $\varepsilon_\perp (= \varepsilon_{zz})$ being of opposite sign to $\varepsilon_\parallel$ and of a magnitude determined by Poisson's ratio σ:

$$\varepsilon_\perp = -2\sigma\varepsilon_\parallel/(1 - \sigma).$$

In semiconductors with tetrahedral bonding $\sigma \sim \frac{1}{3}$, giving rise to the result that

$$\varepsilon \sim -\varepsilon_\parallel.$$

The total strain can be resolved into two components, one purely axial and the other hydrostatic, giving the following relations:

$$\varepsilon_{ax} = \varepsilon_\perp - \varepsilon_\parallel \sim -2\varepsilon_\parallel$$

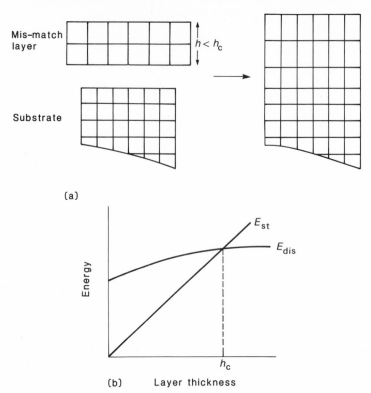

(a)

(b)   Layer thickness

Fig. 13.3 (a) A diagram of the distortion of an epitaxial layer grown on a substrate of different lattice constant. (b) The energy density per unit area stored in an overlayer by virtue of strain and the energy density for a network of dislocations, both as function of layer thickness, leading to the concept of a critical thickness below which dislocations are not nucleated. (After O'Reilly 1989.)

and

$$\varepsilon_{vol} (= \Delta V/V) = \varepsilon_{xx} + \varepsilon_{yy} + \varepsilon_{zz} \sim - \varepsilon_{\|}.$$

This form of analysis of strain in the overlayers allows us to make contact with, and use the results of, a wealth of literature on the measurements of bulk semiconductors under hydrostatic and uniaxial stress. All the data on deformation potentials ($\partial E/\partial \varepsilon_{vol}$ or $\partial E/\partial \varepsilon_{ax}$ where E is some energy, such as the energy gap or the $\Gamma - X$ energy separation) obtained from pressure studies of bulk materials or from transport and optical data in multilayers (cf. Chapters 9 and 10) can be used in the study of strained overlayers.

One important practical parameter is the critical layer thickness h_c. A strained overlayer is thermodynamically stable for thicknesses $h < h_c$, while for $h > h_c$ it is energetically preferable to relieve the strain via the formation of dislocations. The strain energy stored per unit area increases with layer thickness as

$$E_{st} = 2G[(1 - \sigma)/(1 - \sigma)]\varepsilon_\parallel^2 h$$

where G is the shear modulus of the overlayer. The precise energy contained in a network of dislocations is a matter of continuing debate. However, the initial energy for the formation of dislocations is quite large, as it involves breaking a line of bonds. The energy stored in the resulting strain field grows, slowing with thickness h as $\ln(h)$. As shown in Fig. 13.3(b), a dislocation-free strained overlayer is stable until some critical value h_c, above which it is energetically favourable to form a network of dislocations. One particular theory (Voisin 1988) leads to the result that h_c is a solution to an implicit equation

$$h_c = \left(\frac{1 - \sigma/4}{1 + \sigma}\right)\left(\frac{b}{4\pi\varepsilon_\parallel}\right)\ln\left(\frac{h_c}{b} + \theta\right)$$

where b is the dislocation Burger's vector ($c.0.4$ nm) and $\theta \sim 1$ is a constant. This equation yields $h_c \sim 9$ nm for $\varepsilon_\parallel \sim 1$ per cent, setting the scale for strains and thickness that can be tolerated in practice. Note that the product $h_c\varepsilon_\parallel$ is approximately a constant over the range accessible by experiment. The precise value of h_c is further complicated by the fact that the above considerations are about static equilibrium, whereas the formation of strained overlayers is a highly non-equilibrium process. It is possible to grow layers without dislocations that exceed the thickness h_c; they are metastable, with a high activation barrier preventing their relaxation. In practice, the measured critical layer thicknesses depend on the growth conditions (substrate temperature and growth rate), with h_c decreasing with increasing growth temperature.

The two materials systems on which the most extensive strain investigations have been performed have been the growth of $In_xGa_{1-x}As$ layers on GaAs (Fritz *et al.* 1985) and of $Si_{1-x}Ge_x$ layers on Si (Bean *et al.* 1984; Kasper *et al.* 1988). This latter materials system constitutes the subject matter of Chapter 14. We use the adjective 'pseudomorphic' to describe the case where the overlayer is strained but maintains the crystal structure of the substrate. The pseudomorphic $In_xGa_{1-x}As$–GaAs system is important for high-speed microwave transistors (Chapter 16, Section 16.3). In this system, a maximum strain of about 7 per cent has been achieved. Low-temperature photoluminescence (PL) is an important diagnostic tool, as the presence of dislocations can be detected because (i) the PL intensity is reduced since the dislocations act as non-radiative recombination centres, and (ii) the reduction of in-built strain causes a shift of the PL emission energy back towards that of bulk unstrained material. Low-dislocation-density $In_xGa_{1-x}As$ can be prepared on GaAs over a range of strains and thicknesses, as long as $h_c\varepsilon_\parallel < 20$ nm%. For the $Si_{1-x}Ge_x$Si system, the corresponding value is $c.100$ nm%.

All the above analysis has concerned a single thin epitaxial overlayer on a thick substrate. In practical situations, this system is not widely encountered. In Fig. 13.4(a) we show the more common configuration, with a single strained overlayer sandwiched between two unstrained layers (e.g.

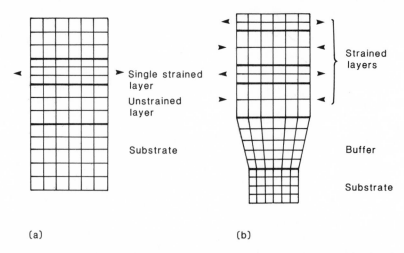

Fig. 13.4 (a) A single strained layer sandwiched between two unstrained epitaxial layers grown on a substrate; (b) a strained-layer superlattice, with alternate layers in tension and compression, on a buffer layer which is itself a (dislocated) strained layer on the substrate. (After O'Reilly 1989.)

GaAs–In$_x$Ga$_{1-x}$As–GaAs) and the theoretical considerations above also apply to this situation. Another common configuration is the strained-layer superlattice (SLS); this has zero net stress, as successive layers are designed to be alternately under biaxial compression and tension. There are two ways of achieving such SLSs. The substrate can be chosen to have the in-plane lattice constant of the superlattice, with the component layers having in-plane lattice parameters on each side (e.g. InAs–GaAs multilayers grown on InP, with approximately equal thicknesses of the InAs and GaAs overlayers). Alternatively, a thick buffer layer can be grown to achieve the same intermediate in-plane lattice constant as the final SLS (Fig. 13.4(b)). The buffer layer will have a high dislocation density because of its lattice mismatch with the substrate, but dislocations do not penetrate the superlattice (cf. Chapter 9, Section 9.6). A zero-net-stress SLS can in principle be grown to an indefinite thickness, provided that each layer is less than its critical thickness. The h_c for layers in such a SLS is four times that of a single strained layer. If the SLS has a non-zero net strain, then there will be a critical thickness H_c for the superlattice, which is related to that of the average alloy on the substrate. For example, if equal thicknesses of In$_x$Ga$_{1-x}$As and GaAs are grown on GaAs, the critical thickness for the superlattice is equal to the critical layer thickness for In$_{x/2}$Ga$_{1-x/2}$As on GaAs, i.e. $H_c(x) = h_c(x/2)$.

13.6 Electronic and optical properties of strained heterojunctions

In Fig. 13.5(a), we show the band structure of an unstrained tetrahedrally bonded direct-bandgap semiconductor, along with the band structure when

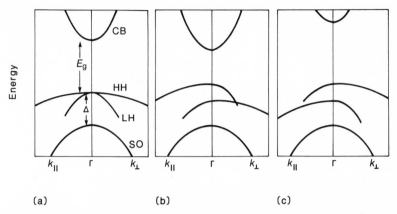

Energy

(a)   (b)   (c)

Fig. 13.5 (a) The typical band structure of an unstrained tetrahedral semiconductor, and the effects of (b) biaxial tension and (c) biaxial compression on the energies of the bands and their dispersion both in the plane of the layers ($k_\parallel$) and perpendicular to the axial plane $k_\perp$ i.e. the direction of growth and of the strain axis. (After O'Reilly 1989.)

the same semiconductor is under biaxial tension and compression (Figs 13.5(b) and 13.5(c)), as occurs in strained epitaxial layers. In each case, the conduction band is approximately parabolic and the dispersion near the centre of the Brillouin zone is given by

$$E(k) = E_g + \hbar^2 k^2 / 2m^*,$$

where both the energy gap E_g (with respect to the top of the valence band) and the effective mass m^* depend on the strain. The valence bands are complicated even in the absence of strain (cf. Chapter 1, Section 1.3). There are heavy-hole and light-hole bands which are degenerate at $\Gamma(k = 0)$, and a spin-split-off band which lies at an energy $\Delta(c.0.35$ eV in GaAs at low temperatures) below the higher bands. Axial strain breaks the cubic symmetry of the semiconductor, and this in turn lifts the degeneracy of the heavy-and light-hole bands at Γ, typically by about 60 meV per % $\varepsilon_\parallel$ strain. The valence band structure also becomes anisotropic (cf. Figs 13.5(b) and 13.5(c)) as the energy band with heavy-hole character along the direction of growth ($k_\perp$) acquires a relatively light-hole character in the direction ($k_\parallel$) perpendicular to growth. (Note that the terminology here refers to growth plane rather than the growth direction, so that $k_\parallel$ is parallel to the atomic growth planes.) The detailed derivation of these results goes back to basic band-structure theory. Thus strain has two major effects on bulk valence bands: the lifting of degeneracy and the introduction of anisotropic band structure.

The introduction of strain into quantum wells has further implications on the energy states, and these are shown in Fig. 13.6. In the absence of strain (Fig. 13.6(a)), the wavefunctions for the carriers take discrete quantum numbers along the direction of growth ($k_\perp$) but have the character of free motion in the planes of the layers ($k_\parallel$). The valence band structure in quantum

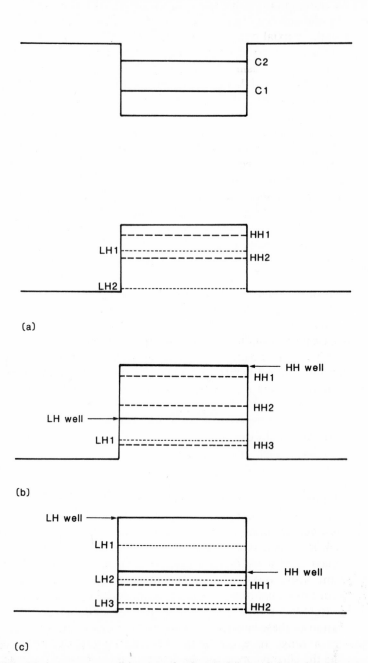

(a)

(b)

(c)

Fig. 13.6 (a) A quantum well in unstrained materials, and the valence band quantum wells under biaxial (b) tension and (c) compression. Note here that from Fig. 13.6 the dispersion in the layers is light-hole-like while the confinement effect mass is heavy-hole-like, and vice versa. (After O'Reilly 1989.)

wells is more complex. First, the quantum confinement energies at $k_\parallel = 0$ are determined by the effective mass in the direction of the growth (and strain). For a layer under biaxial compression (e.g. Ge on Si or InGaAs on GaAs), the heavy-hole band edge shifts up with respect to the light-hole band edge. This has the effect of increasing the energy splitting $E_{HH1} - E_{LH1}$ between the highest confined light- and heavy-hole energies (Fig. 13.6(b)). For a layer under biaxial tension (e.g. GaSb on AlSb) the situation is reversed, and one can even reach the situation where the light-hole state has the highest energy (Fig. 13.6(c)). Strained quantum wells can be used to tailor the energy separations $E_{HH1} - E_{HH2}$ and $E_{HH1} - E_{LH1}$ independently through the well widths and axial strain respectively.

The considerations just given apply to the energy of confinement. The inplane dispersion is a separate issue, but the diagrams in Fig. 13.5 are a guide. Heavy confined states have a light in-plane effective mass, and vice versa, at least for $k_\parallel = 0$. Away from $k_\parallel = 0$, the situation is more complex. The very definition of heavy, light, and spin–orbit split valence bands comes from a diagonalization of a complex Hamiltonian (see Appendix 3), and the breaking of the translational symmetry in the k direction means that this diagonalization must be repeated for each $k_\parallel$. The combination of well width, strain, and the complexity of the near degeneracy of the hole states results in a wide range of possible valence sub-band structures, as shown in Fig. 13.7. We use a strain-induced band-edge splitting parameter S to describe the shift of the heavy-hole band with respect to the mean valence band energy (the light-hole and split-off bands are coupled by the strain), and the value of S (in meV) is derived from a principal deformation potential b by

$$S = -b\varepsilon_{ax}.$$

The hydrostatic strain shifts the conduction band edge with respect to the average valence band edge by

$$\Delta E_g = a\varepsilon_{vol}.$$

The values of a lie in the range -6 to -8 eV, while the values of b are in the range -1.5 to -2 eV. The results in Fig. 13.7 use S as a parameter. Note that the highest valence band sub band can have (a) a low effective mass, (b) a camel's back structure, (c) an electron-like effective mass, and (d) a low zone-centre mass over a restricted energy and $k_\parallel$ range. Light-hole masses have important applications in optoelectronic devices (see Chapter 18). Calculations for finite quantum wells are even more difficult, and the range of energy band structures that can be achieved is wider still.

The verification of these valence band structures of quantum wells has been the subject of a range of experiments. The energy gap can be measured optically, while the effective mass can be determined by transport or cyclotron resonance experiments. In addition, the use of pressure as a tool to enhance or suppress the influence of strain has assisted in the interpretation of the data. The conditions leading to equally spaced Landau levels include a free-electron band structure with a constant effective mass. Non-uniform Landau level spacings that originate from highly non-parabolic valence bands can be seen in magneto-optic experiments (Heiman *et al.* 1987).

The advantages of strain are considerable. In the Drude model for carrier transport, the mobility $\mu = e\tau/m^*$ has an overall $(1/m^*)^2$ dependence, when $\tau \sim 1/m^*$ is assumed for a parabolic band in two dimensions. Traditionally, p-type transistor devices have been slower because of the higher mass and

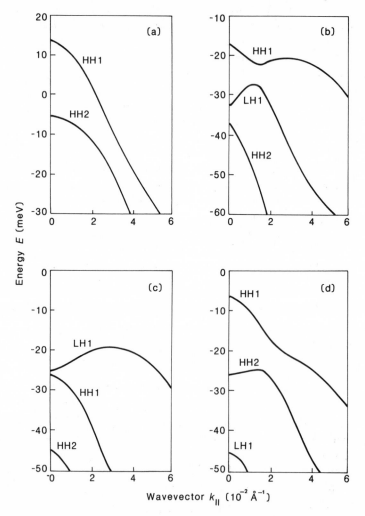

Fig. 13.7 The valence sub-band structure for a 10 nm GaAs infinite quantum well with different biaxial strains: (a) $S = 20$ meV, compression; (b) $S = -20$ meV, tension; (c) $S = -12$ meV, tension, (d) $S = 0$, unstrained. The symbol S, which is defined in the text, is an energy measure of the strain obtained with an appropriate deformation potential. The zero of energy is the mean of the bulk light- and heavy-hole band edges.
(After O'Reilly 1989.)

shorter scattering time for holes. Strain can split the bands apart, reduce the level of scattering, enhance the mobility, and increase the speed of device operation. Among the *n*-channel FETs, the pseudomorphic high-electron- mobility transistor, based on strained InGaAs on GaAs, is the fastest for a given device size. In optical devices the conditions for population inversion, non-linear saturation effects, etc. are all made easier to reach by a low-mass valence band as induced by strain. Strain-layer lasers have a low threshold current and high differential efficiency for converting electrical into optical power. Strain is no longer a nuisance, but an exploitable phenomenon (see Chapters 16 and 18).

13.7 The InAs–GaSb system

The bandgap profile shown in Fig. 13.1(c), with no overlap of bandgaps of the two constituents, is exemplified by the InAs–GaSb system. This materials combination exhibits a larger potential discontinuity than can be achieved with the GaAs–AlGaAs system. The electronic structure of a single heterojunction is unusual. Optical experiments suggest that the misalignment is $c.0.15$ eV, but the precise value depends sensitively on the exact stoichiometry of the layers at the heterojunction. The overlap in energy of the InAs conduction band and the GaSb valence band results in electrons from GaSb spilling over into InAs, leaving a dipole layer formed from parallel and adjacent 2D electron and hole gases in InAs and GaSb respectively. Transport measurements suggest that the hole and electron gas densities are not identical, which is an indication of extrinsic electronic structure associated with a non-ideal interface.

The structure that has been most studied is an InAs quantum well with GaSb barriers, (Fig. 13.8(a)). The two-carrier nature of the transport, when both electrons and holes are present, results in extra structure in the magnetotransport (Fig. 13.8(b)) as proper quantum Hall plateaux and resistance zeros require the Fermi level to be between the electron and hole Landau levels. The analysis of transport data for single InAs quantum wells of differing thickness is shown in Fig. 13.8(c). The electron and hole mobilities are the highest reported for either InAs or GaSb. The most striking feature, and one that had been predicted theoretically, concerns the sharp drop in carrier concentration once the well width is reduced below about 6 nm. With reference to Fig. 13.8(a), as the InAs well shrinks, the energy of the lowest electron bound states in the well rises and will eventually exceed the energy of the conduction band edge in GaSb, at which point there will be no further electron transfer from GaSb into InAs. Thus we have a metal–semiconductor transition as a function of the InAs well thickness.

13.8 II–VI systems and other multilayers

The growth of very-high-quality II–VI semiconductors has always lagged behind that of the III–V and IV semiconductors, in part because of the lesser

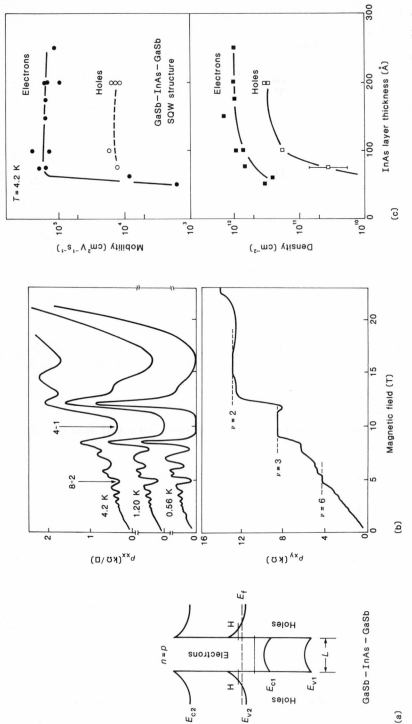

Fig. 13.8 (a) The band structure for an InAs quantum well in GaSb, showing the electron and hole bound-state levels; (b) the magnetotransport data, and (c) the mobilities and carrier densities. (After Esaki © 1986 IEEE.)

total effort, but more particularly because it has proved difficult to eliminate defects and to introduce controllable levels of dopant during growth. In addition, the quality of substrates is not as high as that available in III–V and group IV materials. The effort has increased strongly over recent years, and some progress on the epitaxial material quality and doping has been made. There are two classes of II–VI systems, the narrow-gap materials typified by the HgTe–CdTe system, with applications as infrared materials, and the wide-gap materials typified by the ZnTe–CdTe system, with applications in the blue part of the spectrum. In each case unusual aspects of the electronic structure of the constituent materials are being used. HgTe is a semimetal, and at about 14 per cent CdTe in HgTe the alloy becomes a semiconductor. Appropriate multilayers of CdTe and HgTe can be used to generate a semiconductor with a bandgap of about 0.12 eV needed for 10 μm radiation detection. The stability of such multilayers is very poor, as Hg can diffuse rapidly at the elevated temperatures associated with either MOCVD or MBE growth and even at room temperature. The current status of this materials system as grown by MBE and MOCVD is reviewed by Wu and Kamath (1991) and Irvine *et al.* (1991). Indeed, this difficult materials system, to which an enormous effort was devoted in the 1970s and 1980s in the quest for infrared materials, is about to be challenged by the GaAs–AlGaAs quantum-well structure in many of the same applications (see Chapter 19).

The recent progress on the wide-gap materials has been much more encouraging, with light-emitting diodes and lasers operating at low temperatures in the green and blue parts of the spectrum being made possible with the use of ZnSe and ZnS quantum wells, whose bandgaps and bandgap offsets enable all the optical physics described in earlier chapters to be shifted from the near infrared (0.76–0.9 μm) well into the visible (0.4 μm). Research effort is likely to intensify to achieve devices which are reliable at room temperature, as many applications are envisaged (Jeon *et al.* 1992).

The IV–VI semiconductor materials, typified by the PbTe–SnTe system, have been grown by a variety of techniques. The materials have narrow bandgaps and have applications as mid-infrared optical devices (Bauer and Clemens 1990; Preier 1990).

The range of materials being grown as epitaxial multilayers is ever increasing, and the rapid expansion of the study of magnetic metallic superlattices is described in Chapter 21. Metal–semiconductor heterojunctions and multilayers such as NiAl–GaAs and CoAl–AlAs–GaAs, and dielectric–semiconductor systems such as GaN–GaAs have also been reported, but these are still relatively primitive systems for exploitation in comparison with those described above.

13.9 GaAs on Si

The ability to grow selected areas of high-quality GaAs on Si would enable the combination of the advantages of digital Si signal processing and memory circuitry with the optical functions of III–V materials. The strain levels are large (cf. Chapter 1, Fig. 1.8). However, as growth techniques improve, this can be tolerated. The most important applications involve bipolar devices, such as lasers, and to date the quality of the epitaxially overgrown III–V materials is insufficient to produce high-quality optical devices. The minority carrier lifetimes are too short. Even when uniform GaAs growth is possible, further problems remain. In particular, local-area growth (Chapter 3) is still in a primitive condition. Epitaxial masks will be required, probably of a deposited oxide or nitride. The deposition and patterning of such mask layers, and their subsequent removal after III–V materials deposition together with the processing of the III–V devices, must all be carried out without degrading the devices already on the Si substrate or the quality of the GaAs–Si interface. The status of GaAs–Si materials and devices was last reviewed by Kroemer (1986).

References

Bauer, G. and Clemens, H. (1990). Physics and Applications of IV–VI compound quantum well and superlattice structures. *Semiconductor Science and Technology*, **5**, S122–30.
Bean, J. C., Feldman, L. C., Fiory, A. T., Nakahara, S., and Robinson, I. K. (1984). Ge_xSi_{1-x}/Si strained layer superlattices grown by molecular beam epitaxy. *Journal of Vacuum Science and Technology A*, **2**, 436–40.
Chiaradia, P., Katani, A. D., Sang, H. W., and Bauer, R. S. (1984). Independence of Fermi-level position and valence-band edge discontinuity at GaAs–Ge(100) interfaces. *Physical Review Letters*, **52**, 1246–9.
Esaki, L. (1986). A bird's-eye view on the evolution of semiconductor superlattices and quantum wells. *IEEE Journal of Quantum Electronics*, **QE-22**, 1611–24.
Fritz, I. J., Picraux, S. T., Dawson, L. R., Drummond, T. J., Laidig, W. D., and Anderson, N. G. (1985). Dependence of critical layer thickness on strain for In_xGa_{1-x}As/GaAs strained layer superlattices. *Applied Physics Letters*, **46**, 967–9.
Grant, R. W., Kraut, E. A., Waldrop, J. R., and Kowalczyk, S. P. (1987). Interface contribution to heterojunction band discontinuities: X-ray photoemission spectroscopy investigation. In *Heterojunction band discontinuities: physics and device applications* (eds F. Capasso and G. Margaritondo), pp. 167–206. North-Holland, Amsterdam.
Heiman, D., Pinczuk, A., Gossard, A. C., Fasolino, A., and Alterelli, M. (1987). Magnetic-field tuning of valence subbands in GaAs/(Al,Ga)As multiple quantum well. In *Proceedings of the 18th International Conference on the Physics of Semiconductors* (ed O. Engstrom), pp. 617–20. World Scientific, Singapore.
Irvine, S. J. C., Gertner, E. R., Bubulac, L. O., Gil, R. V., and Edwall, D. D. (1991). MOVPE growth of HgCdTe. *Semiconductor Science and Technology*, **6**, C15–20.
Jeon, H., Ding, J., Nurmikko, A. V., Xie, W., Grillo, D. C., Kobayashi, M., *et al.* (1992). Blue and green diode lasers in ZnSe-based quantum wells. *Applied Physics Letters*, **60**, 2045–7.
Kasper, E., Herzog, H. J., and Schaffler, F. (1988). Si/Ge multilayer structures. In *Physics, fabrication and applications of multilayers structures* (eds P. Dhez and C. Weisbuch), NATO ASI Series B, Physics, Vol. 182, pp. 229–38. Plenum, New York.

Kroemer, H. (1983). Heterostructure devices: a device physicist looks at interfaces. *Surface Science*, **182**, 543–76.

Kroemer, H. (1986). MBE growth of GaAs on Si: problems and progress. In *Proceedings of the Heteroepitaxy on Silicon Symposium*, pp. 3–14. Materials Research Society, Pittsburgh.

Lakrimi, M., Lopez, C., Martin, R. W., Summers, G. M., Sundaram, G. M., Dalton, K. S. H., et al. (1992). Piezoelectric control of doping and band structure in the cross gap system GaSb/InAs. *Surface Science*, **263**, 575–9.

Madelung, O. (1991). *Semiconductors: group IV elements and III–V compounds*. Springer-Verlag, Berlin.

Margaritondo, G. (1986). Do we understand heterojunction band discontinuities? *Surface Science*, **168**, 439–51.

Matthai, C. and Williams, R. H. (1991). Aspects of the physics of heterojunctions. In *Physics and technology of heterojunction devices* (eds D. V. Morgan and R. H. Williams), pp. 1–32. Peter Peregrinius, London.

O'Reilly, E. P. (1989). Valence band engineering in strained layer structures. *Semiconductor Science and Technology*, **4**, 121–37.

Pearsall, T. P. (ed). (1982). *GaInAsP alloy semiconductors*. Wiley, New York.

Ploog, K. and Graf, K. (1984). *Molecular beam epitaxy of III–V compounds*. Springer-Verlag, Berlin.

Preier, H. (1990). Physics and applications of IV–VI compound semiconductor lasers. *Semiconductor Science and Technology*, **5** S12–30.

Shen, T.-H., Elliott, M., Williams, R. H., Woolf, D. A., Westwood, D. I., and Ford, A. C. (1992). Control of semiconductor interface barriers by ∂-doping. *Applied Surface Science*, **56–58**, 749–55.

Tersoff, J. (1987). The theory of heterojunction band lineups. In *Heterojunction band discontinuities: physics and device applications* (eds G. Margaritondo and F. Capasso), pp. 3–57. North-Holland, Amsterdam.

Woolf, D., Sobiesierski, Z., Westwood, D. I., and Williams, R. H. (1992). The molecular beam epitaxial growth of GaAs/GaAs(111)B: doping and growth temperature studies. *Journal of Applied Physics*, **71**, 4908–15.

Voisin, P. (1988). Heterostructures of lattice-mismatched semiconductors: fundamental aspects and device perspectives. In *Quantum Wells and Superlattices in Optoelectronic Devices and Integrated Optics*. (ed. A. R. Adams) *SPIE* **861**, 88–95.

Wu, O. K. and Kamath, G. S. (1991). An overview of HgCdTe MBE technology. *Semiconductor Science and Technology*, **6** C6–9.

14 Silicon and silicon heterojunctions

14.1 Introduction

Si dominates the electronics industry, being associated with more than 90 per cent of the added value in components and subsystems (Sze 1988), and if it were a direct-gap material as required for optical applications, the III–V and other semiconductors would have remained curiosities. Compared with all other semiconductor materials, Si substrates come in large areas. They have high mechanical stability, modest electrical insulation, and a good thermal conductivity. The ease of forming a thin stable high-resistance thermally grown oxide which is capable of withstanding high electric fields and of being patterned by a variety of methods is also a strong advantage. However, the extra speed of carriers in III–V materials, the negative differential mobility of GaAs at high fields, and light emission from direct-gap III–V materials (as light-emitting diodes and as laser diodes) has given semiconductors other than Si a chance. A further advantage, peculiar to GaAs, is that very-high-resistance substrates are available, and this property is maintained during the thermal, chemical, and other cycles associated with microwave device fabrication. During equivalent processing of Si a low level of mobile carriers is introduced into the substrate, and their response to rapidly changing electric fields introduces an extra reactive impedance to the microwave signals and degrades the performances of devices at high frequencies. The advent of heterojunctions has continued to give high-performance III–V devices a lead over their Si counterparts where there is direct competition (as in high-speed microwave and low-density logic applications). This advantage was further helped by the availability of lattice-matched combinations of materials. The technology of heterojunctions has made possible new generations of devices and new device functions in III–V materials. With time, the incorporation of strain has been mastered, and some effects of strain, particularly on the hole band structure, have been deliberately exploited to give superior device performance.

The epitaxial growth of Si presents its own difficulties owing to the higher growth temperatures ($c.800$ °C compared with $c.600$ °C for the III–V materials) and the high temperatures associated with obtaining atomic beams of Si. The absence of any other group IV semiconductors with the same lattice constant has meant that strain will always feature in any Si-based multilayers; the difference in lattice constant between Si and Ge is 4.2 per cent. Furthermore, the limited intersolubility of Si and Sn appeared to limit the range of materials that might be prepared from that combination. These factors have certainly slowed down the development of Si-based multilayers through, for example, the lack of familiarity of Si–Ge alloys compared with GaAs–AlAs alloys. The mastery and exploitation of strain in recent years has led to a vast expansion of research into Si-based multilayers, and some of the device advantages are starting to be realized in the laboratory.

In this chapter, we cover the Si–SiGe materials system, concentrating on the materials issues, the physics of the energy band alignments peculiar to strained Si and Si–Ge multilayers, and the prospects for eventual device exploitation.

Before turning to heterojunctions, it is worth noting that epitaxial layers of Si, deposited on Si substrates, are widely used in the Si industry (Pearce 1988). Si epitaxy has been developed in the context of bipolar transistor devices, as the greater degree of control over doping profiles afforded by epitaxy (compared with ion implantation) can be used to enhance the performance of these devices. This is in the same spirit as described for III–V materials in Chapter 2, Section 2.1. The present-day techniques are the same chemical vapour and molecular beam deposition techniques. There are a number of technical differences in the layout of the MOCVD reactors and the MBE chambers to account for the higher growth temperature and the use of low-energy ion implantation to assist with dopant incorporation, but other-wise the techniques are the same. The quality of the epitaxial layers must be very high to satisfy the needs for large area integrated circuit fabrication, and certainly higher than is demanded of III–V materials for optical or microwave devices.

14.2 Buffer layers as substrates for Si$_{1-x}$Ge$_x$ overlayers

In the last chapter we introduced the concept of a critical thickness for strained overlayers. For 1 per cent lattice mismatch, corresponding to deposition of the alloy Si$_{1-x}$Ge$_x$ with $x = 0.2$ on Si, the critical thickness is of order 100 nm. In most practical devices, we need to be able to grow somewhat thicker layers containing a higher fraction of Ge. Furthermore, the large strains distort the band structure and modify the very electronic properties being sought. As a result, considerable effort has been devoted to the production of a substrate more suitable for the growth of useful device-grade SiGe–Si overlayers (Kasper *et al.* 1988).

The dominant strategy for achieving high-quality overlayers is the prepara-tion of a suitable buffer layer (or 'virtual substrate') between them and the Si substrate. In the spirit of a zero net strain superlattice (cf. Chapter 13, Section 13.5), a thick buffer layer is grown with the average Ge content of the subsequent overlayers. In practice, such a buffer layer provides a good starting surface for subsequent overgrowth, except for a high density of threading dislocations which penetrate the epitaxial layer and which may propagate further into the overlayers of interest. In recent years, the strategy has been to grow a layer of Si$_{1-x}$Ge$_x$ with the x starting at a low value (say

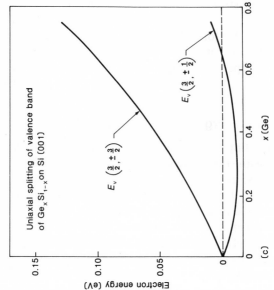

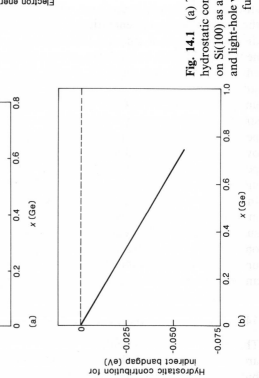

Fig. 14.1 (a) The uniaxial splitting of the conduction band minima and (b) the hydrostatic contribution to the indirect bandgap of coherently strained Si_xGe_{1-x} on Si(100) as a function of the Ge fraction; (c) the uniaxial splitting of the heavy- and light-hole valence band edges of coherently strained Si_xGe_{1-x} on Si(100) as a function of the Ge fraction. (After People © 1986 IEEE)

0–5 per cent) and increasing gradually to the required value (typically 30 per cent) during the deposition of an epitaxial buffer layer that is several micrometres thick. The strain in such a layer relaxes during growth, and the density of threading dislocations at the top of such a layer can be less than 10^{-3} of that nearer the bottom of the layer, i.e. at the interface with the Si substrate. All the highest-quality transport results (e.g. enhanced electron mobility) have been reported with this type of graded-composition buffer layer.

14.3 Si and Si$_1$ – $_x$Ge$_x$ energy bands under strain

Si and the Si$_1$ – $_x$Ge$_x$ alloys ($x < 0.85$) have the same generic conduction band structure, namely with the lowest conduction band states in the $\langle 001 \rangle$ directions and about 85 per cent of the way to the zone boundary in the case of pure Si (cf Chapter 1, Fig. 1.3, and Chapter 5, Fig. 5.3(b)). This is in contrast with the simple direct-gap III–V materials. Here stress reduces the symmetry of the conduction band minima and reduces the degeneracy of the lowest-lying energy levels. If the alloy is grown on a Si substrate, the net biaxial strain is a combination of uniaxial and hydrostatic strain, and the movement of the conduction band minima of Si$_1$ – $_x$Ge$_x$ energy bands under strain under each of these conditions is shown in Fig. 14.1, as a function of the Ge composition, generating a valley degeneracy of four. The splitting of the valence band degeneracy under biaxial strain contains a uniaxial component (Fig. 14.1(c)) and a hydrostatic contribution (Fig. 14.1(b)). The overall effect of the coherent strain on the bandgap of Si$_1$ – $_x$Ge$_x$ energy bands under strain is shown in Fig. 14.2(a) where the fundamental (indirect) gap of the unstrained bulk Si$_1$ – $_x$Ge$_x$ alloys is compared with that of the coherently strained Si$_1$ – $_x$Ge$_x$ alloy overlayer. Note (i) the changeover for more than 85 per cent Ge, where the L nature of the conduction band minima in Ge takes over, and (ii) the spread in data because of the uncertainty in the deformation potentials used in Fig. 14.1. In Fig. 14.2(b), we show the lowest-energy indirect bandgaps for three situations that are often encountered in practice. In the lowest curve we show the bandgap of the alloy grown on a Si substrate, in the middle curve the converse case of the bandgap of Si grown on an alloy substrate with Ge content x_s, and in the upper curve the growth of the alloy on a substrate having a composition with half the fraction of Ge. The uncertainties are reproduced, and some data are given for the first case to give an impression of the accuracy that can be achieved.

14.4 Energy band alignments at Si–Si$_1$ – $_x$Ge$_x$ heterojunctions

The energy band alignments for Si and Si$_1$ – $_x$Ge$_x$ alloys at the heterointerfaces are more complicated than in the III–V case, not only because of the strain but also because of the conduction band degeneracy. The precise values are still being worked out, but typical good estimates are shown in Fig. 14.3

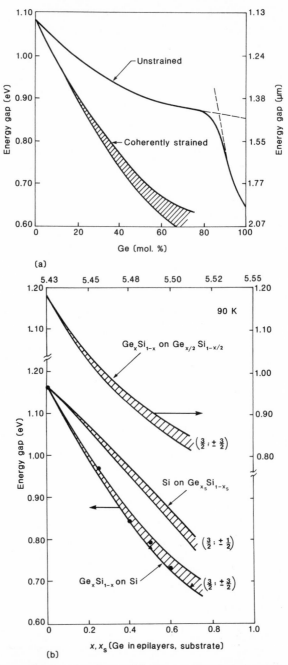

Fig. 14.2 (a) Lowest-energy bandgap of Si_xGe_{1-x} alloy in bulk unstrained layers and in coherently strained overlayers on Si; (b) the lowest-energy bandgap for Si_xGe_{1-x} alloy on Si, Si on Si_xGe_{1-x} alloy, and Si_xGe_{1-x} alloy on a lower-Ge-content $Si_{x/2}Ge_{1-x/2}$ alloy. (After People © 1986 IEEE.)

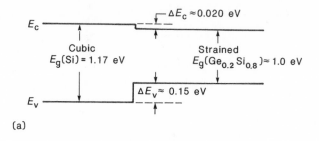

(a)

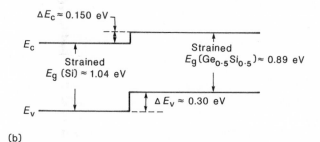

(b)

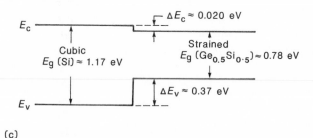

(c)

Fig. 14.3 The heterojunction band alignments for three common situations in strained Si_xGe_{1-x} alloys: in (a) and (c) the substrate is (001) silicon, while in (b) it is $Si_{0.75}Ge_{0.25}$. This shows the variety of band alignments that can be achieved with strain in the epitaxial overlayer and/or the substrate (or lower epitaxial layer). (After People © 1986 IEEE.)

for three cases: a $Si_{0.8}Ge_{0.2}$ alloy on (001)Si (Fig. 14.3(a)), a $Ge_{0.5}Si_{0.5}$ − Si heterojunction grown on a (001)$Ge_{0.25}Si_{0.75}$ substrate (Fig. 14.3(b)), and the same heterojunction grown on (001)Si (Fig. 14.3(c)). We note that the bandgap offset in the conduction band is very small when material is grown on a Si substrate. This makes it difficult to obtain a 2DEG (Chapter 6) in unstrained Si, although it is possible to achieve it in the reduced net-strain structure in Fig. 14.3(c). In contrast, a large valence band offset is always achieved, and studies of the 2D hole gas (2DHG) in Si and $Si_{1-x}Ge_x$ alloys have been possible. The accumulating experimental evidence is leading to refinements of the band alignments, but a consensus comparable to that in the GaAs–AlGaAs system is not yet available.

14.5 Transport and optical properties of Si–SiGe heterojunctions

The two principal device applications of III–V heterojunctions have been associated with quantum confinement: (i) the modulation-doped heterojunction leading to high-mobility 2DEGs and 2DHGs (Chapters 5 and 16), and (ii) the blue shift in the interband optical absorption edge for lasers and modulators because of the quantum energy of confinement of electron and hole states in a quantum well (Chapters 10 and 18). Similar studies have been undertaken in the Si–SiGe system.

The larger valence band offsets made the observation of enhanced hole mobility through modulation doping easier. The Si layers were p-doped, either right up to the heterojunction or set back by *c*.10 nm. The 2DHG forms in the alloy and mobility improvements are seen (Fig. 14.4(a)), although alloy scattering sets a relatively modest upper limit on the gains that can be achieved. Even these modest improvements are important in the context of bipolar transistors (Chapter 16). More recent data with a heterojunction grown on a graded-composition thick buffer show that high mobilities can be achieved in a 2DEG, and these results are given in Fig. 14.4(b). The enhancement in mobility at room temperature is much greater than 2 and is of real device significance in the context of field-effect devices (Chapter 16). At low temperatures the enhancement is more dramatic, as in the case of III–V systems. These cleaner systems allow measurements of the quantum Hall effect, and the Shubnikov–de Haas data can now be revisited to obtain further detailed information on spin-splittings etc., answering questions that were left unresolved in the 1970s (cf. Chapter 5, Section 5.4). Typical recent data are shown in Figure 14.5, where the extra structure in ρ_{xx} is due to the remaining fourfold degeneracy of the electron gas, which in turn could be further reduced by uniaxial stress applied in the plane of the 2DEG.

It is now only a matter of time until ballistic and other quantum interference effects are seen in Si. The various scattering length scales are shorter in Si than in GaAs, but in the past this has been associated largely with scattering from defects and impurities in amorphous thermally grown oxide. The single-crystal system, represented by Si at a Si–SiGe heterojunction, is more promising for these investigations. One problem associated with strain is the limitation that it places on various forms of device/structure processing. In particular, the exposure of a narrow area by means of wet-chemical or dry-plasma etching provides new and undesirable opportunities for lateral strain relief. The techniques of patterned gating, ion implantation, etc., which do not leave free surfaces and edges to strained layers, must be used.

The various luminescence (Fukatsu 1994) and other optical absorption experiments have been performed in SiGe–Si quantum-well structures. The results are unsurprising, given the known optical absorption properties of Si, the quantum size effects associated with Si–SiGe quantum wells, and the fact that the valence band offset is larger than the conduction band offset. The

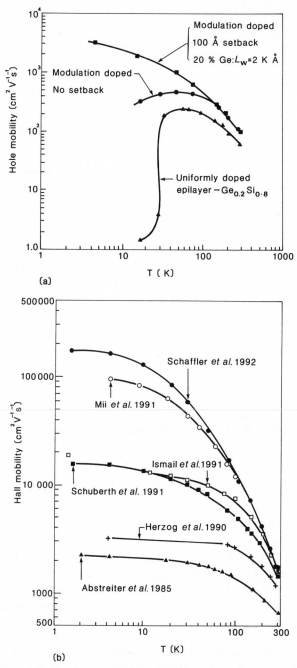

Fig. 14.4 (a) The enhanced hole mobility in SiGe at a p-modulation-doped Si–SiGe heterojunction (after People © 1986 IEEE); (b) the enhanced electron mobility in strained Si at an n-modulation-doped SiGe–Si heterojunction (after Schaffler *et al.* 1992.)

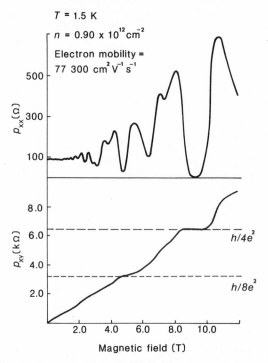

Fig. 14.5 Magnetotransport data for a 2DEG in Si at a Si–SiGe heterojunction (After Abstreiter 1992.)

infrared inter-sub-band absorption in p-doped quantum wells has been investigated; the rationale is that the complexity of the valence bands should not require the light to be polarized strictly perpendicular to the plane of the wells, as in GaAs conduction band quantum wells, and this eases the problems associated with getting the infrared light into the quantum wells (Karunasiri *et al.* 1991; People *et al.* 1992). This approach has been successful, but the inter-sub-band matrix elements are weaker, and the heavier mass and greater scattering of holes during transport results in a poorer collection efficiency. The technologies associated with patterning the GaAs surface with gratings seems superior in terms of ultimate absorption efficiency (Chapter 19).

The search for a light-emitting device in Si has been less successful. The constant need is for a direct-gap semiconductor. In the III–V materials the s-like symmetry of the states at the bottom of the conduction band, coupled with the p-like symmetry of the electron states at the top of the valence band, makes for very efficient light emission from the relaxation of excited carriers. In contrast, the relative efficiency of light emission from the conduction band minima in Si (at 85 per cent of the way to the zone boundary in the $\langle 001 \rangle$ ΓX directions) is only 10^{-6}, as phonon processes must accompany the electron–hole recombination. It was hoped that folding the bandstructure with the use of a short-period superlattice (by direct analogy with folding encountered in

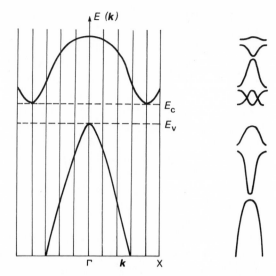

Fig. 14.6 Zone-folding in a short-period superlattice in an attempt to achieve optical emission.

the Esaki–Tsu approach to negative differential resistance) would help matters by bringing the conduction band minimum to Γ in one of the folded energy bands (Fig. 14.6). In practice, the low-lying states in the directions not folded by the superlattice that dominate the optical absorption process have the same indirect-gap character as in bulk Si. A three-dimensional superlattice structure is required! The subject of light emission from quantum dots formed as the microstructure of porous silicon has given the subject a new lease of life. (Cullis and Canham 1991).

14.6 Device prospects from Si-based multilayers and Si–SiGe heterojunctions

In Chapters 16–19 we concentrate on the III–V heterojunction devices, which have set the pace in the applications of heterojunctions as features in device design. Here we briefly review how Si multilayers and heterojunctions will have similar advantages in some Si devices (Meyerson 1994). It is important to note that conventional Si technology for memory and logic circuits is extremely sophisticated, and some of the technologies that have been developed (e.g. trench capacitors), or isolated devices (as in Si-on-sapphire), or some of the process steps (particularly etching) may not be directly compatible with the use of a Si–SiGe multilayer as the starting substrate material incorporating parts of the device through doping profiles etc. (Sze 1988). As is the case with the III–V devices, it has been in specialist application areas that the first prototype devices have been fabricated. Heterojunction bipolar transistors in III–V materials are now ubiquitous (see Chapter 16), as the advantages to the gain are exponential (exp($\Delta E/kT$) where ΔE is the energy gap difference at the emitter–base junction). This improvement in gain can be traded off to some extent against other design features, such as incorporating higher p-type doping in the SiGe alloy base to keep the base layer thin and with low lateral resistance. The fastest Si-based transistors are heterojunction bipolar transistors, with an upper cut-off frequency of 94 GHz at 77 K (Schäffler 1994) (see Chapter 16 for details of this concept). Small very-high-speed signal processing circuits are likely to benefit from heterojunction technology. The heterojunction gives further advantages of higher speed and lower noise to FETs in III–V materials, and the same will be true for Si-based high-electron-mobility transistors. Thermal SiO_2 is a better dielectric, capable of withstanding higher gate voltages, and the trade-off between SiO_2 and SiGe in terms of higher mobility remains to be carried out.

With Si epitaxy, it is possible to prepare bulk unipolar diodes, by analogy with the planar doped barrier structures in GaAs (Chapter 2, Section 2.6). Ion implantation has already been used to make structures that demonstrate the rectifying diode properties (Shannon and Goldsmith 1985) and hot electron transistor structures (cf. Chapter 7, Section 7.2) (Shannon and Slater 1982), and epitaxial growth allows tighter control over the doping profiles.

Further Si devices including avalanche photodetectors, solar cells, IM-PATT, and other microwave diodes may all exhibit advantages from heterojunction design. This may not be just a matter of heterojunction energy steps and barriers *per se*, but some of the peculiarities associated with the Si electronic structure and its modifications under strain may provide additional advantages. These device aspects have not yet been researched.

A Si-based light-emitting diode and laser diode are not available yet–further physics research is required, although the general direction remains clear, namely engineering the bandgap to become direct.

References

Abstreiter, G. (1992). Engineering the future of electronics. *Physics World*, 5 (3), 36–9, and references cited therein.

Cullis, A. G. and Canham, L. T. (1991). Visible light emission due to quantum size effects in highly porous silicon. *Nature*, 353 335–8.

Fukatsu, S. (1994). Luminescence investigation on strained $Si_{1-x}Ge_x$/Si coupled quantum wells. *Solid State Electronics*, 37, 817–23.

Karunasiri, R. P. G, Park, J. S., and Wang, K. L. (1991). Si/GeSi multiple quantum well infrared detector. *Applied Physics Letters*, 59, 2588–90.

Kasper, E., Herzog, H. J., and Schäffler, F. (1988). Si/Ge multilayer structures. In *Physics, fabrication and applications of multilayer structures* (eds. P. Dhez and C. Weisbuch). NATO ASI Series B, Physics, Vol. 182, pp. 229–38. Plenum, New York.

Meyerson, B. S., (1994) High speed silicon-germanium electronics. *Scientific American*, 270, (3), 42–7.

Pearce, C. W. (1988). Epitaxy. In *VLSI technology* (ed. S. M. Sze) (2nd edn), pp. 55–97. McGraw-Hill, New York.

People, R. (1986). Physics and applications of Ge_xSi_{1-x}/Si strained-layer heterostructures. *IEEE Journal of Quantum Electronics*, **QE-22**, 1696–710.

People, R., Bean, J. C., Sputz, S. K., Bethea, C. G., and Perticolas, L. J. (1992). Broadband 8–14 μm normal incidence hole intersubband quantum well infrared photodetector in pseudomorphic Ge_xSi_{1-x}/Si strained layers operating between 20 K and 77 K. *Applied Physics Letters*, **61**, 1122–4.

Schaffler, F. (1994). Strained Si/SiGe heterostructures for device applications. *Solid State Electronics*, **37**, 765–71.

Schaffler, F., Tobben, D., Herzog, H.-J., Abstreiter, G., and Hollander, B. (1992). High electron mobility Si/SiGe heterostructures: influence of the relaxed SiGe buffer layer. *Semiconductor Science and Technology*, **7**, 260–6.

Shannon, J. M. and Goldsmith, B. J. (1985). Bulk unipolar camel diodes formed using indium implantation into silicon. *IEEE Electron Device Letters*, **EDL-16**, 583–5.

Shannon, J. M. and Slater, J. A. G. (1982). Monolithic hot electron transistor in silicon with $F_T > 1$ GHz. *Japanese Journal of Applied Physics*, **22** (Suppl. 22–1), 259–62.

Sze, S. M. (ed) (1988). *VLSI technology* (2nd edn). McGraw-Hill, New York.

15 Vibrational, thermal, and other properties

15.1 Introduction

So far, we have concentrated on the electronic and optical properties of semiconductor multilayers. In this chapter, we consider the vibrational properties of the host lattice, including the effects of the multilayers themselves. The lattice vibrational dispersion relations for bulk GaAs and AlAs are given in Chapter 1, Fig. 1.6. Such phonon spectra are derived from theories of varying complexity, in an attempt to reproduce experimental data in great detail, but a simple model involving bond-stretching and bond-bending forces is sufficient to generate a semiquantitative understanding (Harrison 1980). We can use such a model to describe the lattice vibrations of a superlattice, and the various acoustic, optic, and interface phonon modes that can arise. We describe a number of optical and transport experiments used to detect phonons. The interactions between electrons and phonons play an important role, already considered in Chapter 7, in relaxing hot carriers that have been excited optically or with an electric field.

The thermal properties of carriers and of the semiconductor lattice are of intrinsic scientific interest and are vital in the design of devices. This applies particularly to devices which operate with high power levels, or are required to maintain a high temperature stability as in metrology applications. We describe a number of thermal phenomena in macroscopic and mesoscopic systems. At low temperatures, the dominant phonon wavelength in a blackbody thermal spectrum for most semiconductors is given by $\lambda = 100$ nm$/T$, for the temperature in Kelvin (Kelly 1982). The phonons viewed as elastic waves are an interesting probe of the materials integrity of multilayers and a monitor of heat dissipation, both of which we discuss.

There are other properties of semiconductors and multilayers that are of importance in their applications to devices—mechanical and elastic properties, to mention two. We conclude this chapter with a brief description of these.

15.2 Phonons in superlattices and microstructures

A close look at Fig. 1.6 in Chapter 1 shows a qualitative similarity between the phonon spectra of GaAs and AlAs; the detailed methods by which these spectra are obtained from experiment and theory are not specific to semiconductors and are treated in several texts (Ashcroft and Mermin 1976; Madelung 1978; Harrison 1980). If one notes that the frequency of long-wavelength optic phonons is given by $\omega \sim \sqrt{(K/M)}$, where K is some interatomic force constant and M is the mass of the atoms, the difference in scale between the two spectra is determined by the difference in mass between the Ga and Al atoms in the unit cells. This implies that the characteristic force constants between Al and As atoms and Ga and As atoms in AlAs and GaAs are not too different. The spectra consist of two sets of dispersion curves because there are two atoms per unit cell in each of the materials. At the lower frequencies, the acoustic branches (where $\omega \to 0$ as $k \to 0$) have a slope determined by the speeds of sound, and there is one branch of longitudinal and two branches of transverse polarization. The high-symmetry directions in which phonon spectra are presented often involve degeneracy of the phonon branches. For acoustic modes, the atoms within a unit cell tend to move in the same direction, with a long-wave modulation on the displacement. The characteristic mass of these modes is related to the sum of the masses of the two atoms. At the higher frequencies, the optic branches ($\omega \to \omega_0$ as $k \to 0$) have a very small slope, the atoms within the unit cell tend to move in opposite directions, and the characteristic mass is the reduced mass $1/\mu = 1/m_{Ga, Al} + 1/m_{As}$. There are two characteristic frequencies for long-wavelength optic phonons (the ω_0 above), corresponding to transverse and longitudinal polarizations of the atomic displacements. Apart from the lower frequency of the optic modes in GaAs, compared with AlAs, the other differences are subtle and are related to the different polarizabilities of the GaAs and AlAs bonds, a level of detail which need not concern us here (Bruesch 1982).

In layered superlattices we have a reduced Brillouin zone for the phonons, just as for the electrons (cf. Chapter 9, Section 9.3). The description for superlattices is nicely encapsulated in a 1D chain model in Figure 15.1. Here we have taken a line of alternating atoms corresponding to five unit cells of GaAs followed by four unit cells of AlAs, employed a single spring constant between adjacent atoms, and considered only longitudinal displacements. We see that, at lower energies, the folding back of the acoustic branch gives many low-lying branches having optic mode behaviour as viewed within the superlattice unit cell and which are therefore detectable in optical experiments (cf. Section 15.4). At intermediate energies, there are modes closely associated with GaAs optic modes. Above these there is a gap to the optic modes associated with AlAs. Note that, strictly, we are dealing with normal modes of the entire superlattice, but in practice the amplitudes of the different modes can be greatly enhanced in one or other section of the superlattice unit cell, particularly for the optic modes. The 3D version of Fig. 15.1 is more complex, with the extra polarizations and the 3D geometry, but the physics can be obtained directly from a combination of the phonon spectra of GaAs and AlAs and the superlattice concept.

One important feature missing from this simple model is the possibility of modes that are localized at the interface. These modes are particular solutions of the dynamical equations of motion, which are not free to propagate in a direction normal to the interface, and they have an exponentially decaying

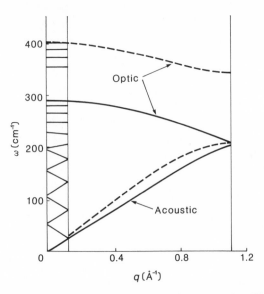

Fig. 15.1 A 1D schematic diagram of superlattice phonon modes. The phonon spectra for GaAs (solid curves) and AlAs (broken curves) are plotted in the large Brillouin zone, and the modes for a superlattice consisting of five layers of GaAs and four layers of AlAs are given in the small Brillouin zone. (After Klein © 1986 IEEE.)

amplitude away from the interface in each normal direction. The existence of such modes, like those of electronic surface and interface states, depends in a subtle manner on the quantitative details of the force constants and mass differences. In cases where we have a single quantum well of GaAs in AlAs, the GaAs optic modes are confined to the vicinity of the quantum well and can be considered localized. These local modes are available to take part in electronic and electromagnetic processes (Ridley 1993*a*, *b*).

In the analysis to date, we have implicitly assumed the continuity of the atomic displacement amplitude and group velocity in all three dimensions as boundary conditions at the interface. For quantum pillars, we need to consider the appropriate boundary conditions of the semiconductor structure with free space. These are in general that both the normal stress and shear stress at the surface are zero. The interest in the vibrational properties of microstructures parallels that of electron states in the same microstructures, namely finite size effects and the discrete separation in energy of the normal modes of vibration (Sottomayor-Torres 1992). The densities of phonon modes in 1D, 2D, 3D, Q1D, and Q2D systems, are given in Fig. 15.2 (compare the electronic case in Chapter 4, Fig. 4.2). The differences are due to the quadratic versus linear dispersion of energy with wavevector for electrons and acoustic phonons respectively. As we shall see in the next section, scattering of electrons or photons by phonons is possible when energy and momentum are conserved, and the changing effective dimensionality limits the scope for these

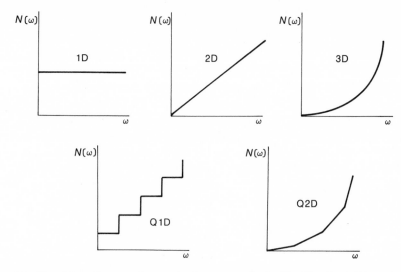

Fig. 15.2 Densities of states of acoustic phonon modes in 1D, 2D, 3D, Q1D, and Q2D systems.

processes as the dimensionality is reduced. Just as the scale of dimensionality for electrons is c.20 nm (cf. Chapter 4), the scale for phonons at low temperatures is set by the 100 nm/T(K) wavelength of dominant phonons, or by the fact that, with sides of c.100 nm, the standing modes of vibrations across a structure are separated by $\hbar\omega \sim h\upsilon/100$ nm ~ 0.1 meV $\sim kT$ for $T = 1$ K (υ is the speed of sound). Experiments in this regime are described later in the chapter; it is an interesting regime where the precise relationship between the discrete (quantum) and continuum (classical) theories of vibration remains to be fully explored.

15.3 Electron–phonon interaction in quantum wells and superlattices

Both acoustic and optic mode vibrations of III–V semiconductors establish a dipole in each unit cell by virtue of relative displacements of the atoms. This dipole interacts strongly with free carriers. In the case of optic modes the dipole density is very large, and the interaction between electrons and optic phonons is very strong. For example, the electric field set up by a longitudinal relative ionic displacement $\boldsymbol{u}$ is given by (Harrison 1980; Ridley 1993a)

$$\boldsymbol{E} = -(e_L^*/V_0\varepsilon_0)\boldsymbol{u} \qquad e_L^{*^2} = MV_0\varepsilon_0^2\omega_L^2(1/\varepsilon_\infty - 1/\varepsilon_s),$$

where V_0 is the volume of the unit cell, ε_0, ε_∞ and ε_s are the permittivities of free space and of the semiconductor at high and low frequencies respectively, M is the reduced mass of the oscillating ions, and ω_L is the frequency of the longitudinal optical phonon. Here e_L^* is a measure of the charge being displaced by the longitudinal optical phonon. For acoustic phonons, it is only the differential motion of the two atoms in the same direction that creates a dipole, and this can be relatively weak at long wavelengths.

The electron–phonon interaction in bulk solids is well understood, as an electron can scatter between a state k and a state $k \pm q$ via the emission or absorption of a phonon of wavevector q. In quantum wells it is a topic of much research and continuing controversy (Ridley 1993a, b). This is in part because the calculation of the scattering matrix elements of such an interaction requires sums of the quantity $\delta E \sim e|\psi(r)|^2 \delta r \cdot \nabla V(r)$, where δr is the displacement of a given atom, normally at r, and this quantity must be summed over every atom in the structure according to the lattice displacement of the appropriate phonon mode. At the same time, the detailed nature of the electron wavefunction enters, as each δE above includes a weighting with the local electron density $|\psi(r)|^2$. $V(r)$ is the one-electron periodic potential for the perfect crystal. In structures with reduced symmetry, for example quantum wells, the detailed displacements of atoms depend sensitively on the model used for the force constants between atoms and the boundary conditions at the interfaces of the quantum well (or the physical boundaries of a microstructure). The possible existence of modes localized at an interface or within a particular layer is also a sensitive function of the model and boundary conditions. Approximations in the electronic structure also limit the reliability of detailed calculations. Some of these issues are reviewed by Klein (1986), Babiker (1992), Das Sarma *et al.* (1992), and Molinari *et al.* (1992).

At high frequencies, the interaction of phonons and light lead to polaritons (Madelung 1978) whose dispersion is shown schematically in Fig. 15.3. The coupled modes have a mixture of optic phonon and light character. These quantities are studied in bulk semiconductors, but become more complicated in quantum wells when one considers the steeply rising light line superimposed on the folded dispersion curve of a superlattice as in Fig. 15.1.

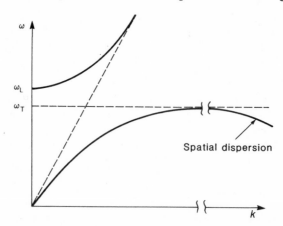

Fig. 15.3 The phonon–polariton interaction, showing the effects on the dispersion curves of interactions between optical vibrational modes and light in a solid.

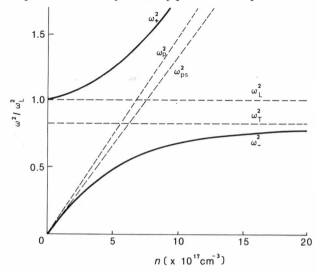

Fig. 15.4 The phonon–plasmon interaction in a semiconductor. At a carrier density of $c.5 \times 10^{17}$ cm^{-3} the plasmon energy is comparable to that of the optical phonon. Both excitations can couple to the electronic charge, and coupled excitation modes described by frequencies ω_+ and ω_- are possible. The creation of the ω_+ modes is very effective in taking energy from hot electrons in doped semiconductors (cf. Chapter 7). At low densities this coupled mode is longitudinal optical phonon like, but at high densities it is transverse optical phonon like.

The other important electron–lattice interaction concerns the collective plasma oscillations of an electron gas, which interact with the optic phonon spectra. As a function of density, at $c.5 \times 10^{17}$ cm^{-3}, the natural plasma frequency $\omega_p^2 = 4\pi n e^2/m^*$ takes a value which is comparable with the optic phonon frequency of 36 meV in GaAs, and a full treatment of the coupled system is a complex matter although the results for the $k \sim 0$ modes are summarized in Fig. 15.4. Note the shifts in the excitation energy that are measurable. The details of the analysis take us beyond the scope of this text, but we note that the effective interaction between a single electron and one of the coupled modes can be exceedingly high, even 10 times higher than the two interactions separately; this is because of the detailed nature of the eigenfunction of the hybrid mode, and an enhancement brought about by the interaction (Kim *et al.* 1978). Half of the energy loss caused to hot electrons injected into doped GaAs is due to the emission of coupled phonon–plasmon modes (cf. Chapters 7 and 16).

15.4 Experimental identification of phonons in superlattices

One of the earliest manifestations of phonons in superlattices was the observation of a dielectric phonon filter (Narayanamurti *et al.* 1979). It is a straightforward matter to calculate the transmission of long-wavelength phonons through a multilayer stack of different acoustic impedances $Z = \rho v$

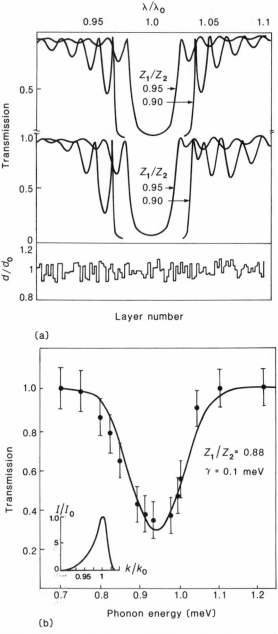

Fig. 15.5 (a) The theoretical transmission of long-wavelength acoustic waves through a 50-layer superlattice of different acoustic impedances (with ratios of 95 and 90 per cent); (b) a comparison between theory and experiment of the transmission intensity, with an impedance ratio of 88 per cent, and a Gaussian broadening of the acoustic phonon energy of 0.1 meV to account for all inhomogeneities. The inset shows the inferred breadth of the incident phonon wavelengths. (After Narayanmurti *et al.* 1979.)

(ρ is the density and v the speed of sound). This is in direct analogy with light transmission through a stack of alternating dielectrics. The results for transmission through a 50-layer stack with difference impedance ratios are given in Fig. 15.5(a) for both an ideal multilayer stack and for a stack with $c.10$ per cent random disorder in layer thicknesses. Superconducting tunnel junctions can be used to emit phonons and detect them after transmission through a superlattice. The results of the analysis are given in Fig. 15.5(b). There is good agreement between experiment and a theory which includes various other phonon-scattering events as an empirical broadening factor.

Subsequent work has concentrated on optical and electrical experiments. The Raman effect is one of the most important probes. A photon of given frequency, polarization, and propagation direction is absorbed, and another photon of different character is emitted. The difference in energy and direction is used to reconstitute the $\omega - q$ relation for excitations of the solid. By setting up to detect the weakly scattered light, it is possible to map out the phonon (and other) excitation spectra of solids. The two regimes of importance to superlattices are shown in Fig. 15.6, with folded acoustic modes and slab modes (modes confined in GaAs quantum wells in a superlattice). The former reinforces the accuracy of simple acoustic models applied to superlattices. In the latter, the polarization of the incident and scattered light is an important further indicator of the type of phonon excitation. See Klein (1986) and references cited therein for more details.

The energy of optic phonon modes $c.30-50$ meV is in the infrared part of the spectrum, and optical absorption of superlattices also provides a method for detecting superlattice and multilayer phonons. The technique is restricted to ultrapure undoped materials, as there are many electronic excitations of larger cross-section, such as inter-sub-band excitation, that mask the creation of phonons.

15.5 Effects of phonons on transport and optical properties

This topic has been dealt with in previous chapters, in particular in chapter 7 with reference to the relaxation of hot carriers by the emission of phonons. The reduction in the rate of carrier cooling (whether the hot carriers were generated via a high electric field or via optical pumping) has been attributed to the restrictions on the number of phonon modes, their energy distribution, and their polarization properties.

The full general expression for the rate at which a hot carrier in an initial state ψ_i scatters by the emission of a phonon into a final state ψ_f is given by Fermi's golden rule:

$$W = (2\pi/\hbar) \int (n_p + 1) f(E_f) [1 - f(E_i)] |\langle \psi_f | H_{ep} | \psi_i \rangle|^2 \delta(E_f - E_i + \hbar\omega_q) \, dS_f,$$

where the $f()$ are the Fermi occupation functions for the initial and final electron states, $n_p = n_p(E_f - E_i)$ is the Bose–Einstein phonon occupation num-

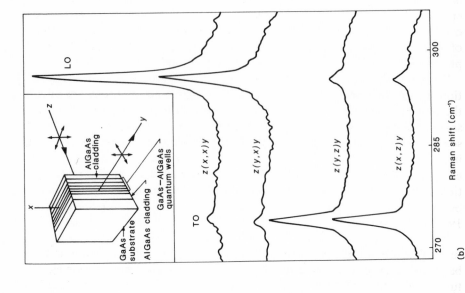

Fig. 15.6 Raman spectra (of inelastically scattered light) used to detect (a) folded longitudinal acoustic modes (after Colvard *et al.* 1985) and (b) longitudinal (LO) and transverse (TO) optical modes in a superlattice (after Zucker *et al.* 1984). In the former, the arrows indicate where the signals would be expected on the basis of bulk elastic constants and an X-ray analysis of the superlattice. In the latter, the symbol $p(q, r)s$ refers to the geometry in which the experimental takes place, namely the incident light travels in the p direction with polarization q, while the scattered light travels in the s direction with polarization r.

ber, E_i and E_f are the initial and final state energies, $\hbar\omega_q$ is the phonon energy, and the integral is taken over all possible final states. The electron–phonon matrix element $\langle \psi_f | H_{ep} | \psi_i \rangle$ was described in Section 15.3. Carriers that are electronically or optically excited relax their excess energy (and/or momentum) via the emission of phonons, most notably optical phonons. Once there is an excess energy of $c.36$ meV, hot carriers are very effective at emitting their energy. This was illustrated in Chapter 7, Fig. 7.8, where electrons injected over a heterojunction recombined with the emission of photons, the spectrum of which had a series of peaks corresponding to $1, 2, 3, \ldots, 10$ optical phonon losses.

Within quantum wells, the modifications to the electron–phonon interaction via the specific electron states and phonon modes alter and complicate the optical relaxation processes. The localized nature of electron states and some phonon modes in quantum wells means that the occupation number of some phonon modes can become appreciable, and the phonon distribution itself may become 'hot' in the same sense of non-equilibrium as is applied to electrons. When combined with modifications to the non-equilibrium occupation functions for the electrons, the precise calculations of cooling rates becomes complex, with little consensus emerging in recent years (Babiker 1992; Das Sarma *et al.* 1992; Molinari *et al.* 1992).

One of the important roles of phonon modes comes in the calculation of tunnelling probabilities, as hinted in Chapter 8. There the calculations were based on elastic tunnelling, but if phonons can be emitted during the tunnelling process, an extra channel for the current is possible. This effect has been observed in that, above the main current peak in the *I–V* characteristics of a high-quality double-barrier diode, a satellite peak can be observed at an overall bias corresponding to an additional $c.36$ meV over some part of the double-barrier diode. Several possibilities are mentioned, for example the tunnelling electron might lose $c.36$ meV in the first barrier and enter the quantum well with an energy corresponding to the bound state in the well. An electron tunnelling via the first excited state of the quantum well might emit phonons and relax into the lower state, and tunnel out from there. These processes add to the current through a resonant tunnel diode, but at biases where it degrades the peak-to-valley current ratio which is an important device figure of merit (Goldman *et al.* 1987).

15.6 Thermal phenomena in multilayers and microstructures

Acoustic phonons are responsible for the transport of heat through an insulating solid in response to a gradient in their occupation number in a temperature gradient (Ashcroft and Mermin 1976; Madelung 1978). If carriers are present, they can dominate the thermal conduction, and we shall return to them below. The low-temperature (below 60 K) expression for lattice thermal conductivity

$$K = C_v v_s \lambda / 3$$

relates the lattice specific heat C_v (which varies with T^3), the speed of sound, and the phonon mean free path λ all of which are determined by the acoustic modes. In high-quality samples (with no defects including isotope variations which act as phonon scatterers), the phonon mean free path is set by the smallest linear dimension of the solid and scaled by a shape factor. This is called the Kasimir limit, which assumes that phonons are scattered, rather than reflected, at the surfaces because of asperities, adsorbed molecules, etc. (Berman 1976). With the advent of multilayers and the techniques for fabricating clean surfaces prior to epitaxy, the number of reflected phonons can rise substantially. This has been seen in highly polished sapphire wafers where the specular reflection coefficient for phonons can exceed 99 per cent after suitable polishing of both sides, with a significant increase in the thermal conductivity of such wafers at low temperatures. The reflection coefficient drops to less than 50 per cent with even gentle abrasion or deposition of overlayer material (Wybourne *et al.* 1984). The lattice-matched heterojunctions encountered in the GaAs–AlGaAs materials do not have a sufficient thermal impedance discontinuity to modify the thermal transport processes. Some of the highly strained systems described in previous chapters might now usefully be investigated for phonon waveguiding and confined thermal transport. Optical phonons are not efficient transporters of thermal energy because of their small group velocity and finite lifetime with respect to decay into two or more acoustic phonons.

With the speed of sound in semiconductors at about 3000 ms^{-1} and with substrates about 1 mm thick, the detection of heat pulses on time scales of *c*.1 µs on the back-face of a substrate allows phonons to be used as a diagnostic tool to monitor the generation and dissipation of heat in device-like structures. The experimental arrangement is qualitatively similar to that shown in Fig. 15.5(a). Furthermore, the different sound velocities of longitudinal and transverse modes, and the anisotropic propagation in a cubic lattice structure, can be used in the analysis. There have been several such studies. A *c*.1 ns pulsed electric field applied to a 2DEG in a HEMT-type structure (Chapter 5) at low temperatures results in heated electrons which emit acoustic phonons as they relax. The time–temperature distribution of a bolometer on the back-face was analysed as a function of the electric field strength of the pulse (Fig. 15.7(b)), and the signal height shows a linear increase with power in the electric pulse, but with a change between two regimes at *c*.5 pW per electron pulse (Fig. 15.7(c)), subsequently analysed to give an indication of the cross-over to optical phonon emission at higher power excitations when the electron temperature reaches about 60 K. With T_e and T_l the electron and lattice temperature respectively, the energy dissipation rate per electron at the lower power end varies as $T_e^3 - T_l^3$ (Fig. 15.7(d)), changing to an $\exp(-\hbar\omega_{LO}/kT_e)$ rate at the higher power end (Fig. 15.7(e)). Other applications have included the analysis of heat dissipation at the corners of a quantum Hall bar (cf. Chapter 5, 5.10(b)) and a preferential emission of phonons from accelerated electrons in the forward-going direction.

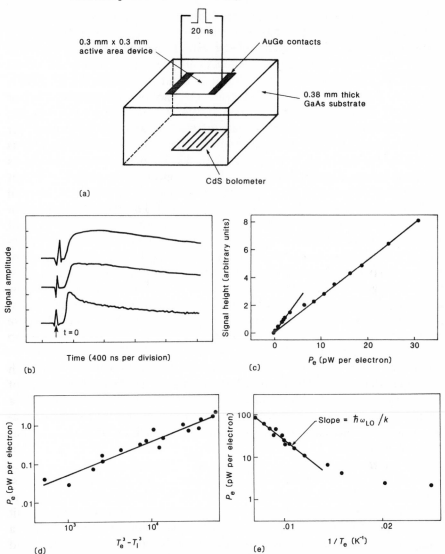

Fig. 15.7 (a) The experimental arrangement for heat pulse studies, where a pulse current generates phonons from Joule heating in a 2DEG. The heat pulses are detected by a CdS bolometer on the back-face of the substrate. (b) The raw data are the detected heat pulse signals at the back-face a GaAs substrate 0.38 mm thick beneath a 2DEG subject to a nanosecond electrical pulse for three different excitation powers (7.5 pW, 50.5 pW, and 600 pW power delivered to each electron, from bottom to top), with the $t = 0$ pulse an electrical pick-up. (c) The signal height versus power delivered to the electrons shows a cross-over between two limited values for about 5 pW per electron. (d) At low levels of excitation, the power per electron varies as $\Delta(T_e^3 - T_l^3)$, as expected from theory. (e) At higher powers, there is an exponential relationship between electron temperature characteristic of optical phonon emission. (After Hawker *et al.* 1992.)

The thermal conductivity of most conducting materials is dominated by the electronic contribution. This has been neatly verified in semiconductor micro-structures formed into free-standing fine wires (of bridge topology) as shown in Chapter 3, Fig. 3.5(a). A layer of GaAs, n-doped to $c.5 \times 10^{17}$ cm^{-3} so as to remain conducting at low temperatures, has been grown on an undoped substrate, and the undercut microstructures formed by a combination of lithography and wet chemical etching (Potts *et al.* 1990*a, b*). If large-area ohmic contacts are made to either side of the bridge, current will flow only through the centre of the free-standing wires. Furthermore, the largest element of the resistance is associated with the fine wires, and this resistance has been measured as a function both of temperature at a fixed very low current and of current at fixed lattice temperature. As the substrate tempera-ture rises from about 0.4 K to about 5 K, the low-current resistance falls by about 10 per cent owing to the removal of the localization and electron–electron scattering effects described in Chapters 4 and 6. The analysis of these data confirms that the conducting channel in the conducting wire has a diameter of about 120 nm, equivalent to the occupation of about 10 laterally quantized 1D sub-bands (Potts *et al.* 1990*a*). Since the material is doped, the electron phase coherence length (cf. Chapter 6) is quite short (*c.*100 nm). Since the energy input is small, the electrons and lattice re-main at the same temperature, and these same data can be inverted for use as an electron thermometer, i.e. $R(T)$ becomes $T_e(R)$. The electron heating experiments obtain $R(I)$ (cf. Fig. 15.8(b)) for fixed substrate temperature. The fall in resistance is attributed to a rise in electron temperature in the wire, and the elementary heat conduction equation for the temperature profile, involving a position-dependent temperature, the thermal conductivity $K(T(x))$, the thermal resistivity $\rho(x,T)$, and the cross-section A of the conducting channel

$$\frac{-\partial\{K[T(x)]\partial T(x)/\partial x\}}{\partial x} = \frac{I^2\rho(x,T)}{A}$$

can be solved analytically for the two important forms of thermal conductiv-ity (Potts *et al.* 1990*b*). If the lattice contribution dominates, and the temperature is low enough, then $K \propto C_v \propto T$ if the phonons are Q1D (i.e. the lateral modes are frozen out, and only the modes along the wire are involved). If the system behaves like a Q3D lattice at higher temperature, then $K \propto C_v \propto T^3$. In fact, the best fit to theory comes with $K \propto T$. Moreover, the constant of proportionality equals that given by the Weidemann–Franz law, namely $\mathscr{L}T/\rho$ where $\mathscr{L} = 2.44 \times 10^{-8}$ $W\Omega$ K^{-2}. This result implies that *all* heat is carried out of the system *solely* by the electrons in the low to modest-current regime (i.e. less than 100 µA per wire). This result was obtained over the entire 0.5–5 K range of substrate temperature. The implication is that the phonons play no role, with the lattice remaining at a fixed temperature and all energy being transported by the electrons. This was verified by repeating the experiments on wires that were partially supported, i.e. where the

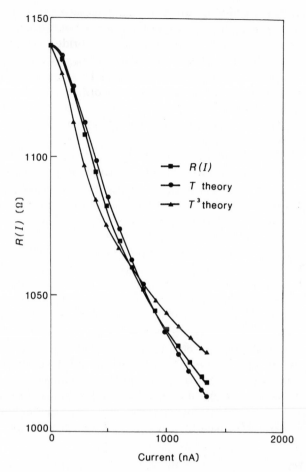

Fig. 15.8 The resistance versus current data compared with theory based on thermal transport by electrons in a free-standing wire (after Potts *et al.* 1992(b)). The theory invokes either a T or a T^3 dependence on temperature for the thermal conductivity. The best fit to the data comes from a linear dependence on temperature.

undercut etch was incomplete and the system was coupled directly to the substrate lattice phonons; the same quantitative results were obtained. This is a new manifestation of circumventing the electron–phonon interaction. As the current was increased, two further regimes of behaviour were observed. For currents in the range 100–300 μA per wire, some phonon emission effects were incorporated in the theory and the improved fit to experiment was used to infer a ballistic phonon mean free path of c.20 μm, i.e. longer than the wires. This provided the first evidence of reduced-dimensionality phonon effects that was the original objective of the heating studies (Potts *et al.* 1992). Finally, at higher currents there was evidence for direct lattice heating of the wires, invalidating the use of the resistance as a measure of electron temperature.

The investigation of the interaction of electrons and lattice in semiconductor and other microstructures, as described in this and earlier sections, is in its infancy. There is a decade or more of further research to be done, both of a basic research nature and also with applications to the design of more efficient devices and those with a reduced dependence on lattice temperature. Better ways of extracting unwanted heat from the active device volume will help is this latter regard.

15.7 Mechanical properties of multilayers and microstructures

While we have been concerned with the electronic and optical properties of semiconductors and multilayers, their mechanical properties place limits on the time–temperature cycles that can be tolerated during device processing. These same properties determine the types of packaging and the range of environments in which devices may operate. Mechanical properties, including brittleness, hardness, dislocation structures, thermal gradients, etc., are complex in their own might and show a strong dependence on the substrate temperature and its doping. Rather than treat the subject in detail, a few pertinent facts are collected here.

The principal elastic constants (Chen *et al.* 1992) are given in Table 15.1. These constants can be inferred from the phonon spectra, and confirm the greater bond strength of Si compared with GaAs and of GaAs compared with InP. Possibly more important is the fracture resistance, i.e. the energy per unit area needed to break bonds in a given crystallographic plane, which occurs when a crack propagates in a brittle material (as all semiconductors are at and below room temperature). Values of fracture resistance (Clarke 1992) are given in Table 15.2. The values in parentheses are measured fracture surface energies, where available, and show a distinct correlation. In contrast with some other materials, there is no evidence that the fracture energy in semiconductors is related in any way to the

Table 15.1 Elastic constants ($\times 10^{11} \, \mathrm{dyn \, cm^{-2}}$)

	Si	GaAs	AlAs	InP
Bulk	9.923	7.69	7.81	7.11
$C_{11}-C_{12}$	10.274	6.63	6.32	4.50
C_{44}	8.036	6.04	5.89	4.56

From Madelung 1991

Table 15.2 Fracture resistance ($\mathrm{J \, m^{-2}}$)

	Si	GaAs	AlAs	InP
{100}	2.13	2.2	2.6	1.9
{110}	1.51(0.81)	1.5(0.76)	1.8	1.3(0.86)
{111}	1.46(1.23)	1.3	1.5	1.1

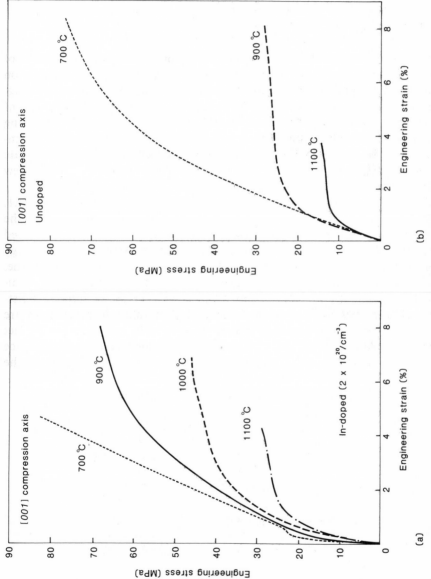

Fig. 15.9 The stress–strain curves of (a) In-doped GaAs and (b) undoped GaAs. (After Guruswamy et al. 1992.)

358 *Vibrational, thermal, and other properties*

strain rate. One intriguing result is that illumination causes increased dislocation mobility, and the hardness of semiconductors increases with light
intensity.

Semiconductors deform in a brittle manner at low temperature and in a
ductile manner at higher temperatures. The transition temperature range for
GaAs is just above room temperature (*c*.30–70 °C). At the higher temperatures, there is evidence of a dependence of the fracture resistance on the strain
rate associated with the shock of dropping or with large accelerations as in
collisions. The transition temperature depends on the level of doping, at least
in Si. The plastic behaviour of semiconductors at the elevated temperatures
that occur during growth and processing is fairly complex, and is beyond the
scope of this text (Faber and Malloy 1992). Figure 15.9 shows that the
yielding of stressed GaAs depends on temperature and on the presence of In
doping. The elastic regime is larger in undoped GaAs, but the In-doped GaAs
is harder. The second linear regime (typically starting from *c*.0.5 per cent)
corresponds to work hardening but becomes less pronounced at higher
temperatures. The equivalent shear stress shows a similar hardening with the
incorporation of In.

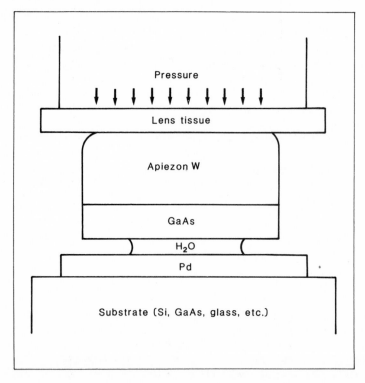

Fig. 15.10 The formation of composite structures using lifted-off epitaxial overlayers
produced with highly selective etches which are rebonded by van der Waals forces to
new substrate materials. (After Yablonovitch 1990.)

The mechanical properties of microstructures and semiconductor multi-layers has not been well investigated. Some isolated studies include the formation of free-standing superlattices, where a thin AlAs sacrificial layer is etched away with HF and the GaAs–AlGaAs epitaxial overlayer is lifted off and deposited on a new substrate, as shown schematically in Fig. 15.10. The thin layers seem remarkably robust (Yablonovitch *et al.* 1987, 1990).

The free-standing GaAs wires discussed in Chapter 3, Section 3.3, and in Section 15.6 reveal a number of incidental mechanical properties. Scanning electron micrographs show that long 40 μm wires sag in the middle. It was found that GaAs doped at 5×10^{17} cm^{-3} or less could be wet etched and yield very smooth side-walls. When doped at 10^{18} cm^{-3}, the side-walls were rough and broken sections of wire were very rigid (Potts *et al.* 1991). This behaviour is also seen in Si_3N_4 wires, while Au–Pd wires bend sharply when broken.

References

Ashcroft, N. W. and Mermin, N. D. (1976) *Solid state physics.* Holt, Rinehart, Winston, New York.

Babiker, M. (1992). Coupling of polar optical phonons to electrons in superlattices and isolated quantum wells. *Semiconductor Science and Technology*, 7, B52–9.

Berman, R. (1976). *Thermal conduction in solids.* Clarendon Press, Oxford.

Bruesch, P. (1982). *Phonons: theory and experiments I.* Springer-Verlag, Berlin.

Chen, A.-B., Sher, A., and Yost, W. T. (1992). Elastic constants and related properties of semiconductor compounds and their alloys. In *Semiconductors and semimetals*, Vol. 37, *The mechanical properties of semiconductors* (eds K. T. Faber and K. Malloy), pp. 1–77. Academic Press, New York.

Clarke, D. R. (1992). Fracture of silicon and other semiconductors. In *Semiconductors and semimetals*, Vol. 37, *The mechanical properties of semiconductors* (eds K. T. Faber and K. Malloy), pp. 79–142. Academic Press, New York.

Colvard, C., Gant, T. A., Klein, R., Merlin, M. V., Fischer, R., Morkoc, H., and Gossard, A. C. (1985). Folded acoustic and quantized optic phonons in (GaAl) As superlattices. *Physical Review B*, 31, 2080–91.

Das Sarma, S., Campos, V. B., Stroscio, M. A., and Kim, K. W. (1992). Confined phonon modes and hot-electron relaxation in semiconductor microstructures. *Semiconductor Science and Technology*, 7, B60–6.

Faber, K. T. and Malloy, K. (ed.) (1992). *Semiconductors and semimetals*, Vol. 37, *The mechanical properties of semiconductors.* Academic Press, New York.

Goldman, V. J., Tsui, D. C., and Cunningham, J. E. (1987). Evidence for LO-phonon-emission-assisted tunnelling in double barrier heterostructures. *Physical Review B*, 36, 7635–7.

Guruswamy, S., Faber, K. T., and Hirth, J. P. (1992). Mechanical behaviour of compound semiconductors. In *Semiconductors and semimetals*, Vol. 37, *The mechanical properties of semiconductors* (eds K. T. Faber and K. Malloy), pp. 189–230. Academic Press, New York.

Harrison, W. A. (1980). *Electronic structure and the properties of solids.* Freeman, San Francisco, CA.

Hawker, P., Kent, A. J., Hughes, O. H., and Challis, L. J. (1992). Changeover from acoustic to optic mode phonon emission by a hot two-dimensional electron gas in the gallium arsenide/aluminium gallium arsenide heterojunction. *Semiconductor Science and Technology*, **7**, B29–32.

Kelly, M. J. (1982). Thermal anomalies in very fine wires. *Journal of Physics C*, **15**, L969–73.

Kendall, D. L., Fleddermann, C. B., and Malloy, K. J. (1992). Critical technologies for the micromachining of silicon. *Semiconductors and semimetals*, Vol. 37, *The mechanical properties of semiconductors* (eds K. T. Faber and K. J. Malloy), pp. 293–337. Academic Press, New York.

Kim, M. E., Das, A., and Senturia, S. D. (1978). Electron scattering interaction with coupled plasmon polar phonons in degenerate semiconductors. *Physical Review B*, **18**, 6890–9.

Klein, M. V. (1986). Phonons in semiconductor superlattices. *IEEE Journal of Quantum Electronics*, **QE-22**, 1760–70.

Madelung, O. (1978). *Introduction to solid state theory*. Springer-Verlag, Berlin.

Madelung, O. (1991). *Semiconductors: group IV elements and III–V compounds*. Springer-Verlag, Berlin.

Molinari, E., Bungaro, C., Gulia, M., Lugli, P., and Rücker, H. (1992) Electron–phonon interactions in two-dimensional systems: a microscopic approach. *Semiconductor Science & Technology*, **7**, B67–72.

Narayanamurti, V., Stormer, H. L., Chin, M. A., Gossard, A. C., and Wiegmann, W. (1979). Selective transmission of high-frequency phonons by a superlattice: the dielectric phonon filter. *Physical Review Letters*, **43**, 2012–15.

Potts, A., Hasko, D. G., Cleaver, J. R. A., Smith, C. G., Ahmed, H., Kelly, M. J., et al. (1990a). Quantum conductivity corrections in free-standing and supported n⁺-GaAs wires. *Journal of Physics: Condensed Matter*, **2**, 1807–15.

Potts, A., Kelly, M. J., Smith, C. G., Hasko, D. G., Cleaver, J. R. A., Ahmed, H., et al. (1990b). Electron heating effects in free-standing single crystal GaAs fine wires. *Journal of Physics: Condensed Matter*, **2**, 1817–25.

Potts, A., Kelly, M. J., Hasko, D. G., Smith, C. G., Cleaver, J. R. A., Ahmed, H., et al. (1991). Thermal transport in free-standing semiconductor fine wires. *Superlattices and Microstructures*, **9**, 315–18.

Potts, A., Kelly, M. J., Hasko, D. G., Cleaver, J. R. A., Ahmed, H., Ritchie, D. A., et al. (1992). Lattice heating of free-standing ultra-fine wires by hot electrons. *Semiconductor Science and Technology*, **7**, B231–4.

Ridley, B. K. (1993a). *Quantum processes in semiconductors* (3rd edn). Clarendon Press, Oxford.

Ridley, B. K. (1993b). Quantum confinement and scattering processes. In *Handbook of Semiconductors*, Vol. 1 (2nd edn) (ed. P. T. Landsberg). North-Holland, Amsterdam.

Sottomayor-Torres, C. M. (1992). Spectroscopy of nanostructures. In *Physics of nanostructures* (eds J. H. Davies and A. R. Long). Institute of Physics, Bristol.

Wybourne, M. N., Eddison, C. G., and Kelly, M. J. (1984). Phonon boundary scattering at a silicon–sapphire interface. *Journal of Physics C*, **17**, L607–12.

Yablonovitch, E., Gmitter, T., Florez, L. T., Harbison, J. P., and Bhat, R. (1987). Extreme selectivity in the lift-off of epitaxial GaAs films. *Applied Physics Letters*, **51**, 2222–4.

Yablonovitch, E., Hwang, D. M., Gmitter, T. J., Florez, L. T., and Harbison, J. P. (1990). Van der Waals bonding of GaAs epitaxial lift-off films onto arbitrary substrates. *Applied Physics Letters*, **56**, 2419–21.

Zucker, J. E., Pinczuk, A., Chemla, D. A., Gossard, A. and Wiegmann, W. (1984). Optical vibrational modes and electron–phonon interaction in GaAs quantum wells. *Physical Review Letters*, **53**, 1280–3.

16 Devices I: field-effect and heterojunction bipolar transistors

16.1 Introduction

Having established the physics and the technology of semiconductor multi-layers, in the following four chapters we consider the way in which that physics is being exploited in commercial and advanced research devices. These chapters are not intended to be complete expositions of the devices, which can be found in standard texts (e.g. Sze 1981, 1990). In this chapter we concentrate on transistors used as switches in digital circuits and as amplifiers in analogue circuits. In the next chapter, we focus on two-terminal devices used as sources, detectors, and mixers of microwaves and millimetre waves. These two chapters cover electronic devices with applications in computation and communications (including radar). In Chapter 18, we concentrate on a range of sources, modulators, and detectors of coherent and incoherent optical signals which find their principal application in optical fibre communication links. In Chapter 19 we describe a wider range of optical devices, used as mid- and far-infrared detectors of radiation and in solar cells.

It will become apparent that new generations of devices result from the use of heterojunctions as a feature in device design. The superior performance of the new devices is a direct consequence of the physics associated with heterojunctions that has been described in Chapters 5–10. In addition to some primary device attributes such as efficiency, power, speed, etc., where hetero-junctions as design elements offer advantages, there is a much wider range of secondary attributes, such as noise, insensitivity to ambient temperature, manufacturability, design tolerance, reliability, etc., which together are equally important, if not more so, to a systems engineer calling for devices. The high-electron-mobility transistor, the heterojunction Gunn diode, the quantum-well laser, and the quantum-well infrared detector are examples from each of the next four chapters where the heterojunctions in the devices offer advantages to the systems that exploit them. We shall also describe the operation of some radically new devices whose principles have been demon-strated, but whose performance advantages are as yet insufficient to warrant the redesign of a system to incorporate them. In time, as the pressure builds up for ever higher performance figures, these radical devices may well represent the new way forward. Furthermore, some of the effects described in Chapters 11 and 12 may come into their own as principles of device operation. These possibilities will be discussed in Chapters 21 and 22.

The structure of this chapter is as follows: in the next section we introduce the basics of transistor action, and hint at the general role of heterojunctions. Thereafter, we describe each of three transistor families in some detail, and we complete the chapter with a discussion of some new transistor concepts.

16.2 Basic FET and bipolar transistor action

There are two principal types of transistor whose action is shown schematically in Fig. 16.1. The FET involves currents flowing parallel to the layers, while the (vertical) bipolar transistor has currents flowing through multilayers.

The FET is basically a metal–insulator–semiconductor capacitor structure. In the case of a Si MOSFET, a positive bias applied to the gate metal attracts electrons in the semiconductor to the semiconductor–insulator interface. If the Si is p-type, and the source and drain contacts are n-type, then a source–drain current will not flow in the absence of a gate bias; at least one of the source–Si or Si–drain p–n junctions will be reverse biased. At sufficient gate bias (the threshold voltage V_{th}) the conduction band edge at the semiconductor–insulator interface is lowered below the Fermi energy of the bulk semiconductor and allows an excess of electrons there. The source–channel–drain path is electron rich throughout, allowing a current to flow. In digital circuits, the switching operation is between two values of gate bias corresponding to no conduction or conduction between the source and drain. In an analogue circuit, the amplification comes from the fact that the source–drain resistance in the conducting state is larger than the gate resistance, so that small changes in input voltage results in larger changes in source–drain voltage. Alternatively, a modest gate current can be used to control a much larger source–drain current.

The elementary equations governing this situation are as follows. If the area of the gate is A (of length L in the source–drain direction and of width W) and the thickness of the insulator is d, then the electron density n_s in the semiconductor is determined by the capacitor relation

$$en_sA = C(V_g - V_{th}) = (\varepsilon_s\varepsilon_0 A/d)(V_g - V_{th}).$$

When conducting, the source–drain bias of a large transistor is sufficient that electrons move at approximately their saturated drift velocity v_s (cf. Chapter 1, Fig. 1.9), and so

$$I_{SD} \sim n_s Wev_s.$$

The important figure of merit is the transconductance g_m, which is the derivative of the (output) source–drain current with respect to the (input) gate voltage and is given by

$$g_m = \partial I_{SD}/\partial V_g = Cv_s/L.$$

For a fast transistor, the time taken by the electrons to traverse the source–drain region is given by $\tau = L/v_s$, and a combination of small structures

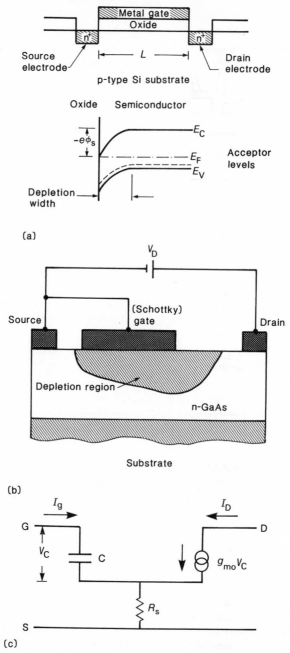

(a)

(b)

(c)

Fig. 16.1 The Si and GaAs FET and bipolar transistor, showing the structure, principle of operation, and equivalent circuit: (a) Si MOSFET, a capacitor structure with source and drain electrodes having access to the variable electron density at the Si–SiO2 interface; (b) the GaAs MESFET, a layer of n-doped GaAs with the depletion layer from a reversed bias Schottky gate controlling the width of the conducting channel; (c) the equivalent circuit

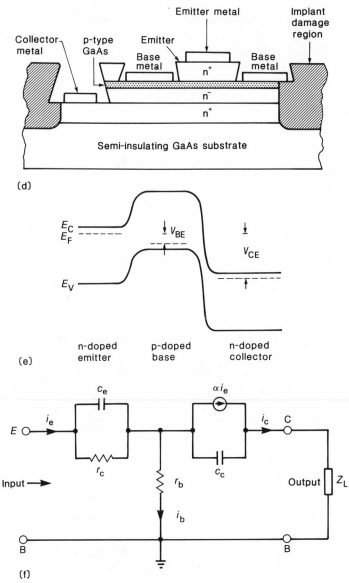

(d)

(e)

(f)

for a MESFET as an effective current source, where C is the source–gate capacitance, R_s is the access resistance between the source and the entrance to the gated region, and g_{m0} is the intrinsic transconductance as described in the text; (d) a side-view of an n–p–n bipolar transistor, emphasizing the complexity of the processing required to access and contact each of the three layers and to isolate devices from each other (the semi-insulating (SI) substrate has no mobile charge to interfere with the high-speed operation of the device); (e) the band profile for a homojunction bipolar transistor with forward-biased emitter and reverse-biased collector; (f) the equivalent circuit for small-signal analysis, where α is the transfer efficiency of electrons from emitter to base and the other symbols have their usual meaning.

and suitable materials is required to minimise τ. There is a more intricate version of this model, taking a two-line approximation to the velocity-field characteristic (cf. Chapter 1, Fig. 1.9), which gives appropriate limiting values for long and short transistors (Weisbuch and Vinter 1991). The equivalent circuit for a FET is shown in Fig. 16.1(c), and standard a.c. circuit theory (at frequency f, or angular frequency $\omega = 2\pi f$) gives an expression for the current gain h_{21}, which is the ratio of drain to gate current, as

$$h_{21} = I_D/I_g = g_m/j\omega C = f_T/jf$$

where $f_T = g_m/2\pi C = 1/2\pi\tau$ is known as the cut-off frequency, giving a measure of the most rapid frequency at which a device can respond to a signal on the gate.

A variant on the MOSFET is the MESFET (see Fig. 16.1(b)), where a thin epitaxial layer on the semiconductor substrate is conducting but a negative bias on the gate forms a depletion region wide enough to prevent conduction. The underlying device operation is the same as that just described for the MOSFET if we transfer the origin of gate voltage to that which depletes the substrate. The current is proportional to the thickness of the undepleted region, which in turn varies approximately as the square root of the gate voltage above threshold (cf. Chapter 2, Section 2.5).

The vertical bipolar transistor consists of two p–n junctions back to back, with the intervening (base) layer sufficiently thin that the actions of the two junctions are coupled. If electrons are injected into a region of p-doped material, they will eventually recombine. If the p layer is thin enough and ends in another p–n junction, many electrons will pass through into the second n region. This condition is required for the bipolar transistor, and is shown in Fig. 16.1(d). The amplifying action in this device comes from the fact that small base currents can lead to a significant lowering of the barrier represented by the p–n junction and a great increase in emitted current. In terms of transferring resistance, the forward-biased emitter–base junction has a rather lower slope resistance than the reverse-biased base-collector junction. The role of heterojunctions will be to modify the various current ratios.

The standard theory for bipolar transistor action (Sze 1981) gives the following relation for the collected current I_c, after being injected into the base as the result of a forward base–emitter bias V_{BE} and successfully transiting the base region (as $I_B < 0.01 I_E$ in practice), as

$$I_c = A (D_n e n_{p0}/w_B)[\exp(eV_{BE}/kT) - 1]$$

where A is the area of the emitter, D_n is the diffusion constant for electrons in the p-doped material of the base, where their equilibrium concentration is n_{p0}, w_B is the width of the base layer, and V_{BE} is the emitter–base bias. If we regard $2D_n/w_B$ as an average diffusion velocity v_d, the expression for the current can be regarded as the product of the diffusion velocity and the number of electrons injected into the base. The high transconductance g_m follows from the exponential dependence of I_c on V_{BE}:

$$g_{\mathrm{m}} = \partial I_{\mathrm{c}}/\partial V_{\mathrm{BE}} = (e/kT)\,I_{\mathrm{c}}.$$

In digital circuits, this means that small swings in input voltage suffice to make large changes in the output current. The total number of electrons in the base can be used to determine an emitter–base capacitance as

$$C = \partial Q_{\mathrm{B}}/\partial V_{\mathrm{BE}} = (e/kT)(en_{\mathrm{p}}w_{\mathrm{B}}/2)$$

where n_{p} is the number of electrons at the entrance to the base. From this we derive the transconductance as

$$g_{\mathrm{m}} = C(2D_{\mathrm{n}}/w_{\mathrm{B}}^2) = Cv_{\mathrm{d}}/w_{\mathrm{B}} = C/\tau$$

where τ is the base transit time for the electrons. High-speed devices require thin base layers.

Fuller descriptions of the device equations, and the way they are used to optimize transitor geometry for device performance, are given in appropriate texts (Sze 1981), but a few general points are appropriate here. At the heart of both transistors, transport of electrons in the active region of the device (under the gate or through the base) determines much of the device performance (Ladbrooke 1986). We have described above the first parameter of importance, the transit time τ. Another feature that determines speed is the time taken to charge various capacitors, the gate in the case of a FET and the emitter–base barrier in the case of the bipolar transistor, which both have the form of an RC time constant. This implies that resistances and capacitances have to be kept small for fast devices. This latter condition is in conflict with the requirements for high-gain devices, and so trade-offs are required in the construction of a transistor.

If we take the region under the gate of a FET and the base layer of a bipolar transistor, there is a further limiting set of performance figures, the Johnson criteria, that place fundamental materials-based limits on device performance (Johnson 1965). They are particularly simple to specify with respect to a sample of semiconductor which is of length L and subject to a maximum applied bias V_{M}. The electric field V_{M}/L cannot exceed the dielectric breakdown strength E_{B}. Furthermore, the maximum velocity of carriers is taken to be the saturated drift velocity v_{s} (neglecting the overshoot effects seen in GaAs which are difficult to access in practice), and so the transit-time cut-off frequency is $f_{\mathrm{T}} = 1/2\pi\tau \sim L/2\pi v_{\mathrm{s}}$, from which it follows that

$$V_{\mathrm{M}}f_{\mathrm{T}} = E_{\mathrm{B}}v_{\mathrm{s}}/2\pi.$$

This implies a direct trade-off between supply voltage and device speed which becomes important at high speeds. Note that the figure of merit $E_{\mathrm{B}}v_{\mathrm{s}}$ takes the value $6 \times 10^{12}\,\mathrm{V\,s^{-1}}$ for GaAs, implying that a 500 GHz device cannot operate with a supply of more than 2 V. If V_{M} is the peak-to-peak voltage swing, the root-mean-square voltage is $V_{\mathrm{M}}/2\sqrt{2}$, and so the maximum power deliverable to a real impedance Z is

$$P = (V_{\mathrm{M}})^2/8Z$$

so that

$$PZ(f_T)^2 = (E_B v_s)^2/32\pi^2.$$

This implies a limit on the output power which is quite severe for high-frequency devices. These two Johnson relations, which contain materials parameters only, limit the performance of all transistor devices, and the various transistors we examine represent different trade-offs of performance factors subject to these limits (Ladbrooke 1986). Heterojunctions are another design tool that can help approach these limits. (Note that the intrinsic materials figure of merit $E_B v_s$ is three times higher for GaAs than for Si, but is in turn 16 times less than that for diamond if electronics in this material were to be possible (see Chapter 21).)

16.3 Heterojunction field-effect transistors

In Fig. 16.2, we make a direct comparison between a GaAs MESFET and the two forms of heterojunction transistor that have attracted the most research and development attention, namely the high-electron-mobility transistor (HEMT) and the pseudomorphic high-electron-mobility transistor (P-HEMT). Note that we have drawn transistors of directly comparable feature size as presented by lithography and other aspects of device processing. Only the structure of the semiconductor multilayers is different. A negative bias on the (Schottky) gate limits the thickness of the conducting path in n-doped GaAs grown on an undoped GaAs substrate, as shown schematically in Fig. 16.3(a). This is a normally on device, and a fuller analysis shows that

$$I_D \sim (V_G - V_T)^2.$$

In the two HEMT structures, the negative gate bias raises the bottom of the conduction band and controls the carrier concentration in the channel under the gate as shown in Fig. 16.3(b). The P-HEMT has a deeper channel because of the thin layer of InGaAs containing the 2DEG, but otherwise the control over the source–drain current is precisely the same.

   In the rest of this section, we describe some of the device results, and relate the superior performance figures from the heterojunction devices to the physics of Chapter 5 in particular. For a given investment in lithography, HEMT devices work faster, as shown in Fig. 16.4. Several factors are playing a role; however, a principal reason is that the access resistance to the active region of the device (i.e. the series resistance of the source contact and the parts of the semiconductor between the source and the gate (Fig. 16.1)) is rather less in a HEMT than in a MESFET. The high mobility of a 2DEG at low electric fields translates to a lower resistance, which leads to smaller RC time constants associated with switching. In the case of the AlGaAs–GaAs HEMT and the MESFET, the saturated drift velocity of carriers in high fields under the gate is not very different. In the case of the pseudomorphic materials, there is a further advantage in that the saturated drift velocity of

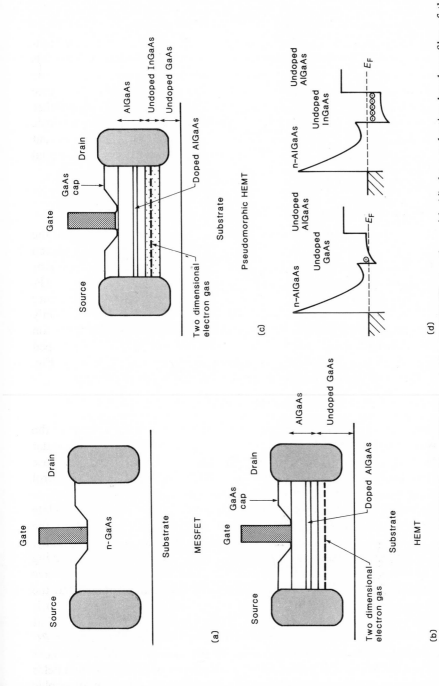

Fig. 16.2 Cross-sections of (a) a MESFET, (b) a HEMT, and (c) a P-HEMT, together with (d) the conduction band profiles of the HEMTs. The basic geometry and operation of each device is the same, but the multilayer semiconductor profile is exploited to obtain lower resistances, higher-speed operation, etc.

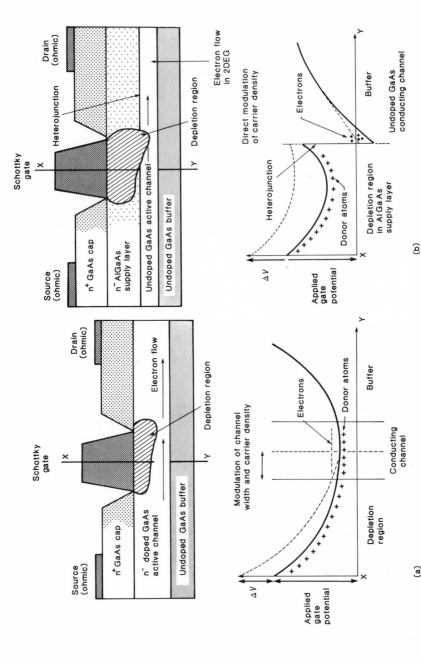

Fig. 16.3 Control by the gate over the carrier density in the channel of (a) a MESFET and (b) a HEMT. The conduction band profile along the line X–Y of the upper panel is shown in the lower panel. (After Tasker 1991.)

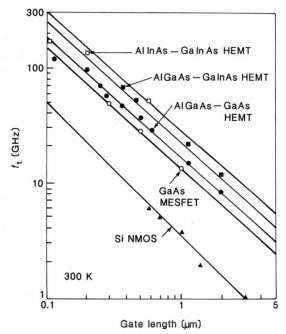

Fig. 16.4 The transit-time-limited upper frequency of various transistor types versus their gate lengths, showing the role of GaAs and multilayers in achieving higher-speed devices for the same commitment to lithography. (After Pearton and Shah 1990 from S. M. Sze (ed) © 1990. Reprinted with permission of John Wiley and Sons. Inc.)

electrons in InGaAs is faster than in GaAs (because of the lower effective mass), and the low-field mobility is also higher than in GaAs. The logical extension of this argument is that in very small transistors (say with about 0.1 µm gate length and very short source-to-gate distances) the contribution of the high mobility to higher speeds is less significant. This indeed has been seen in the modest difference in speed between ultrasmall MESFETs and HEMTs above 100 GHz. The second specific advantage of HEMT devices is a reduced level of noise when they are used as amplifiers of microwave signals. Again, this is related to the reduction in access resistance. The two relevant circuit formulae that govern the maximum frequency f_{max} for which a transistor can deliver power gain and the noise figure NF (with thermal noise in the parasitic resistances being the dominant noise mechanism) are given by (Weisbuch and Vinter 1991; Hollis and Murphy 1990)

$$f_{max} = f_T/2[G_0(R_g + R_s) + 2\pi f_T C_{dg} R_g]^{1/2}$$

and

$$NF = 10\log\{1 + K(f/f_T)[g_m(R_g + R_s)]^{1/2}\}.$$

respectively, where $G_0 = \partial I_d/\partial V_{DS}$ is the output conductance, g_m was defined above, R_g is the resistance of the metal gate, C_{dg} is the parasitic capacitance between gate and drain, R_s is the access resistance, and K is an empirical constant. A reduction in R_s is an important contributor to the improvement in both figures of merit. HEMTs are in wide use as microwave amplifiers, exploiting their low-noise properties. For specialist applications, a reduction in temperature reduces the resistance levels even further so that the devices are both faster and quieter at 77 K.

The above discussion focuses on the basic improvements. There are many further design aspects for optimization, of which the carrier density in the 2DEG is the most important. In physics-based structures (cf. Chapter 5), there was a spacer layer of undoped AlGaAs between the GaAs and the doped AlGaAs, which had the effect of reducing Coulomb scattering to very low values. This scattering largely determines the inverse of the low-temperature mobility (cf. Chapter 5, Fig. 5.7), but it is less significant at room temperature. The doping–thickness product and the position of the centre of the doped AlGaAs layer plays an important role in HEMT devices. The spacer layer is reduced to a minimum, typically $c.5$ nm, and the doped layer is made as thin as possible, even reducing to monolayer doping. These variations increase the efficiency of transferring electrons from the AlGaAs layer to the GaAs, without leaving a parallel conducting channel in the AlGaAs. A related feature is the optimum value of Al in the AlGaAs. Above about 25 per cent Al, the donor levels from the silicon dopants become deep levels (Jaros 1982) and are less efficient at transferring their electrons to the channel. Furthermore, the remaining deep donors are very susceptible to optical excitation, imparting an unwanted photosensitivity to the devices. The conduction band disconti-nuity needs to be as large as possible, so that carriers remain confined in the GaAs. In Section 16.5 we shall see devices which exploit the fact that electrons heated by the source–drain voltage transfer back into the AlGaAs layer from which they came, giving rise to a negative differential resistance. In practice the Al concentration is held to about 25–30 per cent, and the band disconti-nuity is raised by incorporating a layer of InGaAs to contain the 2DEG.

The dominance of Si in digital circuits is almost complete, but for special applications HEMT devices have advantages and a role. In digital circuits the key parameters are the speed with which a particular device can be switched on and off, and the power consumed in the process. One can regard the charging of a gate capacitance to induce a swing from V_{on} to V_{off} as the determinant of speed. How fast can the charge be delivered, and at what energy dissipation? In practice these questions depend on the details of device design and circuit layout, but the high g_m/C and f_T increase the speed, while the lower access resistances lessen power dissipation. Heterojunctions allow charge to be confined in precise regions, and with very thin layers between gate and source–drain channel, the gate capacitance of a HEMT dominates the capacitance of the device to a greater degree than a MESFET, so reducing delays arising from parasitic capacitances. At any time, small-scale integration of GaAs-based devices will always outperform comparable Si circuits in terms of higher speed and lower power consumption.

One problem encountered with the HEMT is that there is a practical limit ($c.\ 10^{12}\,\text{cm}^{-2}$) on the carrier charge density in the channel, as attempts to raise this value result in leakage between the channel and the gate. This is qualitatively similar to having only a layer of equivalent bulk doping only 10 nm thick. A MESFET layer can tolerate a larger charge density in the channel, and so the MESFET still has an edge in power applications. This limitation is one of dimensionality, which becomes more severe when contemplating devices based on Q1D transport.

There is an enormous technical literature on details of all aspects of HEMT device and circuit design that takes us beyond the present text (Linh 1987; Nakanishi 1990; Pearton and Shah 1990). One feature of the physics has important implications for manufacture. It is now possible to produce analytical models that compute the microwave behaviour for a device, given its geometry and the details of the semiconductor multilayers. Such models can be inverted to infer (i) tolerances that must be placed on each manufacturing step to achieve a give yield and performance, (ii) the identification of errors in these steps in the case of poor devices, (iii) the internal properties of devices from other manufacturers, and (iv) the conditions for adopting a 'get it right first time' philosophy to circuit design and manufacture (Ladbrooke *et al.* 1993).

At the time of writing, leaking performance figures for III–V HEMT devices *Electronics Letters* (IEE) and the *IEEE Electron Devices Letters* of recent years, were as follows.

1. Microwave (AlGaAs–InGaAs–GaAs): $f_T > 300$ GHz with a $g_m = 1$ S mm^{-1} length gate with a mushroom cross-section; 7.3 dB gain at 140 GHz, 8.6 dB at 94 GHz, with 1.3 dB noise figure; noise figure below 0.5 dB over 10–20 GHz.

2. Digital: switching speed, $c.4$ ps.

The present status of Si-based FETs is that modulation doping and high carrier mobility have been demonstrated (cf. Chapter 14), and high-speed transistor operation has been achieved using holes in SiGe or electrons in Si. The bandgap offsets are small compared with those encountered in III–V materials, and this will always limit the usefulness of Si heterojunctions. Their application will depend on integrated circuit demonstrations of superior performance, and this is awaited. The advances in Si MOS technology continue at a pace such that the competition for heterojunction Si FETs remains daunting.

The HEMT is regarded, along with the quantum-well laser of the next chapter, as a first generation of quantum semiconductor device, in that it operates with quantum-scale electron states and is in widespread commercial exploitation in both domestic and professional communication and computational systems.

16.4 Heterojunction bipolar transistors

The first transistors were bipolar, and even with the advent of field-effect devices (which in Si technology dominate the memory market), they continue to play a prominent role. The levels of current that can be handled are greater than in field-effect devices—the problem of current drive in thin layers was hinted at in the previous section. High current levels translate to high speed and to high-power devices. The critical step of lithography for defining short gate lengths in field-effect devices is replaced here by precision preparation of the thin base region, and the technologies for uniform growth of wafers have been deployed to keep bipolar technology competitive. The full fabrication process for bipolar transistors is more complex than for field-effect devices, and this factor has been important in ceding ground to field-effect devices.

Heterojunctions have become pervasive for bipolar transistors in III–V materials. The various advantages of a wider-gap emitter layer are first order, and they are fully realized in practice. With respect to standard transistor theory, the current gain at d.c. (the ratio of the collector current to the hole current flowing from the base to the emitter) is given by

$$\beta = (n_e v_e / p_b v_p) \exp(\Delta E_g / kT)$$

where n_e and p_b are the electron and hole densities in the emitter and base respectively, v_e and v_p are the electron and hole effective velocities, and ΔE_g is the difference in bandgap between emitter and base. This last factor relies on heterojunctions. This expression is not exact, but it contains all the important physics (Asbeck 1990); Ashbun & Morgan (1991). A thin highly doped base is important in bipolar transistors, as described in Chapter 16, Section 16.2. Heavy doping results in an effective narrowing of the bandgap, typically by about kT, so that in a homojunction bipolar transistor there is always a significant contribution to the gain from the exponential factor. This contribution has to be offset against some reverse injection of hole current from the base to the emitter as a consequence of heavy base doping. With respect to Fig. 16.5(a), a heterojunction between AlGaAs and GaAs at the emitter–base interface gives rise to simultaneous improvements as follows: (i) the difference in bandgaps is now $c.10$ kT and the exponential factor can lead to $\beta \sim 1000$ at room temperature (10–100 is sufficient in practice); (ii) the doping in the base can be increased significantly, while relying on the heterojunction to stop the reverse flow of holes, and the lower base resistance translates to a higher speed device; (iii) the doping in the emitter can be increased without having to compromise the relative effect of bandgap narrowing, and in turn this leads to a higher current-handling ability; (iv) by grading down the Al composition over the last few nanometres, a notch in the conduction band-edge can be eliminated, so improving the emitter efficiency (Fig. 16.5(b)). In the trade-offs associated with transistor design, heterojunctions can be used as a further design tool. For example, a thin base increases the d.c. current gain (as there is a smaller probability of electron–hole recombination in the base) but will slow down the device (because of a larger lateral base resistance) unless the p-doping is increased, as allowed by heterojunctions, to offset any rise in lateral base resistance. A further advantage of heterojunctions is that the bipolar transistor can operate over a

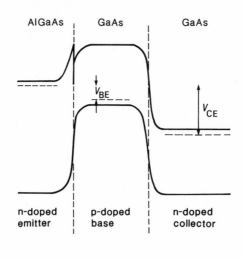

(a)

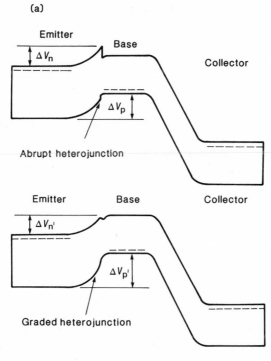

(b)

Fig. 16.5 (a) The energy band profile of a heterojunction bipolar transistor (compare with Fig. 16.1(e)); (b) the effect of a graded-composition layer at the emitter–base interface in improving the electron injection efficiency of the emitter without compromising the blocking action for hole injection from the base. (After Asbeck 1990 from S. M. Sze (ed) © 1990. Reprinted with permission of John Wiley and Sons Inc.)

much enhanced range of temperatures–higher because the heterojunction prevents runaway thermal-based currents, and lower because all the doping densities are above the levels at which carriers freeze out at low temperatures.

The wide-gap emitter is not the only role of bandgap engineering in improving bipolar transistors. If the composition of the base is graded over about 0.2 μm from a modest (about 10 per cent) Al content at the emitter–base heterojunction down to no Al at the base–collector interface, the quasi-electric field represented by the falling conduction band edge can be used to keep accelerating the injected hot electrons (Fig. 16.5(b)). A wider-gap collector can also be used to reduce hole injection from the base to collector when forward biased (i.e. turned off), and to allow higher voltages to be applied to the collector junction without breakdown or leakage.

Although these gains in performance are all realized in research devices, the market for GaAs-based heterojunction bipolar transistors (HBTs) is modest. The yield of circuits that include such devices (in millimetre-wave and microwave applications in particular) is modest, as the multiple processing steps, even with epitaxial multilayers, are difficult and critical. The structure shown in Fig. 16.1(d) gives an indication of the complexity; the various layers have to be contacted and isolated, and the etching, ion implantation (for both contacting and isolation), metallization, and annealing all have to be correct and precise for a working device. One additional problem of bipolar devices in III–V materials that is not encountered with Si is that of light pollution. Electron–hole recombination in the base can give rise to light emission. The multilayer structure acts as a waveguide and the light can be reabsorbed in adjacent devices in high-density circuits, altering the operation of the second device! The multilayer and bipolar nature of this transistor in III–V materials makes it the natural choice in many optoelectronic circuits, where it is used to drive lasers, light-emitting diodes, modulators, and detectors (cf. Chapter 18).

The main electronic applications of the III–V HBT have been in microwave and millimetre-wave amplifying circuits, as at any time the f_T of these devices comfortably exceeds that which can be achieved with Si (typically 100 GHz versus 20 GHz at present). Recent advances with Si–SiGe HBTs (see below) are narrowing this advantage. In digital circuits, the low values of the base resistance and the short electron transit time from emitter to collector imply fast device operation (on–off switching speeds of less than 10 ps have been demonstrated). The larger bandgap in AlGaAs compared with Si implies that larger drive voltages are required for III–V HBTs. In a number of analogue circuits, the uniformity of the turn-on voltage of the emitter–base junction, the absence of trapping effects present in field-effect devices (where hot electrons are injected into the substrate or into the AlGaAs layer of a heterojunction device), and a low $1/f$ noise all attract systems designers who use the HBT in various types of amplifier and in high-speed digital-to-analogue converters. Compared with Si bipolar transistors, the lower base resistance in the III–V HBTs leads to lower power consumption.

Layer	Composition	Doping n (cm^{-3})	Thickness (µm)
Contact	Ga InAs	$n = 1 \times 10^{19}$	0.15
Emitter	Al In As	$n = 1 \times 10^{19}$	0.1
Emitter	Al In As	$n = 5 \times 10^{17}$	0.15
Spacer	GaInAs	$n = 5 \times 10^{17}$	0.02
Base	GaInAs	$p = 5 \times 10^{18}$	0.15
Collector	GaInAs	$n = 1 \times 10^{16}$	0.6
Subcollector	Ga InAs	$n = 1 \times 10^{19}$	0.7
Substrate	InP	Semi-insulating	

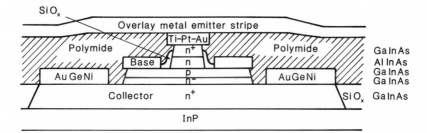

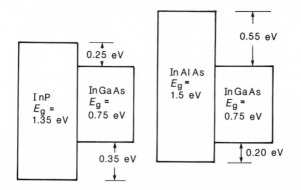

Fig. 16.6 The InGaAs–InAlAs HBT, showing the multilayer structure, the physical structure, and the energy band alignment compared with the InGaAs–InP version. (After Mishra *et al.* © (1988 IEEE.)

In$_{0.53}$Ga$_{0.47}$As and In$_{0.52}$Al$_{0.48}$As are both lattice matched to InP, and provide useful complementary materials for both HEMT and HBT devices. HBTs with InGaAs as the base and InAlAs or InP as the wide-gap emitter have been demonstrated (Fig. 16.6). The particular advantages of these combinations are (i) the relative maturity of the growth of these materials (spurred on by requirements for optical devices (see Chapter 18), for which this device is particularly suited), (ii) an electron mobility 1.6 times higher than in GaAs implying faster devices, (iii) a smaller bandgap (0.75 eV) compared with the 1.42 eV of GaAs, implying a lower turn-on voltage, leading to a reduced power consumption and lower voltage supply, and (iv) advantages of InP as a substrate (e.g. better thermal conductivity).

At the time of writing, state-of-the-art HBTs have the following performance figures (again taken from the IEE and IEEE journals of recent years).

1. Microwave (InGaAs–InP): $f_T > 165$ GHz, $f_{max} > 100$ GHz.

2. Digital: switching speed, less than 10 ps.

The performance of Si-based HBTs is impressive: using an Si–Ge alloy as the base, transistors of about 100 GHz have been reported. Such a figure is a factor of about 3 beyond that predicted as the upper limit of Si homojunction bipolar transistors. So far, only discrete devices and simple circuits have been reported, and the yield and lifetimes of integrated circuits that incorporate laterally patterned strained layers are awaited with interest. If ion-implantation techniques, rather than etching, are used to fabricate devices and so preserve the overall planarity of the multilayer structure, the Si HBT may establish its place. At present the isolation of doped SiGe alloys by oxygen implantation presents a problem, as SiO_2 preferentially forms in the base, leaving segregated Ge which is inadequate as a isolation material.

As with all devices, in this text we have focused on the role of heterojunctions in improving device performance and their role in the optimization of device operation. Fuller texts are devoted to further details of the engineering of HBT circuits which cover important practical details far beyond our possible scope.

16.5 Hot electron transistors

For ever-faster devices, the relatively heavy mass of holes is a limiting factor in bipolar devices. In addition, the voltages required for turning the emitter and collector junctions in bipolar transistors on and off are comparable with the bandgaps of the semiconductors. In principle, a unipolar device can circumvent these problems, using only electrons and conduction band offset voltages. The history of the hot electron transistor mirrors and follows that of the FET: the latter was proposed in the 1920s, and finally fabricated in the 1960s, while the former was proposed as the metal–base transistor in the 1960s but to this day only a few discrete devices operating at 77 K have been reported with both high speed and high gain (Luryi 1990). It was suggested that a metal–base transistor (Fig. 16.7(a)) would allow a very thin base with

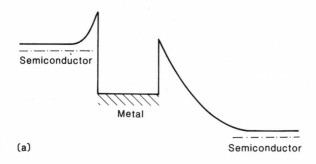

(a)

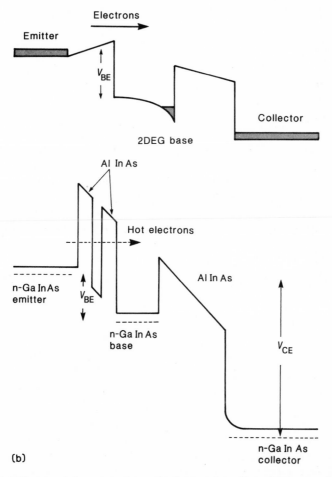

(b)

Fig. 16.7 (a) The metal–base transistor; (b) the band profiles of two promising forms of monolithic semiconductor hot electron transistor, one with a 2DEG base and the other with In-based materials that allow higher energy electron injection from the emitter than can be accomplished with GaAs. (After Yokoyama *et al.* 1988.)

acceptably low lateral resistance. In practice, this failed for two reasons: (i) the materials science for the overgrowth of metals on semiconductors, and more particularly semiconductors on metals, has not been mastered to the stage where electron scattering in the metal layers is small enough not to degrade the transfer efficiency of emitted electrons; (ii) the potential barriers of two back-to-back Schottky diodes present serious problems, such as the efficient quantum reflection of electrons reaching the base–collector barrier. During the 1980s, a number of hot electron spectroscopy studies were undertaken on all-semiconductor structures (cf. Chapter 7, Section 7.3) in which the combination of high speed and high gain were still not met. The high base doping required for a low base resistance scattered the injected hot electrons too effectively, and a base transport factor above 50 per cent could not be achieved in a GaAs–AlGaAs system with a base thickness of $c.0.1$ μm. To achieve a higher base transfer efficiency, the base doping had to be reduced to a level that seriously slowed down the device through the increased RC time constants. Two ways around this problem have been pursued: (i) a materials option and (ii) a structure option (Fig. 16.7(b)). The hot electron spectroscopy studies suggest a dominant energy loss process of of order 60 meV each 200 nm. One cannot inject hot electrons into GaAs with greater than about 0.3 eV energy, as intervalley transfer into the slow resistive satellite valleys will slow down the device. By moving to an InGaAs base, the satellite valleys are more than 0.5 eV above the Γ minimum, and electrons injected with more than 0.4 eV can sustain several energy loss processes during their base transit and still be collected. This process has led to discrete devices known as resonant tunnelling hot electron transistors (RHETs) that also use an AlGaAs–GaAs (or InAlAs–InGaAs) double-barrier structure as the emitter–base barrier, thus injecting electrons with a relatively narrow spread of energies (Fig. 16.8). This is an example of novel band-structure engineering that will become common in future device generations. The collector barrier is of interest, as by modifying its shape the collector–base characteristics can be greatly improved, with a reduction in the dependence of the turn-on voltage on the injected current (Mori *et al.* 1991). Devices with $f_T > 120$ GHz have been reported, but again at 77 K, as thermionic currents become appreciable above that temperature.

The second alternative is to eliminate the main scattering mechanism encountered by the injected electrons, namely the emission of 3D plasmon modes (coupled to optic phonons). A 2DEG base has been demonstrated to have high gain (cf. Chapter 7, Fig. 7.6) in a structure compatible with high-speed device operation, although this aspect has not yet been demonstrated.

Why have researchers persisted with the hot electron transistor in the face of difficulties in both science and technology? In terms of the Johnson criteria (cf. Section 16.2), the turn-on voltages are set by heterojunction band offsets, and these are quite small, implying that fast transistors ($f_T \sim 1$ THz are feasible, whereas FETs and bipolar transistors run out at about 400 GHz. The

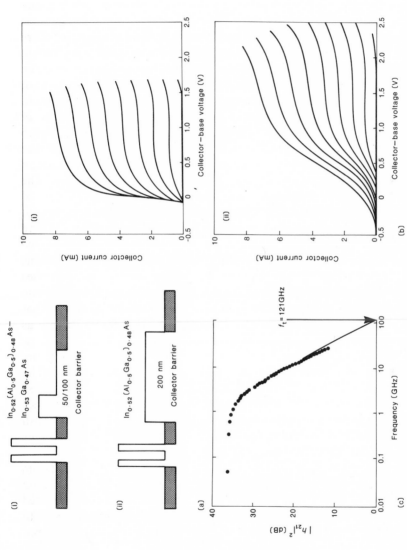

Fig. 16.8 The resonant tunnelling hot electron transistor and the role of band-structure engineering in both the emitter barrier (to control the energy of the injected electrons) and the collector barrier (to reduce the probability of avalanching in the collector barrier layer). Sharper turn-on and flatter saturation characteristics come with the use of a composite collector layer. These are >100 GHz transistors with 2.0 μm × 2.5 μm emitter–base junctions, i.e. very fine lithography in FETs has been traded for precision depth processing to make low resistance contacts at three levels. (After Mori *et al.* 1991.)

RC time constant associated with charging the base–emitter and base–collector capacitances through the lateral base resistance is given by (Luryi 1990)

$$RC = \tau_B = e_s L^2 / L_i \mu \sigma$$

where $L_i (\sim 0.1\,\mu m)$ is the sum of the thicknesses of the emitter and collector barriers, L (~ 1 μm) is a typical lateral feature size of the transistor, μ is the mobility of the electrons, and σ is the sheet charge density of the base. For a competitive transistor, we need $\tau \sim 1$ ps, which implies $(\mu\sigma)^{-1} < 1\,k\Omega/\square$, a value that is attainable in the transistor above with the 2DEG base. If the ohmic contact resistances can be reduced (see Chapter 17, Section 17.5), a terahertz transistor is possible. The small voltages almost certainly dictate 77 K operation, as the thermionic leakage currents which vary as $\exp(-e\Delta V k T)$ become intolerable. The motion of electrons from the emitter layer to the collector layer will be quasi-ballistic, and in this case further analysis of the ultimate speed of the device is required. The *lateral* version of the hot electron transistor (cf. Chapter 7, Fig. 7.12) works by switching injected electrons between base and collector, and this process is determined by the degree of collimation in energy and direction of the emitted current. However, since tunnelling barriers in that structure are thin and have a high capacitance, the sheet resistivity must be down by a factor of about 30 from the value cited above for a vertical transistor, and again this is just possible using the high mobility at low temperatures. The ballistic motion and the absence of electron–hole recombination reduce the intrinsic noise mechanisms, and the hot electron transistor may offer low noise as a bonus.

16.6 Other novel transistor concepts

It would be hoped that, with the range of fabrication technologies now available, novel transistor structures might emerge. The resonant tunnelling hot electron transistor in the previous section is one example. Real-space transfer of hot electrons is a new physics concept peculiar to heterojunctions, as described in Chapter 7, Section 7.5.2. Under a source–drain bias, electrons can be heated to achieve energies of about 0.3 eV, sufficient that they scatter across a heterojunction into a layer where a different field is available to sweep them to another terminal. This transfer out of the channel represents a diminution of the current between source and drain and a form of negative differential resistance. As shown in Fig. 7.14, unusual transistor characteristics can be obtained, relying on this new physics. The negative differential resistance FET (NERFET), shown in another configuration in Fig. 16.9(a)), makes great demands on the contacting technology, just as did the hot electron transistor. The inventors argue that the high-speed performance may be limited only by the time of flight of hot electrons over the high-field region (i.e. from emitter to collector barrier in Fig. 16.9(a)). A few devices have been made (see Luryi (1990) and references cited therein) of modest feature size

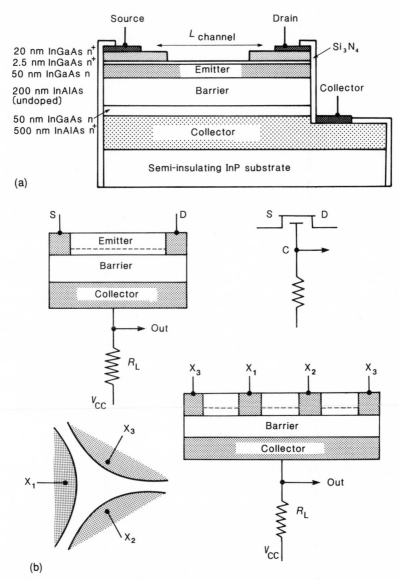

Fig. 16.9 (a) The charge-injection transistor (cf. also Chapter 7, Fig. 7.14), where the source–drain layer acts as an emitter of hot electrons which reach the collector via real-space transfer over the barrier, and (b) the basis of NORAND logic using a three-surface-terminal version. (After Luryi *et al.* 1990.)

(about 2 μm minimum), giving $f_T \sim 60$ GHz, $f_{max} \sim 18$ GHz, and an output conductance of 1000 mS mm in InGaAs–AlGaAs–GaAs structures, figures now exceeded by the field-effect devices subject to a much greater investment. The real future for these devices may come from the new circuit arrangements and concepts. For example (Fig. 16.9(b)), if the top surface is patterned into three equivalent areas X_j, each acting as a source, there are six possible input states corresponding to each X_j having a high or a low bias. By symmetry, we have the same amount of hot electron injection into the collector layer if any one of the three X_j are different from the other two. This means that the output has the logical form

$$\text{Out}(\{X_j\}) = (X_1 \cap X_2 \cap X_3) \cup (\text{not}X_1 \cap \text{not}X_2 \cap \text{not}X_3)$$

(where $\cup$, $\cap$ and notA are the logic functions AND, OR, and NOT A). The novel application of this structure is that if X_1 and X_2 are considered the logic levels of interest, then the NORAND device acts as a NOR if $X_3 = 0$ and as an AND if $X_3 = 1$. To date relatively simple logic elements have been built from groups of transistors, but this new logic structure may become important in future, when area on a chip surface is at a premium in very high performance circuits. One of the new multigate transistors would perform the function of several conventional devices.

The term multifunctionality has been coined to cover the properties of transistors that have complex internal structure which is exploited to reduce circuit complexity. The resonant tunnelling hot electron transistor has been the most widely researched to date. It was originally introduced as a flip-flop circuit containing only one transistor as opposed to two or more conventional devices. The basic physics of this device as a circuit element can be seen in Fig. 16.10(a). The non-monotonic nature of the emitter–base current–voltage characteristics, following from the resonant tunnel diode as emitter–base junction, means that in a circuit there can be two stable states for a load attached to the base. In Fig. 16.10(b) we show schematically the form of an exclusive NOR gate. Several full circuits have been demonstrated with RHETs, and typical results on adder circuits include a fourfold reduction in the number of transistors (from 28 to 7), a reduction in the voltage drive level, and a 10-fold reduction in power consumption to exhibit the same functionality in Si bipolar technology, which represents the fairest comparison (Imamura *et al.* 1992).

If a resonant tunnel diode replaces the emitter layer of the charge-injection transistor of Fig. 16.9, and two contacts are made to the top surface and one to the substrate, the bias on one top contact can alter the *I–V* characteristics through the other. This type of structure has been used to frequency multiply, parity check, and a number of other functions, again with a reduced device count in the circuit (Sen *et al.* 1987).

The resonant tunnel element has also been incorporated into the emitter–base barrier of a heterojunction bipolar transistor, and qualitatively similar operation occurs to that in the RHET. A whole range of structures have been suggested for the base layer (including double barriers, superlattices, parabolic quantum wells, etc. (Fig. 16.11)) for investigating electron tunnelling, and the possible use of multiple internal states for novel logical applications as in the RHET.

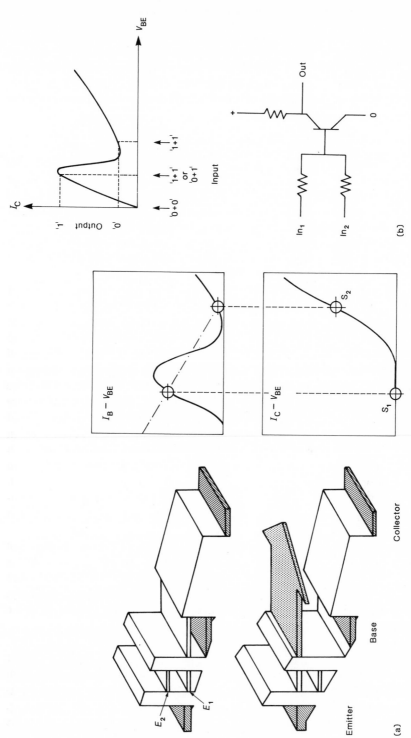

Fig. 16.10 (a) The bistable nature of the current in a base load in a RHET, and (b) the use of a RHET as an exclusive NOR gate. (After Yokoyama *et al.* 1988; Weisbuch and Vinter 1991.)

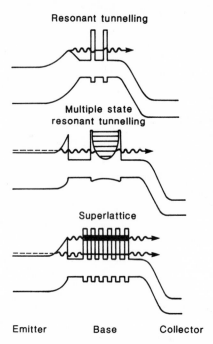

Fig. 16.11 Advanced transistor concepts based on multilayer structures in the base of a bipolar transistor. (After Capasso *et al.* 1990 from S. M. Sze (ed) © 1990. Reprinted with Permission of John Wiley and Sons Inc.)

16.7 The competition between transistor families

FETs in Si are the dominant devices in memory circuits. Bipolar devices dominate high-speed computational circuits. With the move to one-chip computers, a technology has been developed that allows both types of circuit to be made on the same wafer. Although field-effect devices were later to emerge, and they have taken part of the bipolar market, neither device has been displaced by the other. This is because the function required differs and is better served by one transistor than the other.

There are complex trade-offs between devices associated with the required function to be performed and the speed, efficiency, power consumption, cost, noise, etc. with which it can be accomplished. The same applies with III–V heterojunction transistors: in general, the devices are more expensive to manufacture, but for specific applications the performance advantages warrant the cost. For very-high-speed computation and communications, the III–V device families win out because of the intrinsic physics advantages of the III–V materials over Si. In an attempt to make comparisons in digital circuit applications, the speed–power product associated with the switching of a given transistor is plotted. Depending on various fabrication technology assumptions, one obtains diagrams such as that shown in Fig. 16.12 where

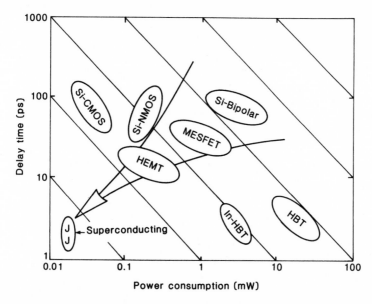

Fig. 16.12 The various transistor families compared in the context of power–speed product for digital circuit applications. The arrow points in the direction of higher-speed lower-power devices as would be required for computation applications.

the boundaries are not sharp and are evolving with time in the direction of lowest speed–power products. Superconducting Josephson junction technology is already established. This technology needs to operate at cryogenic temperatures. The stress associated with different thermal expansion of metals and oxides when cycling the chips between 300 K and 4 K has provided a major challenge. The loss of information above the superconducting transition temperature is a major bar to the use of Josephson junction computers in other than highly structured applications. Otherwise, the diagram shows the high speed, but high power consumption, of HBTs traded off against the lower speed but lower power consumption of Si field-effect devices, and the compromise that is represented by HEMT technology.

References

Asbeck, P. M. (1990). Bipolar transistors. In *High-speed semiconductor devices*, (ed. S. M. Sze), Wiley, New York, pp. 335–97.

Ashburn, P. and Morgan, D. V. (1991). Heterojunction bipolar transistors In *Physics and technology of heterojunction devices* (ed D. V. Morgan and R. H. Williams), pp. 201–30. Peter Peregrinus, London.

Capasso, F., Sen, S., and Beltram, F. (1990). Quantum-effect devices. In *High-speed semiconductor devices* (ed. S. M. Sze), pp. 465–530. Wiley, New York.

Hollis, M. A. and Murphy, R. A. (1990). Homogeneous field-effect transistors. In *High-speed semiconductors devices*, (ed. S. M. Sze), pp. 211–81. Wiley, New York.

Imamura, K., Takatsu, M., Mori, T., Adachihara, T., Muto, S., and Yokoyama, N. (1992). A full adder using resonant-tunneling hot electron transistors (RHETs). *IEEE Transactions on Electron Devices*, **39**, 2707–10.

Jaros, M. (1982). *Deep levels in semiconductors*. Adam Hilger, Bristol.

Johnson, E. O. (1965). Physical limitations on frequency and power parameters of transistors. *RCA Review*, **26**, June 163–77.

Ladbrooke, P. H. (1986). Comparison of transistors for monolithic microwave and millimetre wave integrated circuits. *GEC Journal of Research*, **4**, 115–25.

Ladbrooke, P. H., Hill, A. J., and Bridge, J. P. (1993). Fast FET and HEMT solvers for microwave CAD. *International Journal of Microwave and Millimeter-Wave Computer-Aided Engineering*, **3**, 387–60.

Linh, N. T. (1987). Two-dimensional electron gas FETs: microwave applications. In *Semiconductors and semimetals*, Vol. 24 (ed. R. T. Dingle), pp. 203–47. Academic Press, New York.

Luryi, S. (1990). Hot-electron transistors. In *High-speed semiconductor devices* (ed. S. M. Sze), pp. 399–461. Wiley, New York.

Luryi, S., Mensz, P. M., Pinto, M. R., Garbinski, P. A., Cho, A. Y., and Sivco, D. L. (1990). Charge injection logic. *Applied Physics Letters*, **57**, 1787–9.

Mishra, U. K., Jensen, J. F., Rensch, D. B., Brown, A. S., Pierce, M. W., McGray, L. G., *et al.* (1988). 48GHz AlInAs/GaInAs heterojunction bipolar transistors. *Technical Digest of the International Electron Devices Meeting*, p. 873. IEEE, New York.

Mori, T., Adachihara, T., Takatsu, M., Ohnishi, H., Imamura, K., Muto, S., and Yokoyama, N. (1991). 121 GHz resonant-tunnelling hot electron transistors having new collector barrier structure. *Electronics Letters*, **27**, 1523–5.

Nakanishi, T. (1990). Metalorganic vapour phase epitaxy for high-quality active layers. In *Semiconductors and Semimetals*, Vol. 30 (ed. T. Ikoma), pp. 105–55. Academic Press, New York.

Pearton, S. J. and Shah, N. J. (1990). Heterostructure field-effect transistors. In *High-speed semiconductor devices* (ed S. M. Sze), pp. 283–334. Wiley, New York.

Sen, S., Capasso, F., Cho, A. Y., and Sivco, D. (1987). Resonant tunneling device with multiple negative differential resistance: digital and signal processing applications with reduced circuit complexity. *IEEE Transactions on Electron Devices*, **ED–34**, 2185–90.

Sze, S. M. (1981). *Physics of semiconductor devices* (2nd edn). Wiley, New York.

Sze, S. M. (ed.) (1990). *High-speed semiconductor devices*. Wiley, New York.

Tasker, P. J. (1991) High electron mobility transistors. In *Physics and technology of heterojunction devices*. (eds D. V. Morgan and R. H. Williams), pp. 146–200. Peter Peregrinus, London.

Weisbuch, C. and Vinter, B. (1991) *Quantum semiconductor structures*. Academic Press, New York.

Yokoyama, N., Imamura, K., Ohnishi, H., Mori, T., Muto, S., and Shibatomi, A. (1988). Resonant tunnelling hot electron transistor. *Solid State Electronics*, **31**, 577–82.

17 Devices II: microwave diodes

17.1 Introduction

The generation and detection of microwave signals for both communications and radar is one area of application where devices with heterojunctions and precision semiconductor multilayers have found rapid deployment. These have been largely two-terminal devices. We begin this chapter with a short introduction to the basic device principles, and then take several diode structures and show how the physics of the preceding chapters has been translated into commercially successful devices. One feature of microwave systems has helped this rapid exploitation, namely that many of the highest performance systems are still hybrid, and there is an immediate role for improved discrete devices in each part of the system. This is particularly so, as the new generation devices have similar radiofrequency impedance properties as the earlier devices, and can therefore be used as instant retrofits. Indeed, the superior properties of the new generation devices have been exploited to improve the overall system specification or to relax the burden on some other critical component of the system. While the discrete devices have improved efficiency or sensitivity, as appropriate, a whole range of secondary attributes are also important. For example the sensitivity of the new devices to ambient temperature is much reduced, and the dynamic range and robustness increased, so that a system can be simplified by dispensing with circuits that compensate for temperature variations or protect formerly delicate components. In monolithic integrated circuit applications, many extra constraints in terms of overall chip compatibility must be satisfied, and it may be some time before some of these discrete devices are deployed in such circuits. The debate between hybrid technology and monolithic integration is active and ongoing, typified by the systems proposed for automotive radar. The hybrid systems are available earlier and will compete on cost until a one-chip solution is found.

In order to focus our treatment, we note that frequencies in the range 1–10 GHz and near 35, 60, 94, 140, and 220 GHz are of interest. With one exception, all these correspond to windows in the atmospheric absorption which permit the relatively efficient communication of signals over long distances (Fig. 17.1). At 60 GHz there is a maximum in the attenuation due to oxygen absorption, and this is convenient for short-range communications that do not suffer from interference if the frequency is re-used at a high spatial density.

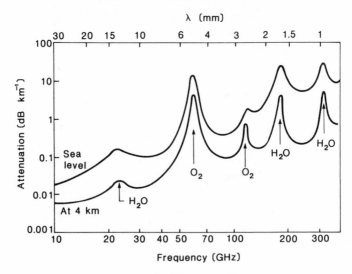

Fig. 17.1 Atmospheric attenuation of millimetre waves at sea level and in the upper atmosphere, showing the effects of water and oxygen. (After Thoren 1985.)

17.2 Microwave sources, mixers, and detectors

In this section we describe the basic principles of some key devices in a microwave system—the source, the mixer, and the detector. This will allow us to trace the relevance of key semiconductor physics in the later sections.

17.2.1 Negative differential resistance and sources

In earlier chapters on hot electrons and tunnelling, attention was drawn to the feature of NDR in a form typified by the schematic current–voltage characteristic in Fig. 17.2(a). Without reference here to the precise origin of the NDR, the important point to note is its role in the conversion of power

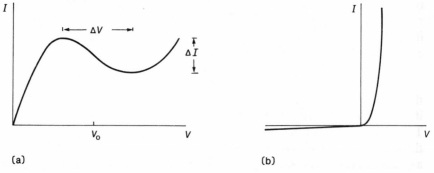

Fig. 17.2 Current–voltage characteristics used in microwave applications: (a) NDR to produce oscillations in an *LRC* circuit; (b) non-linearity for mixing and detection of microwaves.

from d.c. to radiofrequency. Suppose that a device with such an *I–V* characteristic is d.c. biased to a voltage V_0 in the middle of the NDR regime. A small extra radiofrequency signal will see this device as having a negative effective resistance, as any increasing voltage will induce a decreasing current and vice versa. In a conventional *LRC* circuit, a time-varying signal, typified by a sinusoidal variation, will grow exponentially in amplitude with an envelope of $\exp(+t/RC)$ in response to a negative effective resistance R. Indeed, any unintentional time varying current (e.g. noise) will also be amplified. If we take a single-frequency component, the signal will result in the device traversing the current–voltage characteristic in ever larger swings, and the amplitude will be limited once the excursions take the device outside the NDR region. A net positive effective average resistance is then encountered. By placing the device inside a resonant cavity (in a waveguide, or across a transmission line), a d.c.-biased device exhibiting NDR will begin to oscillate, and the largest amplitude will be at the resonant frequency. The engineering of intrinsically wide-band oscillators represented by NDR devices so as to produce a precise output frequency is the subject of electrical engineering texts (Ishii 1989). NDR devices are the starting point for oscillators that act as sources of microwave power. The physics is important for determining the range of frequencies that in principle can be achieved, the efficiency of d.c. to radiofrequency conversion, noise levels, and many other parameters that enable engineers to choose between devices for a given application.

The two principal solid state microwave sources are the Gunn and IMPATT diodes. Heterojunctions can be used to improve the performance of the former (but not the latter) device, as described later in this chapter. The improvements relate to efficiency, noise levels, temperature stability, robustness, and ease of manufacture.

17.2.2 Microwave mixing and detection

A strongly non-linear (rectifying) current–voltage characteristic is at the heart of any device used for detecting microwave signals, as in a straightforward radar application (Fig. 17.2(b)). If we have an incoming signal $V_{sig}\cos(2\pi ft)$ incident on a diode whose non-linear current–voltage characteristic is Taylor expanded about a d.c. bias point (V_0, I_0) as

$$I = I_0 + \alpha(V - V_0) + \beta(V - V_0)^2,$$

then the current induced by this signal has an extra d.c. term $\beta V_{sig}^2/2$ which, if β is sufficiently large, may be detectable across the terminals of the device. The design of diodes is first concerned with the maximization of β, although there are theoretical limits to that, but more particularly with other important aspects of the diodes that improve their utility. Examples include getting the device to operate at zero bias, i.e. without the need for a biasing circuit, or reducing the temperature dependence of the output voltage, or increasing the dynamic range over which the Taylor expansion above is an accurate representation of the diode characteristics. Note that in this (so-called) square-law regime, the induced current is linear in the *power* of the incident radiofrequency signal, and this eases the determination of the incident signal power. For larger powers the square-law approximation breaks down and the output voltage at the terminal is linear in the voltage of the incoming signal, while at the largest input powers the output voltage tends to saturate.

In communication systems, an audio or video signal is impressed on a radiofrequency or microwave signal using a non-linear device such that the sum and difference frequencies are formed, and one of these is transmitted as the radio or microwave signal using diode sources. The incoming signal must be detected in such a way that the audio or video signal is retrieved. Mixing devices are required in both the impression and the extraction of the carrier. Again, a non-linear current–voltage characteristic is required, so that if (say) the incoming signal is of the form

$$V_{\text{sig}} = A[1 + m\cos(2\pi pt)]\cos(2\pi ft),$$

where f is the carrier frequency, p is the modulation (signal) frequency, and m is the degree of modulation, then the response of a diode with the non-linear characteristics described above will again be of the form

$$I = I_0 + \beta A^2 m\cos(2\pi pt) \ + \text{higher frequency terms,}$$

and with a low-pass filter the signal at frequency $p(<<f)$ is recovered. Note that the same figure of merit β appears in mixing and detection.

For several decades the conventional mixer–detector diode has been the metal–semiconductor Schottky diode, for which the current–voltage characteristics, when thermionic emission over a potential barrier is the dominant current mechanism, are given by

$$I = I_0[\exp(eV/nkT) - 1] \qquad I_0 = \text{Area} \times A^{**}T^2\exp(-e\phi_B/kT)$$

where A^{**} is the Richardson factor from thermionic emission theory, ϕ_B is the Schottky barrier height, and $n \geq 1$ is the ideality factor which is a measure of the role of other current mechanisms including tunnelling (see Chapter 2, Section 2.5.4, for the details). The current increases rapidly ('turns on') once $V > \phi_B$. The usual form for presenting β is as a curvature coefficient:

$$\gamma = \frac{\partial^2 I/\partial V^2}{\partial I/\partial V} = \frac{e}{nkT} \ \sim 40V^{-1} \text{ at 300 K.}$$

Since the slope $\partial I/\partial V$ is increasing with V, the Schottky diode is a better detector if biased, typically to a value $V \sim \phi_B$.

Among the other important considerations for mixer and detector diodes are (i) noise, (ii) sensitivity, (iii) dynamic range, and (iv) the sensitivity of the device output to temperature variations. When mixing high frequencies (GHz) with audio (kHz) or video (MHz) frequencies, the intrinsic noise level associated with the mixer diode at these lower frequencies serves as a floor below which the mixed signals cannot be detected. At high input powers the device output saturates, setting an upper limit to the dynamic range. The very high sensitivity to temperature of thermionic emission currents ($\sim T^2 \exp(e\phi_B/kT)$) is undesirable in systems that operate between -40 and $+80\ ^\circ\mathrm{C}$, the temperatures encountered on Earth. In the sections below we shall compare the performances of the new devices to the Schottky diode in these and other respects such as robustness, manufacturability, etc.

17.3 Heterojunction Gunn diode microwave source

The NDR at the heart of a Gunn diode is explained in the context of Fig. 1.10 of Chapter 1. In a high electric field, electrons in GaAs (or InP) originally in the Γ valley with high mobility scatter (or transfer) into satellite valleys with lower mobility. The higher the field, the more efficient is the transfer. The scattered electrons move more slowly, so that a rising electric field results in a lower net current, i.e. NDR. The instabilities associated with a piece of GaAs in the NDR regime were first discovered by Gunn (1963), and the transferred-electron devices that generate microwaves bear his name. In the language of dielectric relaxation, a local electron density imbalance n decays exponentially with a time constant $\tau_r = \rho\varepsilon$, where ρ is the resistivity and ε the dielectric constant. If the resistivity is effectively negative, a charge imbalance grows and a uniform field over the n layer spontaneously deforms to create higher-field domains, as described more fully below. Under bias the charge imbalance moves with a drift velocity v close to the saturated drift velocity, and so a GaAs layer of thickness L has associated with it a characteristic time L/v, during which the change imbalance can build up by a factor $\exp(L/v\tau_r)$. For instabilities to occur we need $L/v\tau_r > 1$ in practical device structures, a condition which can be expressed in the form

$$n^- L > \varepsilon v/e|\mu| \sim 10^{12}\ \mathrm{cm}^{-2}\ \text{(for GaAs and InP)}$$

where n^- is the doping of the GaAs layer and μ is the negative differential mobility.

The conventional Gunn diode has the form of an n^+–n^-–n^+ multilayer, as shown in Fig. 17.3(a). The outer layers act as contact layers, and the field is dropped predominantly over the n^- layer. We consider a device under a steady d.c. bias with an impressed radiofrequency bias. As the radiofrequency field increases, the total field tends to be highest near the cathode, where there are extra space-charge gradients associated with the n^+ – n^- interface (cf. Chapter 2, Section 2.5.2). This part of the structure reaches the critical field E_p (at the peak of the velocity–field characteristics in Fig. 1.9 of Chapter 1) for intervalley transfer before the rest of the structure, and the electrons that transfer move more slowly. The electrons injected before E_p was reached propagate in the Γ valley at a higher velocity, and a depletion region forms ahead of the layer of transferred electrons. Note that the field near the cathode will fall because of the high-field domain, and so faster electrons are

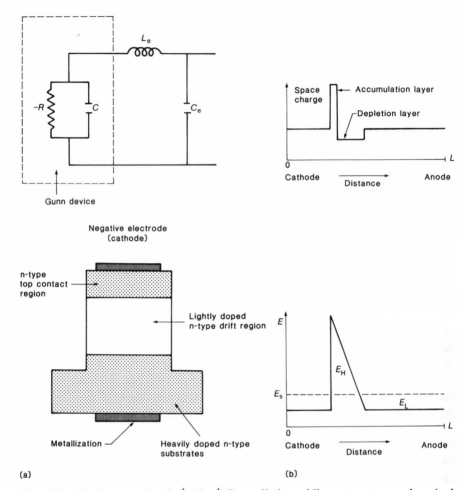

Fig. 17.3 (a) A conventional $n^+-n^--n^+$ Gunn diode multilayer structure and equivalent-circuit model for a packaged device and (b) the space-charge and electric field distributions in a long device in the transit time mode of operation (*GEC Plessey Semiconductors microwave products handbook* 1994.)

injected into the n^- region which catch up with these transferred electrons, enhancing the domain field and the charge density contained in it (Fig. 17.3(b)). This domain will drift down the n^- region as discussed above, and a natural frequency $f = v/L$ is associated with such structures. This mode of operation is known as the transit time mode, and the average electric field across the structure exceeds that for intervalley transfer. The current waveform consists of narrow spikes, indicating a high harmonic content and a relatively low power efficiency at the fundamental frequency (Fig. 17.4(a)). The radiofrequency field across the device is small, indicating a low imped-

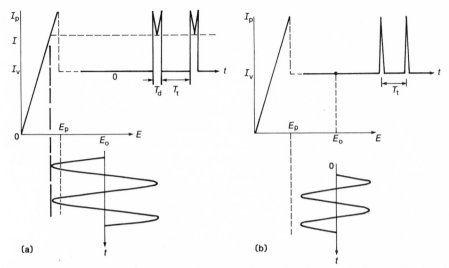

Fig. 17.4 The current and electric field waveforms for the transit time and delayed domain modes of operation of a Gunn diode (*GEC Plessey Semiconductors microwave products handbook* 1994.)

ance (a few ohms), and the transit time frequency f is a strong function of both temperature and applied bias for a fixed length. For this reason this mode of operation is not widely used.

If the average field is reduced, during parts of the radiofrequency cycle the electric field across the device falls below that required for intervalley transfer. As soon as the field rises above E_p a new domain is formed which drifts down the device, reaching the anode just when the field is about to fall below E_p. There will be a small delay (Fig. 17.4(b)) before the field rises at the cathode to allow a new domain to build up, during which time the accumulation of charge discharges at the anode. In this mode of operation the frequency is lower than in the transit-time mode and is given by $f = 1/(T_t + T_d)$, where T_t is the transit time and T_d the delay time. As a result the current waveforms consist of broader spikes, indicating lower high harmonic content and higher efficiency in the fundamental mode of operation. The radio frequency field across the device is high and the device has a higher impedance. The operating frequency of this device is determined via T_d and this is controlled by the resonant frequency of the external circuit which can be very stable as a function of temperature, bias, etc. This device can be tuned to operate over a greater bandwidth than the transit-time device, and the delayed-domain mode is the one most commonly used in commercial applications. There are other modes of Gunn diode operation (Sze 1981), but these are of less practical interest.

Some further aspects of the device performance are summarized here, as the heterojunction Gunn diode with the engineered cathode can improve on these.

The output power of a Gunn diode is determined by the device area and doping level of the drift region. The upper limit on output power is mainly determined by the amount of d.c. power that the device and its encapsulation can tolerate with appropriate heat sinks as a part of the circuit design. In practice currents of about 1 A and biases of about 5 V are applied in the 1–100 GHz range of device operation. The diode itself is basically a broad-band NDR device, and random noise is required to start it oscillating. This becomes more difficult at low temperatures and in high-Q cavities. At high currents heat-sinking is the limitation, while at low currents non-ohmic behaviour in the contacts can limit the linearity in output power with current. Diodes that operate at above 70 °C require lower doping levels in the transit region to ensure sufficiently low powers to maintain a reasonable device lifetime. There are further special considerations that apply to devices that are operated in pulse mode. The frequency-modulated and amplitude-modulated noise associated with the device is a complex issue, as the final results cannot always be disentangled from the contribution of the circuit. The drift of frequency with temperature can also be a problem in practical applications. In general devices that have to operate across a broad range of external parameters (such as temperature or tuning range), or with other restrictions on voltage or current, may function with suboptimum performance. If a device is required to operate at low temperatures (say −40 °C in the polar regions), some of the carriers in the n⁻ region freeze out and a higher starting voltage is required for high-field domains to form, and sometimes this bias is close to the bias at which peak power is extracted from the diode, implying that the device will only operate effectively over a very narrow range of applied biases.

The role of heterojunctions has been to introduce a new generation of higher-performance Gunn diodes. As the systems requirements have become more stringent, and demands are made for higher frequency and higher output powers, the basic Gunn diode structure has reached its limits of performance. For example, there is a rapid fall-off in output power in GaAs devices at frequencies of about 60 GHz, above which one has to use the device in the less efficient second-harmonic mode of operation. The requirement of a device to operate over the −40 to +80 °C range of ambient temperatures poses real problems—the range of biases for which useful power can be extracted shrinks to zero at the lower temperatures.

The main problem is the process of heating the electrons so that intervalley transfer can take place. The acceleration of the electrons in conventional n^+–n^-–n^+ structures occurs within the n^- layer, and the 'dead zone' within which electrons are heated to about 0.3 V can be as much as about 0.25 µm out of a drift region of about 1.5 µm for millimetre-wave devices, acting all the while as a parasitic resistance. This is a cause of a reduction of efficiency which becomes worse as the device is shortened to achieve higher-frequency operation. If the electrons could be injected hot into the drift region, many perceived limitations on conventional Gunn diodes could be circumvented.

The electrons cannot be injected too hot, as the drift velocity of GaAs decreases until reaching the saturated value (Chapter 1, Fig. 1.9). An early method of hot electron injection was to remove the top n^+ layer and form a Schottky barrier injector by depositing metal directly on the n^- layer. In practice the Schottky barrier height of about 0.6 V is too high, and the structures are annealed to degrade the effective barrier height by intermixing at the metal–semiconductor interface until a barrier of about 0.3 V is formed. This has never been other than a 'green-fingered' method of achieving high-performance Gunn diodes.

With the advent of multilayers, both the planar doped barrier diode (cf. Chapter 2, Section 2.6) and the graded AlGaAs injector (Chapter 7, Fig. 7.1) have been used for injecting hot electrons into the n^- region of a Gunn diode, and the multilayer structures are compared directly with the original structure in Fig. 17.5. In both, there is an attempt to raise the energy of the electrons injected into the drift region to an energy of about 0.3 eV. We shall concentrate on the graded-gap structure and make comments on the planar

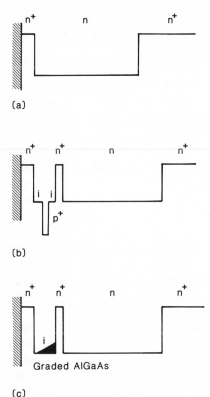

Fig. 17.5 Comparison of (a) the conventional Gunn diode structure with two forms of hot electron injector multilayer structure, namely (b) a planar doped barrier and (c) a graded AlGaAs layer (*GEC Plessey Semiconductors microwave products handbook* 1994.)

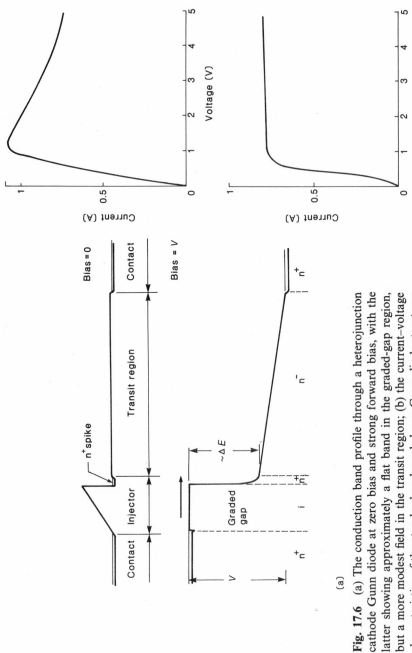

Fig. 17.6 (a) The conduction band profile through a heterojunction cathode Gunn diode at zero bias and strong forward bias, with the latter showing approximately a flat band in the graded-gap region, but a more modest field in the transit region; (b) the current–voltage characteristics of the standard and graded-gap Gunn diode structures (*GEC Plessey Semiconductors microwave products handbook* 1994.)

doped barrier structure below. As discussed in Chapter 7, a layer of AlGaAs with the Al content increasingly linearly from zero to about 30 per cent over 50 nm can be used as a source of hot electrons; note that this distance is already five times shorter than the 0.25 μm referred to above. In order to achieve hot electron injection, a bias is applied over the graded-gap layer in order to remove the quasi-electric field. This field is too high if dropped uniformly over the whole structure, and so the working version of the heterojunction Gunn diode has a thin layer (about 10 nm) of heavily doped GaAs immediately after the heterojunction where the Al content returns to zero. The purpose of this doped layer, which is depleted of its mobile carriers, is to provide a charge layer which takes up some of the electric field (as is seen through the application of Gauss's theorem). The layer is sufficiently thin that the probability of significant electron energy loss is small, as inferred from the data on hot electron spectroscopy in Chapter 7. Thus, under bias (Fig. 17.6(a)), the composite cathode has the effect of launching hot electrons with about 300 meV of energy into the drift region, whose field is modestly above the critical field for intervalley transfer. The current–voltage characteristics for a graded-gap structure (Fig. 17.6(b)) show that the actual field value is not sensitive to bias. Simulations have been carried out on three structures: a conventional n^+–n^-–n^+ diode with only a 1 μm drift region, a structure with a graded-gap launching pad, and one with the composite graded gap and n^+ doping spike. The simulations involve tracking about 10 000 electrons launched into the drift region with appropriate initial energy and allowing for intervalley scattering, drift, phonon emission, and other processes before equilibrating with cold electrons in the anode contact layer. The results in Fig. 17.7 show the electric field profiles as a function of time, and how the combination of hot electron injection into a correct field in the transit region leads to oscillations in a region that is too short to sustain conventional bulk Gunn oscillations.

The heterojunction Gunn diode has more than double the output power and efficiency of a conventional Gunn diode over the 60–100 GHz range in both the fundamental and second-harmonic modes of operation (Fig. 17.8(a)). The electrons in the cathode remain cold until they are given all their energy at the heterojunction, and the competing processes of energy relaxation, particularly by phonon emission, are eliminated. The phonon-emission process is very temperature dependent, and the heterojunction Gunn diodes have less than 30 per cent the variation in output power with temperature than conventional Gunn diodes (Fig. 17.8(b)). The excess energy given to the electrons at the heterojunction is equivalent to a temperature of about 3000 K, and whether the cathode is at 300 K or 400 K is relatively less important. The heterojunction Gunn diode is quieter with respect to noise generation than a homojunction Gunn diode. A principal source of noise, associated with phonon emission processes making uncertain the exact position in the diode where the intervalley transfer takes place, is eliminated. Intervalley transfer takes place close to the cathode. Typical performance figures are shown in the Table 17.1.

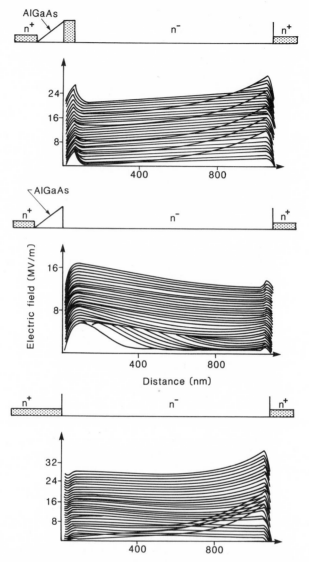

Fig. 17.7 Simulations of the electric field profile as a function of time for three structures as described in the diagram, each with a 1 μm drift length. Whereas oscillations build up in structure (a) static high-field domains form at the cathode or anode of the other structures. (After Couch *et al.* 1988.)

Other attractive systems attributes include the fact that close-to-peak output power is achieved over a wide range of applied biases (about 2 V) compared with almost no range in a conventional Gunn diode (Couch *et al.* 1989). The heterojunction device has a higher impedance and is easier to match in circuits. The manufacture is simple, as one good wafer, and simple

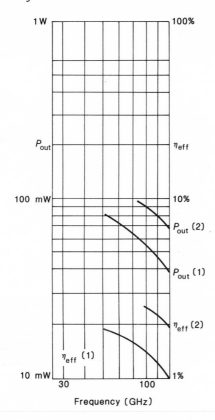

(a)

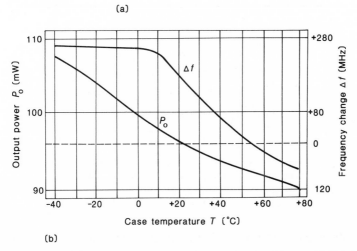

(b)

Fig. 17.8 (a) The efficiency of the conventional (1) and graded-gap (2) Gunn diodes; (b) the drift in frequency and output power of a 60 GHz graded-gap Gunn diode over the −40 to +80 °C temperature range (*GEC Plessey Semiconductors microwave products handbook* 1994.)

Devices II: microwave diodes

Table 17.1 Typical performance of graded-gap GaAs Gunn diodes

Frequency (GHz)	Temperature (°C)	V_{on} (V)	V_{op} (V)	I_{op} (mA)	Power (mW)	Efficiency (%)	Noise (dBc Hz^{-1})
90	−40	3.9	4.9	680	58	1.75	
90	+25	3.2	4.7	660	50	1.6	−86
90	+80	3.1	4.8	640	42	1.4	
94	+25	3.0	4.5	600	60	2.0	−88
60	+25	3.0	4.5	600	120	5.0	−90
35	+25	3.0	4.5	1300	350	5.5	−95

V_{on} = threshold voltage for microwave power.
V_{op}, I_{op} = operating voltage and current.

Source: *GEC Plessey Semiconductors microwave products handbook*, 1994

ohmic contact technology has ensured a high yield. The lifetime tests have shown that heterojunction diodes are very stable and reliable. In contrast, a degraded Schottky contact has a lower reliability and lifetime. All the benefits can be related back to the heterojunction physics. The critical regions of the device are within regions of the single-crystal semiconductor far from metallurgical interfaces. Full use of band-structure engineering of Gunn diodes has not yet been made. A spatial variation in the doping in the drift region has been shown to improve the output power of a conventional Gunn diode, and this design aspect needs to be incorporated with the heterojunction design elements (Ondria and Ross 1987).

The planar doped barrier injector has been used to generate 120 mW of power at 94 GHz, 20 per cent more than the best result from the graded-gap structure and double that offered routinely in a commercial product (Beall *et al.* 1989). The other important device attributes (noise, temperature stability, reliability, etc.) have not been reported upon.

17.4 Heterojunction IMPATT diodes

The most powerful and efficient (but noisy) solid state source of microwave power over the 1–100 GHz range is the impact avalanche transit-time (IMPATT) diode (Sze 1981). We outline the basis of its operation, the potential role for heterojunctions, and the failure to date of heterojunctions to significantly improve the performance. The operation of the device can be seen with respect to a heterojunction version in Fig. 17.9(a). The device is d.c. biased and subject to an additional radio frequency voltage. Under bias, one region of length L_a may be under a sufficiently high field to induce impact ionization and avalanching. This region is followed by a drift region of length L_d where the electric field is lower but still large enough to ensure that electrons travel with their saturated drift velocity v_s. If α is the ionization rate, which is assumed (for convenience) to be the same for electrons and holes, and which typically varies as $\exp(-1/E^2)$ for local electric field E, the current in the device can be written as

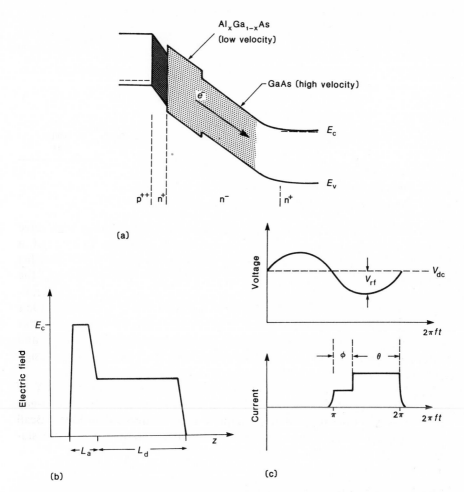

Fig. 17.9 (a) A multilayer structure and (b) an electric field profile for an IMPATT structure which has an n⁺ GaAs layer for avalanching (shown in black) followed by a double-velocity structure in the drift region; (c) the idealized voltage and current waveforms for this structure. (After Kearney *et al.* 1993.)

$$dJ/dt = (2Jv_s/L_a)\left[\int_0^{L_a} \alpha(z, t)dz - 1\right] + 2J_sv_s/L_a,$$

where J is the total current density and J_s is the sum of the electron and reverse hole saturation current densities. During the positive part of the radio frequency cycle, the expression in square brackets becomes positive and the current increases exponentially, i.e. avalanching occurs. At a time $t \sim T/2$ where $T = 1/f$ is the radio frequency period, a pulse of electrons is injected into the drift region, giving rise to a current in the external circuit

$I(t) \sim nev_s(t)$, where n is the mobile space-charge density and $v_s(t)$ is its instantaneous velocity. The d.c. power supplied and radio frequency power generated are given by

$$P_{dc} = (V_{dc}/T) \int_0^T I(t)\,dt$$

and

$$P_{rf} = (V_{rf}/T) \int_0^T \sin(2\pi ft)I(t)\,dt$$

respectively. By choosing $L_d \sim v_s/2f$, the induced current is about $180°$ out of phase with the voltage, ensuring that the diode exhibits close to maximum negative terminal resistance, and a high efficiency

$$\eta = |P_{rf}/P_{dc}| \sim (2/\pi)(V_{rf}/V_{dc})$$

Heterojunctions and multilayers might be introduced to refine the shape of the current pulse. If it is assumed that carriers have a lower saturated drift velocity $v_s(x)$ in $Al_xGa_{1-x}As$ than in GaAs, the waveform is shown as in Fig. 17.9(c) and the efficiency can be recalculated for this shape to give (Kearney *et al.* 1993)

$$\eta = (V_{rf}/V_{dc})[\cos(\phi + \theta) + \gamma - 1 - \gamma\cos(\phi)]/[\theta + (1 - \gamma)\phi],$$

where $\gamma = [v_s(x = 0) - v_s(x)]/v_s(x = 0)$, $\phi = 2\pi f L_1/v_s(x)$, $\theta = 2\pi f L_2/v_s(x = 0)$, and $L_d = L_1 + L_2$. The thickness L_1 of the $Al_xGa_{1-x}As$ layer can be chosen to maximize η, giving larger values than can be achieved with only one material in the drift region provided that the materials have sufficiently different values of v_s: (η is maximized when $\gamma \to 1$). This argument assumes that the voltage ratio prefactor is unchanged when going to a double-layer drift region, but various arguments indicate that the ratio is degraded. If the average field in the avalanche region is E_c and that in the drift region is $E_c/2$, then $V_{dc} \sim E_c(L_a + L_d/2)$. If ΔE is the maximum change in the field during the radio frequency cycle, then $V_{rf} \sim \Delta E(L_a + L_d)$. Now ΔE cannot be too large, otherwise the field in the drift region will fall below that required to obtain saturated drift transport (say $\Delta E \sim E_c/4$)). This gives an upper bound of

$$V_{rf}/V_{dc} = \frac{0.5(1 + L_a/L_d)}{1 + 2L_a/L_d},$$

which falls off as the ratio L_a/L_d increases. In $GaAs–Al_xGa_{1-x}As$ structures the overall drift region will be shorter than in a comparable homojunction device, and in terms of the efficiency equation above the gain from the shaping of the current pulse is partly offset by the degradation in voltage ratio. The gains to IMPATT diodes from heterojunctions have proved modest at best (Kearney *et al.* 1993). The effective series resistance of the alloy layers tends to be higher, and these layers have a poorer thermal conductivity, hampering the extraction of excess heat, and the difference in saturated drift velocities is small ($\gamma \sim 0$). There are other forms of band-structure engineering of the drift and the avalanche regions, but these have not proved successful with devices in the GaAs–$Al_xGa_{1-x}As$ materials system. In contrast, $In_{0.53}Ga_{0.47}As$ which is lattice matched to InP, has only about 60 per cent bandgap of InP, so that the alloy is attractive for the avalanche region and the saturated drift velocity of InP is high. Predictions of 40 per cent efficiency at 10 GHz and 20 per cent at 80 GHz have been made, but not as yet realized in practice. IMPATT diodes operate on the dielectric breakdown properties of semiconductors, and there does not appear to be the range of values in the constituents of semiconductor multilayers to make significant improvements on existing state-of-the-art Si and GaAs IMPATT devices.

17.5 Resonant tunnel diodes (RTDs) and quantum-well injection transit-time (QWITT) diodes

The principles of resonant tunnelling in double-barrier diodes and the design of current–voltage characteristics were covered in detail in Chapter 8. In this section we concentrate on the device performance of these structures as sources of millimetre-wave radiation. To date they represent the fastest of the purely electronic devices: in the GaAs–AlAs system, they have generated 420 GHz oscillations (Brown *et al.* 1989), while in the InAs–AlSb system 712 GHz oscillations have been reported (Brown *et al.* 1991), both in the fundamental mode and at room temperature. These frequencies are much higher than those obtained with Gunn, IMPATT, and other semiconductor diode sources. However, the power levels are very small (0.2 μW), but a recent improved structure (the quantum-well injection transit-time (QWITT) diode) is able to increase the efficiency of radiofrequency power generation at the expense of the upper frequency of operation (Javalagi *et al.* 1992) as described below. The use of non-linear devices (such as the quantum-barrier varactor diode described in Section 17.8) to triple the frequencies obtained from Gunn diode offers an alternative way of producing these levels of output power at these very high frequencies.

Using a simple circuit analysis of the device as a negative resistance of magnitude R in parallel with a diode capacitance C and in series with a resistance R_s (Fig. 17.10(a)), we obtain the maximum frequency above which the diode does not exhibit NDR as

$$f_c = 1/(2\pi RC)\sqrt{(R/R_s - 1)}$$

This frequency can be greater than 500 GHz because the capacitance can be made much smaller than in an Esaki diode (the reverse-biased degenerately doped p^+–n^+ junction with $f_c \sim 100$ GHz) (Esaki 1976). The energy width ΔE of the resonant tunnelling transmission peak defines an intrinsic cavity delay time $\tau \sim h/\Delta E$ (but note the controversy about the precise definition of tunnelling time in Chapter 8, Section 8.3.4.2), and the traversal time of the depletion region at the anode must also be taken into account, but both of these are less than 10^{-12} s. Experimental evidence, in terms of direct measurements of the switching speeds (Whitaker *et al.* 1988) and infrared reflectivity of these structures, suggests that terahertz switching speeds should be possible (Sollner *et al.* 1983).

A simple estimate of the maximum power and efficiency that can be generated in an oscillator, valid for $f \ll f_{\mathrm{c}}$, is given in terms of the current and voltage swings ΔI and ΔV in the NDR region (Fig. 17.10(b)) as

$$P_{\max} = (3/16)\Delta I \Delta V$$

and

$$\eta = \frac{P_{\max}}{(I_{\mathrm{p}} - \Delta I/2)(V_{\mathrm{p}} + \Delta V/2)}$$

where I_{p} (V_{p}) is the peak current (voltage) across the device. $P_{\max}$ turns out to be a few milliwatts, although the intrinsic efficiency can be 10–20 per cent

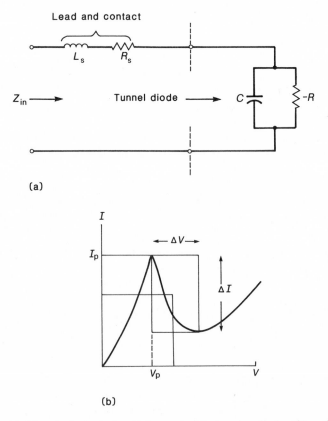

Fig. 17.10 (a) The equivalent circuit for a double-barrier diode; (b) estimation of output power and efficiency from the d.c. current–voltage characteristics.

(and 50 per cent has been reported for the QWITT diode described below). Although the peak current densities are quite high with very thin barriers $c.10^9 \mathrm{A\ m^{-2}}$, the power is fundamentally limited by the small value of ΔV (0.2–0.3 V across the diode) set by the conduction band offsets and the energies of quasi-bound levels in the central GaAs layer in the GaAs–AlAs materials system. In the regime $f \ll f_c$, power levels of 60 µW at 56 GHz (in fundamental mode, $\eta \sim 2$ per cent) and 18 µW at 87 GHz (second harmonic) have been reported in GaAs–AlAs materials. Key results are shown in Fig. 17.11. In particular, the output power is less than that predicted by theory, but does follow the same pattern. The upper cut-off frequency is limited in practice by the R_s term; if the series resistance of the contact layers and the ohmic contacts could be reduced by a factor of 3 (a very tall order), 1 THz operation of GaAs–AlAs devices would be possible (Brown *et al.* 1988).

Output power is not the only consideration in practice. An oscillator must be placed in a circuit that prevents any form of spontaneous oscillations, particularly at lower frequencies. The condition that must be satisfied is (Kidner *et al.* 1990)

$$L_s/(|R|^2 C) < R_s/|R| < 1$$

where $|R|$ is the magnitude of the negative resistance (typically a few tens of ohms), L_s is the series inductance of the device ($c.1$ nH), and R_s is the series resistance. Since $|R|$ is not large, C must be kept small for high-frequency operation, which limits how large the devices can be made. Small devices both limit the power available to the microwatt level and also provide technological problems for routine manufacture.

The Esaki tunnel diode based on a reverse-biased $p^+ - n^+$ junction has retained a niche market as it is a very low noise device. A systematic analysis of noise measurements on resonant tunnel diodes has not been made, although preliminary reports suggest this device is also relatively quiet.

One way of improving the device output power is to increase the voltage swing ΔV. This can be done by expanding the voltage in the depletion region before the anode, so that only a small fraction of the terminal voltage is used to take the double-barrier structure into and out of the resonant tunnelling condition. Once this region is somewhat longer than the double barrier itself, the device acts in a different way (Fig. 17.12(a)). The double barrier acts as a quantum-well injector of electrons into a drift region, and the transit mode of operation has some similarity with that of an IMPATT diode, although the detailed physics is quite different. The transit time of the electrons over the depletion region becomes appreciable, and limits the upper frequency of operation to $c.300$ GHz. The results achieved using the InGaAs–InAlAs materials system are impressive: at 2 GHz a 50 per cent d.c. to radio frequency conversion efficiency with an output power of 20 mW per diode is obtained (Javalagi *et al.* 1992). The 1 mW barrier for output power from AlAs–GaAs-based structures has been broken, and $c.4$ mW should be attainable. The designers of short-range communication links and personal mobile communi-

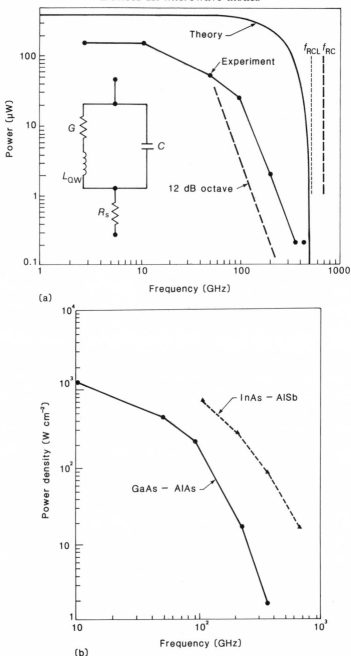

(a)

(b)

Fig. 17.11 A collage of experimental results of output power versus frequency on double-barrier diodes in (a) GaAs–AlAs (Brown *et al.* 1989) and (b) InAs–AlSb (Brown *et al.* 1991). The former shows the extent to which practical devices fall short of theoretical predictions of idealized diode structures. (Reprinted with permission of MIT Lincoln Laboratories, Lexington, MA.)

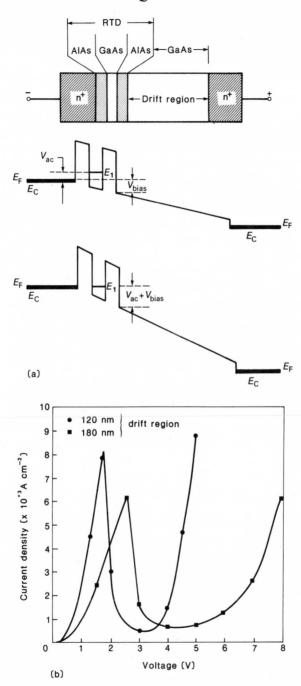

Fig. 17.12 (a) The multilayer structure of the QWITT diode and the principle of its operation; (b) the static current–voltage characteristics. (After Kesan *et al.* © 1987 IEEE; Javalagi *et al.* 1992.)

cations systems are likely to be attracted by the combination of modest powers and high efficiencies.

17.6 Planar doped barrier diodes as mixers and detectors

The physics of this n–i–p–i–n diode was discussed in Chapter 2, Section 2.6, and here we concentrate on the device aspects (Kearney and Dale 1990). The electrostatic barrier in a Schottky diode is fixed in a limited range of values by the combination of metal and semiconductor, and in practice the formation of the intimate contact is difficult in manufacture—even a monolayer of oxide can alter the electronic properties. In contrast, in the planar doped barrier we have control over the doping–thickness product of the thin p^+ layer, the lengths of the two intrinsic regions on either side, and the level of doping in the contact layers. Together, these determine the height of the electrostatic barrier formed when the acceptors trap their extra electron. In particular, the height can be varied over a wide range ($c.0.1$–0.7 eV and more) by the specification of the p^+ layer. The ratio of the lengths of the two undoped regions (typically more than $10:1$ in mixer/detector applications) defines the asymmetry of the triangular barrier and hence the asymmetry of the current–voltage characteristics. Note that there are applications where a non-linear but antisymmetric current–voltage characteristic is required (for example when pumping a mixer at approximately half the frequency of the signal to be detected), and in this case the p^+ layer is placed exactly in the middle of the i region. In contrast with obtaining two balanced Schottky diodes in an antiparallel pair, the symmetric planar doped barrier solution is much simpler and more precise. We show a typical structure in Fig. 17.13(a) and the range of room temperature current–voltage characteristics that can be achieved in Fig. 17.13(b). A suitable choice of structure can be made to move the point of maximum curvature towards the origin, and so the device can operate as a mixer or detector under conditions of zero bias, which is not possible with most Schottky diodes. In this case the barrier tends to be relatively low, and the reverse breakdown voltage is smaller than in a Schottky diode. Since the principal current mechanism is thermionic emission, and the current–voltage characteristics are qualitatively similar to those of a Schottky diode, it is found that all the important radio frequency parameters of a planar doped barrier diode involved in matching it into a circuit are similar to those of a Schottky diode. Planar doped barrier diodes have been tested in circuits and systems as direct replacements for Schottky diodes, and the following features have been noted.

1. The dynamic range of the planar doped barrier diodes is among the largest of all microwave diodes, such that they are able to detect and mix signals with input powers are low as −53 dBm. The linearity (terminal voltage proportional to received power) tends to break down for input powers above 10 dBm, although the devices do not saturate until even higher levels of received power (Fig. 17.14(a)).

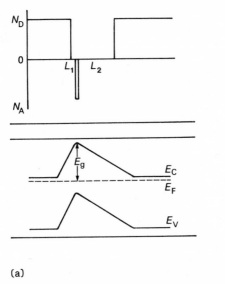

(a)

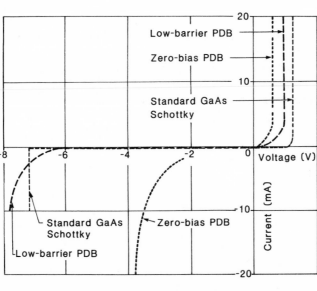

(b)

Fig. 17.13 (a) The multilayer doping profile and the conduction band profile for a planar doped barrier (PDB) diode structure; (b) the room temperature current–voltage characteristics of zero-bias and low-barrier planar doped barrier diodes compared with Schottky diodes, showing the tailorability of the former (after Kearney and Dale 1990; *GEC Plessey Semiconductors microwave products handbook* 1994.)

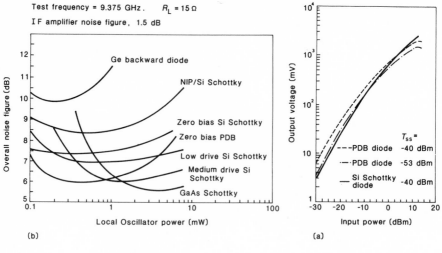

Test frequency = 9.375 GHz. R_L = 15 Ω

IF amplifier noise figure, 1.5 dB

(b)

(a)

F_0 = 9.2 GHz

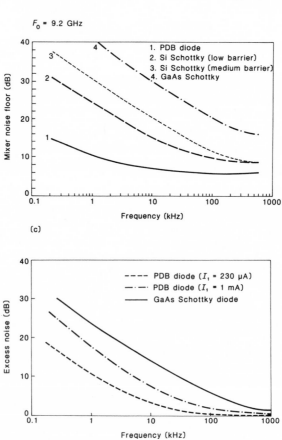

1. PDB diode
2. Si Schottky (low barrier)
3. Si Schottky (medium barrier)
4. GaAs Schottky

(c)

(d)

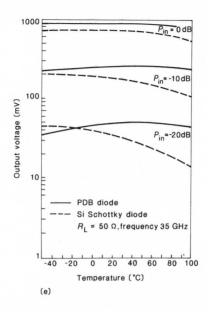

(e)

Test frequency 9.375 GHz. Pulse frequency 1000 pps. R_L=15Ω

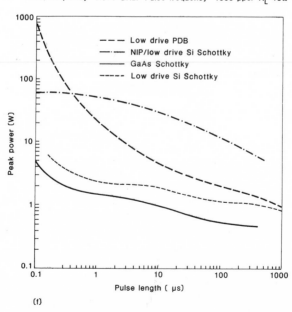

(f)

Fig. 17.14 The superior microwave properties of the planar doped barrier (PDB) diode compared with the Schottky diode including (a) dynamic range, (b)–(d) lower noise levels, (e) reduced sensitivity to temperature, and (f) resistance to pulse-power burn-out. (After *GEC Plessey Semiconductors microwave product handbook* 1994.)

2. For levels of local oscillator power up to 1 mW, the overall noise figure of a zero-bias planar doped barrier mixer diode is lower than that of a Schottky diode when tested in the same cavity, and mixing has been achieved with local oscillator power as low as 0.2 mW (Fig. 17.14(b)).

3. Measurements of the balanced mixer noise floor and of the $1/f$ noise levels close to carrier both indicate the superiority of planar doped barrier over Schottky diodes (Figs 17.14(c) and 17.14(d)). The abscissa is the difference frequency.

4. The variation in the terminal voltage of a planar doped barrier diode over the temperature range -50 to $+100$ °C is much smaller than that of a Schottky diode. The strong temperature dependence ($T^2 \exp(-\phi_B/kT)$) of thermionic currents is accentuated in a Schottky diode whose barrier height reduces with temperature. In the planar doped barrier diode the rise in thermionic current is partly counteracted by increased spill-over of charge into the i layers at the n^+–i contact interfaces at increased temperature, which increases the effective barrier height (Figs 17.14(e) and 2.15(b) of Chapter 2).

5. When subject to intense pulses of microwave power, the local electric fields can induce breakdown of the metal–semiconductor interface owing to the formation of metallic filaments in the semiconductor, while lower levels of continuous power result in thermal degradation (and eventually melting) of the diode. In a test circuit where pulses of varying duration, repeated 1000 times per second, were applied, the planar doped barrier diodes could withstand nearly 1000 times the power in very short pulses, although the thermal heating (with pulses of duration $c.1$ ms implying d.c. exposure) degradation was comparable (Fig. 17.14(f)). The devices that burned out when subject to pulsed power showed evidence of dielectric breakdown over the approximately 0.1–0.2 µm of nominally undoped GaAs between the two n^+ contact layers.

All these advantages for the planar doped barrier diode can be traced to the absence of the metal–semiconductor contact of a Schottky diode, where the polycrystalline microstructure of the metal is a source of noise and instability. In the planar doped barrier diode, the device is all within a single-crystal semiconductor, and the ohmic contacts are well away from the active part of the device. The total thickness of GaAs between the two n^+ contacts is typically 0.1–0.2 µm, and the transit time of carriers corresponds to an upper-frequency cut-off of $c.300$ GHz; this is down by an order of magnitude on the Schottky diode where the depletion region in the semiconductor can be as small as 0.01 µm. The Schottky diode continues to be the diode of choice for experiments in the 0.5–2.5 THz region.

17.7 The asymmetric spacer layer tunnel (ASPAT) diode

The physics and design of this type of diode was discussed in Chapter 8, Section 8.2.2. If the p^+ layer of a planar doped barrier is replaced with a thin

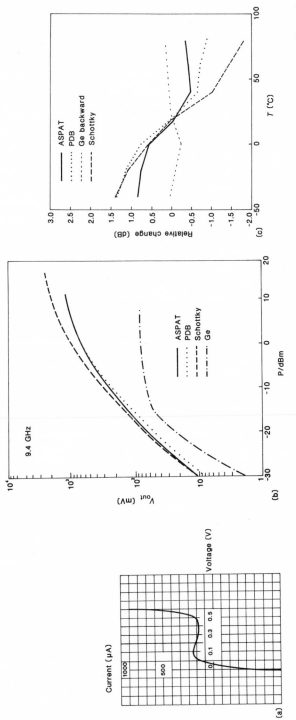

Fig. 17.15 (a) The current–voltage characteristics of a Ge backward diode (equivalent to the Esaki diode), the temperature stability of whose reverse-bias characteristics is exploited in microwave detection; (b) the sensitivity of the ASPAT diode compared with other microwave detector diodes; (c) the sensitivity to ambient temperature of the terminal voltage for fixed input power for the various diodes. Note how the ASPAT improves on the planar doped barrier (PDB) and Schottky diodes, but does not reach the Ge backward diode. (After *GEC Plessey Semiconductors microwave product handbook* 1994; Syme 1993.)

AlAs tunnel barrier, comparable asymmetric current–voltage characteristics are achieved (cf. Fig. 8.3(a)) (Syme 1993). The current is now controlled by the tunnelling through the AlAs barrier, and tunnelling is intrinsically less sensitive to temperature than thermionic emission. The ASPAT diode has been developed as a replacement for the Ge backward diode, a reverse-biased p–n tunnel diode in Ge which has retained a niche market by virtue of its very weak sensitivity to ambient temperature. Diodes that rely on thermionic currents require complex circuitry to compensate for large variations (factors of 3 or more) in current for fixed voltage over the −40 to +80 °C range, whereas diodes that use tunnel currents can dispense with such circuitry, reducing cost. The particular form of the current–voltage characteristics of the Ge backward diode (Fig. 17.15(a)) allows mixer/detector action at zero bias, but the output saturates at a relatively low power (c. −20 dBm) (Fig. 17.15(b)) once the radio frequency signal takes the device into the NDR regime above about 0.1 V. At zero bias, the ASPAT diode matches the sensitivity of Schottky diodes and planar doped barrier diodes over the range −50 to +10 dBm of input power where they have been compared (see Fig. 17.15(b) for a part of the comparison), and has a reduced sensitivity to ambient temperature over the −40 to +80 °C range (Fig. 17.15(c)), where the terminal voltage for a fixed input power of −20 dBm is measured for a variety of diodes. Preliminary evaluations also suggest that this is a particularly quiet diode, as any hot-electron-induced noise associated with thermionic currents is absent, and the electrons are not strongly excited anywhere. This combination of properties makes the ASPAT a probable competitor component in future portable low-power radar and communication systems.

17.8 Other microwave diodes and applications

While the diodes described above represent the major components of a microwave system, there are further microwave applications of multilayer semiconductor diodes. We describe two in this section.

17.8.1 Quantum-barrier varactor diodes

One element in the tuning of resonant microwave circuits is a varactor diode, a variable capacitor whose precise value is determined by an applied bias. In other contexts, a non-linear element is required for frequency multiplication to achieve microwave power at high frequencies. In this latter application a device with antisymmetric current–voltage characteristics and symmetric capacitance–voltage characteristics is required. In such a case only odd harmonics are obtained, with applications as 3×, 5×, and 7× elements, and the circuit does not have to be refined to handle any unwanted 2×, 4×, etc. elements, as is the case when Schottky diodes are used as frequency-multiplication devices. By placing an AlGaAs barrier in the centre of the undoped region (cf. Chapter 8, Fig. 8.3(a)) one can achieve the desired result. In Fig. 17.16 we show the structure of the quantum barrier varactor (QVB) diode

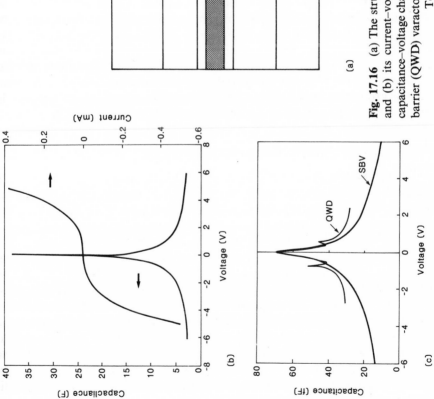

Fig. 17.16 (a) The structure of a quantum single-barrier varactor diode, and (b) its current–voltage and capacitance–voltage characteristics; (c) capacitance–voltage characteristics of a single-barrier (SBV) and a double-barrier (QWD) varactor diode. (After Rydberg et al. © 1990 IEEE 1990; Tolmunen and Frerking 1991.)

0.4 μm GaAs 3.4 × 10^{18} cm^{-3}

530 nm GaAs 10^{17} cm^{-3}

5 nm GaAs undoped

20 nm Al$_{0.7}$Ga$_{0.3}$As

5 nm GaAs undoped

530 nm GaAs 10^{17} cm^{-3}

2 μm GaAs 3.4 × 10^{18} cm^{-3}

(a)

(b)

(c)

(Fig. 17.6(a)) and its current–voltage and capacitance–voltage characteristics (Fig. 17.6(b)). In trebling the frequency to reach the range 210–280 GHz, this device has delivered more than 2 mW at over 5 per cent efficiency, and a 0.2 per cent efficiency has been estimated on the basis of 5× multiplication to 310 GHz. The power-handling limit is established by the size of the electric fields inducing avalanching in the diode (Rydberg *et al.* 1990; Reddy and Neikirk 1993).

Double-barrier structures have been investigated in the same varactor context, and a comparison of the capacitance–voltage characteristics for single-and double-barrier structures is shown in Fig. 17.16(c) (Tolmunen and Frerking 1991). The extra structure in the capacitance–voltage curve at the resonant tunnelling condition is an undesirable feature.

17.8.2 Resonant tunnel mixer diodes

Mixer diodes rely on non-linear current–voltage characteristics to generate sum and difference frequencies. The devices are inherently lossy, as input power is diverted into unwanted harmonics and dissipated in parasitic impedances. The non-linearity of the characteristics of a double-barrier diode near the current peak is suitable for mixing, and have the added advantage that power might also be extracted from the oscillations of the double-barrier structure. A reduction in conversion loss has been demonstrated for this device, with a conversion gain of more than 1 dB at 18 GHz. Recent effort has focused on the complex problem of establishing the regions of stability and maximum gain of resonant tunnelling mixers as a function of the bias on the device, the frequency of operation, and the level of incoming and local oscillator power (Hayes *et al.* 1993).

17.9 Comparison of microwave diode sources

In Fig. 17.17, we compare the output power as a function of frequency for the range of diode sources considered in this chapter. We see that over the 10–100 GHz range the output powers vary by nearly six orders of magnitude. In earlier figures in this chapter, the performances of the various mixer and detector diodes have been compared. Apart from the primary indicator of output power or transfer function, one must consider the host of other device attributes (efficiency, noise level, sensitivity to ambient temperature, manufacturability, yield, lifetime, robustness, etc.) as well as the ease with which the diodes can be matched into circuits efficiently. The new generation heterojunction devices offer power and sensitivity, but the many advantages in these latter aspects commend themselves to systems designers.

In this chapter we have considered only the first generations of heterojunction devices, and research continues on more sophisticated multilayer structures. There have been proposals and demonstrations of devices where superlattices have been used in both the avalanche and drift regions of devices to exploit the narrow bands of energy of the electrons in superlattices. The

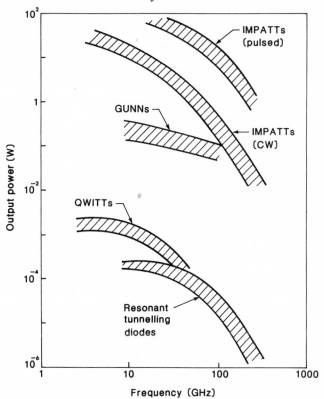

Fig. 17.17 A comparison of the output power of various semiconductor diode sources.

results to date are modest (Christou and Varmazis 1986), but a future edition of this text may make further reference to such work. With two-terminal devices, it should be possible to set up a calculus of variations to obtain the optimum thickness doping level and composition of each layer in a multilayer semiconductor device structure, with a given application in mind, so that specific trade-offs of efficiency, noise, temperature stability, etc. are made at the design stage.

References

Beall, R. B., Battersby, S. J., Grecian, P. J., Jones, S., and Smith, G. (1989). W-band GaAs camel-cathode Gunn devices produced by MBE. *Electronics Letters*, **25**, 871–3.

Brown, E. R., Goodhue, W. D., and Sollner, T. C. L. G. (1988) Fundamental oscillations up to 200 GHz in resonant tunnelling diodes and new estimates of their maximum oscillation frequency from stationary-state tunneling theory. *Journal of Applied Physics*, **64**, 1519–29.

Brown, E. R., Sollner, T. C. L. G., Parker, C. D., Goodhue, W. D., and Chen, C. L. (1989). Oscillations up to 420 GHz in GaAs/AlAs resonant tunneling diodes. *Applied Physics Letters*, **55**, 1777–9.

Brown, E. R., Soderstrom, J. R., Parker, C. D., Mahoney, L. J., Molvar, K. M., and McGill, T. C. (1991). Oscillations up to 712 GHz in InAs/AlSb resonant-tunneling diodes. *Applied Physics Letters*, **59**, 2291–4

Christou, A. and Varmazis, K. (1986). Superlattice GaAs mixed tunnelling avalanche transit time device structure. *Applied Physics Letters*, **48**, 1446–8

Couch, N. R., Beton, P. H., Kelly, M. J., Kerr, T. M., Knight, D. J., and Ondria, J. (1988). The use of linearly graded composition AlGaAs injectors for intervalley transfer in GaAs: theory and experiment. *Solid State Electronics*, **31**, 613–16.

Couch, N. R., Spooner, H., Beton, P. H., Kelly, M. J., Lee, M. E., Rees, P. K., and Kerr, T. M. (1989). High-performance, graded AlGaAs injector, GaAs Gunn diodes at 94 GHz. *IEEE Electron Device Letters*, **10**, 288–90.

Esaki, L. (1976). Discovery of the tunnel diode. *IEEE Transactions on Electron Devices*, **ED-23**, 644–7.

Gunn, J. B. (1963). Microwave oscillations of current in III–V semiconductors. *Solid State Communications*, **1**, 88–91.

Hayes, D. G., Higgs, A. W., Wilding, P. J., and Smith, P. J. (1993). Conversion gain at 18 GHz from resonant tunnelling diode mixer operated in fundamental mode. *Electronic Letters*, **29**, 1370–2.

Ishii, T. K. (1989). *Microwave engineering* (2nd edn). Harcourt Brace Jovanovich, New York.

Javalagi, S., Reddy, V., Gallapalli, K., and Neikirk, D. (1992). High efficiency microwave diode oscillators. *Electronics Letters*, **28**, 1699–701.

Kearney, M. J. and Dale, I. (1990). GaAs planar doped barrier diodes for mixer and detector applications. *GEC Journal of Research*, **8**, 1–12.

Kearney, M. J., Couch, N. R., Stephens, J., and Smith, R. S. (1993). Heterojunction impact avalanche transit-time diodes grown by molecular beam epitaxy, *Semiconductor Science and Technology*, **8**, 560–7

Kesan, V. P., Neikirk, D. P., Streetman, B. G. and Blakey, P. A. (1987) A new transit time device using quantum-well injection. *IEEE Electron Device Letters*, **8**, 129–31.

Kidner, C., Mehdi, I., East, J. R., and Haddad, G. I. (1990). Power and stability limitations of resonant tunnel diodes. *IEEE Transactions on Microwave Theory and Techniques*, **MTT-38**, 864–72.

Ondria, J. and Ross, R. L. (1987). Enhanced TED mmW device performance using graded doping profiles. *17th European Microwave Conference Digest*, pp. 673–80, European Microwave Management Committee.

Reddy, V. K. and Neikirk, D. P. (1993). High breakdown voltage AlAs/InGaAs quantum barrier varactor diodes. *Electronics Letters*, **29**, 464–6.

Rydberg, A., Gronqvist, H., and Kollberg, E. (1990). Millimeter- and submillimetre-wave multipliers using quantum-barrier-varactor (QBV) diodes. *IEEE Electron Devices Letters*, **11**, 373–5.

Sollner, T. C. L. G., Goodhue, W. D., Tannenwald, P. E., Parker, C. D., and Peck, D. D. (1983). Resonant tunneling through quantum wells at frequencies up to 2.5 THz. *Applied Physics Letters*, **43**, 588–90.

Syme, R. T. (1993). Microwave detection using GaAs/AlAs tunnel structures. *GEC Journal of Research*, **11**, 12–23.

Sze, S. M. (1981). *Physics of semiconductor devices* (2nd edn). Wiley, New York.

Thoren, G. R. (1985). Advanced applications and solid-state power sources for millimeter-wave systems. In *Millimeter Wave Technology III* (ed. J. C. Wiltse).

Proceedings of the Society of Photo-Optical Instruments and Optical Engineering (SPIE), **544**, 2–9.

Tolmunen, T. J. and Frerking, M. A. (1991). Theoretical performance of novel multipliers at millimeter and submillimeter wavelengths. *International Journal of Infrared and Millimeter Waves*, **12**, 1111–33.

Whitaker, J. F., Mourou, G. A., Sollner, T. C. L. G., and Goodhue, W. D. (1988). Picosecond switching time measurement of a resonant tunneling diode. *Applied Physics Letters*, **53**, 385–7.

18 Devices III: lasers, modulators, and detectors

18.1 Introduction

In this chapter we discuss the components required to generate, control, and detect optical signals at wavelengths of order 0.8–1.5 μm. The growth of communication systems based on optical fibre as the transmission medium has been spectacular over the last two decades. In 1993 alone, 80 000 route kilometres of undersea cable was programmed for installation. The bandwidth is enormous when the carrier frequency is of order 300 THz, compared with the $c.100$ GHz in the last chapter, although one trades off propagation in free space there for that within a fixed link here. The optimum wavelengths for transmission in silica fibres are 1.3 μm and 1.55 μm corresponding respectively to a wavelength where two non-linear effects cancel each other and to a minimum in the absorption, both of which permit long-haul communications without significant distortion or weakening of the optical signals (Tsang 1985). This has led to the development of high-performance lasers at both wavelengths. With the discovery that erbium-doped fibres can act as lasers operating at 1.5 μm while pumped at 0.98 μm (Barnes *et al.* 1989), methods are being sought of pumping fibres with semiconductor lasers at a comparable wavelength. In the first part of this chapter, we describe the subsequent generations of lasers with reference to the GaAs–AlGaAs system (bulk GaAs lases at a wavelength of 0.86 μm determined by the bandgap), and then we apply some of the principles behind these lasers to those operating at longer wavelengths.

The storage and reading of data by optical means is another major area of application. At present GaAs-based lasers (costing less than £0.1 each!) are used. The optical systems for writing and reading digital information are limited by diffraction effects, and so ever-shorter wavelengths are required for increased data storage density. GaAlInP-based red lasers are available. Most recently the effort to produce semiconductor laser diodes that operate in the visible has paid off with initial reports of (low-temperature) lasers in the blue and green (Haase *et al.* 1991, Jeon *et al.* 1992).

As with microwave sources, the aim is to make increasingly efficient lasers and to reduce the level of current at the threshold of laser operation. In Fig. 18.1, we show how the threshold current density for GaAs-based lasers has dropped over the last 20 years as new design principles have been incorporated. As with the microwave devices discussed in the last chapter, the new generations of lasers have other desirable properties, of which a reduced sensitivity to ambient temperature is most important. Another is the ability to tune the wavelength of the laser action, and quantum confinement of

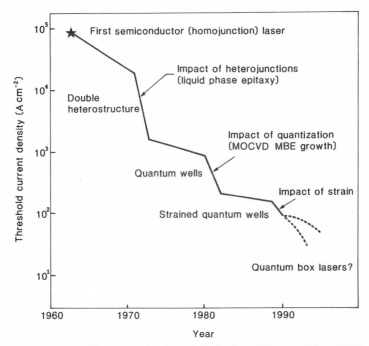

Fig. 18.1 The threshold current density of GaAs-based lasers. (After Weisbuch and Nagle 1990; Weisbuch 1993.)

electron and hole states is particularly effective in this respect. The incorporation of strain has led to further improvements in laser operation, and the current controversy is over any possible role that Q1D or Q0D systems might play in further improvements in laser performance. We air this controversy.

Having generated a monoenergetic beam of (coherent) light, we may wish to modulate the intensity so as to convey analogue or digital information. This can be done by modulating the current fed to the laser, or by passing the light through another structure whose optical properties (particularly the refractive index and absorption coefficient) can be modified by applied biases. We describe both options. The properties of quantum wells in electric fields play a prominent role here. The present state of the art corresponds to modulation at 30 GHz, comparable with the carrier frequency of microwaves!

Finally, we need to detect and demodulate the optical signal, and here we are concerned with the sensitivity and signal-to-noise ratio, just as in a microwave system. We leave other optical devices, such as solar cells, and infrared detectors, etc., to the next chapter.

18.2 Basic principles of double-heterojunction semiconductor lasers

The elementary optical processes of interest in this and the following chapter are indicated in Fig. 18.2, namely absorption, spontaneous emission, and

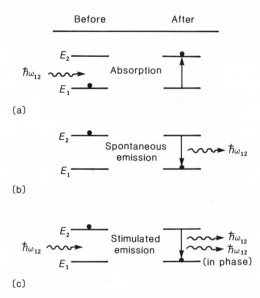

Fig. 18.2 The basic optical processes in a direct-gap semiconductor: (a) absorption, (b) emission, and (c) stimulated emission of a photon.

stimulated emission. The absorption processes have been discussed in Chapters 1, 4, and 10. Light-emitting diodes using p–n junctions in direct-gap semiconductors have been in widespread use: the annihilation of an electron–hole pair can lead to the spontaneous emission of light. The condition for the dominance of stimulated emission over spontaneous emission is that the occupancy of higher-energy initial states is greater than that of lower-lying final states for the optical transition (Loudon 1973). We shall show below how quantum confinement makes this condition easier to achieve.

The first semiconductor lasers were generated with forward-biased p–n junctions in GaAs (Sze 1985). At the interface it is possible to obtain the non-equilibrium condition of population inversion (higher electron population of higher energy states, with lower and incomplete occupation of lower energy states), as shown in Fig. 18.3(a). Since the laser action involves the interaction of an electron density with a photon field, we are also concerned with the extent to which light can be confined in the region where the population inversion is achieved, i.e. the plane of the p–n junction. In practice, there is a small (about 1 per cent) change of refractive index as the free carrier concentration on either side of the junction is reduced, and the optical field (light intensity) is spread out over several micrometres. The semiconductor diode acts as its own cavity, with two ends cleaved or etched and polished to act as mirrors and the other two sides roughened. Ohmic contacts are made to the top and bottom, and light is emitted through the polished ends.

The first ever device application of heterojunctions came in the double-heterostructure laser, where the n and p contact layers were made of

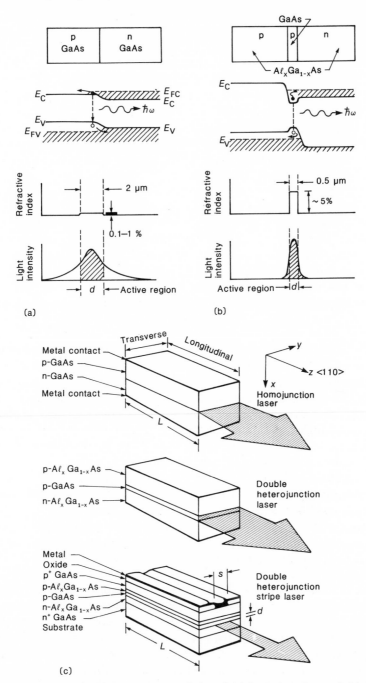

Fig. 18.3 A comparison of some characteristics of (a) homojunction and (b) double-heterojunction semiconductor lasers; (c) a comparison of the laser diodes. (After Sze © 1985 Reprinted with permission of John Wiley and Sons. Inc.)

$Al_xGa_{1-x}As$, and the 0.5 μm GaAs junction region was left p-doped (cf. Casey and Panish 1978). The net gain to the laser operation was twofold: the heterojunctions provided barriers to the current flow in such a way as to increase the density of both electrons and holes in the junction region, and furthermore the difference in refractive index became about 5 per cent with a greatly increased confinement of light in the active region (Fig. 18.3(b)). The two effects resulted in a 100-fold reduction of the threshold current density (Fig. 18.1), and double-heterostructure lasers were the first practical semiconductor lasers to be exploited commercially. In addition, by patterning the top contact into a narrow stripe (Fig. 18.3(c)), the light can be arranged to come from a narrow part of the end facet.

We shall map the improvements of lasers via the reduction of the threshold current density for the onset of stimulated emission. There are two important considerations associated with this figure of merit (Weisbuch and Vinter 1991, and references cited therein). First, the threshold condition itself is that light should make one round trip in the cavity under the condition that losses due to absorption, scattering, etc. are just matched by the optical gain in the active volume. Secondly, the relationship between the current and the optical gain must be established. The threshold condition applied to light of intensity I_0 in a cavity of length L, with end-mirror reflectivities R_1 and R_2, can be written as

$$I_0 R_1 R_2 \exp(2\Gamma g_{th} L)\exp[-2L(\Gamma\alpha_i + (1-\Gamma)\alpha_c)] = I_0$$

where Γ is the optical confinement factor described below, $\alpha_i(\alpha_c)$ are the internal loss coefficients in the active (optical confining) layers, and g_{th} is the volume gain of the active layer at threshold, calculated below. This equation can be rewritten in the form

$$\Gamma g_{th} L = \Gamma\alpha_i + (1-\Gamma)\alpha_c + (1/2L)\ln(1/R_1 R_2)$$

Note that typical values for the right-hand side are 40–80 cm^{-1}, with the cavity loss playing a major role c.30 cm^{-1} for $L = 300$ μm and $R_1 = R_2 = 0.3$ without any special coatings.

The optical confinement factor plays an important role in laser design. It is a measure of the fraction of the photon field within the electronically active region, and is given by

$$\Gamma = \int_{-d/2}^{d/2} |E(z)|^2 dz / \int_{+\infty}^{-\infty} |E(z)|^2 dz$$

which is calculated using standard techniques of wave propagation in stratified media; the difference in dielectric constant plays an important role here. In the double heterostructure, Γ can be quite large. Below about 100 nm in the GaAs–AlGaAs system the optical wave is less well confined between the two heterostructures. In the case of quantum wells (see next section), where the active layers are only a few nanometres thick, Γ can become very small and other aspects of the laser design must be modified to compensate.

The second factor is the relationship between gain and current. We think of gain as the opposite of absorption—as an optical wave passes through a region, if the conditions can be met for the absorption to be negative, the amplitude of the optical wave will increase exponentially. For a particular transition frequency, we can write

$$g(\omega) = \alpha_0(\omega)[f_c(\omega) - f_v(\omega)]$$

where $\alpha_0(\omega)$ is the absorption loss in the host medium, and $f_c(\omega)$ and $f_v(\omega)$ are the occupancies of the conduction and valence band states associated with the transition of frequency ω. In equilibrium $f_c = 0$ and $f_v = 1$, the gain is negative, and the material is absorbing. If a region of material is pumped so that $[f_c(\omega) - f_v(\omega)] > 0$, which is a condition of population inversion, then the optical wave will grow in amplitude. In the case of a semiconductor solid the spectral gain curve is given by

$$g(\omega) = [e^2/(2n_r\varepsilon_0 m_0^2 c\,\omega)] \int M^2 \rho_j(E,E')[f_c(E) - f_v(E')]\delta(E - E' - \hbar\omega)\mathrm{d}E\mathrm{d}E'$$

It has the same structure as the optical absorption given in Chapters 1 and 10, where n_r is the refractive index, M is the matrix for interband transitions, and $\rho_j(E)$ is the joint density of states between conduction and valence bands for a transition of energy E. For a given level of injection, f_c and f_v are the values of the Fermi–Dirac distribution functions in the conduction and valence bands. We seek the condition that $f_c(E) - f_v(E') > 0$. There is an equivalent way of expressing this condition in terms of the quasi-Fermi energy for electrons in the conduction band and holes in the valence band:

$$E_{\mathrm{Fc}} - E_{\mathrm{Fv}} > \hbar\omega$$

We use this expression to calculate the gain–current relation. A given density of electrons and holes are injected, from which we calculate the quasi-Fermi energies and the spectral gain curve incorporating the interband matrix elements. Now the relation between the injected carrier densities and the injection current is obtained by assuming a fixed recombination time, which allows us to write $J = eNd/\tau$ (where N is the injected carrier density, τ is the relaxation time, and d is the thickness of the active region), or we calculate the time for each density according to the rate of spontaneous recombination. The results are shown schematically in Fig. 18.4 for both a double-heterostructure laser, where the densities of states vary as $\sqrt{E}$, and the quantum-well laser discussed in the next section where the density of states is a constant (i.e. $\sim E^0$). The key role of the density of states in the gain curve will be discussed further in the next section. For an energy less than the gap, there are no transitions, and at very large energies the injected current will not greatly influence the occupancy of the electron and hole states so that the gain will be negative. There is a peak in the gain–energy curve, and with exponential growth of the optical wave amplitude and intensity lasing action occurs at the energy of the maximum gain. In practice there is a threshold level of injection

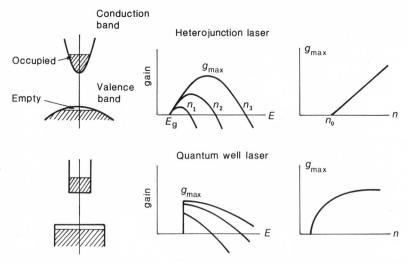

Fig. 18.4 The calculation of gain for a semiconductor laser. (After Weisbuch and Vinter 1991.)

before gain becomes positive. In the case of 3D density of states, the volume gain is approximately linear with injection current as shown. We write

$$g_{max} = g = A(J - J_0)$$

where A is a differential gain and J_0 is known as the transparency current. At this current density, light can just pass through the material without alteration, i.e. it is the threshold for achieving gain and below it no gain is achieved. By combining the threshold condition for matching gain and loss with the gain equation, we can obtain the threshold current density as

$$J_{th} = J_0/\eta + \alpha_i/\eta A + (1 - \Gamma)\alpha_c/\Gamma\eta A + (1/\eta\Gamma A)(1/2L)\ln(1/R_1 R_2)$$

where η is the internal quantum efficiency representing the fraction of total recombination transitions that are radiative ($c.100$ per cent in GaAs, but less than 100 per cent in poorer-quality materials). The threshold current decreases as the thickness of the active region decreases (hence the factor of $c.100$ between the p–n junction and the double-heterojunction lasers). The optimum thickness for the active volume of a double-heterostructure laser in the GaAs–AlGaAs system is about 100 nm: above this value there are more electron and hole states to be inverted, and the differential gain A decreases for this reason, and below this value Γ decreases (as d^2) whereas J_0 decreases only as d and A increases as d^{-1}, where d is the active-layer thickness.

18.3 The single- and multiple-quantum-well laser

With reference to Fig. 18.1, the next steep drop in threshold current density came with the introduction of quantum wells of thickness 20 nm or less as the active volume of the laser. In Fig. 18.5, we exhibit the energy band profile and

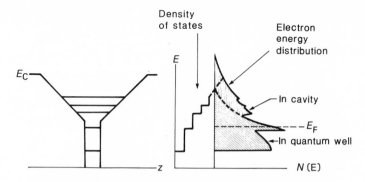

Graded index separate confinement quantum well laser

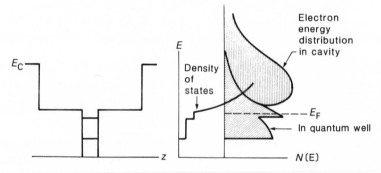

Separate confinement heterostructure quantum well laser

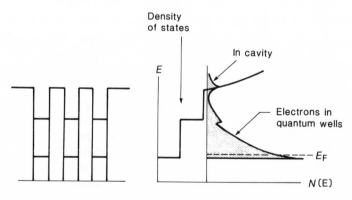

Multiple quantum well laser

Fig. 18.5 The energy band profile, densities of states, and electron distribution for various quantum-well laser structure: GRINSCH, graded-index separate-confinement heterostructure; SCH, separate-confinement heterostructure. (After, 'Weisbuch and Vinter 1991.)

the density of states in energy for various quantum-well laser structures that have been investigated.

Using a very thin GaAs active layer has some advantages.

1. The electron and hole levels have quantum energies of confinement, so that the interband transitions are not tied to the bandgap of GaAs (0.86 μm) but can be tuned over a modest range $c.0.67$–0.86 μm by varying the thickness of the wells. There are other effects, such as charge density corrections to the precise energy levels, but these are matters of fine detail.

2. The density of states becomes 2D, and the contribution of injected carriers to the gain is highly effective. In 3D we see (Fig. 18.4) the gain peak move away from the bandgap as the density of injected carriers increases, and all the carriers at lower energies are ineffective in contributing to the gain. The peak in 2D stays at the effective gap, and the differential gain above threshold is higher for carriers in 2D than in 3D. A further effect of the 2D density of states is that within an energy range of kT there are $c.10^{12}$ states cm^{-2} in a quantum well, but $c.10^{13}$ states cm^{-2} in a 1000 nm double-heterostructure laser, and threshold is achieved sooner with the reduced number of states to be inverted.

3. The optical confinement factor Γ is very small for a single quantum well, but further bandgap engineering is used to achieve separate confinement by heterojunctions (SCH) of the optical field. The separate confinement is achieved by using an intermediate Al composition or by grading the Al profile (cf. the implications of Fig. 18.3), leading to comparable but still small values of Γ.

4. Fig. 18.5 also shows the importance of the graded index in increasing the electron population in the quantum wells relative to the optical cavity.

The overall advantages are summarized in Fig. 18.6(a), where a comparison is made of the modal gain (i.e. the product Γg) for a 80 nm double-heterostructure laser and a 12 nm quantum-well laser with graded-index layers providing the optical confinement. Both the lower threshold current density and the steeper gain curve as the current increases can be seen. Note that it is possible to achieve lasing between the first excited electron and hole states in the quantum well at higher current densities. Note also that the steeper rise in modal gain above threshold means that improvements in mirror reflectivity at the ends has a rather larger relative effect in reducing threshold currents. Fig. 18.6(b) shows the variation of the gain with quantum-well thickness. The 40 nm well is essentially 3D like in its gain–current relation, and the transparency condition is reached for lower currents as the quantum-well width decreases, until very thin wells are reached. By then the quasi-Fermi energies are so high in the quantum wells that some carriers are not confined in the quantum well, but only in the optical confinement cavity. In practice in SCH quantum-well lasers only about 20 per cent of the electrons are in the

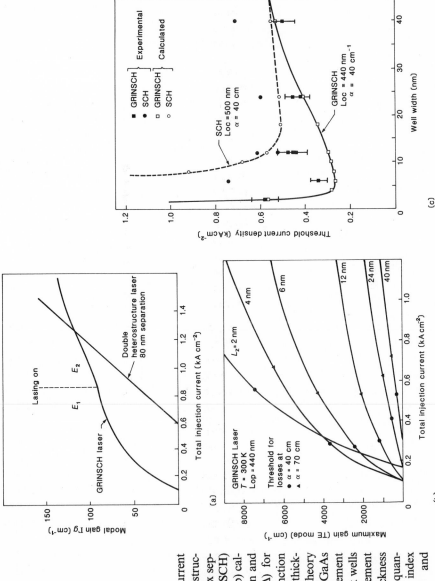

Fig. 18.6 (a) The gain–current relations for a double heterostructure laser and a graded-index separate-confinement (GRINSCH) single-quantum-well laser; (b) calculations of the volume gain and threshold loss rates as a function of the quantum-well layer thickness; (c) a comparison of theory and experiment in GaAs–AlGaAs lasers with separate confinement wells and with graded-index wells for the optical confinement (GRINSCH). L_{oc} is the thickness of the optical cavity, i.e. the quantum wells plus graded index layers. (After Weisbuch and Vinter 1991.)

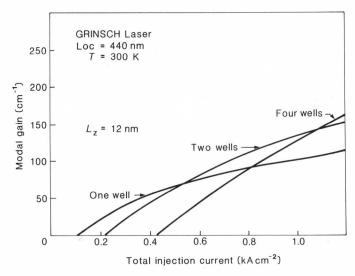

Fig. 18.7 The modal gain versus current relationship for one-, two-, and four-well. GaAs–AlGaAs quantum-well lasers (each well is 12 nm thick) with 18 per cent Al barriers for electronic confinement and a maximum of 40 per cent Al for optical confinement. (After Weisbuch and Vinter 1991.)

quantum well at threshold. Given these refinements, the agreement between theory and experiment is close to being quantitative, as seen in Fig. 18.6(c).

If several quantum wells are placed in parallel in the active region (the so-called multiple-quantum-well laser) and it is assumed that (i) the current injection into each well is the same and (ii) Γ is the same within each well, the product Γg can be scaled up from the calculations on a single well. From the results in Fig. 18.7, there is no advantage in having several quantum wells, in low-loss systems, as the low transparency current and the low number of states to be inverted will win out. At higher losses, the steeper part of the gain–current curve can be used rather than the saturated curve for a single quantum well, and multiple-quantum-well lasers win out.

18.4 The strained-quantum-well laser

The third steep drop in threshold current density comes with the incorporation of strain into the quantum well, as with the addition of InGaAs layers to the GaAs–AlGaAs system (Adams 1986). In Chapter 13 we described the modifications to the valence bands of semiconductors under the biaxial strain encountered in highly strained layers. Whereas in unstrained layers the high degeneracy of the valence bands can be partly lifted by virtue of the reduction of symmetry associated with a thin layer (Fig. 13.5), biaxial strain can be used to move the heavy and light-hole quantum wells and energy levels with respect to each other (cf. Fig. 13.6). In particular, under biaxial tension the light-hole

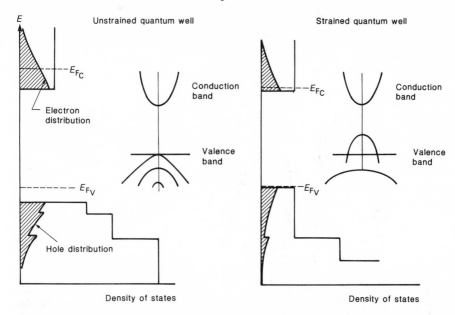

Fig. 18.8 A comparison of energy levels, densities of states, and band-filling in unstrained and strained quantum wells as active layers for lasers. (After Weisbuch and Vinter 1991.)

quantum well deepens and the highest confined state has light transverse hole character with a low density of states. The effect of strain on the density of states is shown schematically in Fig. 18.8.

The density of electron and hole states adjacent to the bandgap is now comparable. In unstrained lasers, the shift in quasi-Fermi level is much greater for electrons than holes ($f_v(E) \ll f_c(E)$). This is because of the higher hole density of states in the valence bands. In order to account for charge neutrality in the active region, the number of holes and electrons are equal, but the density of hole states is higher. With strained layer lasers and a given hole density, the shift in quasi-Fermi energy for the hole states is greater and the gain is correspondingly increased.

More recently, it has been noted that the threshold current reaches a minimum value for non-zero values of both compressive and tensile stress in structures with InGaAs quantum wells and InGaAsP barriers (Fig. 18.9(a, b)). The original compressive strain was exploited as it split the valence bands and reduced the in-plane densities-of-states mass of the heavy-hole sub-band. The tensile strain brings the light-hole states to the highest energy and leads to a further reduction in effective mass. In addition to reduced threshold current densities, higher differential and total gains have been reported in tensile rather than compressive strained lasers in this system. Note (i) the absence of lasing at small tensile stress where the bands are degenerate or indirect, and (ii) a rise on threshold current for larger strains. There are competing

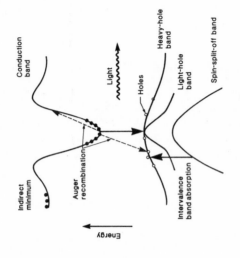

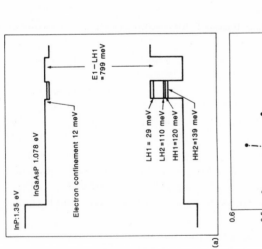

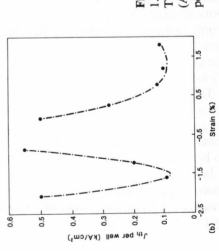

Fig. 18.9 (a) The multilayer structure and (b) the threshold current density for 1.5 μ quantum-well lasers as a function of strain in the quantum well (after Thijs *et al.* 1991); (c) two examples of non-radiative recombination processes (Auger recombination and intervalence band absorption (IVBA)), which compete with the radiative processes in these materials systems. (After Adams and O'Reilly 1992.)

non-radiative transitions associated with intervalence band absorption, shown schematically in Fig. 18.9(c), which reduce the η factor from near unity in GaAs structures to much lower values in quaternary alloys. Auger recombination reduces the relaxation time τ, which must be offset by higher carrier injection describes. These processes occur for given sets of separations between bands and tend to be more prevalent in the case of longer-wavelength (1.5 μm) lasers; we shall return to this in Section 18.10.

18.5 General features of quantum-well lasers

As with the devices discussed in previous chapters, efficiency, power, and speed are important parameters, but there are many other secondary factors which make the new devices attractive. The temperature dependence of the threshold current density is important in practical systems. By convention it is described empirically by

$$J_{th}(T_2) = J_{th}(T_1)\exp[(T_2 - T_1)/T_0],$$

where T_1, T_2 are two adjacent temperatures; a high value of T_0 is desirable in systems applications. The features that enter the evaluation of T_0 include the Fermi–Dirac occupation factors, the densities of states in the quantum wells and the optical confining wells, and the thermal excitation of carriers in addition to their injection, leading to a greater spill-over of carriers into the optical confining layers. The value of T_0 increased from a single quantum well with a separate square-well confinement structure, through a single quantum well with graded composition layers, to a multiple-quantum-well structure. With reference to Fig. 18.5, the 3D density of states is further away in energy from the carriers associated with population inversion. As we go to strained quantum wells, with their stronger confinement and lower band-filling from the symmetric electron and hole densities of states (cf. Fig. 18.8), the value of T_0 increases even further.

The principal mechanism leading to the catastrophic failure of semiconductor lasers during operation is the degradation of the mirror ends of the structure as a result of high non-radiative recombination processes. Since these processes are less efficient in quantum-well structures, these lasers prove to be more reliable.

In communication systems, one wants to modulate the light output at high speed. An upper limit to the modulation frequency is set by the characteristic frequency with which a laser can respond to altered current levels, and is given by (Yariv 1989)

$$\Omega = \sqrt{(AP_0/\tau_p)}$$

where A is the differential gain, P_0 is the average laser power, and τ_p is the lifetime of a photon in the laser cavity. One can only decrease τ_p by increasing cavity losses, and so the increase in A at large gain in quantum-well lasers is important (cf. Fig. 18.6(a)).

The use of arrays of quantum-well lasers to pump other solid state lasers has resulted in a new generation of high-efficiency narrow-linewidth frequency-doubled visible power lasers. The high internal quantum efficiency, lower transparency current, and smaller internal losses are the important factors contributed by quantum-well lasers.

18.6 Vertical-cavity surface-emitting lasers

In all the lasers discussed so far the light is emitted from the edges. There has always been a desire for surface-emitting lasers, which would allow free-space communication from chip to chip between specific locations and 2D arrays of high-power light sources. Earlier attempts introduced mirrors or gratings to turn the edge-emitted light through 90°, but recently major advances have been made by using GaAs–AlGaAs multilayers (grown at the same time as the laser structure) as integral but distributed mirrors (Scherer *et al.* 1989). The multilayer mirrors (known as Bragg reflectors) act on the light (in the same way that superlattices act on electrons) to reflect it at certain wavelengths, including that of the laser. The GaAs and AlGaAs layers are each typically a quarter-wavelength thick. The structure is shown in Fig. 18.10, together with a photograph of an array of such devices. It is straightforward to place ohmic contacts top and bottom, and there is no need for cleaving and polishing end facets. The yield is high. The three main features that determine their operation are (i) the high reflectivity of the mirrors, (ii) the means of lateral confinement of the active region (i.e. the equivalent of the stripes in

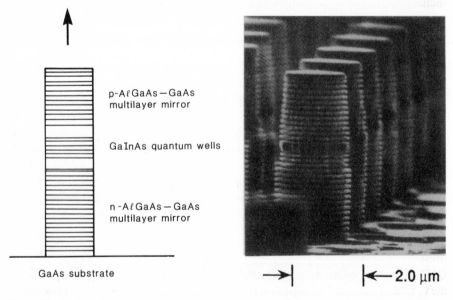

p-AℓGaAs–GaAs
multilayer mirror

GaInAs quantum wells

n-AℓGaAs–GaAs
multilayer mirror

GaAs substrate

$\rightarrow$| |←2.0 μm

Fig. 18.10 (a) The multilayer structure for a surface-emitting laser featuring distributed multilayer mirrors; (b) its physical realization. (After Scherer *et al.* 1989.)

Fig. 18.3(c)), and (iii) the the limitations imposed by laser heating from carrier injection through the resistive mirrors.

Whereas the reflectivity at a semiconductor–air interface is about 30 per cent, and can only be improved with mirror coatings, it is straightforward to achieve reflection in excess of 99 per cent by using a multilayer stack. In turn, this allows a 100-fold reduction in the cavity length to 1–2 μm which is consistent with epitaxial-layer overgrowth. Surface-emitting lasers have comparable losses and gains to edge-emitting lasers but are much more compact. The need to confine the current to achieve a narrow beam presents problems. Appropriate etch patterning of the substrate can be used to confine the optical wave, and a lateral p–n junction will confine the current. An alternative is to etch through the laser structure itself, which confines both the current and the light. However, the dry-etching technique favoured for this processing step induces large and undesirable surface leakage currents. Whereas one would expect a steady reduction in threshold current (down to *c*.0.1 μm diameter where Γ would start to fall steeply), the threshold current rises for diodes of diameter more than 5 μm because of the parallel surface channel for the current. Damage induced during ion implantation is an attractive alternative for current and optical confinement. The current flowing to the cavity can heat up the mirrors through which it passes. This can lead to severe heating during continuous-wave operation. At present the devices are about 10 per cent efficient in electrical to optical power conversion (Wang *et al.* 1991). Submilliampere threshold currents can be achieved with strained active layers. The surface-emitting geometry has many attractions for future systems, including a smaller beam divergence making it easier to couple.

18.7 Quantum-wire and quantum-dot lasers

In Chapter 11 we described electronic and optical experiments on pillar and wire structures where two of the three spatial dimensions were on the scale of less than 0.1 μm required for lateral quantization to become effective. Here we concentrate on the possible device implications of such structures. Quantum-wire lasers have been made using the structures shown in Fig. 2.4, where a multiple-quantum-well laser structure is grown on a V-groove substrate. The wells tend to be thicker at the bottom of the groove and to taper to relatively thin wells up the sides owing to different growth rates. The widening shape at the bottom of the well contains a number of Q1D bound states. The advantages in going from a $\sqrt{E}$ to an E^0 density of states in the case of the quantum well are enhanced with the $1/\sqrt{E}$ density of states in the subbands of a Q1D structure, and lasing has been achieved in these structures. At present the results are primitive, but indicate some problems. The various Q1D sub-bands in the structures examined are separated by only about 12–15 meV, and the thermal occupation of the different sub-bands at room temperature, means that lasing on a given sub-band transition is less efficient than might be the case if this energy separation could be increased to more than

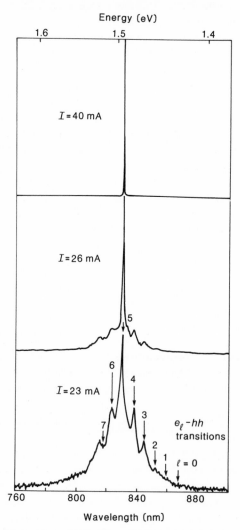

Fig. 18.11 The optical spectra below, at, and above threshold for a four-quantum-wire laser of length 2.5 mm operating at room temperature in a pulsed mode (cf. Chapter 2, Fig. 2.4). (After Kapon *et al.* 1992.)

30 meV. The lasing does not occur on the lowest sub-band (Fig. 18.11) but on the third or fifth excited levels in laser structures of slightly different widths, or on two excited levels in some cases. The spectrum below threshold allows the effects of quantum confinement in two spatial dimensions to be verified in the number and separation of the Q1D sub-bands. A real challenge will be to grow a number of wires of sufficiently equal cross-section to achieve a narrow line spectrum output. One will also need many wires in parallel and at a sufficiently high spatial density to obtain adequate output power. The

spontaneous emission spectrum below threshold and the lasing spectrum above threshold give evidence of both TE and TM modes, a feature not encountered in quantum-well lasers where the TE modes dominate. The preliminary results are encouraging further work on quantum-wire structures (Arakawa 1994). It has also been suggested that a reduction of the hole mass by strain could also lead to a 1D density of states distribution for holes at the wire widths currently obtainable.

In contrast, the quantum-dot structures seem less suited to laser action. In this case millions of identical dots are needed to create a sufficient active volume in the device, and this has not been achieved to date. The processing of dots by etching and regrowth introduces defects that degrade the optical emission intensity by providing alternative non-radiative defect channels (Andrews *et al.* 1990). Furthermore, the dynamics of relaxation can be inhibited by a bottleneck effect, namely that if there are too few states for the excited carriers to relax into, the efficiency of the stimulated emission process seems to be unacceptably degraded (Benisty *et al.* 1991). The quantum-dot laser seems far off, but provides a mix of contemporary technology and physics challenges.

18.8 Quantum-well optical modulators

In Chapter 10, we described the optical properties of quantum wells and demonstrated the large variations in optical absorption (particularly associated with the excitonic absorption peak near the band edge—the so-called quantum-confined Stark effect) which could be induced by the application of an electric field across the quantum well. In this section we describe the practical application of these electric-field-dependent optical properties in optical modulators, optical switches, and other devices. There was rapid progress on these components in the mid-1980s, but the pace has slowed more recently. If we place typically 50 quantum wells in the i region of a reverse-biased p–i–n structure (Fig. 18.12), we can obtain shifts in optical absorption as shown in Chapter 10, Fig. 10.8(b) and 10.9, as the biasing is increased. We take precautions with the doping profile to ensure that the field across each well is the same, as depletion effects modifying the local electric fields in some wells would broaden out the optical absorption. The main early thrust in the development was the need to obtain a high on–off ratio of the transmitted signal intensity in a device capable of high-speed operation. The earliest results gave a 2:1 contrast for a device with an *RC* time constant of 2 ns. Since then there have been improvements in the materials quality, structure, design and the operation in a two-pass reflection mode using the same type of distributed mirror described above for surface-emitting lasers (Fig. 18.13(a)). The results include a 100:1 contrast ratio with the reflectivity changing between 43 and less than 0.4 per cent at 860 nm (Fig. 18.13(b)). An additional important figure of merit is the insertion loss (the power attenuation from unwanted absorption in the structure), which is at the level of about 3.5 dB for this device (Whitehead *et al.* 1989).

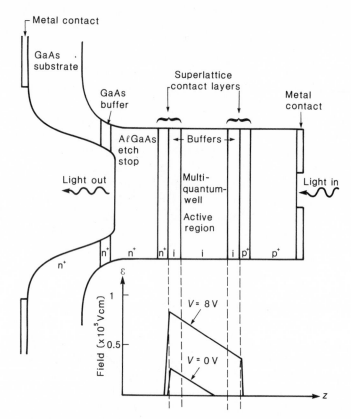

Fig. 18.12 Schematic diagram of a p–i–n multiple-quantum-well modulator. The device is about 4 μm thick and 600 μm in diameter. The lower inset shows the electric field over the active region. (After Miller *et al.* 1985.)

At the outset, it is important to note that the optical properties are very wavelength specific, but laser systems can be finely tuned and the width and depth of the quantum wells carefully chosen to develop components for an optical system. The competing modulator devices made of $LiNbO_3$ rely on electric-field-induced changes in refractive index. The effect is weak, and therefore long optical interaction lengths (centimetres as opposed to micrometres) are required. These devices are much less lossy and also much less wavelength specific (Tamir 1988; Kawano 1993). The great advantage of the multiple-quantum-well modulator is the opportunity to integrate devices on a chip, with part of the wafer forward biased to give laser action and another section reverse biased as the modulator. High-speed electronic signal processing in III–V materials (cf. Chapter 16) can be integrated with the laser and modulators to produce whole circuits (Fig. 18.14).

One qualitatively new device has emerged from the optical properties of quantum wells, namely the self-electro-optic device (SEED) (Miller *et al.*

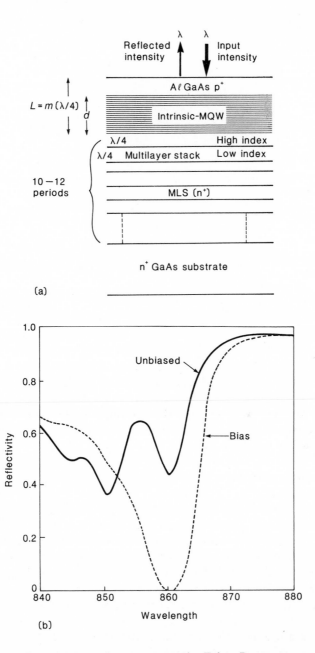

Fig. 18.13 (a) The structure of an asymmetric Fabry–Perot resonant structure containing a multiple-quantum-well structure above a multilayer mirror structure; (b) its performance as a reflectivity modulator showing the biased (broken) and unbiased (solid) reflectivities as a function of wavelength. (After Whitehead *et al.* 1989.)

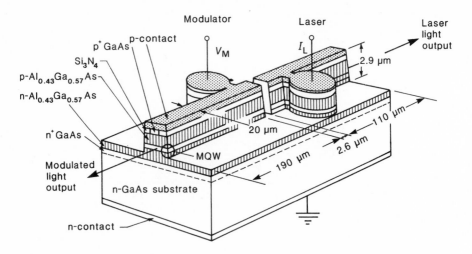

Fig. 18.14 Elements of an optoelectronic integrated circuit. (After Tarucha and Okamoto 1985.)

1984). The principle is shown in Fig. 18.15(a). It relies on the fact that a reverse-biased p–i–n multiple-quantum-well structure is a good photodetector (see next section). The structure is placed in series with a load resistor and a d.c. bias, so that the modulator structure is under a high electric field. Light at a wavelength slightly shorter than the excitonic peak at zero bias (say 1.45 eV in Fig. 10.8(b) of Chapter 10) is incident on the structure, which exhibits weak absorption since it is under bias. As the light intensity increases, so does the absorption, and hence the current flowing through the device, lowering the bias on the modulator as the current drops bias over the load resistor. This effect shifts the exciton absorption peak back towards its zero bias value and provides a positive feedback in the further increase in optical absorption of the incident light. If the response is sufficiently non-monotonic, the system can become unstable at intermediate currents, switching between a stable low-current low-transmission state and a stable high-current high-trans-mission state (Fig. 18.15(b)). The switching power varies inversely as the switching time (cf. 400 ns at 3.7 mW), and the switching energy of $c.4$ fJ μm^{-2} is much lower than encountered in other forms of optical bistability. The application of SEED-type devices can be enhanced by incorporating the p–i–n structure as the emitter–base barrier of a heterojunction bipolar transistor, so that the photocurrent is then subject to the gain considerations of the transistor.

All-optical switching used for routing in a multiple optical fibre network is an attractive prospect. The switching characteristics have been used to demonstrate optical logic functions and memory, with further investigations to establish whether it has a role, if not in optical computing (where the power budget makes optics uncompetitive with electronics), then in optical signal

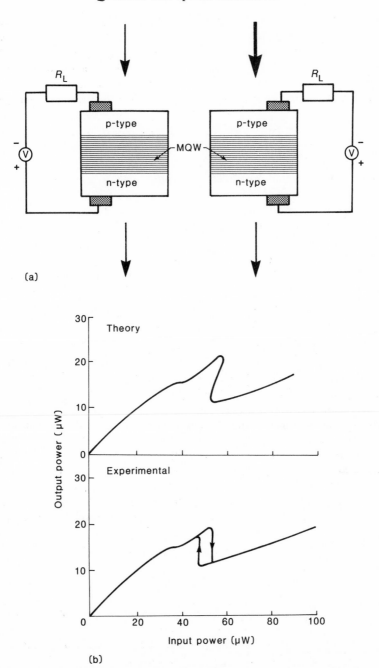

Fig. 18.15 (a) A multiple-quantum-well modulator configured as a self-electro-optic device, where the photocurrent modifies the distribution of voltage drops from that established by the battery in the absence of incident light; (b) early bistability results. (After Miller *et al.* 1984.)

processing more generally. In the latter context, 2048 element arrays of pairs of SEED devices have been made, each a logic element, with 40 pW required to hold a device in the high-current state and subnanosecond switching speeds (Boyd *et al.* 1990). SEEDs are under intensive technological development at present, with optical oscillators (Giles *et al.* 1990) and high-contrast waveguide directional couplers (Keyworth *et al.* 1990) having been demonstrated.

18.9 Photodetectors

When an optical signal has been generated with a laser and its intensity modulated either directly with the drive current of the laser or by passage through an optical modulator, it must be detected after passage through an optical fibre. We begin by outlining the two most common types of detector, the p–i–n photodiode and the avalanche photodiode (Sze 1981). We then show the role of semiconductor multilayers in enhancing the sensitivity, signal-to-noise, and other attributes of semiconductor photodetectors. The operation of a Si p–i–n diode under reverse bias is shown in Fig. 18.16(a). A fraction $1-R$ of the incident optical power is transmitted into the semiconductor, where it is absorbed with a characteristic inverse length α as electron–hole pairs are created and then separate under the applied field. If we denote by L_p the characteristic length over which the photocarriers can travel before recombining (($L_p \sim \sqrt{D_p \tau_p}$) where D_p is the diffusion constant and τ_p is the excess carrier lifetime), then the overall efficiency of the photodetector, in terms of number of electrons out per photon in, can be written as

$$\eta = (1 - R)[1 - \exp(-\alpha L)/(1 + \alpha L_p)]$$

for a width L of the i region. In conventional structures the optimum trade-off between total amount of light absorbed (i.e. efficiency) and speed occurs when $L \sim 1.5/\alpha$, and the maximum speed at which the photodetector can respond to a changing optical signal is determined by the transit time of the excess carriers. If their saturated drift velocity is v_s, the the modulation bandwidth is given by

$$f_{3\,\mathrm{dB}} \sim 0.4 v_s/L.$$

In Si acting at 0.8 μm and $\alpha \sim 1000$ cm^{-1} and $L \sim 15$ μm, $\eta = 75$ per cent and $f_{3\mathrm{dB}} \sim 2.5$ GHz.

The measure of sensitivity of a photodetector is the r.m.s. modulated input power required to produce a signal-to-noise ratio of unity in a 1 Hz bandwidth, which is called the noise-equivalent power (NEP). A fuller analysis than we have space to describe yields (Sze 1981)

$$\mathrm{NEP} = \sqrt{2}\sqrt{(eI_B + eI_D + 2kT/R_{\mathrm{eq}})}/R_\lambda$$

where I_B is the current arising from the background current, I_D is the thermally generated dark current, and R_λ is the responsivity (output current

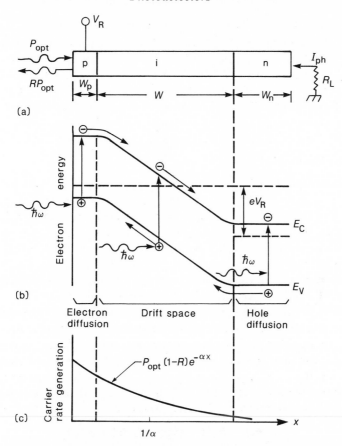

Fig. 18.16 The p–i–n photodiode structure, showing (a) the multilayer structure, (b) the energy bands with the diode under reverse bias, and (c) the variation with depth of the optically generated carriers. The symbol $1/\alpha$ is a measure of the length over which light penetrates and is absorbed with the creation of electron–hole pairs. (After S.M. Sze © 1981. Reprinted with permission of John Wiley and Sons. Inc.)

per unit input signal power). The expression takes into account the fact that both the background current and the dark current have shot noise associated with them, and that thermal noise is generated by the external equivalent circuit, modelled by its resistance R_{eq}. The background and dark currents must be minimized and the responsivity enhanced to improve the signal-to-noise ratio.

If the device is reverse biased to the extent that avalanching takes place in the intrinsic (or usually low-doped) region, then the device has internal gain, and the responsivity is enhanced, as is the noise, not just from the extra current but from the random and therefore noisy nature of the multiplication process itself. By taking this into account, the average multiplication factor per carrier M translates into an excess noise factor

$$F(M) = M\{1 + (\alpha_p/\alpha_n - 1)[(M - 1)/M]^2\}$$

where α_p and α_n are the hole and electron ionization rates respectively. (If hole multiplication is dominant, these two terms are interchanged). In GaAs α_p and α_n are comparable and $F(M) \sim M$, but in Si the ratio α_n/α_p can be as large as about 30, and one can achieve almost 100 per cent quantum efficiency between 0.6 µm and 1 µm with $F(M)$ as low as 4 for $M \sim 100$ (Sze 1981). In these devices, the undoped (or low-doped) region is generally longer than the equivalent i region in the non-avalanching devices. Thus, while avalanching is good for picking up weak signals, a penalty is slower speed.

Heterojunction technology has played an important role in improving the performance of photodetectors. GaAs-Al_xGa_{1-x}As p–i–n diodes are more efficient than Si in the 0.64–0.85 µm range. The direct bandgap of GaAs is exploited to enhance the optical absorption, and the wider-gap Al_xGa_{1-x}As forms a transparent contact region that allows the light to reach the absorption region without prior attenuation (Miller *et al.* 1978). The lattice-matched $In_{0.53}Ga_{0.47}$As–InP system is ideal for longer wavelengths; at 1.3 µm this device has been proved to be at least 80 per cent efficient with a modulation bandwidth of 20 GHz (Bowers and Burrus 1987). This system has the added advantage that the substrate is transparent and back-illumination is an optional configuration.

The most impressive heterojunction device is the separate absorption and multiplication avalanche photodiode (SAM–APD) shown schematically in Fig. 18.17. In this device, the light is absorbed in the $In_{0.53}Ga_{0.47}$As layer which is a low-field region. This reduces the amount of tunnelling across the main bandgap, helping to reduce the dark current. The carriers (holes in this case) then enter the InP region, which is under a high electric field, and undergo rapid avalanche multiplication. The wider bandgap of InP prevents tunnelling across the gap. The ionization rate for holes in InP is three times that for electrons, and this helps to reduce the excess noise factor $F(M)$

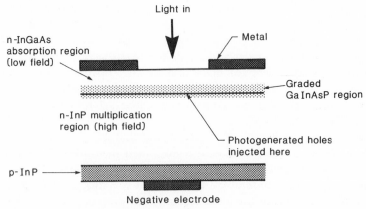

Fig. 18.17 Schematic diagram of the SAM-APD for 1.3–1.6 µm operation.

above; this is the reason for using hole injection. The optimization of this device is complex; the electric field in the InGaAs layer must be neither too large to induce unwanted multiplication in that layer, nor too small that the slow hole diffusion limits the speed of the device response. It is also necessary to grade the composition of the heterojunction interface in order to prevent holes from becoming trapped at the bandgap discontinuity there; this is achieved with a quaternary alloy $Ga_xIn_{1-x}As_yP_{1-y}$, with x going from 0.47 to zero and y going from zero to unity over the graded region (Matsushima *et al.* 1982). SAM–APDs have been made which have low dark currents, improved signal-to-noise ratios, and gain–bandwidth products of about 30 GHz, and in the range 1.3–1.6 μm they have replaced avalanche photodiodes based on Ge (Campbell *et al.* 1988; Tarof *et al.* 1990).

The advantage of a difference in the electron and hole ionization rates was referred to above in the context of reduced excess noise. Early experiments on GaAs–AlGaAs multiple-quantum-well structures suggested that the ratio of the two rates was different, with the holes taking advantage of the lower band discontinuities at heterojunctions (Capasso *et al.* 1982). This would have led to a new generation of low-noise avalanche photodiodes. However, as the band offset ratio evolved from the earlier 85:15 per cent to the present 60:40 per cent (cf. Chapter 2), further experiments revealed that the avalanche ratio was in fact closer to unity. In the lattice-matched system of $In_{0.53}Ga_{0.47}As$–$In_{0.52}Al_{0.48}As$–InP, the ionization rates are different for the two carriers, and $F(M) \sim 2.7$ for $M \sim 10$ has been measured for a system with a bandwidth of 9.3 GHz (Kagawa *et al.* 1990).

18.10 Optical devices in other materials and at other wavelengths

In earlier sections much of the description of laser and modulator performance was based on the GaAs–AlGaAs systems working in the wavelength range from the bandgap of GaAs at 0.86 μm to about 0.67 μm which can be spanned by quantum confinement corrections to electron and hole energies in quantum wells. Since the improvements to avalanche photodetectors have been modest in the GaAs–AlGaAs system, in the last section we described significant advances in the 1.3–1.6 μm range using materials which were lattice-matched to InP substrates. While most band-structure engineering concepts and devices have been first demonstrated in the GaAs–AlGaAs system, there are several factors driving the research in other wavelength ranges, some of which we describe here. We defer to the next chapter all reference to materials, devices, and systems associated with infrared detection in the 3–5 μm and 8–14 μm ranges.

The optical fibre communications systems exploit desirable properties of the fibre, namely minima in the distortion and absorption of input signals after transmission over hundreds of kilometres at wavelengths around 1.3 μm and 1.55 μm respectively. Here the $Ga_xIn_{1-x}As$ materials system is in widespread use in quantum-well structures, with the barrier layers being either InP or, more recently, the quaternary alloy materials $Ga_xIn_{1-x}As_yP_{1-y}$, and with MOCVD as the almost universal method of crystal growth. At present, the emphasis is on the optimization of design of structure and layers, with the freedom to chose the alloy compositions in the barriers and wells to obtain

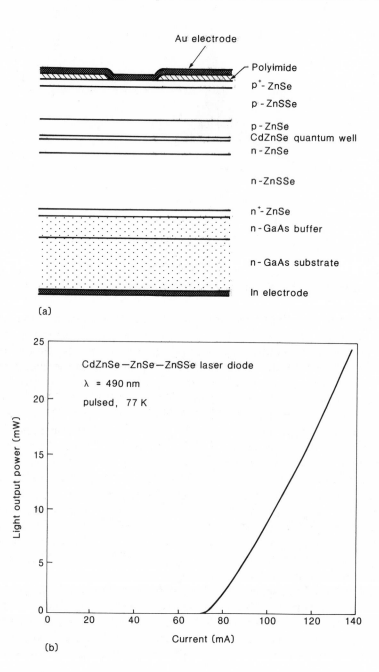

(a)

(b)

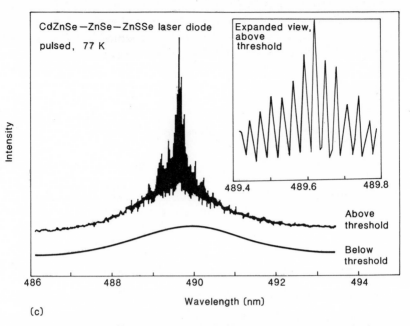

Fig. 18.18 (a) The structure, (b) the light–current relation, and (c) the output spectrum of a ZnSe quantum-well laser. (After Haase *et al.* 1991.)

appropriate wavelength of operation but with the incorporation of the correct amount of strain (cf. Section 18.4). The number of variables (composition, thickness, and layer number) poses problems for the qualification of layers grown for device applications, and the precise properties at the interfaces (cf. Chapter 2). Special problems have been posed by this materials system. For example the optical confinement factor Γ for the InGaAs–InP quantum-well system is poorer than that for the AlGaAs–GaAs system. The diminished bandgap and effective masses help, but very large carrier-heating effects associated with Auger recombination and intervalence band absorption (cf. Fig. 18.9(b)) are more severe here. At one point in the development of lasers with square wells of intermediate bandgap as the optical confining layer, lasing was obtained from these layers rather than from the quantum wells! The results in Fig. 18.9(a) indicate that these difficulties have been overcome. The band structure of the strained InGaAs quantum wells with InGaAsP optical confining layers on InP substrates (see Fig. 18.9(b)) show how much further the design of multilayers has developed from the GaAs–AlGaAs lasers to achieve the required device performance.

The 0.4–0.8 µm range spans the visible spectrum, and there are many potential applications for compact cheap visible laser sources. The GaAs– AlGaAs quantum lasers reach into the red end of the spectrum. Blue and green lasers would have applications in displays. The density of data stored and read by optical means is always limited by diffraction optics associated with the

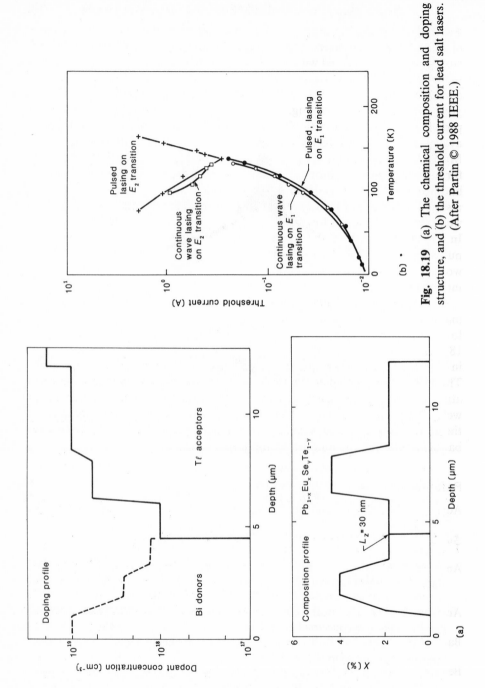

Fig. 18.19 (a) The chemical composition and doping structure, and (b) the threshold current for lead salt lasers. (After Partin © 1988 IEEE.)

wavelengths used. One cannot use a pixel of diameter less than the wavelength of light. Blue–green lasers would allow a quadrupling of the density of data storage of those achieved with red lasers. In this area, recent progress has been made with II–VI materials based on ZnS–ZnSe, CdTe–ZnTe, and related combinations. The bandgaps and offsets are ideally suited to achieve blue–green optical devices (cf. Fig. 1.9). Progress has been hampered by severe problems intrinsic to the II–VI materials and the methods of crystal growth available. The stoichiometry is very hard to achieve by MBE and MOCVD. Most dopants are amphoteric (i.e. will go randomly onto either lattice site and form donors and acceptors in comparable numbers), and in particular it is hard to achieve high p-doping. Steady improvements in recent years have resulted in reports of light emission and lasing action in ZnSe-based quantum wells, but so far only at low temperatures (Haase *et al.* 1991; Jeon *et al.* 1992). In Fig. 18.18 we show the structure and some results of the first report of pulsed lasing giving 100 mW optical power per facet at 0.490 µm. Further work is refining and improving on this with continuous-wave operation and increasing temperatures—room temperature operation has yet to be achieved.

At very long wavelengths (3–30 µm), a range of lead salts are used for making lasers. Because of the low bandgaps involved, the devices operate at low temperatures, although pulsed operation is possible up to 235 K. In Fig. 18.19(a) we show the doping and composition profile of a laser structure, and in Fig. 18.19(b) we show the threshold current as a function of temperature. The bifurcation at higher threshold currents for the continuous-wave operation corresponds to lasing at two different quantum-confined states in the wells. Again, as with the II–VI compounds, the progress is slow because of the materials quality and is made more difficult here because of the low-energy bandgap required for long-wavelength operation.

References

Adams, A. R. (1986). Band-structure engineering for low-threshold high-efficiency semiconductor lasers. *Electronics Letters*, **22**, 249–50.

Adams, A. R. and O'Reilly, E. P. (1992). Semiconductor lasers take the strain. *Physics World*, **5**(10), 43–7.

Andrews, S. R., Arnot, H., Kerr, T. M., Rees, P. K., and Beaumont, S. P. (1990). Radiative recombination in free-standing GaAs/AlGaAs quantum boxes. *Journal of Applied Physics*, **67**, 3472–9.

Arakawa, Y. (1994). Fabrication of quantum wires and dots by MOCVD selective area growth. *Solid State Electronics*, **37**, 523–8.

Barnes, W. L., Morkel, P. R., Reekie, L., and Payne, D. N. (1989). High quantum efficiency Er^{3+} fibre lasers pumped at 980 nm. *Optics Letters*, **14**, 1002–4.

Benisty, H., Sotomayor-Torres, C. M., and Weisbuch, C. (1991). Instrinsic mechanism for the poor luminescence properties of quantum-box systems. *Physical Review B*, **44**, 10945–8.

Bowers, J. E. and Burrus, C. A. (1987). Ultrawide-band long-wavelength p–i–n photodetectors. *Journal of Lightwave Technology*, **5**, 1339–50.

Boyd, G. D., Fox, A. M., Miller, D. A. B., Chirovsky, L. M. F., D'Asaro, L. A., Kuo, J. M., *et al.* (1990). 33 psec Optical switching of symmetric self-electro-optic effect devices. *Applied Physics Letters*, **57**, 1843–5.

Campbell, J. C., Tsang, W. T., Qua, G. J., and Johnson, B. C. (1988). High-speed InP/InGaAsP/InGaAs avalanche photodiode grown by chemical beam epitaxy. *IEEE Journal of Quantum Electronics*, **QE-24**, 496–500.

Capasso, F., Tsang, W. T., Hutchinson, A. L., and Williams, G. F. (1982). Enhancement of electron impact ionization ratio in a superlattice: a new avalanche photodiode with a large ionization rate ratio. *Applied Physics Letters*, **40**, 38–40.

Casey, H. C. and Panish, M. B. (1978). *Heterostructure lasers.* Academic Press, New York.

Giles, C. R., Wood, T. H., and Burrus, C. A. (1990). Quantum-well SEED optical oscillators. *IEEE Journal of Quantum Electronics*, **QE-26**, 512–17.

Jeon, H., Ding, J., Nurmikko, A. V., Xie, W., Grillo, D. C., Kobayashi, M., *et al.* (1992). Blue and green diode lasers in ZnSe-based quantum wells. *Applied Physics Letters*, **60**, 2045–7.

Kagawa, T., Kawamura, Y., Asai, H., and Naganuma, M. (1990). InGaAs/InAlAs superlattice avalanche photodetector with a separated photoabsorption layer. *Applied Physics Letters*, **57**, 1895–7.

Kapon, E., Hwang, D. M., Walther, M., Bhat, R., and Stoffel, N. G. (1992). Two-dimensional quantum confinement in multiple quantum wire lasers grown by MOCVD on V-grooved substrates. *Surface Science*, **267**, 593–600.

Kawano, K. (1993). High speed Ti:LiNbO$_3$ and semiconductor optical modulators. *IEICE Transactions in Electronics*, **E76C**, 183–90. (IEICE, Tokyo).

Keyworth, B. P., Cada, M., Glinski, J. M., Springthorpe, A. J., Rolland, C., and Hill, K. O. (1990). Multiple quantum well directional coupler as a self-electro-optic-effect device. *Electronics Letters*, **26**, 2011–13.

Loudon, R. (1973). *The quantum theory of light.* Clarendon Press, Oxford.

Matsushima, Y., Akiba, S., Sakai, K., Kushiro, Y., Noda, Y., and Utaka, K. (1982). High-speed-response InGaAs/InP heterostructure avalanche photodiode with InGaAsP buffer layers. *Electronics Letters*, **18**, 945–6.

Miller, D. A. B., Chemla, D. S., Damen, T. C., Gossard, A. C., Weigmann, W., Wood, T. H., and Burrus, C. A. (1984). Novel hybrid optically bistable switch: the quantum well self-electro-optic-effect device. *Applied Physics Letters*, **45**, 13–15.

Miller, D. A. B., Chemla, D. S., Damen, T. C., Gossard, A. C., Wiegmann, W., Wood, T. H., and Burrus, C. A. (1985). Electric field dependence of optical absorption near the band gap of quantum-well structures. *Physical Review B*, **32**, 1043–60.

Miller, R. C., Schwartz, B., Koszi, L. A., and Wagner, W. R. (1978). A high-efficiency GaAlAs double heterojunction photovoltaic detector. *Applied Physics Letters*, **33**, 721–3.

Partin, D. L. (1988). Lead salt quantum well lasers. *IEEE Journal of Quantum Electronics*, **QE-24**, 1716–26.

Scherer, A., Jewell, J. L., Lee, Y. H., Harbison, J. P., and Florez, L. T. (1989). Fabrication of microlasers and microresonator optical switches. *Applied Physics Letters*, **55**, 2724–6.

Sze, S. M. (1981). *Physics of semiconductor devies*, (2nd edn), Wiley, New York.

Sze, S. M. (1985). *Semiconductor devices: physics and technology.* Wiley, New York.

Tamir, T. (ed) (1988). *Guided wave optoelectronics.* Springer-Verlag, Berlin.

Tarof, L. E., Knight, D. G., Fox, K. E., Miner, C. J., Puetz, N., and Kim, H. B. (1990). Planar InP/InGaAs avalanche photodetectors with partial charge sheet in device periphery. *Applied Physics Letters*, **57**, 670–2.

Tarucha, S. and Okamoto, H. (1986). Monolithic integration of a laser diode and an optical waveguide modulator having a GaAs/AlGaAs quantum well double heterostructure. *Applied Physics Letters*, **48**, 1–3.

Thijs, P. J. A. (1992). Progress in quantum well lasers: applications of strain. *Conference Digest of the 13th IEEE International Semiconductor Laser Conference*, Takamatsu, Japan, September 1992, pp. 1–5. IEEE, New York.

Thijs, P. J. A., Tiemeijer, L. F., Kuindersma, P. I., Binsma, J. J. M., and van Dongen, T. (1991). High performance 1.5 μm wavelength InGaAs–InGaAsP strained quantum well lasers and amplifiers, *IEEE Journal of Quantum Electronics*, **27**, 1426–39.

Tsang, W. T. (ed) (1985). Lightwave communication technologies. *Semiconductors and Semimetals*, vol. 22 (Parts A–E), Academic Press, New York.

Wang, Y. H., Hasnain, G., Tai, K., Wynn, J. D., Weir, B. E., Choquette, K. D., and Cho, A. Y. (1991). Molecular beam epitaxy growth of AlGaAs/GaAs vertical cavity surface emitting lasers and the performance of PIN photodetector/vertical cavity surface emitting laser integrated structures. *Japanese Journal of Applied Physics*, **30** (12B), 3883–91.

Weisbuch, C. (1992). Perspectives of low-dimensional semiconductor heterostructures. In *New concepts for low-dimensional electronic devices* (eds. G. Bauer, F. Kucher, and H. Heinrich), pp. 3–14. Springer–Verlag, Berlin.

Weisbuch, C. and Nagle, J. (1990). On the impact of low-dimensionality in quantum-well, wire and dot semiconductor lasers. In *Science and engineering of one- and two-dimensional semiconductors* (eds S. P. Beaumont and C. M. Sotomayor Torres), NATO ASI Series B: Physics, Vol. 217, pp. 309–16. Plenum, New York.

Weisbuch, C. and Vinter, B. (1991). *Quantum semiconductor structures*. Academic Press, New York.

Whitehead, M. A., Rivers, A., Parry, G., Roberts, J. S., and Button, C. (1989). Low-voltage multiple quantum well reflection modulator with an on : off ratio > 100 : 1. *Electronics Letters*, **25**, 984–5.

Yariv, A. (1989). *Quantum electronics* (3rd edn). Wiley, New York.

19 Devices IV: infrared and solar devices

19.1 Introduction

In the last of these chapters devoted to the applications of multilayer semiconductor structures, we concentrate on two further optical devices, namely infrared detectors and solar cells. The former exploit inter-sub-band absorption in quantum wells, introduced briefly in chapter 10, Section 10.2.6, while the latter use multilayers to increase the absorption from the solar spectrum. Both devices have specialized applications, where the cost of high performance is justified. Thermal imaging and night vision, based on infrared radiation in the 8–14 μm atmospheric window not affected by water vapour, is becoming increasingly exploited outside the military arena for which it was developed, for example in see-through-smoke applications associated with firefighting, thermal imaging of people buried in collapsed buildings, surveillance and security, etc. The applications have in the past withstood the expense of less stable and convenient materials such as HgCdTe alloys, exploiting a principal bandgap of $c.10$ μm for an 18 per cent CdTe–82 per cent HgTe alloy (cf. Chapter 1, Fig. 1.8). The prospect of replacing this material with the rather more mature GaAs–AlGaAs system is very attractive, and the pace of development has been particularly rapid in recent years (Levine 1993). In contrast, solar cells for domestic and commercial applications must be cheap before being highly efficient. In space and in other applications where access is difficult (e.g. weather or other monitoring from remote locations, and in some military arenas) one requires a compact, light, and highly efficient method for capturing solar energy and converting it into electricity (Bube 1993).

19.2 Long-wavelength infrared detectors

Broadband detectors of thermal radiation in the 3–5 μm range working at 200 K and in the 8–14 μm range working at 80 K, whose performance approached the theoretical limits set by the optical aperture, already existed as discrete devices 10 years ago. At the time, it was noted that further improvements in systems performance would come through larger numbers of elements in 2D arrays with integrated signal processing. We shall be concerned with devices that detect the infrared photons, but there are others which detect the heat (in, say, a temperature-sensitive resistor) or detect the radiation field using the concepts of Chapter 17 suitably modified to handle 10 μm wavelengths. These others are not of interest to us here.

19.2.1 Principles of detector action (using HgCdTe)

The three relevant optical absorption processes for infrared detection by photoconduction are shown in Fig. 19.1(a)—intrinsic absorption across the bandgap, extrinsic absorption from donor levels in the gap, and free-carrier absorption (Elliott and Gordon 1993). We shall concentrate here on the first of these. The detectors work via the excitation of carriers and their collection at the terminals as a photocurrent in the geometry shown in Fig. 19.1(b). Extrinsic absorption devices tend to have a low optical absorption coefficient. They also need to be cooled below 77 K as there is a very high probability that ionized donors recapture the free electrons, leading to very short lifetimes. Free-carrier absorption is confined to 4 K detection of submillimetre waves in InSb.

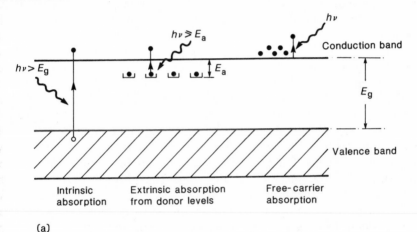

(a)

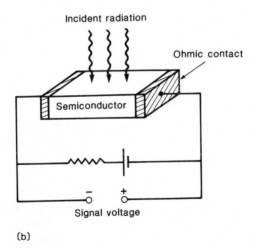

(b)

Fig. 19.1 (a) Optical absorption processes in a semiconductor for infrared detection; (b) the geometry of the photoconductive device. (After Elliott and Gordon 1993.)

Of the intrinsic photoconductors, the alloy $Hg_{1-x}Cd_xTe$ is most widely used in the 8–14 μm band. The energy gap is $c.0.1$ eV and photoexcited carriers have a lifetime of several microseconds at 77 K, while alloys appropriate for the 3–5 μm band have lifetimes in excess of 20 μs at 200 K. The limits to performance are set primarily by noise, which may be in the incoming signal itself, in the detector, or in the subsequent electronic amplifiers. Except in special applications (as in radio astronomy) where detectors are cooled to very low temperatures, amplifier noise is no problem, and in any event low-noise HEMT devices are now being used in that context. The noise in the radiation comes from the ambient background when warm objects are being detected. In the limit where $\hbar\omega >> kT_b$, as is the case at 300 K for wavelengths shorter than about 30 μm, there is a simplified expression for the noise equivalent power (the r.m.s. value of the radiant power incident on the detector which gives an r.m.s. signal equal to the r.m.s. noise from the detector):

$$P_N = E_\lambda \sqrt{(2A\phi_B B/\eta)}$$

where E_λ is the photon energy of the signal radiation, A is the detector area, η is the quantum efficiency, which is assumed constant up to the cut-off wavelength (in practice some bandgap beyond which the detector does not absorb), B is the electrical bandwidth of the receiver, and ϕ_B is the total background photon flux to which the detector can respond. This last is obtained by assuming that the background radiation is black-body like, and the Planck distribution function of the radiation is integrated up to the cut-off wavelength. One also considers the field of view associated with a particular camera optics. The background-limited detectivity is the inverse of P_N for a unit area and bandwidth:

$$D_{BL}^* = \frac{\sqrt{(\eta/\phi_B)}}{E_\lambda}$$

The other important figure of merit is the responsivity, which is the r.m.s. voltage or r.m.s. signal current per unit r.m.s. radiant power incident on the detector. Over the past 20 years, HgCdTe alloys have dominated the invest-ment in materials for infrared photodetectors. The relevant bandgap is shown for 77 K and 300 K in Fig. 19.2(a). A photon incident on n-type material generates an electron–hole pair, and under bias the drift of the minority holes towards the positive electrode is controlled by the ambipolar mobility ($v_d = \mu_a E$):

$$\mu_a = (n-p)\mu_h\mu_p/(n\mu_n + p\mu_p)$$

At low fields the average drift length $\mu_a E\tau$, where τ is the recombination lifetime, is much less than the detector length l and recombination occurs in the bulk of the material. The concentrations of electrons and holes are those of thermal equilibrium, supplemented by those generated by the background radiation:

$$n_b = p_b = \eta \phi_B \tau / t$$

where t is the thickness of the device and η is the quantum efficiency of the absorption process. The signal radiation generates $\eta \phi_s \tau$ excess carriers and a signal current of

$$i_s = \eta \phi_s e [\mu_p E \tau (b + 1)/l]$$

where $b = \mu_e/\mu_p$. The quantity in square brackets is the net charge flow in the external circuit due to a single absorbed photon, and is known as the signal gain G. When b is large, so too is G. The small-signal voltage responsivity is just this current per unit signal power multiplied by the sample resistance, giving

$$R_V = \frac{\eta V (1 + b) \tau}{(E_\lambda l w t)(nb + p)}$$

$$= \frac{\eta V \tau}{(E_\lambda l w t) n} \qquad b \gg 1, \, bn \gg p$$

where V is the bias voltage and w is the width of the detector. At low bias ($\mu_a E \tau \ll l$) the responsivity increases linearly with bias and minority carrier lifetime. At higher biases, the opposite condition ($\mu_a E \tau \gg l$) holds, and a further analysis of non-uniform excess carrier distributions must be made. In the limit where the time taken for a minority carrier to drift the length of the device is much smaller than the recombination time, the maximum voltage responsivity takes the form

$$R_V = \frac{\eta l}{(2 E_\lambda w t)(n_e \mu_p)}$$

where n_e is the actual electron density. For typical values in HgCdTe, namely $\eta = 0.6$, $l = w = 50$ μm, $t = 8$ μm, $n_e = 5 \times 10^{14}$ cm^{-3}, $E_\lambda = 0.1$ eV, and $\mu_p = 450$ cm^2 V^{-1} s^{-1}, we obtain $R_V \sim 10^5$ V W^{-1} or $R_I \sim 3$ A W^{-1}. Experiments approach to within 50 per cent of these performance figures, and a detectivity D^* of 10^{11} cm Hz$^{1/2}$W^{-1} at 77 K operation have been achieved (Borello *et al.* 1977). A summary of typical results calculated for a 50 μm square detector at 80 K as a function of bias is given in Fig. 19.2(b), where the peak wavelength of response is 12 μm, and D^* is enhanced by a factor of about 4 over D^*_{BL} corresponding to a device with an angle of view of only 30°.

These results are over a decade old, and only modest progress has been made on intrinsic materials properties and the quality of epitaxial films since then. We compare HgCdTe and quantum-well infrared detectors in Section 19.2.4.

19.2.2 Detection using inter-sub-band absorption

The principle of infrared detection using inter-sub-band absorption is very simple. Quantum wells in an n–i–n structure (Fig. 19.3(a)) are arranged to have ground and first-excited electron states with a separation of $c.0.12$ eV,

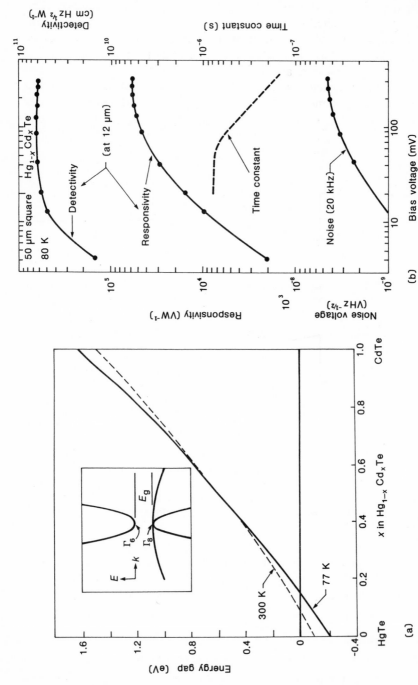

Fig. 19.2 (a) Band structure of HgCdTe near the principal gap (after Hansen *et al.* 1982); (b) a summary of device parameters for a typical HgCdTe detector (after Capocci 1978).

which is equivalent to the energy of infrared radiation of 10 μm wavelength. The quantum wells are doped, so that the ground state is occupied. The position of the first excited state must be such that electrons excited into that state can be swept out of the system by a modest applied field. This leaves three possibilities, as illustrated schematically in Fig. 19.3(b): (i) this state is near the top of the barrier, and electrons tunnel out from the top of the quantum well; (ii) the excited state is in the continuum above the barrier and so electrons are free to move under the electric field; (iii) the excited state is

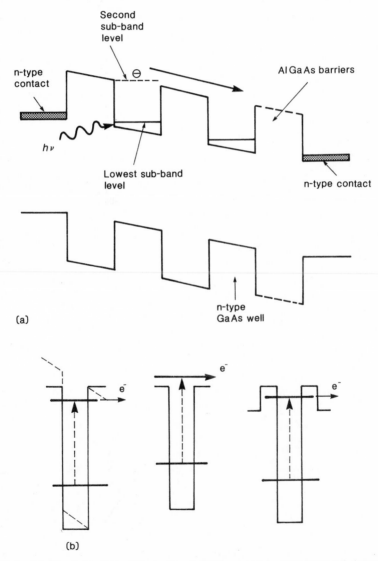

Fig. 19.3 (a) The overall structure of a multiple-quantum-well infrared detector; (b) three quantum-well configurations used for inter-sub-band absorption.

Table 19.1 Quantum-well infrared detector structures

Sample	L_w (Å)	L_b (Å)	x	N_D (10^{18} cm^{-3})	Doping type	Periods	Inter-sub-band transition
A	40	500	0.26	1.0	n	50	B–C
B	40	500	0.25	1.6	n	50	B–C
C	60	500	0.15	0.5	n	50	B–C
D	70	500	0.10	0.3	n	50	B–C
E	50	500	0.26	0.42	n	25	B–B
F	50	50	0.30	0.42	n	25	B–QC
		500	0.26				

L_w, quantum-well width; L_b, barrier width; x composition in Al$_x$Ga$_{1-x}$As; N_D doping density; B–C, bound-to-continuum transition; B–B, bound-to-bound transition; B–QC, bound-to-quasi-continuum transition.

Source: Levine 1993.

in the quasi-continuum, or is like an excited state in a double-barrier diode, with a finite lifetime before emerging. Each of these possibilities has been examined for the various physical properties that go to make a good infrared detector, and in Table 19.1 we give the relevant parameters of six structures which have been compared. In Chapter 10, Section 10.2.6, we described the fact that light normally incident on the top surface of a wafer has the wrong polarization vector to be absorbed, and a 45° bevel is used with back-illumination (cf. Fig. 10.11(b)). Note that this is a vertical as opposed to a transverse device as far as the photocurrent is concerned. In Fig. 19.4 we show the normalized responsivity of the structures and the absolute bias-dependent peak responsivity of all six structures. This latter quantity, after correction for the oblique incidence, is

$$R_p = (e/h\nu)\eta G$$

where η is the total quantum efficiency for unpolarized incident light of frequency ν and G is the optical gain of the structure, which is the same as in the previous section. Note in Fig. 19.4(a) that the responsivity is intrinsically narrow band for structures working with bound-to-bound and bound-to-quasi-continuum state transitions ($\Delta\lambda/\lambda \sim 10$ per cent), whereas structures that excite into the continuum are broader band ($\Delta\lambda/\lambda \sim 25$ per cent). For most applications the broader-band detection is desirable. The responsivity is linear with bias for structures exciting into the continuum or quasi-continuum. If the excited state is bound, a finite bias must be applied to induce an appreciable tunnelling probability for the excited electrons to tunnel out with an appreciable probability, and so the responsivity is non-linear. Note also that the responsivity is in the range 0.3–0.5 A W^{-1} for most structures, rising above 0.7 A W^{-1} for the structure with a particularly low barrier ($x \sim 0.10$), somewhat lower than for HgCdTe detectors.

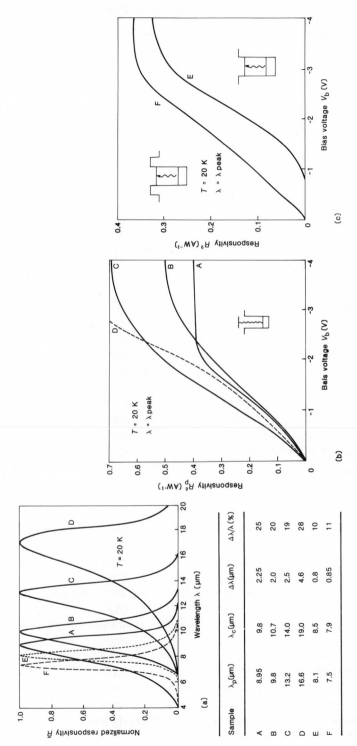

Fig. 19.4 (a) Normalized responsivity of six quantum well structures; (b) (c) the bias-dependent absolute responsivity of these structures. (After Levine 1993.) Table : λ_p = peak, λ_c = cutoff, $\Delta\lambda$ = FWHM wavelengths.

Sample	$\lambda_p(\mu m)$	$\lambda_c(\mu m)$	$\Delta\lambda(\mu m)$	$\Delta\lambda/\lambda$ (%)
A	8.95	9.8	2.25	25
B	9.8	10.7	2.0	20
C	13.2	14.0	2.5	19
D	16.6	19.0	4.6	28
E	8.1	8.5	0.8	10
F	7.5	7.9	0.85	11

Optical gain is determined by the relation between the noise current in a unit frequency range and the dark current, and so can be measured experimentally. The values of G obtained using

$$i_n = \sqrt{(4eI_d G \Delta f)}$$

are shown in Figs 19.5(a) and 19.5(b). Another measure of the gain is the ratio of the hot carrier recapture mean free path (i.e. the distance travelled before carriers relax into a second quantum well nearer the anode) to the overall superlattice length (2–3 μm), and this allows the former quantity to be extracted and displayed. Note that the unipolar nature of these structures eliminates effects associated with recombination (for which here recapture is the nearest equivalent phenomenon) and its associated noise. Furthermore, the velocity and mean free paths of the electrons are considerably greater than can be achieved with HgCdTe detectors where hole transport is the limiting factor. The saturation of the optical gain at +2 V because of this recapture is what limits the responsivity of quantum well detectors, the only exception being for the structure with very low (10 per cent Al) barriers. The data collected so far allow the quantum efficiency to be extracted from the formula above for the responsivity, and this is shown in Figs 19.5(c) and 19.5(d). The qualitative similarity of the results suggests that the excitation, escape, and recapture processes for all the structures are comparable.

All these data are a prelude to determining the detectivity of quantum-well infrared detectors, the figure of merit used to compare different types of detector (even if not always the most critical device parameter):

$$D_\lambda^* = R_p \sqrt{A}/i_n.$$

The data for these structures, typical of quantum-well detectors, are plotted as a function of their cut-off energy (the energy of the long-wavelength cut-off in Fig. 19.4) in Fig. 19.6, which is fitted by the expression

$$D^* = (1.1 \times 10^6)\exp(E_c/2kT)\text{cmHz}^{1/2}\text{W}^{-1}.$$

Note that for $\lambda_c = 10$ μm at 77 K, we have $D_\lambda^* = 1.2 \times 10^{10}$ cm Hz$^{1/2}$W^{-1} which is approximately a factor of 10 down on the value for HgCdTe. The data have all been obtained with the 45° bevelled back-face, but surface gratings have already improved on this geometry and crossed gratings to capture both polarizations of incident light should be able to improve the prefactor by nearly an order of magnitude, making the two directly comparable.

19.2.3 *Quantum wells versus HgCdTe for infrared detectors*

Given 30 years of development of HgCdTe to 5 years for quantum wells, the former is more advanced as a technology. The fact that the preparation of large-area arrays of narrow-gap II–VI semiconductors is not without problems has spurred the effort on quantum wells. Despite some important limitations of quantum wells, it seems likely that they will come to play a

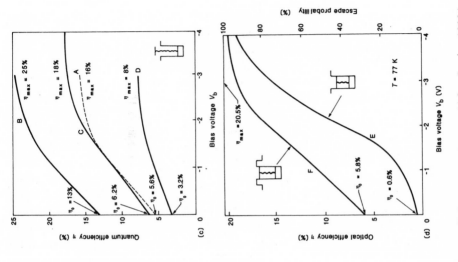

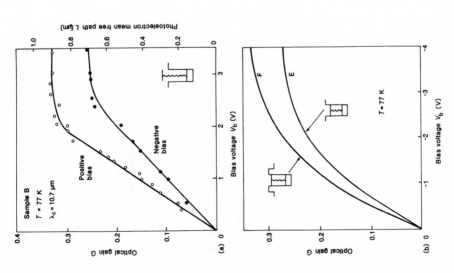

Fig. 19.5 (a) (b) Optical gain and (c) (d) quantum efficiency of the quantum-well detector structures. (After Levine 1993.)

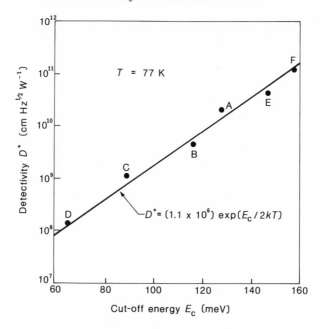

Fig. 19.6 Detectivity D^* (at 77 K) versus cut-off energy E_c for GaAs quantum-well infrared detectors. (After Levine 1993.)

competing role in systems over the next few years. In this section we describe the advantages and disadvantages of the two systems.

The HgCdTe alloys have been the subject of much materials research, and are grown on CdTe and GaAs substrates by all the main epitaxial techniques. The detectivities are very high and background-limited performance is achieved at temperatures above 77 K, in part because the intrinsically long lifetime of minority carriers associated with band-to-band transitions. The response speed, power consumption, and operating temperature are important systems attributes in addition to detectivity. Detectivities of quantum wells at 77 K can be achieved in HgCdTe at above 100 K, somewhat easing the refrigeration requirements which are critical in some space applications. The shortcomings are also considerable. The materials are mechanically very soft and are very sensitive to elevated temperatures such as those associated with epitaxial growth and device processing. Various trapping centres associated with native defects, the weakness of the Hg–Te bond and the clustering effects of particular ions in the alloy lead to spatial inhomogeneities which can become intolerable. There has not been much progress in the last decade on these problems. In staring-array applications, i.e. where camera images are required, grey scales are important, and this places very tight tolerances on the uniformity of the layers over large areas, which are more easily met by III–V quantum wells than by II–VI epitaxy. In practice, discrete devices and linear arrays in HgCdTe have produced very high performance figures, but 2D arrays have been less successful.

The dark current in quantum wells is always several orders of magnitude higher than that in HgCdTe for the same cut-off frequency and operating temperature, and this degrades the detectivity D^*. This is because the quantum-well devices are majority-carrier devices, and the dark current is associated with thermally assisted tunnelling of carriers already present in the conduction band. Some strategems can be adopted with the band structure at low temperatures, but most of these are ineffective above 77 K. The uniformity over large areas and the compatibility of quantum wells with integrated III–V signal processing technology (cf. Chapter 16) may more than compensate the detectivity problem in the development of infrared video cameras for example. The overall performance of 128×128 focal plane arrays in infrared imaging cameras using quantum wells already exceeds that of the HgCdTe equivalent (Kozlowski *et al.* 1991). Here it is the uniformity of D^* rather than D^* itself which is more important.

In contrast with results in earlier chapters where the new generation devices have shown clear superiority over earlier devices, here we have two systems that have their own materials and physics limitations. In common with more general experience (do not do it in GaAs if you can do it in Si), quantum wells in GaAs are likely to supersede HgCdTe in due course. This is particularly true for systems performance if the limitations associated with the GaAs quantum-well structure can be overcome with sophisticated signal processing in large arrays. In addition, the lessons learned from GaAs–AlGaAs may yet be applied to other heterojunction combinations such as Sn–Ge or InAs– InAsSb to achieve even better results than can be achieved with GaAs.

19.3 Solar cells

Solar cells have long made use of heterojunctions to improve the efficiency with which solar energy is converted into electricity (Bube 1993). We shall describe the basic principles and recent achievements using multilayer semiconductor structures. As far as the atmosphere attenuates the solar energy, we refer to air-mass 1 (AM1) as the ground-level spectrum when the sun is directly overhead. In Fig. 19.7 we show the solar spectrum under conditions of air-mass 0, i.e. in space at a distance from the Sun comparable with the Earth, and air-mass 1.5 when the light is incident at $c.45°$ to the normal, a typical average value on the Earth's surface. The former is close to the blackbody spectrum corresponding to a surface temperature of 5800 K, and the latter shows strong absorption bands due to water vapour, dust, etc. The total average incident power at ground level is $c.\ 0.8\ \text{kW m}^{-2}$, about half that at the top of the atmosphere.

There are various types of solar cell, such as those in the desert collecting solar energy from very large large areas or panels on buildings. In these

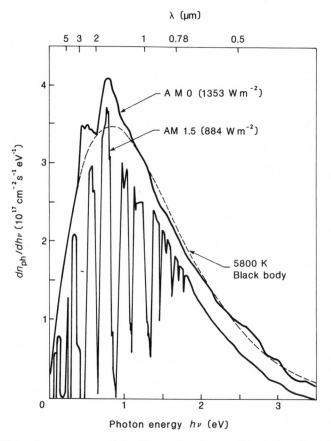

Fig. 19.7 The solar spectrum of the Sun under the conditions described in the text. (After Henry 1980.)

terrestrial applications extraction of extra energy by the last few per cent conversion efficiency is not commercially viable. In contrast, in space applications a combination of high efficiency and light weight is at a premium.

19.3.1 Principles of solar cell action

The basic semiconductor solar cell action relies on the separation of photoexcited electrons and holes within a p–n junction, thus involving only that light which has an energy in excess of the principal bandgap. The equivalent circuit is given in Fig. 19.8(a), which shows a diode with a parallel internal resistance and a series resistance including the effects of the contacts. The light current source is at opposite polarity to the internal diode. In the dark, the current–voltage characteristics are those of a diode, but under illumination we regard the optical absorption as generating carriers at a rate

$$\mathrm{d}n(x)/\mathrm{d}t\big|_{\mathrm{gen}} = \alpha L_0 \exp(-\alpha x)$$

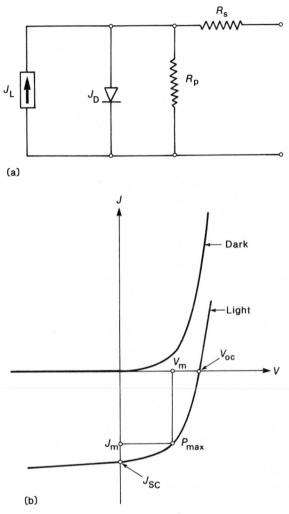

(a)

(b)

Fig. 19.8 (a) The equivalent circuit of a solar cell as a current source; (b) the relationship between current density and applied bias for the solar cell diode in the dark and the light.

where $\alpha(\lambda)$ is the absorption coefficient and $L_0(\lambda)$ is the incident phonon flux which are both wavelength dependent. Some of the carriers recombine on a length scale set by a diffusion length L_D and a minority carrier lifetime τ, such that $L_D^2 = D\tau$ where D is the Einstein diffusivity given by $D = \mu kT/e$. We integrate the generated current over all wavelengths of incident light and define a light current J_L. We introduce a collection function $H(V)$ to distinguish between the measured short-circuit current J_{sc} and J_L. The relation

$$J_{sc} = H(0)J_L$$

defines the quantum efficiency of the cell, incorporating the effects of recombination during diffusion and at the interfaces. The important measured properties of a solar cell are shown in Fig. 19.8(b). An illuminated solar cell will generate an open-circuit voltage and a short-circuit current. The voltage and current density for maximum power P_{max} are denoted by V_m and J_m respectively. An important figure of merit is the fill factor f, defined as

$$f = P_{max}/J_{sc}V_{oc} = J_m V_m/J_{sc}V_{oc},$$

which gives a quantitative measure of how square the J–V characteristic is in the fourth quadrant. Typical values of interest are $V_{oc} = 0.5$–0.8 V, $J_{sc} = 10$–40 mA cm^{-2}, $f = 0.6$–0.8, and $\eta = 6$–20 per cent (Bube 1993).

Solar cells incorporating internal p–n junctions have been made in single-crystal, amorphous, and polycrystalline Si, single-crystal GaAs, Cu$_x$S/CdS, InP, and CdTe. InP and CdTe have bandgaps that allow absorption up to about 1 μm, and applications make trade-offs between efficiency, cost, and lifetime. The ultimate systems performance includes many further considerations, for example the way that the solar energy can be concentrated (focused) onto the semiconductor to increase the light flux (sometimes by as much as 10^3). These further aspects will not concern us here.

19.3.2 Multilayer semiconductor solar cells

One early use of heterojunctions is exemplified in Fig. 19.9, where the p-layers of a GaAs solar cell include Al$_x$Ga$_{1-x}$As. The wider-gap alloy, even if quite heavily doped to lower series resistance, will act as a window and let light of energy less than its bandgap enter to be absorbed at the GaAs layer. Light of higher energy that is absorbed within a diffusion length of the p–n junction will also contribute to the photocurrent. The spectral response is defined as the total photocurrent divided by the product of the electronic charge and the incident flux at a particular wavelength (i.e. the number of carriers out per photon in). Fig. 19.9 shows how the normalized spectral response spans an ever-increasing range of photon energy with increasing alloy bandgap. In addition, higher radiation can be tolerated if the semiconductor is thick as well as of wide bandgap. The heterojunction as a design element has been taken much further, both in monolithic devices with many absorbing layers, and with hybrid structures involving mechanical stacking. A recent example, consisting of a two-terminal AlGaAs–GaAs heterojunction cell, each semiconductor with its own p–n junction, mechanically mounted in series with an InGaAsP p–n junction cell, is shown in Fig. 19.10(a). The relevant band-gaps are 1.93 eV (AlGaAs), 1.52 eV (GaAs), and 0.95 eV (InGaAsP). The external quantum efficiency (carrier out per photon in) of the component layers and the overall current–voltage characteristics of the two-terminal three-junction cell under one-sun AM 0 illumination are shown in Figs 19.10(b) and 19.10(c) respectively. Because of its structure, the cell can be operated with most

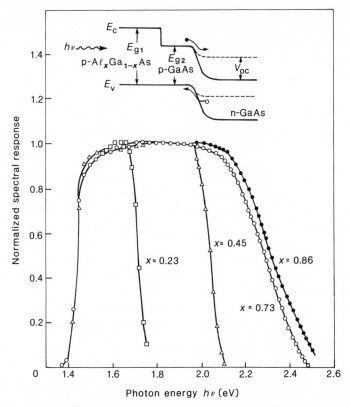

Fig. 19.9 Normalized spectral response of heterojunction AlGaAs–GaAs solar cells. (After Hovel and Woodall 1973).

combinations of p–n junctions either operating or short-circuited. Table 19.2 shows the efficiencies—and fill factors FF obtained under the varying conditions. The overall power conversion efficiency of 25.2 per cent is one of the highest reported in this type of cell.

Progress with a fully monolithic structure is hampered by the difficulty of growing relatively thick and very-high-quality epitaxial overlayers of the

Table 19.2 Solar-cell parameters of an AlGaAs–GaAs–InGaAsP three-junction solar cell at 25°C under one-sun illumination,

J_{sc} (mA cm^{-2})	V_{oc} (V)	FF	η (%)	AlGaAs	GaAs	InGaAsP
15.63	2.88	0.77	25.20	M	M	M
15.63	2.36	0.78	21.20	M	M	—
15.81	1.40	0.75	12.00	M	—	—
15.63	0.99	0.80	9.09	—	M	—
15.97	0.52	0.72	4.39	—	—	M

M, measured;—short circuit.

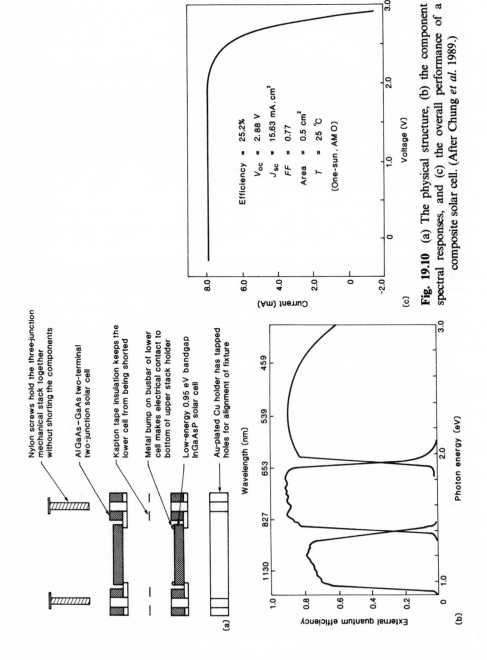

Fig. 19.10 (a) The physical structure, (b) the component spectral responses, and (c) the overall performance of a composite solar cell. (After Chung et al. 1989.)

Nylon screws hold the three-junction mechanical stack together without shorting the components

AlGaAs–GaAs two-terminal two-junction solar cell

Kapton tape insulation keeps the lower cell from being shorted

Metal bump on busbar of lower cell makes electrical contact to bottom of upper stack holder

Low-energy 0.95 eV bandgap InGaAsP solar cell

Au-plated Cu holder has tapped holes for alignment of fixture

(a)

Wavelength (nm)

External quantum efficiency

Photon energy (eV)

(b)

Efficiency = 25.2%

V_{oc} = 2.88 V

J_{sc} = 15.63 mA.cm^2

FF = 0.77

Area = 0.5 cm^2

T = 25 °C

(One-sun, AM O)

Current (mA)

Voltage (V)

(c)

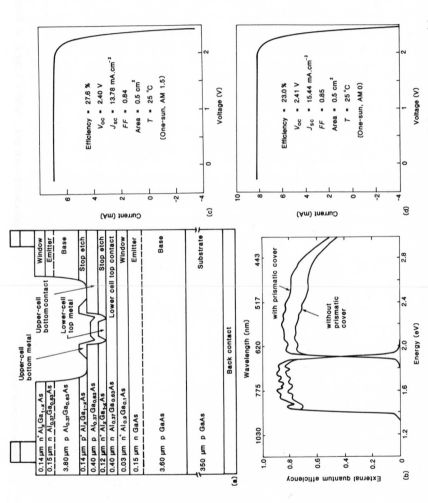

Fig. 19.11 (a) A monolithic cascaded AlGaAs–GaAs solar cell, (b) the spectral response of the components with and without a prismatic cover, and the current–voltage characteristics under (c) AM 1.5 and (d) AM 0 illumination. (After Chung *et al.* 1992.)

various materials. The quality of GaAs grown strained on InGaAs is insufficient. In a cascade structure the energy band profiles need carefully tailoring, so that in a two-terminal device the internal electric fields are optimally distributed. The state of the art in the GaAs–AlGaAs system is the two-junction structure shown in Fig. 19.11(a), grown by MOCVD. The optimum connection between the two cells is achieved by short-circuiting the additional reverse-biased p–n^+ junction with a metal interconnect grid set out across the structure, part of which is shown in the figure. The part of the solar cell area that is lost in this process is compensated by placing an array of glass prisms over the grid to deflect incident light only onto the remaining active regions. In Fig. 19.11(b) we show the spectral response of both parts of the solar cell with and without this prismatic cover, and in Figs 19.11(c) and 19.11(d) we show the two-terminal current–voltage characteristics of the monolithic integrated cascade solar cells under AM 0 and AM 1.5 illumination. The interesting point is the decrease in power-conversion efficiency from 27.6 to 23.0 per cent as the intensity increases from AM 1.5 to AM 0. This is attributed to a photocurrent mismatch: under AM 0 conditions (cf. Fig. 19.7), more blue light is absorbed in the upper cell, generating a higher photocurrent than is generated in the lower cell and distorting the overall band profile. This mismatch can be improved by further widening the AlGaAs bandgap to allow more light to enter the GaAs layer for absorption there. Table 19.3 shows how the two cells are operating either separately or in tandem.

Table 19.3 Monolithic cascaded AlGaAs–GaAs solar cell.

Condition	Component cell	J_{sc}	V_{oc}	FF	η
AM 1.5 (three-terminal)	Upper cell	13.92	1.41	0.82	16.00
	Lower cell	13.78	1.00	0.82	11.30
AM 0 (three-terminal)	Upper cell	17.34	1.41	0.82	14.60
	Lower cell	15.44	1.00	0.82	9.33
AM 1.5 (two-terminal)		13.78	2.40	0.84	27.60
AM 0 (two-terminal)		15.44	2.41	0.85	23.00

The full range of band-structure engineering has not yet been applied to solar cell design. Fine tuning the potential profile while photocurrent is flowing has been suggested, but none of the advantages of more interfaces have yet been realized in terms of increased solar energy conversion efficiency.

The competition from other materials is intense. Recent work on Si devices claims to have achieved 35 per cent efficiency, with higher current densities but lower voltages (Li *et al.* 1992). In general, III–V materials are more resistant to radiation damage than Si, which accounts for their use in space applications.

19.4 Concluding remarks on devices

In this and the three preceding chapters we have described the successful application of multilayer semiconductor structures to the improvement of device performance. Only in avalanching microwave sources and solar cells has there been no notable demonstrated advantage of using many different finely tuned layers within the device. This is mainly because both devices are tied to bulk properties and are relatively intolerant of any extra defects introduced as a part of the multilayer growth techniques.

It is important to remember that, while we have focused on particular figures of merit such as efficiency or speed, in many cases the take-up of the new generation of devices by systems designers is because of a host of other attractive properties. One feature prominent throughout chapters 16–18 is the reduced sensitivity to operating temperature of the transistors, optical devices, and microwave devices that incorporate multiple heterojunctions. Another is the reduction in noise. Even if the improvements in the primary efficiency are modest, a composite figure of merit associated with all these other attributes is very much improved. The better performances achieved by the new devices can be traded off against some other limiting feature within an overall systems design. Thus improved signal-to-noise properties of a microwave detector diode can be used to increase the radar range or to reduce the power consumption of a system operating over the existing range.

Just as the very high mobilities associated with the 2DEG do not survive to room temperature and to high electric fields, the other figures of merit quoted have to be regarded critically in the context of the overall system application. Single-crystal solar cells are uneconomical for domestic terrestrial use, but come into their own in demanding environments.

The first generation of devices that make serious use of semiconductor multilayers have been described in this and the preceding chapters. There is room for many further refinements. As control over the composition, doping, and crystalline quality increases for an increasing range of materials, there is scope for repeated iterations of device improvement.

The recent state of the art across all semiconductor devices is described in Volume 4 (*Device physics*) of the second edition of the *Handbook on semiconductors* (Hilsum 1993), and the role of heterojunctions and multilayers in most device families is emphasized.

References

Borello, S., Kinch, M., and LaMont, D. I. (1977). Photoconductive HgCdTe detector performance with background variations. *Infrared Physics*, **17**, 121–5.

Bube, R. H. (1993). Solar cells. In *Handbook on semiconductors*, Vol. 4, *Device physics* (eds C. Hilsum), pp. 797–839. Elsevier, Amsterdam.

Capocci, F. A. (1978). Private communication to C. T. Elliott cited by Elliott and Garden 1993.

Chung, B.-C., Virshup, G. F., Hikido, S., and Kaminar, N. R. (1989). 27.6% efficiency (1 sun, air-mass 1.5) monolithic $Al_{0.37}Ga_{0.63}As$/GaAs two-junction cascade solar cell with prismatic cover glass. *Applied Physics Letters*, **55**, 1741–3.

Chung, B.-C., Virshup, G. F., Klausmeier-Brown M., Ladle Ristow, M., and Wanlass, M. W. (1992). 25.2% efficiency (1 sun, air-mass 0) AlGaAs/GaAs/InGaAsP three-junction, two-terminal solar cell. *Applied Physics Letters*, **60**, 1691–3.

Elliott, C. T. and Gordon, N. T. (1993). Infrared detectors. In *Handbook on Semiconductors*, Vol. 4, *Device physics* (ed. C Hilsum), pp. 841–936. Elsevier, Amsterdam.

Hansen, G. L., Schmit, J. L., and Casselman, T. N. (1982). Energy gap versus alloy composition and temperature in $Hg_{1-x}Cd_xTe$. *Journal of Applied Physics*, **53**, 7099–101.

Henry, C. H. (1980). Limiting efficiencies of ideal single and multiple energy gap terrestrial solar cells. *Journal of Applied Physics*, **51**, 4494–500.

Hilsum, C. (ed). (1993). *Handbook on semiconductors*, vol. 4, 2nd edn, *Device Physics*. Elsevier, Amsterdam.

Hovell, H. J. and Woodall, J. M. (1973). $Ga_{1-x}Al_xAs$–GaAs p–p–n heterojunction solar cells. *Journal of the Electrochemical Society*, **120**, 1246.

Kozlowski, L. J., Williams, G. M., Sullivan, G. J., Farley, C. W., Anderson, R. F., Chen, J., *et al.* (1991). LWIR 128 × 128 GaAs/AlGaAs multiple quantum well focal plane array. *IEEE Transactions on Electron Devices*, **38**, 1124–30.

Levine, B. F. (1993). Device physics of quantum well infrared photodetectors. *Semiconductor Science and Technology*, **8**, S400–5.

Li, J., Chong, M., Zhu, J., Li, Y., Xu, J., Wang, P., *et al.* (1992). 35% efficient nonconcentrating novel silicon solar cell. *Applied Physics Letters*, **60**, 2240–2.

20 Amorphous semiconductor multilayers

20.1 Introduction

So far, all the materials physics and devices have used single-crystal multilayer semiconductor structures. Great effort has been made in the growth to achieve precise single-crystal properties and sharp interfaces between adjacent layers. During the fabrication of devices, the process steps attempt to preserve the crystalline integrity. In the past 20 years, hydrogenated amorphous Si has developed rapidly from a research material (first doped to make p–n junction devices in 1976) to the host material for the electronics of displays, with a $4 billion turnover in 1991–1992 (Le Comber 1992). It is now possible to obtain photocells for powering calculators, watches, etc., photoconductors for a wide range of sensing applications, thin-film transistors for driving liquid-crystal displays, and high-voltage thin-film transistors for driving printers. The films of amorphous Si suitable for devices are prepared by plasma-enhanced chemical vapour deposition from SiH_4. In the early days, the role of hydrogen was underestimated in this device-grade material, but it is now known to constitute anywhere in the 2–16 at. % range (Street 1991) (this reference is a comprehensive introduction to hydrogenated amorphous Si, including an introduction to multilayers). The role of hydrogen is to saturate all the unsatisfied dangling bonds in a continuous random network. Studies of ideal random networks (shown schematically in Fig. 20.1) indicate that a dangling bond is electronically active with energy levels in and near the gap. The formation of Si–H bonds saturates these bonds and moves the relevant states far from the gap. Only states associated with unsatisfied Si–H bonds and with traditional donors and acceptors remain in the gap, and the behaviour of the latter is similar to that in crystalline semiconductors. The nearest-neighbour distance between Si atoms is the same as in the crystal, and the density of amorphous Si is only a few per cent less than that of single-crystal Si. There is some evidence that the Si–H bonds are metastable, and that movement of hydrogen takes place between different bonds (Street 1991). The electronic structure of an amorphous semiconductor differs from its crystalline counterpart; despite the absence of long-range order, and a band structure in the conventional sense, there is still a principal bandgap. The band-edge densities of states do not have the sharp $\sqrt{E}$ energy dependence, but rather an exponential tail brought about by electron states localized in regions of lower potential energy within the amorphous network; these states are spread over a range with an energy scale $E_0 \sim 0.1$ eV associated with it (Fig. 20.1(b)). If the Fermi level in an amorphous semiconductor is in this energy region, the

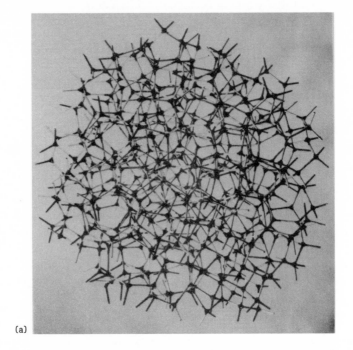

(a)

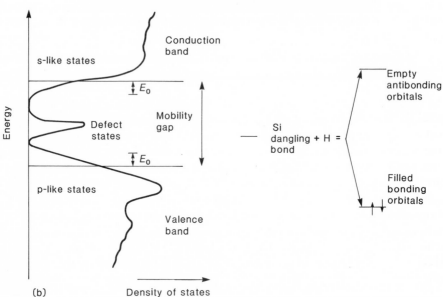

(b)   Density of states

Fig. 20.1 (a) A continuous random network (D. S. Boudreaux, Allied Chemical Corporation, private communication); (b) a schematic diagram of the electronic structure including the effect of the formation of a Si-H bond. The meaning of E_0 is explained in the text.

low-temperature mobility and hence conductivity vanish, but remain finite if the Fermi energy is well into the band. There is a sharp transition between the two regimes at an energy E_c, and one has a mobility gap as shown. The weakening of some Si–Si bonds results in the energy spread of localized states being wider in the valence band than the conduction band.

In this chapter, we describe the extension of the growth techniques to the preparation of multilayers, with components that now extend to include amorphous Si:H (written as aSi:H), aGe:H aSiN$_x$, aSiC$_x$, aSiO$_x$ and materials where fluorine takes the place of hydrogen. This chapter treats the materials preparation and their assessment, the physics of quantum confinement, tunnelling and hot electron injection, and some device applications that are being developed. A principal reason for this type of research is that the combinations of materials open up new regimes of bandgap offsets which are not easily available in crystalline materials. Also, new aspects of the quantum physics of solids have been elucidated in Chapters 5–13, and it is hoped that comparable multilayer structures might probe new regimes of the amorphous state.

20.2 Amorphous semiconductor multilayers

The state of the art of materials, physics, and devices using multilayer amorphous materials is less well developed than that of their crystalline counterparts, but advances are anticipated over the coming decade. Until the mid-1980s, it was thought impossible to prepare sharp interfaces with a sufficiently low defect density, but this has now been done with the aid of hydrogenation.

20.2.1 Preparation

The elevated temperatures of both MBE and MOCVD (see Chapter 2) allow the atoms impinging on the surface to move to take up a crystallographic position. If the substrate is kept at room temperature, this crystalline order is lost and amorphous films are prepared instead. The chemical vapour deposition technique is favoured here, as it allows deposition over large areas. The chemical reaction to break up silane can be induced by a number of methods, but the favoured technique is to use a radiofrequency coil to heat the gas or generate a plasma in the gas, both of which assist the break-up of the compound (Fig. 20.2). Intense illumination, which is an alternative to a radiofrequency plasma, also helps the cracking process. The growth rates achieved are those of MOCVD, typically nanometres per second. As with MOCVD, the composition of the incident gas can be changed to incorporate other materials and appropriate species for doping. The gas flow rates in the reactor are such that sharp changes in composition can be achieved within the time that one layer is grown. A highly schematic diagram of the bonding structure of a thin overlayer of aSi:H on aSiO$_2$ is shown in Fig. 20.3(a).

There are only limited possibilities for monitoring growth other than from the gas flow and the results of calibration runs. One technique is to measure

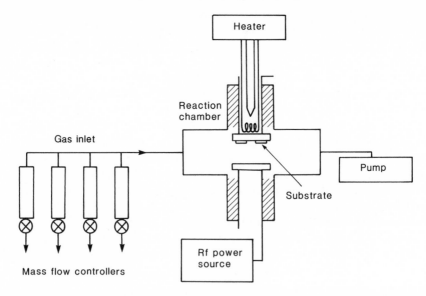

Fig. 20.2 The plasma-enhanced chemical vapour deposition method for preparing aSi-H. (After Street 1991.)

the reflectance at a particular wavelength, as exemplified in Fig. 20.3(b) for 632.8 nm (He–Ne laser) light during the growth of aSi:H–aSiO$_2$:H (2 nm/ 3 nm) films, using pulses of SiH$_4$ and a mixture of 2 per cent SiH$_4$ in N$_2$O during the growth. The inset shows the gross interference fringes associated with film thickness on the scale of the wavelength, but fine structure is also shown in the main diagram. This latter has been closely analysed after growth using transmission electron microscopy, Rutherford back-scattering, and other techniques. The initial steep rise (A) in reflectivity is due to the growth of a c.1 nm file of Si oxide, as the plasma oxidizes the underlying Si, while the more gradual rise (B) is associated with plasma-deposited oxide. The plasma is turned off for 10 s (C) to flush out the N$_2$O with SiH$_4$, and the peak (D) once the plasma is turned on again is associated with layers of Si with macroscopic roughness on the 1–1.5 nm height and 10 nm lateral scales, and by extra hydrogen at the interfaces. The bulk films correspond to SiH$_{0.1}$ and SiO$_{1.9}$H$_{0.09}$, and these values tally with the average composition of Si and oxygen in the superlattice given the layer thicknesses. These findings are corroborated by the results of post-growth materials assessment, to which we now turn.

20.2.2 *Materials assessment*

The techniques outlined in Chapter 2, Section 2.4 for characterizing crystalline multilayer structures for their electrical, optical, and structural properties also apply to amorphous systems. In some cases there is a loss, for example the lack of sharp Bragg spots in X-ray analysis. In optical techniques, the

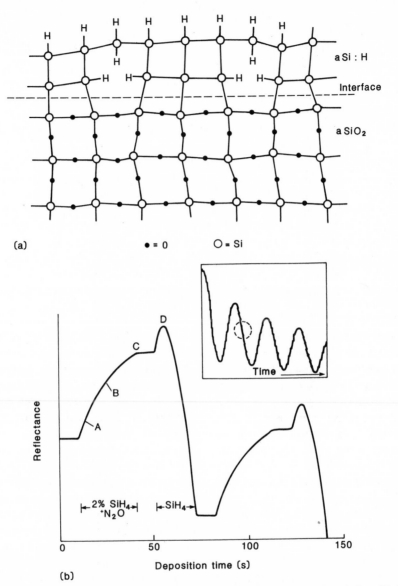

Fig. 20.3 (a) A schematic amorphous semiconductor heterojunction (after Abeles 1989); (b) optical reflectance of aSi:H–aSiO$_2$:H (2 nm/3 nm) films monitored during the growth. (After Abeles *et al.* 1987.)

k-selection rules and considerations of symmetry-allowed transitions do not apply, and further information about the overall system, such as the valence band density of states, can be obtained. The infrared optical properties of multilayer films with different thicknesses of constituent layers are used (in conjunction with data on thick films) to provide further details on the hydrogen coordination of Si in the middle of aSi:H layers and at interfaces in structures such as those described in the previous section. The technique of core-level photoemission (at $c.100$ eV), used in Chapter 13, Section 13.3, to determine bandgap offsets, is also used here for the same purpose, and also to determine the different chemical environments of Si in SiO_x, Si:H, or at an interface (from emission at $c.20$ eV). The energy band diagram for this system is shown in Fig. 20.4(a); the only additional information used is the 1.7 eV optical gap in bulk aSi:H since the conduction band edge is not resolved in the photoemission experiment.

In the inset to Fig. 20.4(b) we show the potential profile over several periods of a superlattice. There are strong built-in fields caused by dipole layers at each interface, which were discovered by an electroabsorption spectroscopy measurement, a variation of experiments described in Chapter 10, Section 10.2.4, to measure the shifts in the excitonic features in quantum wells subject to an applied field. The minimum energy for the absorption edge corresponds to a zero field in the well, and a finite bias must be applied in the amorphous multilayers to achieve this condition. It corresponds to an interface dipole of $c.170$ mV at each heterojunction, whatever the thickness of the aSi:H layer. Such dipole layers are associated with residual charged dangling bonds and other defects, and do not occur in their crystalline counterparts unless doping dipoles are introduced deliberately.

The amorphous state is generally metastable, and annealed superlattices will relax their structure and eventually crystallize. We can also anticipate that the interfaces will become less abrupt during the annealing process. Inelastic light-scattering experiments can be used to probe the excitations of the structure, and the frequency of the various vibrational modes (cf. Chapter 15, Section 15.4). In aSi:H–aGe:H superlattices the average frequency and frequency spread associated with the Si–Ge modes have been monitored during annealing, and the increase in intensity of this peak above 340 °C suggests interdiffusion of the Si and Ge, with complete intermixing to form a homogeneous alloy by 640 °C (Kumar and Trodahl 1991). In contrast, in aSi:H–aSiN$_x$:H there is no evidence of intermixing, and X-ray and light-scattering data suggest that the interfaces may even become more abrupt around 650 °C (Santos *et al.* 1991). In this latter example, some hydrogen is given off and the superlattice period contracts slightly. It seems that a driving force associated with the tendency to form Si_3N_4 overcomes any diffusion process.

20.3 Physics in amorphous semiconductor multilayers

In this section, we describe the results of optical and transport experiments on various multilayer structures of amorphous semiconductors. In general, the quality of the structures is poorer than their crystalline counterparts, so that excitonic effects in quantum wells and room temperature resonant tunnelling, for example, are relatively difficult to achieve.

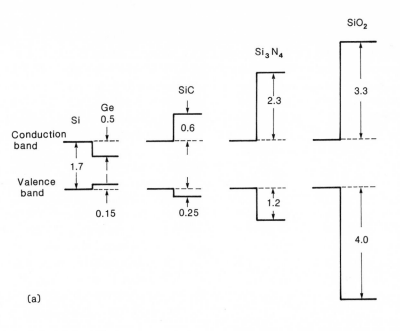

(a)

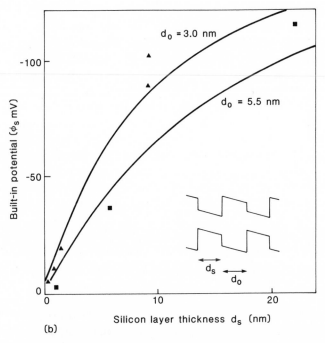

(b)

Fig. 20.4 (a) The energy band diagram for several heterojunctions (after Abeles 1989); (b) the built-in potential field (ϕ_s per period) in a superlattice consisting of a thickness d_0 of aSiO$_x$ and a thickness d_s of aSi:H (after Abeles *et al.* 1987).

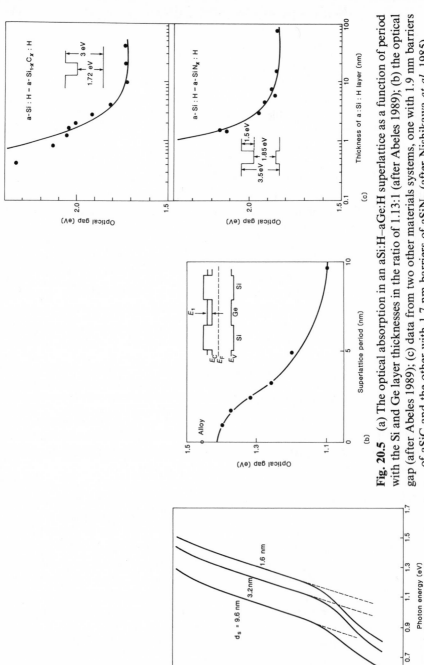

Fig. 20.5 (a) The optical absorption in an aSi:H–aGe:H superlattice as a function of period with the Si and Ge layer thicknesses in the ratio of 1.13:1 (after Abeles 1989); (b) the optical gap (after Abeles 1989); (c) data from two other materials systems, one with 1.9 nm barriers of aSiC and the other with 1.7 nm barriers of aSiN$_x$ (after Nishikawa *et al.* 1985)

20.3.1 Optical evidence for quantum size effects

There have been many optical studies of the bandgap of amorphous semiconductor multilayers, and typical results are given in Fig. 20.5 for the dependence of the optical gap in an aSi:H–aGe:H superlattice with a fixed ratio of 1.13 between the thickness of the Si and Ge layers but a varying period. With 10^3 cm^{-1} as the absorption used to define the gap E_g (or alternatively by fitting the absorption data to the semi-empirical expression $\sqrt{(\alpha E)} \propto (E - E_g)$), we obtain the results in Fig. 20.5(b) for the variation in energy gap with superlattice period. In the inset, we show the energy band profile; the solid line is that of theory (cf. Chapter 10, Fig. 10.1) for an electron in the finite potential well, based on the valence and conduction band offsets shown in Fig. 20.4(a). The data in Fig. 20.5(a) show the exponential tail of localized states, which is characterized by an energy width E_0 (cf. Fig. 20.1(b)). The value of this parameter and the temperature dependence of luminescence intensity, which varies as $\exp(-T/T_0)$, both rise in value with decreasing well width, suggesting that the localized state distribution broadens in thin wells. This type of result has been obtained in several materials systems, two of which are shown in Fig. 20.5(c) along with Krönig–Penney calculations of the lowest-energy quantum-confined states, using a value of $0.4m_e$ for the effective mass of aSi:H. All these data provide optical evidence of quantum size effects in amorphous multilayers.

20.3.2 Transport and tunnelling studies

The conductivities of the superlattice structures whose specifications and optical properties are given in Figs 20.5(a) and 20.5(b) are shown in Fig. 20.6. Again, a simple theory gives a good account of the variation in conductivity. Motion along the layers is assumed to take place in the Ge quantum wells with the conductivity determined by the lowest confined state, while superlattice motion takes place in the perpendicular direction with electrons tunnelling through amorphous silicon layers. The Fermi energy is pinned by defects in the aGe:H layers at 0.44 eV below the Ge band edge. The effective masses are assumed isotropic at $0.3m_e$ for Si and $0.4m_e$ for Ge, and the mean free path of electrons in aGe:H is 0.8 nm. A series of photoemission experiments on aSi:H layers on aSiO$_x$:H show no evidence for a shift in the conduction band edge of the overlayer down to the very thinnest layers. This result implies a hole mass in these systems at least 10 times higher than the electron mass (Abeles 1989).

The correct effective mass for electrons in the aSi:H system is a continuing problem. We do not have band-structure calculations as a guide, and the anisotropy associated with the conduction band valleys in crystalline Si does not obviously apply. In the interpretation of the optical data in the previous section we assumed an effective mass of $0.4m_e$. Most recent detailed experiments on the current–voltage characteristics of Schottky barriers using aSi:H have been interpreted using a tunnelling effective mass of $0.09m_e$ (Shannon

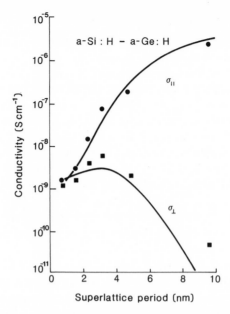

Fig. 20.6   The anisotropy of the dark conductivity of the superlattices whose optical
gap is shown in Fig. 20.5(b). (After Abeles 1989.)

et al. 1993). No systematic study has been made of the tunnelling of electrons
through a single barrier in an entirely amorphous semiconductor system. This
repeats the earlier situation in the GaAs–AlGaAs system, where resonant
tunnelling in double-barrier diodes received much more attention than conventional tunnelling through single-barrier diodes, although the latter has
been essential in the refinement of simulations for tunnelling (cf. Chapter 8).

   Several studies have been made of resonant tunnelling in double-barrier
diodes in amorphous semiconductor materials. In Fig. 20.7 we show (a) the
structure of aSi:H–aSiN$_{1.33}$:H double barrier structures with phosphorus-
doped contact layers, (b) their energy band profile, (c) their current–voltage
characteristics and the calculations of transmission coefficient of a 4 nm well
structure under bias, and (d) the resonant energy levels versus well thickness.
The quality of the results in terms of peak-to-valley ratio and peak current
densities are comparable with those achieved in the early days of crystalline
double-barrier structures, and quite inferior to those now available in crystals.
No negative differential resistance is achieved at room temperature, and the
modest non-linearities in the current–voltage data, even at 77 K, indicate that
leakage currents (i.e. the valley currents in Chapter 8) are much larger in this
system. All this indicates that the microstructure of the interfaces and thin
barriers is not homogeneous on the shorter quantum length scale in these
materials, where the effective Bohr radius is *c.*2 nm in Si as opposed to *c.*10
nm in GaAs. A calculation with the effective mass of the tunnelling electrons
as the variable can be used to correlate features in the current–voltage data

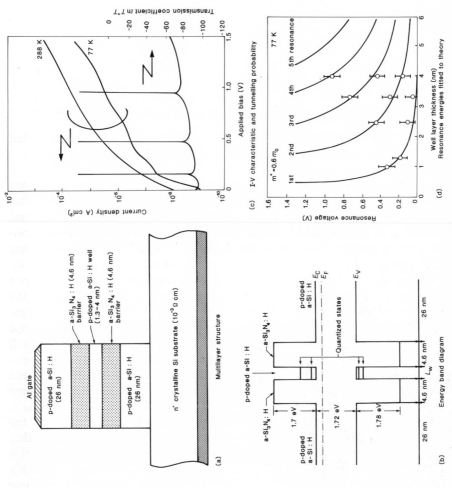

Fig. 20.7 Resonant tunnelling in aSiH–aSiN$_{1.33}$ double-barrier structures. (After Miyazaki *et al.* 1987.)

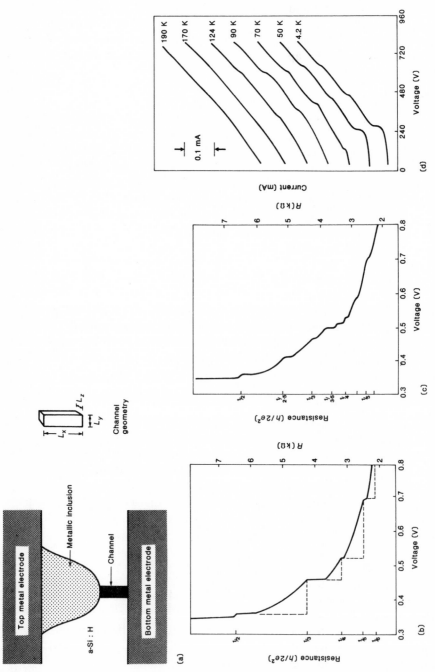

Fig. 20.8 (a) Schematic diagram of a 1D filament; (b) plateaux in the current–voltage characteristics which (c) split with magnetic field and (d) persist to high temperatures. (After Hajto *et al.* 1991.)

with the quantized energy levels in the wells, and the best fit here is with $m^* = 0.6m_e$. This value is higher than others inferred from optical and electrical measurements, and reinforces the point that effective masses in amorphous multilayer structures are not established experimentally, and an adequate theory is also lacking. Given the absence of band structure in amorphous materials, the appropriate mass for electrons in transport and optical experiments need not be the same.

20.3.3 Ballistic motion

One of the more unusual results in amorphous semiconductor multilayers has been the evidence for ballistic motion of the form seen in Q1D channels (cf. Chapter 6). In this case, a $c.100$ nm layer of heavily p-doped aSi:H is sandwiched between two metallic layers and subjected to a short voltage pulse, during which the dielectric breakdown field of the film is exceeded. This leads to a permanent reduction in the resistance of typical films from about $10^6 \, \Omega$ to about $10^3 \, \Omega$, probably by the formation of some conducting filament through the film as shown schematically in Fig. 20.8(a), although its precise nature is unknown. A voltage pulse of reverse polarity can restore the high resistance, and so the structure is proposed as a memory element. The low-temperature current–voltage characteristics give evidence of the plateaux (Fig. 20.8(b)) associated in Chapter 6 with ballistic motion in short channels, with the number of steps doubling in a magnetic field (Fig. 20.8(c)) with spin-splitting. The plateaux persist up to very high temperatures ($c.170$ K), much higher than seen in any of the other III–V systems ($c.10$ K), suggesting that the filament here is very thin so that the energy separation of excited states in the channel is correspondingly larger.

20.3.4 Graded-gap structures

In earlier chapters, graded-gap layers have been incorporated into Gunn diode cathodes, the base region of transistors and phototransistors, avalanche photodiodes, and solar cells. In each case, the gradient in the bandgaps and the quasi-electric fields set up by the slopes of the conduction and valence band edges have been used to improve device performance. Graded-gap layers have also been investigated in the aSi–Ge:H materials system, with typically 16 layers each 50 nm wide and with a slightly increased or decreased Ge gas flow rate for each new layer (Conde *et al.* 1989). Bandgap differences of up to $c.0.6$ eV over 0.4 μm have been prepared. Semitransparent Pd dots were used to form Schottky diodes, with either the wide-gap or the narrow-gap end at the metal–semiconductor interface. Optical pulses generate electrons and holes as carriers, and the product of their mobility and lifetime (before being trapped in some deep level) could be measured for both carriers by a time-of-flight technique for the electrons, and with a fit of the bias dependence of the photocurrent for the holes (cf. Street 1983). Typical results for electrons and holes are shown in Fig. 20.9 as a function of the effective optical gap (extracted from the optical absorption of the graded-gap structure). The

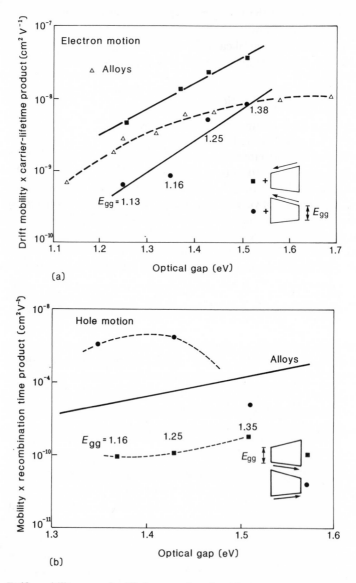

Fig. 20.9   Drift mobility – carrier lifetime product for (a) electrons and (b) holes in the graded-gap aSi–Ge:H materials system. (After Conde *et al.* © 1989, IEEE.)

bandgap profiles with respect to carrier motion are shown in the insets. Where the bias is such that the quasi-electric field gradient is assisting the transport, the mobility–lifetime product exceeds that obtained with a uniform alloy, whereas if the imposed electric field has to overcome the quasi-electric field, the product is less than that obtained in the uniform alloy. Further studies have been made of the transport as a function of temperature to extract activation energies of the traps that act as sources of electrons and holes, the quantum efficiency, and other transport parameters. The results are as encouraging for device applications where amorphous materials are used (e.g. solar cells) as they have been when exploited in crystalline semiconductors.

20.4 Devices and applications

Although thin films of amorphous Si are in widespread use, the incorporation of amorphous semiconductor multilayers into thin-film devices is an early stage. The aSiN$_x$:H materials system has been incorporated into solar cell design to increase the collection efficiency of blue light. Enhanced efficiency pulsed-mode operation of p–i–n light-emitting diodes in aSiC:H–aSi:H quantum well structures have been reported, where the quantum wells trap the photoinjected carriers before the built-in fields separate the carriers and reduce the chance of recombination. While there has been no systematic transport study of 2DEGs at amorphous heterojunction interfaces, the results in Fig. 20.10 should encourage such research. Thin-film transistors have been

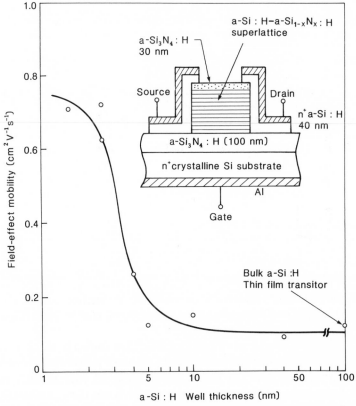

Fig. 20.10 Electron field-effect mobility as a function of well thickness in thin-film transistors with aSi:H–aSi$_{1-x}$N$_x$:H superlattices. (After Hirose and Miyazaki 1989.)

made using an aSi:H–aSi$_{1-x}$N$_x$:H superlattice as the active material. The electron field-effect mobility at low electric fields (less than 2×10^5 Vm^{-1}) is plotted as a function of thickness of the aSi:H quantum well. For wells of thickness c.2.5 nm, there is a fivefold enhancement of mobility. The precise reasons are not yet understood.

The vapour phase deposition technique used for preparing device-grade amorphous Si films is well suited to the growth of multilayers, and as the physics results indicate improving performance of device structures, the amorphous multilayers are likely to find device applications following the route described in Chapter 16–19.

References

Abeles, B., Yang, L., Eberhardt, W., and Roxlo, C. B. (1987). Structure of hydrogenated amorphous silicon/silicon oxide superlattices. In *Proceedings of the 18th International Conference on the Physics of Semiconductors*, (ed O. Egstrom), pp. 731–8, World Scientific, Singapore.

Abeles, B. (1989). Amorphous semiconductor superlattices. *Superlattices and Microstructures*, **5**, 473–80.

Conde, J. P., Shen, D., Chu, V., and Wagner, S. (1989). a-Si:H,F ⇔ a-Si,Ge:H,F graded-bandgap structures. *IEEE Transactions on Electron Devices*, **ED-36**, 2834–8.

Hajto, J., Owen, A. E., Snell, A. J., Le Comber, P. G., and Rose, M. J. (1991). Analogue memory and ballistic electron effects in metal–amorphous silicon structures. *Philosophical Magazine*, **63B**, 349–69.

Hirose, M. and Miyazaki, S. (1989). Superlattics and multilayers in hydrogenrated amorphous-silicon devices. *IEEE Transactions on Electron Devices*, **36**, 2873–6.

Kumar, S. and Trodahl, H. J. (1991). Raman spectroscopy studies of progressively annealed amorphous Si/Ge superlattices. *Journal of Applied Physics*, **70**, 3088–92.

Le Comber, P. G. (1992) Applications of amorphous silicon devices. *Physica Scripta*, **T45**, 22–8.

Miyazaki, S., Ihara, Y., and Hirose, M. (1987). Resonant tunneling through amorphous silicon–silicon nitride double barrier structures. *Physical Review Letters*, **59**, 125–7.

Nishikawa, S., Kakinuma, H., Fukuda, H., Watanabe, T., and Nihei, K. (1985). Optical properties of a-Si:H/a-Si$_{1-x}$C$_x$:H and a-Si:H/a-SiN$_x$:H superlattice. *Journal of Non-Crystalline Solids*, **77–78**, 1077–80.

Santos, P. V., Hundhausen, M., Ley, L., and Viczian, C. (1991). Structure of interfaces in a-Si:H/aSiN$_x$:H superlattices. *Journal of Applied Physics*, **69**, 778–85.

Shannon, J. M. and Nieuwesteeg, K. J. B. M. (1993). Tunneling effective mass in hydrogenated amorphous silicon. *Applied Physics Letters*, **62**, 1815–17.

Street, R. A. (1983). Measurements of depletion layers in hydrogenated amorphous silicon. *Physical Review B*, **27**, 4924–32.

Street, R. A. (1991). *Hydrogenated amorphous silicon*. Cambridge University Press.

21 Towards 2000

21.1 Introduction

Contemporary semiconductor science and technology has a momentum that allows one to make predictions on the progress over the next few years based on the experience of the last 20 years. We are concerned in this chapter with the future of materials, technology, structures, physics, and devices in mainstream multilayer semiconductor science and technology. In the concluding chapter, we consider some of the radical alternatives to this mainstream that are being pursued in research and development. In conventional ultra-large-scale-integration Si technology, the current developments in the various fabrication steps (lithography, etching, metallization, etc.) will ensure a steady progress in the memory and computational capability through this decade and into the next century. Extrapolations from the 1970s and 1980s are a reasonable guide for the 1990s, as far as the technology is concerned (but see Section 21.8 for a more global picture). The science in Si technology is increasingly concerned with the effects of the intrinsic small scale of the devices, particularly including those of high electric fields, and we review this area more fully in section 21.7. The advances in multilayer semiconductor science have relied on the technology for fabricating the test structures, and the interdependence of science, technology, and devices is becoming more complete, a principal reason for the structure of this book. The science and technology of low-dimensional semiconductors is having a large impact on the science of metrology, and we include a discussion of this in the penultimate section of this chapter.

21.2 Materials

To date, the GaAs–AlGaAs multilayer materials system has been dominant in the demonstration of new physics and new devices. Where there is new physics, as with the complex many-body states in the fractional quantum Hall regime, a combination of the extreme materials purity achieved with GaAs–AlGaAs and ultralow temperatures are required. The physics of quantum size effects, tunnelling, and hot electron injection was worked out in this materials system. Other materials combinations with different band offsets and effective masses are now required to investigate new regimes of these phenomena. Early examples described in previous chapters include the AlSb–InAs system with its radically different energy band alignment and Si–SiGe with the combination of the indirect gap implying multiple conduction band minima and strain which reduces the degeneracy of these minima. Other materials systems are now used to extend the range of the existing phenomena to higher temperature, and to achieve greater confinement energies in quantum wells or larger peak-to-valley ratios in tunnel diodes as examples of device relevance. As new materials become available, this trend will continue. The search for quantum-well lasers in the blue in II–VI compounds and in the infrared in the lead salt compounds (Chapter 18, Section 18.10) are examples. The performance of both device types is far from optimal, and materials properties are a limiting factor at present. Major improvements in the doping of II–VI compounds are being sought, and with the impetus of low-temperature lasing in the blue (Haase *et al.* 1991; Jeon *et al.* 1992) the research effort in these materials is set to grow over the coming years, leading to room-temperature optical devices in the blue and green.

Growth research in III–V materials has several aims.

1. The use of patterned substrates to impose structure within quantum wells has only just begun. Quantum-wire lasers are prepared in V-grooves or on ridges that are separated by about 1 μm (see Chapter 2, Section 2.3), and this spacing between quantum wires will need to be reduced 10-fold. Only then can the advantages of Q1D physics be exploited in systems with sufficient active volume to produce useful intensities of output light. Growth on vicinal substrates with steps of several unit cells height every 10 nm along the surface are required.

2. The strained quantum-well lasers are grown with ternary materials for the quantum wells and quaternary for the confining barriers (Chapter 18, Section 18.4). The precise conditions for the growth of the highest-quality interfaces, particularly in terms of the changes in gas flows in a MOCVD reactor, are yet to be established.

3. The quaternary system InGaAlAs is being used in the fabrication of high-speed devices because of the range of conduction band discontinuities that can be achieved (Chapter 16, Section 16.6). The ultimate limits on this materials system are not established.

In the quest for ever greater electronic device performance, diamond as a host material has the greatest value of the key transistor figure of merit (cf. Chapter 16, Section 16.2, and Fig. 21.1). The product of the saturated drift velocity and the dielectric breakdown strength is 50 times that of Si and 16 times that of GaAs in terms of drive voltage and speed of FETs and bipolar transistors. Faster devices which deliver higher output power can be envisaged with electronics based on C and SiC. The thermal conductivity of diamond is greater than that of copper at room temperature, making it an ideal heat sink (Chapter 1, Fig. 1.8). Multilayers of diamond (Collins 1989) and SiC (Ivanov and Chelnokov 1992) offer the prospects of higher-power, higher-temperature, and higher-speed electronics. The quality of epitaxial films of diamond is adequate for heat sinking, but not yet for electronic devices. Doping of

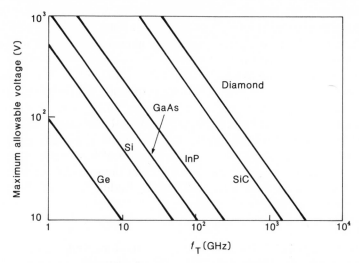

Fig. 21.1 The drive voltage and transit time upper-frequency limits of transistors imposed by the host material. (After Geis *et al.* 1988.)

diamond to produce shallow donors and acceptors still remains difficult, so that operation is possible only at elevated temperature and not at room temperature. Optically, we have the prospects of solid state devices in the ultraviolet.

The real change anticipated in growth studies in the coming years is a vast extension of the range of high-quality materials grown in multilayer forms. Multilayers with metallic components have been prepared, and the range of materials properties that can be tailored through the layer thicknesses include (i) the electric resistance with a metal insulator transition in the Nb–Cu system as the individual layers shrink below about 1 nm (Fig. 21.2(a)), (ii) a giant magnetoresistance in Co–Cu multilayers (Fig. 21.2(b)), and (iii) elastic constants with the biaxial modulus in Au–Ni multilayers increasing strongly as the repeat distance shrinks below about 2 nm (Fig. 21.2(c)). Some of these multilayers, which are prepared by relatively crude sputtering techniques, exhibit the greatest magnetic anomalies, greater than obtained in the same system prepared by molecular beam epitaxy. The reasons for this are not known. A summary of the state of the art of metallic superlattices is given by Schuller (1988).

The possibilities of coercivity engineering in magnetic materials are as wide as those of bandgap engineering in semiconductors, with multilayers designed to have desired *B–H* curves for the particular application which may be in the area of magneto-optical data storage or in magnetic systems more generally (den Broeder *et al.* 1992). Other secondary properties, such as the temperature dependence of the *B–H* relation, the speed of switching, etc., can also be improved in tailored structures, just as was the case with multilayer devices using semiconductors.

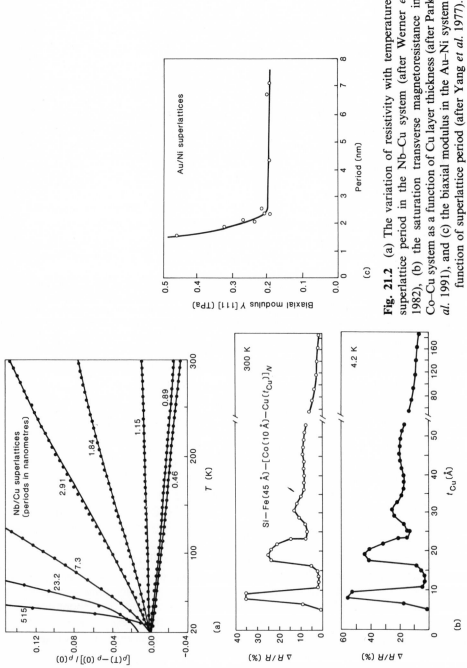

Fig. 21.2 (a) The variation of resistivity with temperature and superlattice period in the Nb–Cu system (after Werner *et al.* 1982), (b) the saturation transverse magnetoresistance in the Co–Cu system as a function of Cu layer thickness (after Parkin *et al.* 1991), and (c) the biaxial modulus in the Au–Ni system as a function of superlattice period (after Yang *et al.* 1977).

Semiconductor–metal (Si–silicide) multilayers have been made of suffi-cient quality to investigate transport through the multilayers, there being only a modest lattice mismatch between the lattice constants of many si-licides (e.g. of Co, Pd, Pt, etc.) and one of the orientations of Si. A problem that remains to be addressed is the fact that there are two or more orientations of silicide on Si (Fig. 21.3). Great care is required to obtain one rather than a mixture of the two, as the interfaces with different orientations have quite dissimilar electric properties; for example the Schottky barrier heights for the A and B orientations of $NiSi_2$ on Si(111) are 0.64 eV and 0.78 eV respectively. High-quality Si can be grown on top of $CoSi_2$, and a wider range of silicides can be formed as buried layers using heavy doses of implanted metal ions followed by annealing to form single crystals. Mixed-fluoride materials ($Ca_x Ba_{1-x} F_2$) are lattice matched to GaAs and offer the prospect of semiconductor–dielectric multilayers with tailored optical properties. In Fig. 21.4 we show a hot electron transistor structure made from the $CoSi_2$–CaF_2 combination of materials. The growth process is a complex sequence of conventional epitaxy and heat treatments. CaF_2 growth proceeds using molecular beams that are ionized and accelerated under modest fields towards the surface. Their energy is sufficient to allow some migration on the surface, which is kept cool enough to prevent agglomeration of subsequent $CoSi_2$ layers. Then Si and subsequently Co layers are deposited and annealed to form the silicide, as co-evaporation of Co and Si leads to clustering of the Co. High-resolution transmission electron microscopy shows high-quality epitaxial films, at least over small areas (Fig. 21.4(a)). The transistor structure is shown in Fig. 21.4(b), and its common-emitter charac-teristics at 77 K are shown in Fig. 21.4(c). With $V_{BE} \sim 6$ V and $V_{CE} \sim 6$ V, the current gain is $c.150$, although leakage currents become unacceptable at elevated temperatures. The transfer efficiency more than 90 per cent was achieved for the collections of hot electrons injected into 1.9 nm of $CoSi_2$ at 77 K.

The prediction is made that, over the coming decade, the range of materials achieving device-grade quality is set to grow enormously as the result of detailed and painstaking research with attention to the details of the kinetics and thermodynamics of the growing crystals. The scope for new physics and new devices using these new materials combinations is explored in Section 21.5. First, we consider two more aspects of new samples for physics and device investigations, namely the technology for making structures and the new structures that will become available.

21.3 Fabrication technologies

The blossoming of multilayer semiconductor science and technology during the 1980s was based on the research into materials and fabrication techno-logies that started in the 1970s and continues until the present day. The trends in that research are set to continue, with the objective of achieving ever-tighter

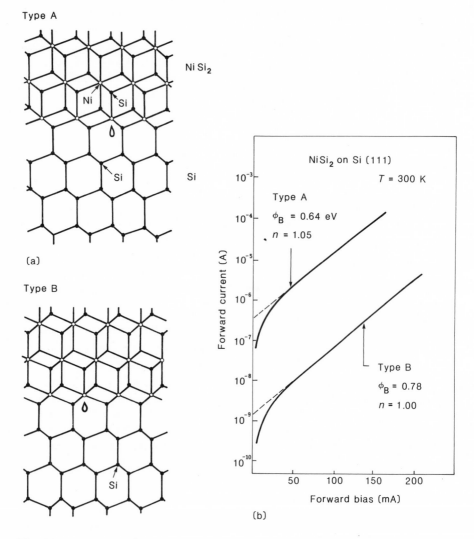

Fig. 21.3 (a) Two orientations of NiSi₂ on Si(111) surfaces; (b) the forward-bias current–voltage characteristics of Schottky diodes with each of these orientations. (After Tung 1984.)

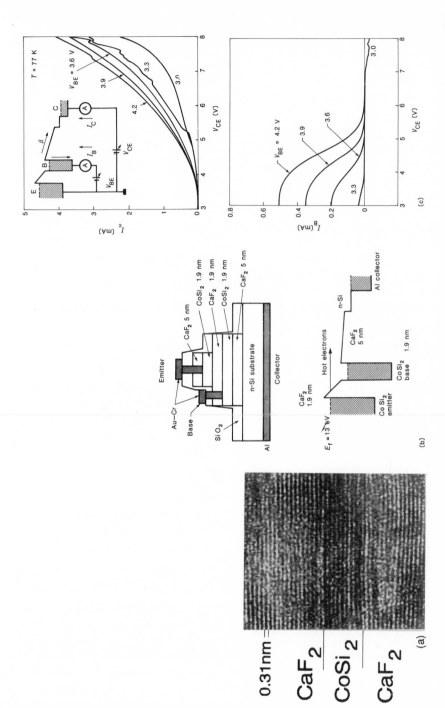

Fig. 21.4 (a) A transmission electron microscope cross-section of a CaF$_2$–CoSi$_2$–CaF$_2$ multilayer structure (after Watanabe *et al.* 1992), (b) a hot electron transistor structure based on these materials, and (c) its 77 K device performance (after Muratake *et al.* 1992.)

control over the dimensions of structures and the retention of materials quality after the fabrication and processing steps are completed.

Since electron beam lithography is already capable of defining nanometre-scale structures, the emphasis is on higher current densities in the beam and clever writing schemes to achieve much higher throughput of large-area wafers with higher densities of ever-finer patterns defined upon them. The trends in optical and X-ray lithography outlined in Chapter 3, Section 3.2, are set to continue. Increased activity is anticipated in combining processing steps along the lines exemplified by the *in situ* focused ion beams used to pattern MBE multilayers during growth (Chapters 2, 3, and 5). The ability to produce device and physics structures that require a non-alloyed ohmic contact technology as the only post-growth processing step represents an attractive and high-yield processing route. The elimination of resists for patterning in this technology is also attractive. There is a need to increase greatly the ion beam current, and hence the speed of the pattern writing.

Research on dry plasma etching is aimed at limiting the surface and subsurface damage induced by the plasma. At present, the first 70 nm of most GaAs-based structures is electrically dead with hydrogen from the plasma passivating the donor and acceptor levels. Etching with rare gas ions is one approach being investigated to eliminate hydrogen.

Although ion implantation is a technology dating back to the 1960s, the greater control over implant energy, the use of low-energy implantation during growth, and the ability to implant in selected areas offers advantages for the fabrication of structures and devices in semiconductor and other multilayers with increasingly tight tolerances of the feature sizes in all three spatial dimensions.

To date, epitaxial metal overlayers have been a research vehicle, but at decreasing structure sizes, and with concerns over high-field and other effects in small structures, patterned single-crystal epitaxial overlayers will assume an increasing importance.

A key trend in all processing technologies is to reduce the time–temperature cycle over which any given step is executed. This reduction allows increasingly precise features to be defined without subsequent degradation of the structure and materials properties through subsequent processing cycles. In this respect the reduction in the number of process steps by *in situ* patterning reinforces this trend.

21.4 Structures

Over the past 20 years, the ability to define and make electrical contacts to a variety of low-dimensional structures—2DEGs, quantum pillars, etc.—has made possible much of the interesting science and device research where some of the important length scales are in nanometres. The central role played by new test structures will continue, and some of those that can be anticipated in the next few years are given here. A collage of some of these structures is shown in Fig. 21.5.

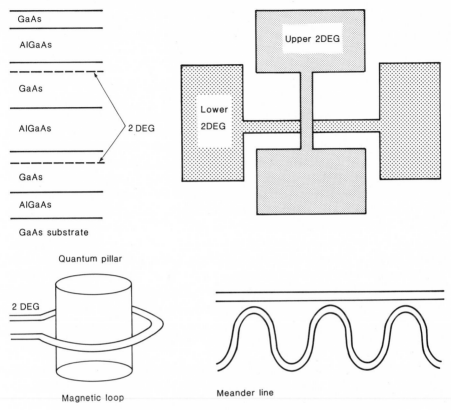

Fig. 21.5 Future structures for physics and device investigations.

Independent contacts to closely adjacent 2DEGs have recently made possible studies of electron–electron interactions which test the relevant theories. Original work used relatively crude depletion of the upper or lower gas with a Schottky gate defined on the top or bottom surface of the wafer. Focused ion beams are used to form localized volumes of damage so that contacts to conducting layers are blocked by this damage. This latter technology is set to produce a wide range of further new structures (Chapter 5, Section 5.8). Adjacent quantum wires will be formed from 2DEGs that are parallel or cross at a predetermined angle, and meander structures with deep submicrometre feature sizes will be possible. These will be used for detailed studies of electron interactions in electronic transport. Furthermore, if electrons in one wire are subject to large electric fields, the transfer of heat to the surrounding lattice will be monitored with electrons in the second wire.

The extension of these ideas from two to three dimensions makes possible the fabrication of new topological structures in which new physics can be investigated, particularly associated with magnetic phenomena. If high-quality conducting channels can be made with focused ion beam doping, then a single or multiple coil threading a Corbino or other type of 2DEG or 1DEG structure is possible.

Periodic patterning of a 2D layer is already achieving a pitch of order 0.2 μm; it is difficult to form more closely adjacent structures with resists and still preserve the intervening resist. As new resist schemes and materials for resists develop, if this pitch can be reduced by a further factor of 5, band-structure effects associated with the in-plane patterning should be readily accessible and exploitable. Such an achievement would represent the start of band-structure engineering in two spatial dimensions. It is already possible to preserve steps on vicinal substrates during epitaxial overgrowth to form a 1D periodic potential in the plane of the 2D layer, with a period comparable to that achieved in the depth direction during epitaxial growth. In these patterning schemes, it is important to achieve large arrays of identical microstructures if the coherence of electrons extending over several repeat distances is to generate real band-structure effects. The inevitable deviations from the ideal structure will enable new regimes of disorder to be investigated. Once new 2D crystals are available and investigated, one will want to proceed to examine arrays where the component microstructures have a specific spatial variation in their feature size to obtain the analogue of the graded-gap layered structures that have found device exploitation. One may want to make superlattices of superlattices in the 2D plane. In each of these schemes, new electronic phenomena will occur only if the highest materials quality of the conducting layer is preserved.

Once the mastery of patterning in two dimensions is achieved, the repeated patterning of many layers will be undertaken, with precise registry between the features in adjacent layers. If the quantum-dot laser is ever to be fabricated, the active volume of the laser for adequate output powers will not be greatly different from those quantum-well lasers already available. It is just that this active volume will need to contain arrays of identical quantum dots precisely ordered in position in all three dimensions. While the feature size for quantum dots in semiconductors will be of order of 20 nm, once this scale is reduced by a further factor of 5–10, the emerging solids will have band structures entirely dependent on the fabrication specifications, i.e. electronic structure to order.

The real challenge, which is still more than a decade away, is the ability to tailor a material in all three spatial dimensions on the scale of 0.1 μm or less. At present, the theory for the photonic bandgap has been developed and exploited with gigahertz frequencies in structures with millimetre feature sizes (Yablonovitch 1987; 1993; Yablonovitch *et al.* 1991). Here, a periodic modulation of the dielectric constant in all three dimensions gives rise to stopbands for light in complete analogy to bandgaps in semiconductors. The implication is that radiative processes can be prevented or controlled over specific energy ranges. The application to practical optical devices, such as lasers, requires a periodicity of a few hundred nanometres in all three spatial dimensions.

Over the coming decade much progress is anticipated in the above agenda as applied to the patterning of semiconductors. As the growth of hybrid material structures advances (e.g. silicides by low-energy implantation of metal ions during epitaxial growth of silicon), the above structures will be fabricated with metal, dielectric, and semiconductor elements.

21.5 Physics

The qualitatively new physics likely to emerge from the new structures just outlined includes extensions of quantum coherence and interference phenomena, in samples where feature sizes are modulated on length scales of 10 nm or less, much less than the coherence length. It is only in such structures that there is any hope of raising these delicate and at present cryogenic phenomena to 77 K and above. The interaction of individual electrons in a current with other electrons in the current, and the environment of the conducting channel, is at a primitive stage. The use of Coulomb interactions in multiple-quantum-wire systems, and in electron gases coupled by tunnel barriers, will enable these studies to be taken much further. The energy transfer between currents in closely coupled 2DEGs is under investigation at present, but in coupled-wire systems the single-electron effects may be brought to prominence. The electron–phonon interaction is also capable of more detailed investigation via thermal experiments in ultrasmall wire systems. The physics of Coulomb blockade is ripe for further exploration in structures with fine-scale modulation. Until now, much microstructural physics has probed only the wave nature of the electrons, and the discrete-charge nature has only recently been receiving attention. Clocking of electrons in time-correlated sequences in different wires will enable further insights to be obtained into the role of charge in the spatial and temporal nature of currents in microstructures, and offer the prospects of a current standard for metrology.

If new band structures can be defined within 2D layers, a whole new regime of 2D band transport phenomena are to be investigated. The altered nature of the density of states will have a profound effect on all the optical and transport properties, and the influence on them of magnetic fields and thermal gradients. The same applies to artificial 3D structures. The microscopic behaviour of electrons and holes in graded-gap structures is not well understood, and further insights might be gained by using 2D arrays of microstructures with carefully tailored gradients in feature size.

The physics from new topological structures promises great interest. If current can be made to loop once or several times around wires on the nanometre scale, the effect of induced magnetic fields on individual electrons in a solid can be monitored. Furthermore, the magnitude of gradients of magnetic fields in both space and time can be greatly enhanced over values obtained from the fields of macroscopic magnets situated far from the samples. In the same manner, higher-temperature gradients can be induced by local heating in the closely adjacent wires above. The Berry phase carried by electrons in the presence of a magnetic field can be investigated on a quantum length scale (Berry 1984).

The move to other materials systems introduces the possibility of exploring nearly all aspects of solid state physics on the nanometre scale in solids that have been tailored at the atomic level in one, two, or all three dimensions. The interaction between localized moments in different layers of magnetic metals separated by non-magnetic intermediate layers has already produced striking new magnetic properties which can be tailored. Superconductivity and metal–insulator phase transitions introduced by temperature, layer thicknesses, or even intensity of optical excitation await investigation, as indeed do all the other solid state phase transitions (ferroelectric, elastic, etc.) The length scales associated with some of these other phenomena, such as the phase coherence length in superconductivity, can be taken from the limit of being much greater than the feature sizes in the microstructural components of the arrays, through approximate equality, to being much smaller. In some cases this tuning of relative length scales can be achieved (as with temperature) in a single sample.

After the developments in Si devices in technologies in the 1960s, the 1970s saw a great expansion in research and development on amorphous Si. During the 1980s, this work on amorphous semiconductors has been concentrated on the science and technology required for large-area electronics applications. After the intensive research on crystalline semiconductor multilayers in the 1980s, the 1990s are likely to see such work followed up in the amorphous semiconductor context.

Lessons learned from the fabrication and investigation of the electronic, optical, magnetic, and other properties of 3D arrays of lower-dimensional microstructures (quantum dots, wires, wells, etc) will pave the way for the ultimate in materials science, namely the ability to place atoms from anywhere in the periodic table one at a time into a 3D solid against some design criterion for some material property. This is the long-term objective of the type of much of the work discussed in this text.

The probes for investigations are developing all the time, and nowhere more so than in the realm of ultrafast optics, with femtosecond time-scale interactions of carriers being investigated. As hinted above, many of these probes, while being fast in the time domain, are uniform in the space dimensions, and the new structures offer large local gradients in the relevant properties. Indeed, even by maintaining the probe intensity, the shrinking of the sample feature size results in the earlier onset of non-linear phenomena, and these (particularly in the optical response) are likely to be the subject of more intensive investigations in the coming years.

21.6 Devices

During this decade, multilayer semiconductor devices as described in Chapters 16–19 will become pervasive, even towards the end within mainstream Si technology (see Section 21.7). The trends to higher efficiency, speed, signal-to-noise ratio, and other device performance figures will continue. Even the two-terminal microwave diodes can be improved until the point is reached where every layer in the diode is chosen to optimize the particular collection of device performance parameters. In this section, we concentrate on the qualitatively new aspects of devices that can be anticipated in the coming years.

First, the increased number of multilayers can enable several device functions to be performed within a single device—the resonant tunnelling hot electron transistor as a one-transistor flip-flop is an early example—but internal multiple stability in a device will become exploited more as the density of devices on the wafer surface reaches limits set by the physics of the device operation, first in very-high-performance applications and subsequently more widely. Resonant tunnel diodes in series connections are suggested as analogue-to-digital converters. The same diodes within three-terminal device structures are used as frequency multipliers. This phase of multifunctional devices comes before a move to 3D integrated circuitry.

As the performance of room-temperature devices is perfected, smaller but important markets for devices operating at higher and lower temperatures are set to emerge. Compared with 300 K operation, the performance of both electronic and optical devices at 77 K is greatly superior because of reduced scattering. Until now, the low temperatures have only been used where essential, such as in cooled infrared detection. The cost of refrigeration is reducing, and the increased performance and lower losses make the over-heads for cooling less prohibitive, and for high-performance applications (including those that use any hybrid semiconductor–superconductor technology) 77 K operation will become attractive. At higher temperatures, in monitoring gas emissions from cars, factories, etc., and in regulating engines the electronics needs to react quickly from a position of close proximity, and here thermally stable devices at 300 °C are required. This may be achieved in III–V technology, but might also involve SiC, C, and other materials as well.

In electronic devices, the move to ever higher speeds raises the question of the feasibility of terahertz electronics. With a saturated drift velocity of 10^5 m s^{-1}, a Fermi velocity in ballistic structures of 2×10^5 m s^{-1}, and a hot electron injection velocity of 10^6 m s^{-1}, the total active length of a terahertz device is between 0.1 and 1 μm, placing tight limits of the technology and other aspects of device performance, particularly if room-temperature operation is sought. Quite a range of resonant tunnelling structures can meet such demands. The real problem is the RC time constants associated with charging and discharging different parts of a device through an external circuit, as happens in transistors. The upper frequency of existing tunnel diodes would exceed 1 THz if the series resistance could be reduced by a factor of 2–3 (Brown *et al.* 1988). Heavier doping in contacts, thinning of substrates, and metal plating up to the edges of the depletion layer would be required. All the

transistor structures in Chapter 16 have upper limits of c. 400 GHz on their
operation, imposed by the smallest feature sizes and lowest RC time constants
compatible with contemporary methods of device fabrication. Only the hot
electron transistor structures (cf. Chapter 16, Section 16.5) have the chance of
achieving 1 THz. Even then the Johnson criteria (Chapter 16, Section 16.2)
imply a maximum of 1 V drive level in GaAs and only 2 mW of power
delivered to a 50 Ω output load. The physics of ballistic transport is yet to be
fully investigated in the context of 1 THz electronics.

The recent rise in the importance of optical signal transmission and
processing (cf. Chapter 18) is likely to continue through this decade. Optical
fibres are becoming fully installed on long-haul intercontinental and trunk
communication routes, and in local area networks where very high data rates
are required, increasing the demand for optical components. There will always
be pressure for optically amplifying repeaters to process the photons rather
than absorb them, process them as charge and re-emit photons. Current
research is aiming at rates in excess of 1 Gbit per channel and a total
bandwidth of hundreds of terahertz, and progress is limited by the components for multiplexing and demultiplexing many channels of information at
the beginning and end of the transmission.

The new physics associated with ballistic motion, single-electron phenomena, electron coherence effects, and other new phenomena (cf. particularly
Chapter 12) has been proposed for very futuristic and radically new devices,
and the performance figures are modest. It remains to focus on devices that
are recognizable modifications and advances on today's mainstream devices
as described in Chapters 16–19 to see whether the limitations on their
performances can be circumvented in some way. It is hard to break with
conventional device technology unless it is in crisis, and that is not foreseen
until at least the second decade of the next century. By that time, some of the
new concepts realized in novel metal–semiconductor–dielectric microstructure
devices will be needed.

21.7 Silicon

How small can Si MOS transistors be made and still operate as recognizable
miniatures of the 1 µm devices in production? With 0.5 µm CMOS circuits
scheduled to become available commercially in memory chips about now,
research on Si devices is already at much smaller length scales. The guiding
principles are scaling laws established 20 years ago by Dennard *et al.* (1974),
and the results are summarized in Fig. 21.6(a). The transistor is scaled down
by a factor $\alpha > 1$ in all spatial dimensions, but the doping and voltages scale
as shown in order to preserve the electric field values from large to small
devices. The interconnects can be scaled in the same way to preserve RC time
constants as in Fig. 21.6(b). A more general scaling allows the electric field to
increase by a factor $\varepsilon > 1$ is shown in Table 21.1. Note that the wiring density
increases as $\varepsilon^2 \alpha$, and the power consumption as ε^2 within this scaling regime,

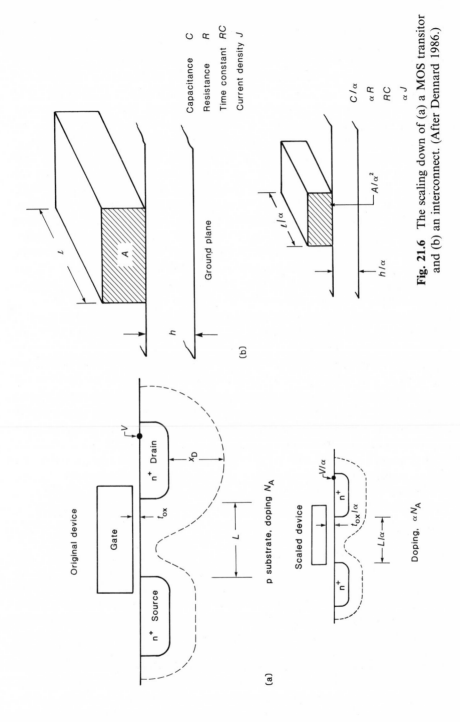

Fig. 21.6 The scaling down of (a) a MOS transistor and (b) an interconnect. (After Dennard 1986.)

Table 21.1 Generalized scaling relationships

Physical parameters	Scaling factor
Linear dimensions	$1/\alpha$
Electric field intensity	$\varepsilon \times 1$
Voltage (potential)	$\varepsilon(1/\alpha)$
Impurity concentration	$\varepsilon\alpha$
Wiring current density	$\varepsilon^2\alpha$
Gate delay	$(1/\varepsilon)(1/\alpha)$
Power/gate	$\varepsilon^3(1/\alpha^2)$

Source: Dennard 1986.

which has worked well for the technologies down to 0.5 μm, while the reduced gate delay speeds up the circuit operation by a factor $\varepsilon\alpha$.

Small circuits with 0.1 μm MOS transistors were demonstrated at 77 K several years ago, and a schematic cross-section of the basic device is shown in Fig. 21.7(a). The transistors behave recognizably as advances on the 0.5 μm devices, but already a number of problems associated with small-scale phenomena occur. The gate oxide is sufficiently thin (less than 5 nm!) that electrons tunnel through from the gate, fluctuations in the polysilicon induce extra scattering in the Si inversion layer, and the reduced gate voltages limit the carrier density and current-handling ability. With short source–drain distances hot electron effects from very high electric fields may degrade the transistor performance, or even lead to dielectric breakdown in the semiconductor. Furthermore, once electrons reach their saturated drift velocity, further bias does not increase speed. The trends would be to reduce the operating voltage, but this must exceed kT/e by a substantial amount to ensure that thermally excited electrons do not play any role in parallel with the bias. A move to low temperatures allows reduced voltages and several other attractive features follow, including a sharpened threshold voltage. Also, one can bias the substrate without leakage to the source–drain in such a way as to reduce the thickness of the depletion regions around the source and drain. The results in Fig. 21.7(b) show that, while there is no intrinsic problem with room-temperature operation, device performance at 77 K is significantly improved. The results in Fig. 21.7(c) at 77 K for the 0.07 μm device do reveal some problems, with a series resistance showing up. The devices are very process dependent, and ever-tighter control over the processing steps is needed.

The move to devices with feature sizes below 0.1 μm will produce problems such as the small number of electrons in the channel (only a few hundred per micrometre gate width) and the statistical fluctuations inherent in the small number involved. The integration of many devices into a circuit cannot tolerate these fluctuations. All the mesoscopic phenomena being investigated in GaAs structures will be encountered in Si research devices over the next decade. Note that 0.1 μm technology will not enter the marketplace until the next century if there is no slowdown in the market.

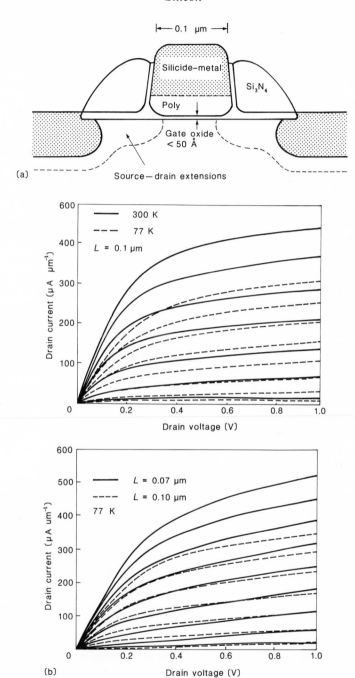

Fig. 21.7 (a) Schematic layout of a 0.1 μm silicon MOSFET; (b) the characteristics at 300 K and 77 K of a 0.1 μm gate length device and at 77 K for a 0.1 μm and 0.07 μm device (After Sai-Halasz *et al* © 1987 IEEE.)

21.8 Metrological applications of low-dimensional semiconductors

Metrology, the science of precision measurements and standards, has been undergoing important developments in recent years. The ability to measure precisely the frequency associated with the Josephson effect has led to its introduction as a voltage standard. Alternatively, through the Josephson relation, i.e. $\omega_J = 2eV/\hbar$ where ω_J is the angular frequency of a supercurrent induced by a bias V applied to a superconducting tunnel junction, we can regard the effect as a method for measuring $e/\hbar$. The quantum Hall effect (see Chapters 4 and 5), discovered von Klitzing *et al.* (1980), measured the quantity h/e^2 to a few parts in 10^6. A decade later the accuracy has been improved by a further factor of 100, and the von Klitzing constant $R_K = h/e^2 = 25\,812.807 \pm 0.005\,\Omega$ was introduced in 1990 as a resistance standard. The attention to detail required for metrological measurements with respect to this resistance standard is reviewed by Hartland (1992).

The recent discoveries of 'electron turnstiles' or 'electron pumps' and the use of Coulomb blockade effects (Chapter 12) for counting individual electrons into and out of circuits that include ultrasmall tunnel junctions, originally of metals (Geerligs 1992), has led to initial investigations of its application as an absolute method for measuring currents (Martinus *et al.* 1994). A set of five ultrasmall tunnel junctions are connected in series. Each junction can also be regarded as a capacitor and each node can be independently biased (Fig. 21.8). An appropriate sequence of biases applied to each node can allow just one electron to be sent through the the set of junctions, and so far an accuracy of 0.5 in 10^6 has been achieved, with photon-assisted tunnelling being conjectured as the principal source of error. Electrons can be clocked in and out of a quantum box in a semiconductor structure of the form shown in Chapter 12, Figs 12.6 and 12.7, by using an a.c. signal applied to the surface gates and a d.c. bias applied to the electron gas so that just one electron enters and leaves in one cycle. This system lends itself to metrological investigations, as the relative importance of the other conduction channels, that might limit the precision, can be monitored more precisely.

The prospect of transferable solid state device standards for current, voltage, and resistance by 2000 is real, and is a matter of great practical convenience. Their mutual consistency must first be established.

21.9 General

We conclude this chapter with some considerations that are not scientific or technological but commercial. The pace of development of electronic circuits has no rival throughout history. Until now there has been an insatiable demand for higher-performance circuits. In the last 2 years, a world recession has slowed down the demand and this has had a serious knock-on effect on research and development. The pause has given users time to realize that the

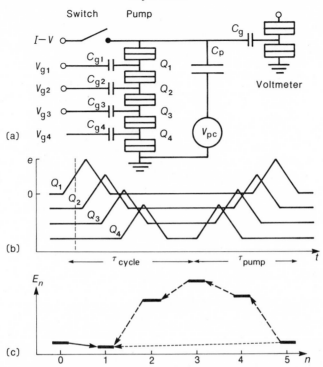

Fig. 21.8 (a) An electron pump consisting of five ultrasmall tunnel junctions (the box symbols), (b) the charge polarizations ($Q_i = C_{gi}V_{gi}$) at the nodes as induced by the gate voltages to obtain the optimum pumping, and (c) the Coulomb energy (solid line) on each node at the instant denoted by the broken line in (b). The broken and dotted lines in (c) indicate parallel transport channels associated with thermal activation or direct tunnelling. (After Martinus *et al.* 1994.)

increased sophistication and performance of the hardware has not been matched in software, and that a technology pause could be tolerated while the further advantages available within today's technology are realized. This is also having a slowing-down effect, so that future demands on hardware may be slower in coming. While this might give science and technology time to address and solve the real problems for the next generations of technology, it also reduces the sums available for investment, exacerbating the problem.

References

Berry, M. V. (1984). Quantal phase factors accompanying adiabatic changes. *Proceedings of the Royal Society of London, Series A*, **392**, 45–57.

Brown, E. R., Goodhue, W. D., and Sollner, T. C. L. G. (1988). Fundamental oscillations up to 200 GHz in resonant tunneling diodes and new estimates of their maximum oscillation frequency from stationary-state tunneling theory. *Journal of Applied Physics*, **64**, 1519–29.

Collins, A. T. (1989). Diamond electronic devices—a critical appraisal. *Semiconductor Science and Technology*, **4**, 605–11.

den Broeder, F. J. A., van Kesteren, H. W., Hoving, W., and Zeper, W. B. (1992). Co/Ni multilayers with perpendicular magnetic anisotropy: Kerr effect and thermomagnetic writing. *Applied Physics Letters*, **61**, 1468–70.

Dennard, R. H. (1986). Scaling limits of silicon VLSI technology. In *The physics and fabrication of microstructures and microdevices* (ed M. J. Kelly and C. Weisbuch), pp. 352–69. Springer-Verlag, Berlin.

Dennard, R. H., Gaensslen, F. H., Yu, H. N., Rideout, V. L., Bassous, E., and LeBlanc, A. R. (1974). Design of ion-implanted MOSFETs with very small physical dimensions. *IEEE Journal of Solid-State Circuits*, **SC-9**, 256–68.

Geerligs, L. J. (1992). The frequency-locked turnstile for single electrons. *Surface Science*, **263**, 396–404.

Geis, M. W., Efremow, N. N., and Rathman, D. D. (1988). Summary abstract: device applications of diamonds. *Journal of Vacuum Science and Technology A*, **6**, 1953–4.

Haase, M. A., Qiu, J., DePuydt J. M., Cheng, H. (1991). Blue-green laser diodes. *Applied Physics Letters*, **59**, 1272–4.

Hartland, A. (1992). The quantum Hall effect and resistance standards. *Metrologia*, **29**, 153–74.

Ivanov, P. A. and Chelnokov, V. E. (1992). Recent developments in SiC single-crystal electronics. *Semiconductor Science and Technology*, **7**, 863–80.

Jeon, H., Ding, J., Nurmikko, A. V., Xie, W., Grillo, D. C., Kobayashi, M., *et al.* (1992). Blue and green diode lasers in ZnSe-based quantum wells. *Applied Physics Letters*, **60**, 2045–7.

Martinus, J. M., Nahum, M., and Jensen, H. D. (1994). Metrological accuracy of the electron pump. *Physical Review Letters*, **72**, 904–7.

Muratake, S., Watanabe, M., Suemasu, T., and Asada, M. (1992). Transistor action of metal (CoSi$_2$)/insulator CaF$_2$ hot electron transistor structure. *Electronics Letters*, **28**, 1002–4.

Parkin, S. S. P., Bhadra, R., and Roche, K. P. (1991). Oscillatory magnetic exchange coupling through thin copper layers. *Physical Review Letters*, **66**, 2152–5.

Sai-Halasz, G. A., Wordeman, M. R., Kern, D. P., Ganin, E., Rishton, S., Zicherman, D. S., *et al.* (1987). Design and experimental technology for 0.1 μm gate-length low-temperature operation FET's. *IEEE Electron Device Letters*, **8**, 463–6.

Sai-Halasz, G. A., Wordeman, M. R., Kern, D. P., Rishton, S. A., Ganin, E., Chang, T. H. P. and Dennard, R. H. (1990). Experimental technology and performance of 0.1 μm-gate-length FETs operated at liquid-nitrogen temperature. *IBM Journal of Research and Development*, **34** 452–65.

Schuller, I. K. (1988). The physics of metallic superlattices: an experimental point of view. In *Physics, fabrication and applications of multilayered structures* (eds P. Dhez and C. Weisbuch), NATO ASI Series B: Physics, Vol. 182, pp. 139–69. Plenum, New York.

Tung, R. T. (1984) Schottky-barrier formation at single-crystal metal–semiconductor interfaces. *Physical Review Letters*, **52**, 461–4.

von Klitzing, K., Dorda, G., and Pepper, M. (1980). Realisation of a resistance standard based on fundamental constants. *Physical Review Letters*, **45**, 494–7.

Watanabe, M., Muratake, S., Fujimoto, H., Sakamori, S., Asada, M., and Arai, S. (1992). Epitaxial growth of metal (CoSi$_2$)/insulator (CaF$_2$) nanometre-thick layered structure on Si(111). *Japanese Journal of Applied Physics*, **31**, L116–18.

Werner, T. W., Banerjee, I., Yang, Q. S., Falco, C. M., and Schuller, I. K. (1982). Localisation in a three-dimensional metal. *Physical Review B*, **26**, 2224–6.

Yablonovitch, E. (1987). Inhibited spontaneous emission in solid state physics and electronics. *Physical Review Letters*, **58**, 2059–62.

Yablonovitch, E. (1993). Photonic band-gap structures. *Journal of the Optical Society of America B*, **10**, 283–95.

Yablonovitch, E., Gmitter, T. J., and Leung, K. M. (1991). Photonic band structure: the face-centred-cubic case employing non-spherical atoms. *Physical Review Letters*, **67**, 2295–8.

Yang, W. M. C., Tsakalakos, T., and Hilliard, J. E. (1977). Enhanced elastic modulus in composition modulated gold–nickel and copper–palladium foils. *Journal of Applied Physics*, **48**, 876–9.

22 Radical alternatives

22.1 Introduction

So far in this text we have dealt with multilayers of conventional III–V and group IV semiconductors, and described what has been the mainstream research in semiconductor science and technology. The rapid advances within this technology have spurred those working in other fields and with other materials to try to match and surpass these achievements. In this closing chapter we introduce some of these radical alternatives to mainstream semiconductor science and technology. While some scientists strive in the vain hope of outperforming Si at what Si is best at, the main advances are likely to be in those areas where the final devices perform functions not available with Si or III–V semiconductors. As mentioned in the early chapters, III–V materials have specific properties, not available in Si, that have allowed them flourish, namely (i) the direct nature of the bandgap, making optical functions efficient, (ii) the turnover in the velocity–field characteristic leading to negative differential resistance and microwave sources, and (iii) the ability of substrates to retain a semi-insulating property after being taken through device processing. Increases in device speed by a factor of 3 in III–V materials compared with Si allow windows of opportunity for III–V digital applications, but in the end Si technology wins out through the greater maturity of its technology. Radical alternative technologies will need distinctive attributes not available with the group IV and III–V materials and devices.

Superconductors, organic materials, and molecular structures are under intense investigation. The first of these exploit the vanishing resistance of metals in the superconducting state to eliminate RC delays in interconnections, and to perform novel device functions with both high speed and minimal power dissipation. For the very highest performance figures (high speed, low noise, etc.), superconducting devices have the edge over their semiconducting rivals (see Chapter 16, Fig. 16.12). The problems come with interfacing these devices to ambient temperature electronics, and the degradation of superconducting devices by repeated thermal cycling during fabrication, use, and maintenance. Superconducting devices have clear niche roles, and there is a debate about superconducting supercomputers in the future. The exploitable properties of both organic materials and molecular structures range far wider than can be achieved with conventional electronic materials. The prospect of truly flexible devices is attractive, but the long-term stability of these materials against temperature, chemical attack, light-assisted chemical processes, etc. is delaying their widespread deployment.

Another radical approach is in the architecture of circuits. The layout of devices in memory and computational chips has become standardized over the last 30 years. Just as the limits to miniaturization have posed many of the problems of basic microscopic physics described in this text, whose answers may circumvent the currently perceived limits, radical architectures are being investigated that allow information to be processed in parallel, or in other ways that differ from the serial computation. For example, since diffraction gratings send out light in the different orders that represent Fourier components, optical devices are particularly suited to this type of signal processing. Biological computing structures are considered in section 22.4.

22.2 Superconducting electronics

For 15 years from about 1970, major research and development was devoted to the superconducting computer (Matisoo 1980). At the outset, the projected device figures of merit (e.g. 50 times faster) were so far in advance of those predicted for conventional Si electronics (cf. Chapter 16, Fig. 16.12) that even problems such as cooling the computer to 4.2 K could be accommodated. By the mid-1980s, the predicted performance gap between superconducting and semiconductor computers had closed to the extent that many companies ceased development work. However, several Japanese companies have continued development and achieved impressive performances in prototypes (Hayakawa 1990). In recent years, new higher-temperature (above 77 K) superconductors have been developed, but the materials constraints imposed by practical devices have not been met so far with these new materials.

At the core of a supercomputer is the Josephson junction device shown in Fig. 22.1(a): the relevant details of the superconducting state will not concern us here, by they can be found in books by Orlando and Delin (1991) and by Ruggeiro and Rudman (1990). A sufficiently thin layer of insulator separates two superconducting layers, and superconducting pairs of electrons can tunnel through this barrier leading to current–voltage characteristics well below the transition to the superconducting state as shown. Provided that the current is below a critical value, no voltage is dropped across the device. If a magnetic field penetrates the insulating layer, the critical current is reduced, varying as shown in Fig. 22.1(b), and reaches zero each time a quantum of magnetic flux (given by $h/2e = 2 \times 10^{-15}$ T m^2) is contained within the oxide layer. In this and in modified forms, the two-terminal device is an extremely sensitive detector of magnetic fields (Clarke 1990). Such fields are easily generated by a d.c. current, and the practical device is shown in Fig. 22.1(c), where the control current reduces the critical current to zero, and there is a voltage (equal to the superconducting gap energy divided by the electronic charge) across the tunnel junction. The use of this structure within circuits for both logic and memory functions takes us into levels of detail beyond the scope of this chapter. The characteristic lateral feature sizes are a few

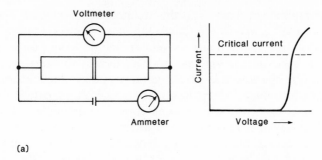

(a)

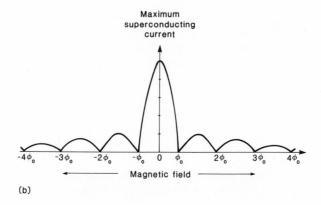

(b)

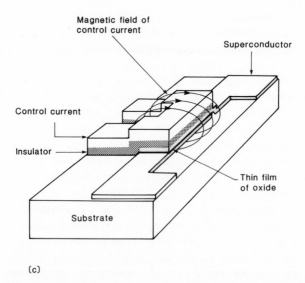

(c)

Fig. 22.1 (a) The Josephson junction diode, (b) the effect on the critical current of a magnetic field, and (c) the use of this junction as a device. Adapted from J. Matisoo, 'The Superconducting Computer' in Scientific American **242**(5), Copyright © by Scientific American Inc. All rights reserved.)

micrometres, and the superconducting materials are lead and niobium, with a native layer of grown oxide as the insulator (this is usually only a few nanometres thick). Under these conditions switching times of $c.10$ ps are achieved. This last figure has now been achieved in GaAs HEMT devices (cf. Chapter 16) working at 77 K. The levels of power dissipated in superconducting devices remain much lower than achieved with semiconductors, although the power consumed in refrigeration reduces the difference. At present, it appears that the improvements to GaAs-based digital electronic circuits with cooling to 77 K provide systems performance comparable to that achieved with superconductors, but using a relatively more mature technology.

Since the discovery of the copper oxide superconductors (with $T_c > 77$ K) in 1986, research on these materials for device applications has been intense (Likharev *et al.* 1990; Donaldson *et al.* 1992). However, the materials requirements for computational devices made through some practical fabrication route have not yet been met. Tunnel insulators must have a very uniform thickness of only about 1 nm. The critical current densities (when a superconductor becomes normal, acquiring a finite resistance) are too low to be practical in many computational and related applications. The major area of concentrated research is on the use of these superconductors as passive microwave components, replacing copper and other materials in waveguides, resonators, and filters (Cracknell 1992). Here use is made of the very small a.c. resistivity in the surface skin depth (the London penetration distance) of these materials compared with that of copper up to $c.100$ GHz; this quantity measures the losses in the passive components, and so the Q of the cavity, filter, etc. These new superconductors are likely to be used in those situations where the increased performance demands it and/or the cost of refrigeration can be accommodated (as in many space-borne applications). Any further applications of high-temperature superconductors at 77 K will lead to the advent of hybrid superconductor–semiconductor devices. Superconducting interconnects will reduce losses, provided that their critical current densities are raised to suitable levels (a factor of 10–100 better than is routine at present). In infrared applications where 77 K is needed for the sensing, it may be most efficient to perform some of the signal processing at the sensing pixel, in which case the choice may be between semiconductor, superconductor, or hybrid technologies.

One area where superconducting materials still win out over semiconductors is in applications where sources of radiation are needed at submillimetre wavelengths (above 300 GHz). The small structures and small RC time constants mean that 1 THz operation is feasible. Arrays of Josephson junctions are proposed as 1 THz sources capable of delivering 10 mW power (Lukens 1990). The extreme non-linearity of the current–capacitance characteristics shown in Fig. 22.1 are routinely exploited at above 300 GHz for astronomical observations of molecular vibrations in gases in outer space (Hu and Richards 1990).

22.3 Organic semiconductors

A distinguishing feature of the world of organic chemistry is the ability to fine tune most electrical, optical, chemical, or other materials properties by incorporating different chemical groups at key points within the overall organic compound. This is no better exemplified than in the field of liquid crystals, whose now exploited properties were enhanced from modest changes in optical properties at low temperatures in materials of poor stability to the striking changes in robust and varied materials that are in widespread use as displays (Bahaddur 1990, 1991, 1992). The subset of organic materials that are semiconductors share this feature of tailorability, and it is widely exploited in the design of materials for particular electronic and optical functions.

We first describe briefly the nature of organic semiconductors (which is quite different in detail from that of the group IV and III–V materials) and the means of doping them (Conwell 1988). Polyacetylene $(CH)_x$ is the simplest form of relevant material, and is shown schematically in Fig. 22.2(a). The polymer is long and flat. Each carbon atom is three-coordinated, with sp^2 bonding (Harrison 1980) between adjacent carbon atoms and to the hydrogen atoms (i.e. the same form as in graphite as opposed to the diamond form of carbon). There is one remaining valence electron for each carbon that sits on p_z orbitals, and the orbitals from adjacent carbon atoms overlap to form a 1D band structure. A slight distortion associated with alternating short and long bonds along the backbone contributes a potential which opens up bandgaps. With one electron per carbon atom (and two per pair of carbon atoms that form the unit cell), we have a filled π (bonding) valence band and an empty π^* (antibonding) conduction band, the gap being about 1.5 eV, i.e. comparable with GaAs. It has been known that very high mobilities can be obtained for electron motion along these chains, and doping is achieved by the introduction of agents that oxidize or reduce the polymer, such as iodine, resulting in conductivities close to copper.

There is much detailed technology associated with the fabrication of these materials and processing them into devices that need not concern us here. In Fig. 22.2(b) we show a transistor structure to scale (except for a channel of 20 μm width and 1.5 m length), and its characteristics are given in Fig. 22.2(c), showing the field-effect modification of the source–drain current for fixed source-drain voltage as the gate-source voltage was varied. Although the devices can be modelled in conventional semiconductor terms, the charge is not stored simply as electrons and holes as in group IV and III–V semiconductors. A carrier (an excess or deficit of charge) on the polyacetylene chain can result in large local distortions, with the combination of charge and distortion called a soliton. The distortion is often large enough to require a recalculation of the energy levels. In Fig. 22.2(d) we show two pieces of $(CH)_x$ where the alternation between double and single bonds is interrupted. This configuration allows for a non-bonding p level to be formed in the middle of the gap, in which the extra charge is stored. As the carrier moves,

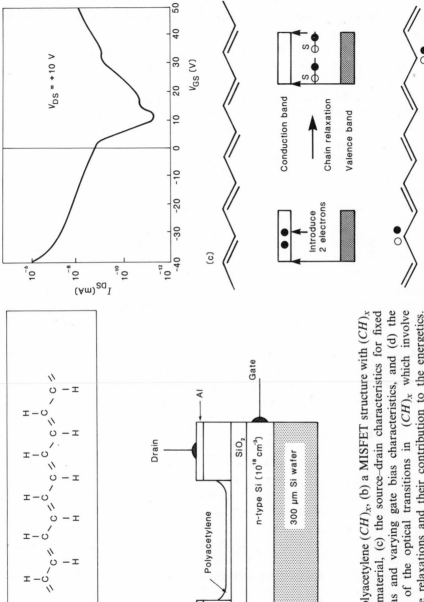

Fig. 22.2 (a) Polyacetylene $(CH)_x$, (b) a MISFET structure with $(CH)_x$ as the channel material, (c) the source–drain characteristics for fixed source–drain bias and varying gate bias characteristics, and (d) the complex nature of the optical transitions in $(CH)_x$ which involve significant lattice relaxations and their contribution to the energetics. (After Burroughes and Friend 1988.)

this local distortion is carried along with it. Instead of the accumulation and depletion arguments of earlier chapters, extra charge is induced in the polymer chains by the applied fields that are associated with the formation of more of these charge solitons. Such distortions also accompany optical excitations of carriers from π to π^* bands, with the extra energy for this distortion being provided by the exciting light in addition to that required for a simple transition between non-distorting energy levels (Fig. 22.2(d)). Optical absorption by charges in the mid-gap states can occur at energy lower than the gap energy resulting in excitations into the conduction bands and further reorganization of the backbone bonding and atomic structure. The polymer transistor performance is very modest compared with the GaAs devices described in Chapter 16, but one would not want to compete directly. Since these polymeric materials are already widely used as insulators, the possibility of active electronics within plastics opens up many new areas of applications not accessible to Si or GaAs. The polymeric backbones can be modified to increase the stability of the bulk or thin film materials against chemical, temperature, light, and other attack that can degrade these materials if such precautions are not taken. The range of mechanical, optical, electrical, and other properties can be modified in concert in a way not available in conventional semiconductors.

The optical properties of some organic polymeric suggest a future as light-emitting devices, again not to compete with the components described in Chapter 18, but rather to complement them, as for example in very-large-area flexible (even foldable) displays. The important factor is to convert current into light (electroluminescence), and while many materials can do this, most are very inefficient. This phenomenon has been reported in polyphenylenevinylene, producing weak yellow-green emission at 0.01 per cent efficiency (Burroughes *et al.* 1990). Chemical tuning of the precursor materials with appropriate side-groups has already increased this to 0.3 per cent, and enabled the wavelength of light emission to be tailored in the visible (Burn *et al.* 1992). These materials now match the as yet relatively inefficient II–VI materials in the blue-green region of the spectrum, and further chemical tailoring is in progress. The variation with pressure and temperature of the optical emission is very high, and may yet be exploited in sensing applications.

The emphasis above is on polymers, i.e. macroscopic samples of materials with rigidity and other desirable properties. Much research is taking place on the electronic and optical properties of small organic molecules, inspired by the trend shown in Fig. 22.3, namely that scaling beyond conventional devices requires us to consider starting with device building blocks of molecular dimensions within two decades. The intimate relationship between atomic and electronic structure is suggested as the basis for novel device functions that might perform on the scale of a few nanometres. Details such as thermal stability of stored charge, the interconnection of dense 3D arrays, and signal input–output have not yet been considered.

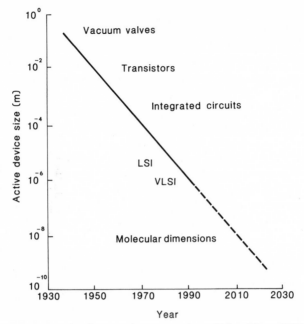

Fig. 22.3 Device feature size versus time. (After Bloor 1991.)

22.4 Bioelectronic models

One growing trend is the examination of information processing in living systems in a search for new concepts for device and signal processing in inorganic systems. The first level of analysis comes from the physiology of the nervous system and the brain. It is appreciated that signals are sent electrically, but often using ionic rather than electronic conduction. The typical speeds at which signals are transmitted are much slower than in a circuit interconnect. With reference to the schematic diagram of a cell shown in Fig. 22.4(a), we note that input signals come in along channels called dendrites, are processed in the soma, and the result is transmitted out via axons. The computational unit of input, processor, and output is called a neurone. At the ends of axons are valves called synapses, which transmit signals into following dendrites. The size of the computational unit is much larger than in modern integrated circuits, but the connectivity and the nature of the computational function are quite different. The very high connectivity of neurones (up to 100 dendrites feeding out to tens of axons) is a major qualitative difference between the biological and the semiconductor computer. So too is the 3D topology of the interconnections. It is clear that modern computers can perform basic arithmetic functions orders of magnitude faster than the biological computer. However, attempts to get semiconductor electronics to perform even simple tasks undertaken by the brain, such as is associated with pattern recognition, quickly consume inordinate amounts of computational

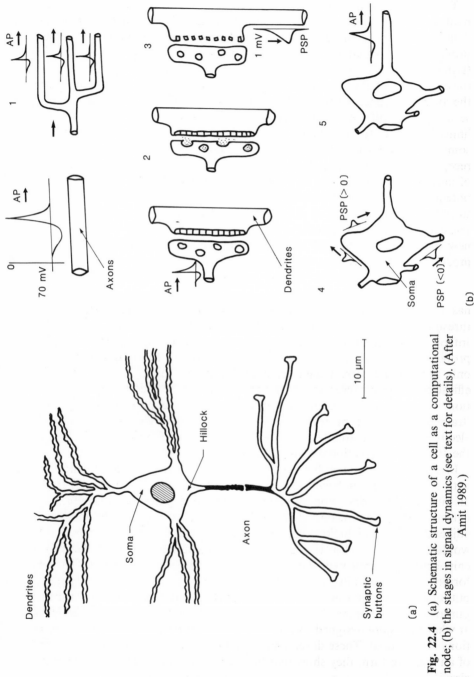

Fig. 22.4 (a) Schematic structure of a cell as a computational node; (b) the stages in signal dynamics (see text for details). (After Amit 1989.)

power and time. There must be radical alternative strategies for such computations.

The dynamic processes in neural communications and computation are relatively simple to describe and model (Fig. 22.4(b)), although the practical details are very complex. First, at any instant a given axon is either propagating a signal or is in a quiescent state resting at some potential (typically a few tens of millivolts). If the axon is transmitting, a potential spike (or action potential (AP)—again a few tens of millivolts) is propagating along the axon; this spike is very stable in shape and amplitude, and is such that a second signal will not propagate at the same time. In fact, repetition rates are limited typically to 500–1000 s^{-1} and in parts of the brain this rate falls in some cases to 30–40 s^{-1}. At the synapse, the transmitted signal causes the receptor side to open up and admit ionic current to the neurone, with amount of current being a measure of the efficacy of the synapse. The post-synaptic potential (PSP), now of amplitude about 1 mV and of either sign diffuses towards the soma, where all the input potentials are summed. When the sum reaches a threshold of some tens of millivolts (thus requiring many inputs), a new pulse is triggered and sent down the output axon. The mathematical model of this process

$$V_{\text{out}} = \text{threshold}(\Sigma\, V_{\text{inputs}})$$

has been closely investigated as a function of the level and shape of the threshold function (i.e. a sharp step function or a smoothed version of it). Interacting collections of such nodes (so-called neural networks) have been proposed as models for learning and for recovering full output data when only part of the correct data are available at the input. In each case, the efficacies of the synapses are the variables that can be used to store information, and they can be established during a process where the network is being 'trained'. At present, prototypes of these systems are being implemented in Si electronics, and the architectures are quite different from those established in the 1960s for conventional computational processors.

In some computations, for example in image processing, one is concerned with a very large number of simple calculations being performed in parallel at different locations. For example an edge in a pixellated image can be detected by taking second differences of a measure of the brightness in adjacent pixels and looking for zeros. Conventional computer architecture would take each pixel in turn and perform that function, but alternatively a chip could be fabricated which would allow these calculations to be performed at each pixel in parallel and simultaneously, and with either a parallel output of all the data or a serial output of the zeros. Chips designed for this purpose (i.e. containing the algorithms in 'hard-wiring') can be very effective at what they were designed for, but are quite inflexible when other computations are required. These developments mimic processes in the visual systems of animals. In turn, they show that the transistors in conventional integrated circuits are not used very efficiently, i.e. they are idle for most of time. There are vast gains to be made if circuits can be designed to make better use of the hardware available. Some have proposed that the perception of continued improvements in computer power will continue without significant change, even after limitations imposed by hardware technology are reached, simply because improved software and design will enhance the usage of the hardware.

The generalization of the parallel pixellated computation just described is to regard each node as a computational machine, and to provide it with an increasingly sophisticated set of time-dependent interconnects. This is the study of cellular automata—each node has a number of internal degrees of freedom obeying prescribed rules analogous to the threshold condition above, and the evolution of the behaviour of a system of such automata is controlled by the weights and time dependences of various interconnects.

It is fair to say that progress has been most rapid where the computational problem is most closely specified. Sceptics suggest that all the current models are too simple to contain anything of the sophistication of living communication systems, while the advocates claim that they will approach the required level of complexity in stages.

Mainstream semiconductor science and technology has a clear future, but radical alternatives are likely to take us even further ahead in more diverse directions in the longer term.

References

Amit, D. J. (1989). *Modelling brain function*. Cambridge University Press.

Bahaddur, B. (1990, 1991, 1992). *Liquid Crystals: Applications and Uses*, Vols 1, 2, 3. World Scientific, Singapore.

Bloor, D. (1991). Breathing new life into electronics. *Physics World*, **4**(11), 36–40.

Burn, P. L., Holmes, A. B., Kraft, A., Bradley, D. D. C., Brown, A. B., Friend, R. H., and Gymer, R. W. (1992). Chemical tuning of electroluminescent copolymers to improve emission efficiencies and allow patterning. *Nature*, London, **356**, 47–9.

Burroughes, J. H. and Friend, R. H. (1988). Polymer transistors. *Physics World*, **1** (November), 24–6.

Burroughes, J. H., Bradley, D. D. C., Brown, A. R., Marks, R. N., Mackay, K., Friend, R. H., *et al.* (1990). Light-emitting diodes based on conjugated polymers. *Nature*, London, **347**, 539–41.

Clarke, J. (1990). SQUIDs: principles, noise and applications. In *Superconducting devices* (eds S. T. Ruggiero and D. A. Rudman), pp. 51–99. Academic Press, New York.

Conwell, E. M. (1988). *Semiconductors and semimetals*, Vol. 27, *Highly conducting quasi-one-dimensional organic crystals*. Academic Press, New York.

Cracknell, D. P. (1992). The application of high temperature superconductors to microwave filters. *GEC Journal of Research*, **9**, 155–65.

Donaldson, G. B., Bowman, R. M., Cochran, A., Kirk, K. J., Pegrum, C. M., and Macfarlane, J. C. (1992). Progress in high T_c magnetic sensors and their applications. *Physica Scripta*, **T45**, 34–40.

Harrison, W. A. (1980). *Electronic structure and the properties of solids*. Freeman, San Francisco, CA.

Hayakawa, H. (1990). Computing. In *Superconducting devices* (eds S. T. Ruggiero and D. A. Rudman), pp. 101–34. Academic Press, New York.

Hu, Q. and Richards, P. L. (1990). Quasiparticle mixers and detectors. In *Superconducting devices* (eds S. T. Ruggiero and D. A. Rudman), pp. 169–96. Academic Press, New York.

Likharev, K. K., Semenov, V. K., and Zorin, A. B. (1990). New possibilities for superconducting devices. In *Superconducting devices* (eds S. T. Ruggiero and D. A. Rudman), pp. 1–49. Academic Press, New York.

Lukens, J. (1990). Josephson arrays as high frequency sources. In *Superconducting devices* (eds S. T. Ruggiero and D. A. Rudman), pp. 135–67. Academic Press, New York.

Matisoo, J (1980) The superconducting computer. *Scientific American*, **242**, 38–53.

Orlando, T. P. and Delin, K. A. (1991) *Foundations of applied superconductivity*. Addison-Wesley, New York.

Ruggiero, S. T. and Rudman, D. A. (eds). (1990). *Superconducting devices*. Academic Press, New York.

Appendix 1
Annotated bibliography

Abram, R. A. and Jaros, M. (eds) (1989). *Band structure engineering in semiconductor microstructures* NATO ASI Series B: Physics, Vol. 189. Plenum, New York.
Contains 32 papers, approximately half of them theoretical, covering the electronic structure and band offsets, and the transport and optical properties of semiconductor multilayers.
Allan, G., Bastard, G., Boccara, N., Lannoo, M., and Voos, M. (ed). (1986). *Heterojunctions and semiconductor superlattices*. Springer-Verlag, Berlin.
Contains 19 papers on theory, experimental physics, technology, and applications.
Balkan, N., Ridley, B. K., and Vickers, A. J. (ed). (1993). *Negative differential resistance and instabilities in 2-D semiconductors*. NATO ASI Series B, Physics, Vol. 307. Plenum, New York.
Contains 30 papers mainly on negative differential resistance for perpendicular transport through heterostructures.
Bastard, G., (1988). *Wavemechanics applied to semiconductor heterostructures*. Editions du Physique, Paris, and Halsted Press, New York.
Eight extensive theoretical chapters with application of quantum mechanics to electronic energy levels in III–V bulk semiconductors and heterojunctions, quantum wells, and superlattices derived from them, together with implications for transport and optical measurements.
Beaumont, S. P. and Sotomayor Torres, C. M. (ed). (1990). *Science and engineering of one- and zero-dimensional semiconductors*, NATO ASI Series B: Physics, Vol. 217. Plenum, New York.
Contains 31 papers on fabrication, transport, and optical properties of zero- and one-dimensional structures and periodic arrays of such structures.
Beeby, J. L. (ed). (1991). *Condensed systems of low dimensionality*, NATO ASI Series B, Physics, Vol. 253. Plenum, New York.
A broad coverage of low-dimensional systems, including growth and characterization of structures, electronic and optical properties, and device applications. There are chapters on multilayers in other than semiconductors and on molecular systems.
Capasso, F. (ed). (1990). *The physics of quantum electron devices*. Springer-Verlag, Berlin.
Contains 12 papers on molecular beam epitaxy, fine lithography, and devices based on resonant tunnelling, hot electron injection, quantum interference, and quasi-one-dimensional physics.
Capasso, F. and Margaritondo, G. (ed). (1987). *Heterojunction band discontinuities: physics and applications*. North-Holland, Amsterdam.

Contains 14 papers on the measurements and theory of band discontinuities, and on the physics of heterojunction devices and band-structure
engineering.

Chemla, D. S. and Pinczuk, A. (ed). (1986). Special Issue on Semiconductor
quantum wells and superlattices: physics and applications, *IEEE Journal of
Quantum Electronics*, **QE-22** (9), 1609–1920.
Contains 32 papers mainly on the electronic and optical properties of
semiconductor multilayers and their applications.

Dhez, P. and Weisbuch, C. (ed). (1988). *Physics, fabrication and applications
of multilayered structures*, NATO ASI Series B: Physics, Vol. 189. Plenum,
New York.
Contains 14 papers, and a larger number of abstracts, about the growth,
characterization, and physics of multilayers in semiconductors, metals, and
other materials, together with applications in electronics, X-ray optics, and
neutron optics.

Dingle, R. (ed). (1987). *Semiconductors and semimetals*, Vol. 24, *Applications
of multiquantum wells, selective doping and superlattices*. Academic Press,
New York.
Eight detailed chapters on the physics of two-dimensional quantized
structures, high-electron-mobility transistors, including their integration
and microwave properties, optical signal processing with quantum wells,
graded-gap and superlattice devices, quantum well lasers, and strained-
layer superlattices.

Grubin, H. L., Ferry, D. K., and Jacoboni, C. (ed). (1988). *The physics of
submicron semiconductor devices*, NATO ASI Series B: Physics, Vol. 180.
Plenum, New York.
Contains 25 chapters covering the general concepts of submicrometre
devices.

Heinrich, H., Bauer, G., and Kuchar, F. (ed). (1988). *Physics and technology
of submicron structures*. Springer-Verlag, Berlin.
Contains 32 short papers on fabrication of microstructures, vertical transport and tunnelling phenomena, quantum interference and mesoscopic
effects, conductance and optical properties of quasi-one-dimensional systems, and the physics and applications of submicrometre devices.

Kelly, M. J. and Weisbuch, C. (ed). (1986). *The physics and fabrication of
microstructures and microdevices*. Springer-Verlag, Berlin.
Contains 34 papers on the physics and engineering of microfabrication, the
physics of microstructures, and perspectives in microfabrication applications, with inputs from physicists, technologists, and device engineers.

Kramer, B. (ed). (1991). *Quantum coherence in mesoscopic systems*, NATO
ASI Series B: Physics, Vol. 254. Plenum, New York.
Contains 11 chapters of lecture notes covering basic electronic properties
and low-dimensional systems, primarily from a theoretical point of view.

Mendez, E. E. and von Klitzing, K. (ed). (1987). *Physics and applications of
quantum wells and superlattices*. NATO ASI Series B: Physics, Vol. 170,
Plenum, New York.
Contains 19 papers on the basic properties and materials growth, electrical
properties, optical properties and applications of semiconductor multilayers, all written by physicists.

Moss, T. S. (ed). (1993–1994). *Handbook on semiconductors*, Vols 1–4 (2nd edn), North-Holland, Amsterdam.

A complete revision of the four volumes of the original 1980 edition, each in excess of 1000 pages. Extensive chapters cover topics in this book, and many other topics in semiconductor science as well, but from a specialist viewpoint.

Reed, M. A. and Kirk, W. P. (ed). (1989). *Nanostructure physics and fabrication*. Academic Press, New York.

Contains 62 papers from a symposium covering the whole range of small structures, their fabrication, and their physics.

Surface Science **142** (1984), **170** (1986), **196** (1988), **228** (190), **267** (1992) and **305** (1994) contain conference proceedings on two-dimensional electron systems, and **174** (1986), **228** (1990), and **270** (1992) contain conference proceedings on modulated semiconductor structures, including general review papers.

Sze, S. M. (ed). (1990). *High speed semiconductor devices*. Wiley, New York.

Contains 10 extensive chapters in three sections: (i) materials, technologies and device building blocks, (ii) field-effect and potential effect transistor devices, and (iii) quantum effect, microwave, and photonic devices.

Weisbuch, C. and Vinter B. (1991). *Quantum semiconductor structures: fundamentals and applications*. Academic Press, New York.

A monograph with chapters on the electronic, optical, and electrical properties of thin semiconductor heterostructure layers, their applications, and the move to one- and zero-dimensional structures.

Appendix 2
Energy, length, and time scales in semiconductor physics

The rescaled hydrogen atom model for a donor electron (Chapter 1) helps to set the scales of energy, length, and time that are of importance in semiconductor physics for devices. The atomic hydrogen Rydberg (the electron–proton binding energy for hydrogen) is scaled for a bulk semiconductor donor by a factor m^*/ε_s^2; here $\varepsilon_s \sim 12$–13 is the static dielectric constant of the semiconductor and m^* is the effective mass of carriers at the bottom of a semiconductor conduction band ($c.0.067m_e$ for GaAs, and a $0.26m_e$ composite of the conduction band masses for Si). Thus $R = 13.6$ eV for the hydrogen atom becomes $R^* = 25.0$ meV for Si and $R^* = 5.3$ meV for GaAs.

Note that 1 meV = 11.6 K = 0.24 THz = 8.07 cm^{-1}. The energy of 1 meV can be obtained in a semiconductor in several different contexts, such as $\hbar\omega_c$, the cyclotron energy in magnetic fields of 1.64 T in Si (with $m^* = 0.19m_e$) and 0.58 T in GaAs. For electrons with $g = 2$, the separation of spin-up and spin-down electrons is $g\mu_B B$, which is 1 meV at 8.64 T. Strain on semiconductors shifts the energy bands, and 1 meV shifts in the Si conduction band structure can be obtained with a stress of 100 N mm^{-2}; note that a Si wafer snaps under bending stresses of approximately 30 times this value. Strong electric fields of 10^6 V m^{-1} give ballistic electrons an energy of 1 meV in 1 nm. For typically thick 0.1 μm layers of SiO_2, the electrons in a 2DEG in a Si MOSFET have their energy shifted by 1 meV for each 1 V applied to the gate.

The corresponding length scale is the effective Bohr radius a^* of the donor electron state which is larger that the atomic hydrogen Bohr radius (0.0529 nm) by a factor of ε_s/m^*, giving $a^* = 2.4$ nm in Si and $a^* = 10.3$ nm in GaAs. Again, this length scale can be achieved in several ways in a solid. The cyclotron radius of an electron in a semiconductor is 25.56 $B^{-1/2}$ nm, and a field of $B = 6.53$ T will give $l_c = 10$ nm in GaAs, while a field in excess of 100 T is required to achieve the equivalent condition in Si.

In terms of $E = h/(2\pi\tau)$, we find that a time of 1 ps is equivalent to 0.66 meV. Many of the transport and optical processes in semiconductors occur on this time scale.

The quantum size and related effects that occupy much of the book are on the 1 meV, 10 nm, and 1 ps scales in energy, distance, and time.

Appendix 3
Valence band structure in the bulk and in quantum wells

A3.1 Bulk valence bands

In Chapters 1, 10, and 13 reference has been made to complications of the electronic structure of semiconductor valence bands in the bulk, in quantum wells, and in strained layers respectively. The Bloch functions associated with electron states near band extrema are relatively simple to describe. Quite generally they are of the form of a plane wave modulated by a function which shares the periodicity of the lattice. In chemical terms, the lowest conduction states near $k = 0$ are such that the periodic function has the spherical symmetry of s-like orbitals placed at each lattice site, and the simple non-degenerate effective mass parabolic band structure is accurate for many purposes. The energy bands have Γ_6 symmetry. The lowest-energy conduction band states are made up of bonding combinations of antibonding orbitals (see Fig. A3.1), which in effect means that alternate atomic sites have opposite signs for the s-orbital amplitude. In contrast, the valence band states are more complicated. The equivalent basis states for building up the periodic parts of the Bloch functions have p-like symmetry and the threefold degeneracy of p_x, p_y, p_z orbitals. Furthermore, there are the two atoms per unit cell whose p orbitals are available to form the Bloch functions. Indeed, the highest-energy valence band states at $\Gamma(k = 0)$ can be described as being antibonding combinations (between adjacent bonds) of bonding combinations of p orbitals within the bonds, as shown schematically in Fig. A3.1 for the case of Si; when we have two different atoms per unit cell, as in GaAs, the situation is more difficult to describe. The results of detailed symmetry analysis of the energy bands in the bulk and in quantum wells is given in outline in the text. Here we give a few more details, referring to Harrison (1980) or Seeger (1991) for a fuller description of the bulk bands, or to Altarelli (1986), Bastard and Brum (1986), or Bastard (1988) for the extension to sub-bands in quantum wells, and to Ekenberg and Altarelli (1984) for the extension to sub-bands at a single heterojunction. Inevitably we make use of group-theoretical results, and those wanting to pursue this aspect further should refer to publications cited in the above references, and also Heine (1960) or Falicov (1966).

Having started with a sixfold degeneracy at $k = 0$, spin–orbit coupling lifts the degeneracy, giving rise to a fourfold degenerate level (labelled Γ_8), which has the properties associated with atomic orbitals having $J = 3/2$, and a twofold degenerate level (labelled Γ_7) with $J = 1/2$. These latter are the spin-split-off states shown in Figs. 1.4 and 10.3, and they have the same symmetry as for the conduction band states and a simple effect mass parabolic band structure. Instead of using atomic s and p basis states, we can construct linear combinations of these of the form $|S\uparrow\rangle$, $|X\uparrow\rangle$, $|Y\uparrow\rangle$, and $|Z\uparrow\rangle$ and $|S\downarrow\rangle$, $|X\downarrow\rangle$, $|Y\downarrow\rangle$, and $|Z\downarrow\rangle$ within a unit cell, and transform them like their atomic counterparts when we map the local tetrahedron of bonds into itself. In what follows their detailed form will not be required. In Table A3.1 we show the appropriate combinations that form the periodic parts of the Bloch functions for the bands at Γ, together with their total and azimuthal angular momentum in this representation, and their energy ($\varepsilon_0 = \varepsilon(\Gamma_6) - E(\Gamma_8)$) is the bandgap energy and $\Delta = \varepsilon(\Gamma_8) - E(\Gamma_7)$ is the spin-split off energy).

	$\lvert iS\uparrow\rangle$	$\lvert\frac{3}{2},\frac{1}{2}\rangle$	$\lvert\frac{3}{2},\frac{3}{2}\rangle$	$\lvert\frac{3}{2},\frac{3}{2}\rangle$	$\lvert iS\downarrow\rangle$	$\lvert\frac{3}{2},-\frac{1}{2}\rangle$	$\lvert\frac{3}{2},-\frac{3}{2}\rangle$	$\lvert\frac{1}{2},-\frac{1}{2}\rangle$
$\langle iS\uparrow\rvert$	$\frac{\hbar^2k^2}{2m_0}$	$-\sqrt{\frac{2}{3}}Ph k_z$	$Ph k_+$	$\frac{-1}{\sqrt{3}}Ph k_z$	0	$\frac{-1}{\sqrt{3}}Ph k_-$	0	$-\sqrt{\frac{2}{3}}Ph k_-$
$\langle\frac{3}{2},\frac{1}{2}\rvert$	$-\sqrt{\frac{2}{3}}Ph k_z$	$-\varepsilon_0+\frac{\hbar^2k^2}{2m_0}$	0	0	$\frac{P}{\sqrt{3}}\hbar k_-$	0	0	0
$\langle\frac{3}{2},\frac{3}{2}\rvert$	$Ph k_-$	0	$-\varepsilon_0+\frac{\hbar^2k^2}{2m_0}$	0	0	0	0	0
$\langle\frac{1}{2},\frac{1}{2}\rvert$	$\frac{1}{\sqrt{3}}Ph k_z$	0	0	$-\varepsilon_0-\Delta+\frac{\hbar^2k^2}{2m_0}$	$\sqrt{\frac{2}{3}}Ph k_-$	0	0	0
$\langle iS\downarrow\rvert$	0	$\frac{P}{\sqrt{3}}\hbar k_+$	0	$\sqrt{\frac{2}{3}}Ph k_+$	$\frac{\hbar^2k^2}{2m_0}$	$-\sqrt{\frac{2}{3}}Ph k_z$	$Ph k_-$	$\frac{1}{\sqrt{3}}Ph k_z$
$\langle\frac{3}{2},\frac{1}{2}\rvert$	$\frac{-1}{\sqrt{3}}Ph k_+$	0	0	0	$-\sqrt{\frac{2}{3}}Ph k_z$	$-\varepsilon_0+\frac{\hbar^2k^2}{2m_0}$	0	0
$\langle\frac{3}{2},\frac{-3}{2}\rvert$	0	0	0	0	$Ph k_+$	0	$-\varepsilon_0+\frac{\hbar^2k^2}{2m_0}$	0
$\langle\frac{1}{2},-\frac{1}{2}\rvert$	$-\sqrt{\frac{2}{3}}Ph k_+$	0	0	0	$\frac{P}{\sqrt{3}}\hbar k_z$	0	0	$-\varepsilon_0-\Delta+\frac{\hbar^2k^2}{2m_0}$

$$\text{(A3.1)}$$

Kane (1957) originally considered only interaction between the periodic functions in Table A3.1, neglecting contributions to other (remote) bands. The first order Hamiltonian takes the form (A3.1) with all other matrix elements zero by symmetry. Note that $k_\pm = (k_x \pm ik_y)/\sqrt{2}$. The sole extra parameter P is a matrix element of the velocity operator between different bands and has the form

$$P = (-i/m_0)\langle S|\mathrm{p}_x|X\rangle = (-i/m_0)\langle S|\mathrm{p}_y|Y\rangle = (-i/m_0)\langle S|\mathrm{p}_z|Z\rangle \qquad \text{(A3.2)}$$

where m_0 is the free-electron mass. In Table A3.2 we list the values of the key band-structure parameters for the main semiconductor compounds. The eigenvalues of matrix (equation A3.1) are as follows (cf. Chapter 10, Section 10.2).

Conduction band:

$$E = E_g + \hbar^2k^2/2m^* + [\sqrt{(E_g^2 + 8P^2k^2/3)} - E_g]/2.$$

Valence bands:

$$E = -\hbar^2k^2/2m \text{ (heavy holes)}$$

$$E = -\hbar^2k^2/2m - [\sqrt{(E_g^2 + 8P^2k^2/3)} - E_G]/2 \text{ (light holes)}$$

Appendix 3

Bottom of conduction band : most bonding
combination of antibonding orbitals

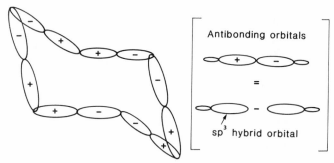

Top of valence band : most antibonding
combination of bonding orbitals

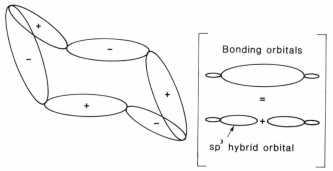

Fig. A3.1 The nature of the states at the bottom of the conduction band and the top
of the valence band from a chemical bond point of view.

$$E = -\Delta - \hbar^2 k^2/2m - P^2 k^2/(3E_g + 3\Delta) \text{ (split-off)} \qquad (A3.3)$$

This model makes explicit the non-parabolicity induced by the interband
matrix elements. It has been checked against more detailed bandstructures
where a more complete set of basis states are used, and is found to be
sufficiently accurate near the zone centre (Γ) for it to be useful.

Kane later included the effect of remote bands using second order perturbation theory. Taking the lowest conduction band and those directly coupled to it
gives a six-band valence Hamiltonian similar to that obtained by Luttinger (1956)

Luttinger further simplified this problem by taking an infinite spin-orbit
energy, giving rise to a four-band Hamiltanion for the Γ_8 states, writing this
in the form

$$E(k) = (\hbar^2/2m_0)[(\gamma_1 + 5\gamma_2/2)k^2 - 2\gamma_2(k_x^2 J_x^2 + k_y^2 J_y^2 + k_z^2 J_z^2)...] \quad (A3.4)$$

where J is the angular momentum associated with the whole state. We can
choose the axis of quantization as the [001] direction and, for
$J_z = \pm 3/2 (\pm 1/2)$, map out the heavy- and light-hole bands as

Table A3.1 Basis Functions at Γ

u_i	$\lvert J, m_j \rangle$	ψ_{jm_j}	$\varepsilon_i \, (k = 0)$
u_1	$\lvert \frac{1}{2}, \frac{1}{2} \rangle$	$i\lvert S\uparrow\rangle$	0
u_3	$\lvert \frac{3}{2}, \frac{1}{2} \rangle$	$-\sqrt{\frac{2}{3}}\lvert Z\uparrow\rangle + \frac{1}{\sqrt{6}}\lvert(X+iY)\downarrow\rangle$	$-\varepsilon_0$
u_5	$\lvert \frac{3}{2}, \frac{3}{2} \rangle$	$\frac{1}{\sqrt{2}}\lvert(X+iY)\uparrow\rangle$	$-\varepsilon_0$
u_7	$\lvert \frac{1}{2}, \frac{1}{2} \rangle$	$\frac{1}{\sqrt{3}}\lvert(X+iY)\downarrow\rangle + \frac{1}{\sqrt{3}}\lvert Z\uparrow\rangle$	$-\varepsilon_0 - \Delta$
u_2	$\lvert \frac{1}{2}, -\frac{1}{2} \rangle$	$i\lvert S\downarrow\rangle$	0
u_4	$\lvert \frac{3}{2}, -\frac{1}{2} \rangle$	$\frac{-1}{\sqrt{6}}\lvert(X-iY)\uparrow\rangle - \sqrt{\frac{2}{3}}\lvert Z\downarrow\rangle$	$-\varepsilon_0$
u_6	$\lvert \frac{3}{2}, -\frac{3}{2} \rangle$	$\frac{1}{\sqrt{2}}\lvert(X-iY)\downarrow\rangle$	$-\varepsilon_0$
u_8	$\lvert \frac{1}{2}, -\frac{1}{2} \rangle$	$\frac{-1}{\sqrt{3}}\lvert(X-iY)\uparrow\rangle + \frac{1}{\sqrt{3}}\lvert Z\downarrow\rangle$	$-\varepsilon_0 - \Delta$

Table A3.2 Key band structure parameters ($E_P = 2m_0 P^2$)

	InP	InAs	InSb	GaAs	GaSb
ε_0(eV)	1.4236	0.418	0.2352	1.5192	0.811
Δ(eV)	0.108	0.38	0.81	0.341	0.752
m_{Γ_6}/m_0	0.079	0.023	0.0139	0.0665	0.0405
E_p(eV)	17	21.11	22.49	22.71	22.88

$$E = (\hbar^2 k_z^2/2m_0)(\gamma_1 - 2\gamma_2) \quad \text{and} \quad E = (\hbar^2 k_z^2/2m_0)(\gamma_1 + 2\gamma_2) \quad \text{(A3.5)}$$

respectively, with heavy- and light-hole masses $m_0/(\gamma_1 - 2\gamma_2)$ and $m_0/(\gamma_1 + 2\gamma_2)$. The axial model is convenient in quantum wells, as one can then obtain motion in the x–y plane with a different effective mass from that which confines the holes in the z direction.

A3.2 Valence bands in quantum wells

The complexity of the bulk valence bands is increased once the cubic symmetry of the bulk lattice is reduced with the introduction of a quantum well.

The simple description of electrons in a box given in Chapters 4, 5, and 10 is strongly justified when applied to semiconductor quantum wells. In Chapter 1, Section 1.4, the electronic wavefunction of an electron weakly bound to a donor ion is a product of the periodic part of a Bloch wavefunction at a band extremum (Γ in the case of donors or acceptors in GaAs) and a relatively smooth envelope function which extends over about 20 nm from the donor ion in the case of GaAs. The mechanics for determining the envelope function is to take the second-order expansion of the band structure about its minimum (i.e. using an inverse effective mass tensor), make the substitution $k \to -i\nabla$, and use it as the kinetic energy operator in a Schrödinger equation. The potential is the perturbing potential, so that

$$-\hbar^2\nabla^2\psi/2m^* + V(r)\psi = E\psi \qquad\qquad (A3.6)$$

in the case of a parabolic band with effective mass m^*, and $V(r) = e^2/(4\pi\varepsilon_s|r|)$ screened with the dielectric constant of the semiconductor host material. This simple result was derived in detail by Luttinger and Kohn (1955). The only serious approximation is that the perturbing potential (the extra charge in the case of a donor ion) is smooth on the scale of a unit cell. In practice this holds except in the immediate vicinity of the ion, and a simple (central-cell) correction can be made if necessary. The same model has been applied extensively to quantum electron states, where the envelope functions are just those electron-in-a-box solutions for an infinite or a finite potential well which in turn is set up by the conduction band discontinuity between two different materials at the heterojunction. Again, although the sharpness of the potential at an interface could be a problem (Burt 1989), the results are sufficiently accurate for comparing experiment and theory in most cases that the equivalent of central-cell corrections are not required.

The treatment of energy levels of holes weakly bound to acceptor ions goes through in the same fashion as for electrons except that the Schrödinger equation becomes a matrix determined by the same $k \to -i\nabla$ substitution in eqn (A3.1) above. The resulting solutions give complex energy levels of predominantly light-hole, heavy-hole, or mixed character. With a natural axis for quantization in the crystal growth direction, the introduction of the quantum-well potential can be made in the generalization of eqn (A3.6) to incorporate the matrix of eqn (A3.1). The terms that couple the k_x and k_y terms with k_z in eqn (A3.1), and hence the derivative terms in the Schrödinger equation (A3.6), mean that numerical solutions must be sought from the outset, leading to the complications of different effective masses determining the in-plane transport and the degree of quantum confinement. The reduced symmetry lifts the remaining degeneracy between the different hole bands, and the various interband matrix elements distort the valence bands leading to the variety of results shown in Chapter 10, Fig. 10.4. Furthermore, the introduction of strain results in further distortions in electronic structure, leading to the even wider variety of results shown in Chapter 13, Fig. 13.7.

The Luttinger simplification of the bulk band structure (eqn (A3.4)) leads to in-plane dispersion in, say, the k_y direction for the heavy hole and light-hole confinement in the k_z direction of

$$E = (\hbar^2 k_y^2/2m_0)(\gamma_1 + \gamma_2) \quad \text{and} \quad E = (\hbar^2 k_y^2/2m_0)(\gamma_1 - \gamma_2).$$

respectively. The relative sizes of the in-plane masses are reversed, being $m_0/(\gamma_1 + \gamma_2)$ for light holes and $m_0/(\gamma_1 - \gamma_2)$ for heavy holes (see Chapter 10).

The detailed calculation of hole states for the (311) GaAs–AlGaAs interface would be of practical interest, as 2D hole gases are formed with Si modulation doping, but results for the extra complexity of rotating the coordinate axes in eqn (A3.1) have not been obtained.

References

Altarelli, M. (1986). Band structure, impurities and excitons in superlattices. In *Heterojunctions and semiconductor superlattices* (eds G. Allan, G. Bastard, N. Boccara, M. Lannoo, and M. Voss, pp. 12–37 Springer-Verlag, Berlin.

Bastard, G. (1988). *Wave mechanics applied to semiconductor heterostructures*. Halsted Press, New York.

Bastard, G. and Brum, J. A. (1986). Electron states in semiconductor heterostructures. *IEEE Journal of Quantum Electronics*, **QE-22**, 1625–44.

Burt M. G. (1989). Exact envelope function equation for microstructures and the particle in a box. In *Band Structure Engineering in Semiconducter Microstructures*, (ed. R. A. Abram and M. Jones), NATO ASI Series B, Vol. 189. Plenum, New York.

Ekenberg, U. and Altarelli, M. (1984). Calculation for hole subbands at a GaAs/Al$_x$Ga$_{1-x}$As interface. *Physical Review B*, **30**, 3569–72.

Falicov, L. M. (1966). *Group theory and its physical applications*. Chicago University Press.

Harrison, W. A. (1980). *Electronic structure and the properties of solids*. Freeman, San Francisco, CA.

Heine, V. (1960). *Group theory in quantum mechanics*. Pergamon, Oxford.

Kane, E. O. (1957). Band structure of indium antimonide. *Journal of Physics and Chemistry of Solids*, **1**, 249–61.

Luttinger, J. M. (1956). Quantum theory of cyclotron resonance in semiconductors: general theory. *Physical Review*, **102**, 1030–41.

Luttinger, J. M. and Kohn, W. (1955). The motion of electrons and holes in perturbed periodic fields. *Physical Review*, **97**, 869–83.

Seeger, K. (1991). *Semiconductor physics: an introduction* (5th edn). Springer-Verlag, Berlin.

Appendix 4
The electrostatics and current–voltage characteristics of an n–i–n structure

In this appendix, we amplify the discussion in Chapter 2, Section 2.5.2, of the electrostatic profile and the current voltage characteristics of an n–i–n (or n^+–n^-–n^+) multilayer structure) following the spirit of the treatment by Luryi (1990) and Grinberg and Luryi (1987). A treatment that is largely analytical is possible. In Fig. A.4.1(a) we show the conduction band profile, the layer geometry, and the electric field profile for a thin symmetric n–i–n structure. We shall consider the carrier profile as being determined by the doping in the n layers, and we shall ignore the contribution from carriers generated in the intrinsic region. We begin by introducing the following dimensionless variables:

concentration	$\nu \equiv n/N_0$	where	$N_0 \equiv N_D$
coordinate	$\xi \equiv x/x_0$	where	$x_0 \equiv L_D = (\varepsilon_s kT/e^2 N_D)^{1/2}$
electric field	$\varepsilon \equiv \mathcal{E}/\mathcal{E}_0$	where	$\mathcal{E}_0 \equiv kT/eL_D = (N_D kT/\varepsilon_s)^{1/2}$
current	$j \equiv J/J_0$	where	$J_0 \equiv eND\mu\mathcal{E}_0 = eN_D D/L_D$ (A4.1)

where L_D is the Debye length, ε_s is the dielectric constant, μ is the mobility and D is the diffusion constant in the i layer. We regard the i layer as extending between $-L/2$ and $+L/2$ (or $-\mathcal{L}/2$ to $+\mathcal{L}/2$ in reduced units of length defined by L_D). Well into the contact layers, we define the zero of potential energy so that $\phi(\pm\infty) = 0$, and the dimensionless carrier concentration can be written as $\nu(\xi) = \exp[\beta\phi(\xi)]$ with $\beta = e/kT$. In terms of this potential, we can write the dimensionless electric field in the form $\varepsilon = -\partial(\beta\phi)/\partial\xi$. The Poisson equation then becomes

$$d\varepsilon/d\xi = -\nu \quad |\xi| \leq L/2, \quad \text{and} \quad \partial^2[\beta\phi(\xi)]/\partial\xi^2 = e^{\beta\phi} - 1, \quad |\xi| \geq L/2. \text{(A4.2)}$$

The drift–diffusion equation for electrons in the i region ($J \sim ne\mu E + eD\Delta n$) becomes

$$\frac{\partial[\varepsilon^2/2 + \partial\varepsilon/\partial\xi]}{\partial\xi} = j. \tag{A4.3}$$

A solution for this equation when $j = 0$, satisfying $\varepsilon(0) = 0$ (by symmetry) and the physical condition $n > 0$, is

$$\varepsilon = -2\gamma \tan(\gamma\xi) \quad |\gamma\xi| < \pi/2 \tag{A4.4}$$

for real γ. This equation can be integrated with respect to ξ to find the potential anywhere in the i region as

$$\beta\phi = \beta\phi(0) - \ln\cos^2(\gamma\xi) \quad |\xi| \leq L/2, \tag{A4.5}$$

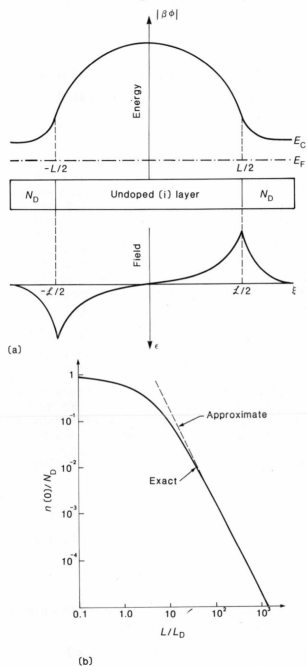

Fig. A4.1 (a) The conduction band profile, the multilayer structure, and the electric field distribution in an n–i–n structure; (b) the carrier density at the middle of the i region of a symmetric n–i–n structure (see text for details). (After Luryi 1990, from S.M. Sze (ed) © 1990, Printed with permission of John Wiley and Sons Inc.)

where $e^{\beta\phi} = v(0) = 2\gamma^2$. Outside the i-layer, the Poisson equation can be integrated twice to give the potential in the form

$$e^{\beta\phi} - \beta\phi - 1 = \varepsilon^2/2 \qquad |\xi| \geqslant L/2. \tag{A4.6}$$

Matching the two solutions for the potential at the interface $|\xi| = L/2$ leads to the closed expression

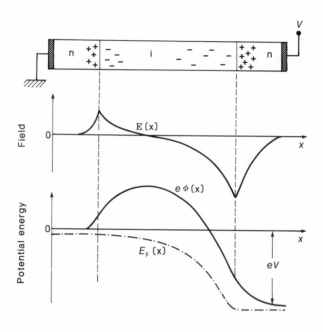

(a)

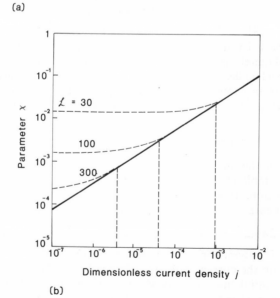

(b)

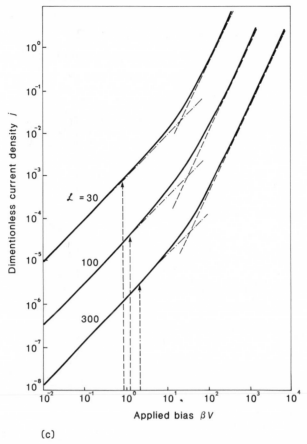

(c)

Fig. A4.2 (a) The n–i–n diode under bias, showing the field, potential, and local reference potential (imref $E_F(x)$); (b) the variation of the χ parameter with the reduced current density j, and (c) the calculated current–voltage characteristics in reduced form. Both the latter are in terms of the reduced length L of the diode. The condition $jL/\chi = 1$ is indicated by the vertical lines in (b) and (c).

$$\cos^2(\gamma L/2) = 2\gamma^2 \exp(1 - 2\gamma^2), \qquad (A4.7)$$

which determines the functional form of $\gamma(L)$ and completely solves the problem. For example in Fig. A4.1(b) we plot $v(0)$, the carrier density at the centre of the i region, as a function of the i layer thickness. There is a simple limiting form when $L \gg 1$ (i.e. $L \gg L_D$ and $n(0) \ll N_D$), as $|\gamma L| \to \pi$ and

$$n(0) \sim 2\pi^2/L^2 \qquad \text{or} \qquad n(0) \sim 2\pi^2 \varepsilon_s k T/e^2 L^2 \quad (\ll N_D). \qquad (A4.8)$$

In the case of a symmetric n–i–n structure, the above analysis can be extended to obtain the current–voltage characteristics. The asymmetric structure involves an intial built-in field which introduces complications. Figure A4.2(a) shows the structure, with its electric field and potential energy distribution. We assume that the mobility and the diffusion coefficient are field independent. Equation (A4.3) can be integrated once to give

$$- \partial \varepsilon / \partial \xi - e^2 / 2 + j\xi - \chi = 0 \qquad (|\xi| \leqslant L/2) \qquad (A4.9)$$

where χ is a j-dependent integration constant. The first form of the Poisson equation (cf. eqn. (A4.2)) allows us to write

$$\nu(\xi) = e^2 / 2 - (j\xi - \chi) \qquad (|\xi| \leqslant L/2). \qquad (A4.10)$$

Using the Boltzmann form of the carrier density $v(\xi) = \exp[\beta\phi(\xi)]$ with eqns. (A4.6) and (A4.9), we can use the constant χ and the current j to write down the values of the potential (ϕ_+) and field ε_+ at the $-L/2$ and $+L/2$ boundaries as

$$\beta\phi_- = \chi - 1 \qquad \beta\phi_+ = \chi - 1 + jL$$

$$\varepsilon_- = - \{2(\exp(\chi - 1) - \chi)\}^{1/2} \quad \varepsilon_+ = \{2[\exp(\chi - 1 - jL) - \chi + jL]\}^{1/2} \quad (A4.11)$$

The solution to eqn (A4.9) is cumbersome, involving the introduction of an auxiliary function U and a new independent variable

$$U(\xi) = \exp\left(- \int^{\xi} \varepsilon \, d\xi/2\right) \qquad \eta = (2/j)^{2/3}(j\xi - \chi) \qquad (A4.12)$$

in terms of which eqn (A4.9) reduces to

$$d^2 U / d\eta^2 - \eta U(\eta) = 0 \qquad (A4.13)$$

which has a Bessel function solution of order 1/3 and argument $z = 2\eta^{3/2}/3$ (Grinberg and Luryi 1987).

In Fig. 4.2(b) we show the dependence of the constant of integration χ on the dimensionless current density j and the reduced thickness L of the i layer. There are two limiting forms of operation: the short-base low-current regime (where $jL << \chi$) and the long-base high-current regimes $(jL >> \chi)$. In Fig. 4.2(c), the reduced current–voltage characteristics $(j/\beta V)$ are plotted, also as a function of L. In the low-current limit, a linear Ohm's law applies with

$$j = [en(0)\mu/(1 + \delta)](V/L) \sim [2\pi^2/(1 + \delta)][\varepsilon_s \mu V/L^3](kT/e) \quad (A4.14)$$

where $\delta = 2^{3/2} L^{-1} e^{0.5}$, while in the high-current regime a Mott–Gurney (1948) law for space-charge-limited current applies with

$$j = (9/8)(\beta V)^2 / L^3 \text{ or } J = (9/8)\varepsilon_s \mu V^2 / L^3. \qquad (A4.15)$$

The detailed explanation of a linear rather than exponential $J-V$ characteristic from a problem with a current over a barrier lies in the fact that the barrier changes shape and height under bias with the barrier maximum (Fig. A4.2(a)) moving towards the cathode under bias.

For an extension of this analysis to treat the capacitance of these n–i–n structures see Grinberg and Luryi (1987), and for an equivalent treatment of the planar doped barrier diode (cf. Chapter 2, Section 2.6) see Luryi (1990).

References

Grinberg, A. A. and Luryi, S. (1987). Space-charge-limited current and capacitance in double-junction diodes. *Journal of Applied Physics*, **61**, 1181–9.

Luryi, S. (1990). Device building blocks. In *High-speed semiconductor devices* (ed S. M. Sze), pp. 57–136. Wiley, New York.

Mott, N. F. and Gurney, R. W. (1948). *Electronic processes in ionic crystals*. Oxford University Press.

Index

absorption
  exitonic 98, 442
  far-infrared 286
  interband 97, 171, 260, 262, 264, 267
  inter-sub-band 125, 260, 265, 268, 270, 272, 454, 457
  intervalence band 435, 449
  intra-sub-band 272
  optical 12–14, 40, 57, 94–5, 97–8, 125, 163, 241, 256, 259–62, 264, 268, 336, 339, 349, 427, 439, 442, 455, 518
acceptor 7, 9, 26, 44–5, 53, 98, 104
acoustic mode 14
Aharonov–Bohm effect 91, 151, 294
AlF$_3$ 80
AlGaAs injector, graded 397
alloys
  quaternary 316–18, 435, 447–8, 492
  ternary 19, 316–18, 492
Al$_x$Ga$_{1-x}$ As alloys 19, 30–1, 50, 57–8, 104, 251, 311, 404–5, 468
amorphous semiconductors 502
  multilayers 476–8, 483, 487, 489
anion (common anion rule) 313
asymmetric spacer layer tunnel (ASPAT) diode 414
avalanche photodetectors 340, 447
avalanching 23, 162–3, 402–3

back-gating 125–7, 157, 185
backward diode 416
ballistic motion 90, 137, 154, 187, 212, 280, 283, 487, 504
band alignment (types I, II, III) 50, 312–13, 331, 333, 491
bandgap offsets 57, 215
band offsets 215, 313, 316
band structure engineering xviii, 447
band structure parameters xviii, 5, 6, 254, 351
Berry phase 501
bioelectronics 519
Bloch functions 2, 7, 57, 84, 95, 198, 254, 256, 260, 528–9, 531
Bloch motion 247
Bloch oscillations 235–7, 242, 247, 271
Bloch theorem 228
Bloch wave vector 228
Boltzmann approximation 47, 49, 116
Boltzmann condition 85

Boltzmann conductivity 87, 122
Boltzmann equation 83
Boltzmann regime 92
Boltzmann transport 82–3, 84
Boltzmann treatment 84–5
bound states 99, 110, 122, 139, 203, 205, 211, 227, 230, 253, 262, 325
Bragg reflection 236
buffer layer 249, 320, 331–3

capacitance–voltage profiling 37, 45
carrier trapping 273
channel (1D) 134, 136, 139, 142, 144–5, 156–7, 178–9, 185, 292–3
characterization 36, 38
coercivity engineering 493
coherence 98, 293–4, 501, 504
  backscattering 296
  length 354, 501–2
  wavefunction 293
conduction band 529
conductivity, minimum metallic 85
contacts
  blocking (Schottky) 71
  ohmic 68, 71, 73, 127, 157, 169–71
CoSi$_2$ 279, 495
Coulomb blockade 300, 303, 305, 508
Coulomb charges 120
Coulomb drag 128
Coulomb interaction 111
Coulomb scattering 104–5, 127
coupled quantum wells 269–70, 273
critical layer thickness 318–20
current–voltage characteristics 47, 49, 54, 162, 165, 183, 185, 190–1, 194–6, 280, 384, 466, 484, 537
cyclotron length 131
cyclotron orbits 76, 88, 148, 150, 212

dark current 444–6, 465
deformation potential 318, 323
density of states (DOS) 5, 22, 80, 81–2, 85, 88, 90, 113, 120, 144–6, 150, 154, 221–30, 259, 427, 430, 433
  3D, 2D, 1D, 0D, Q1D, Q2D 109, 119, 139, 256, 428, 430
  joint DOS 12–14, 94, 256, 427
  in superlattices 230